Powder Technology

Fundamentals of Particles, Powder Beds, and Particle Generation

Powder Technology

Fundamentals of Particles, Powder Beds, and Particle Generation

edited by

Hiroaki Masuda
Ko Higashitani
Hideto Yoshida

CRC Press
Taylor & Francis Group
Boca Raton London New York

CRC Press is an imprint of the
Taylor & Francis Group, an **informa** business

This material was previously published in *Powder Technology Handbook, Third Edition.* © CRC Press LLC, 2006.

CRC Press
Taylor & Francis Group
6000 Broken Sound Parkway NW, Suite 300
Boca Raton, FL 33487-2742

First issued in paperback 2019

© 2007 by Taylor & Francis Group, LLC
CRC Press is an imprint of Taylor & Francis Group, an Informa business

No claim to original U.S. Government works

ISBN-13: 978-1-4200-4410-2 (hbk)
ISBN-13: 978-0-367-38980-2 (pbk)

Library of Congress Cataloging-in-Publication Data

Powder technology. Fundamentals of particles, powder beds, and particle generation / [edited by] Hiroaki Masuda and Ko Higashitani.
    p. cm.
Includes bibliographical references and index.
ISBN 1-4200-4410-9 (acid-free paper)
1. Powders. I. Masuda, Hiroaki, 1943- II. Higashitani, Ko, 1944-

TP156.P3P638 2006
620'.43--dc22   2006270123

Visit the Taylor & Francis Web site at
http://www.taylorandfrancis.com

and the CRC Press Web site at
http://www.crcpress.com

Preface

Particulate, or powder, technology is a fundamental engineering field that deals with a variety of particles, from submicroscale grains and aggregates to multi-phase colloids.

The applications of powders and particles are rapidly expanding into more diverse technologies, from the information market—including mobile phones, copy machines, and electronic displays—to pharmaceuticals, biology, cosmetics, food and agricultural science, chemicals, metallurgy, mining, mechanical engineering, and many other fundamental engineering fields. Fueling some of the latest developments, nanoparticles are the focus of promising research leading to more effective applications of various particles and powders.

Drawing from the third edition of the renowned *Powder Technology Handbook*, this book is solely focused on analyzing the fundamental properties and behavior of particles and powder beds. This individual volume allows the reader to concentrate on the most useful analytical methods for observing, measuring, modeling, and predicting them.

Substantially revised, updated, and expanded, this volume highlights new information and developments from areas including surface properties and analysis, particle motion in fluids, mechanical properties of a powder bed, and the design and formation of composite particles.

While the book consolidates some sections from the last edition, it also incorporates the innovative work and vision of new, young authors to present a broader and fully up-to-date representation of the technologies.

We hope this volume will serve as a strong guide for understanding particle and powder behavior and encourage readers to apply the techniques, as well as the knowledge, to their future research and applications, particularly for scientists studying nanoparticles.

Special acknowledgment is given to all contributors and also to all the original authors whose valuable work is cited in this handbook. We would also like to acknowledge Dr. Matsusaka of Kyoto University, for his collaboration and editing, and we are grateful to our editorial staff at Taylor & Francis Books for their careful editing and production work.

Hiroaki Masuda
Ko Higashitani
Hideto Yoshida

Preface

Editors

Hiroaki Masuda is a professor in the Department of Chemical Engineering at Kyoto University, Japan. He is a member of the Society of Chemical Engineers, Japan and the Institute of Electrostatics (Japan), among other organizations, and is the president of the Society of Powder Technology, Japan. His research interests include electrostatic characterization, adhesion and reentrainment, dry dispersion of powder, and fine particle classification. Dr. Masuda received his Ph.D. degree (1973) in chemical engineering from Kyoto University, Japan.

Ko Higashitani graduated from the Department of Chemical Engineering, Kyoto University, Japan in 1968. He worked on hole pressure errors of viscoelastic fluids as a Ph.D. student under the supervision of Professor A. S. Lodge in the Department of Chemical Engineering, University of Wisconsin–Madison, USA. After he received his Ph.D. degree in 1973, he moved to the Department of Applied Chemistry, Kyushu Institute of Technology, Japan, as an assistant professor, and then became a full professor in 1983. He joined the Department of Chemical Engineering, Kyoto University, in 1992. His major research interests now are the kinetic stability of colloidal particles in solutions, such as coagulation, breakup, adhesion, detachment of particles in fluids, and slurry kinetics. In particular, he is interested in measurements of particle surfaces in solution by the atomic force microscope and how the surface microstructure is correlated with interaction forces between particles and macroscopic behavior of particles and suspensions.

Hideto Yoshida is a professor in the Department of Chemical Engineering at Hiroshima University, Japan. He is a member of the Society of Chemical Engineers, Japan and the Society of Powder Technology, Japan. His research interests include fine particle classification by use of high-performance dry and wet cyclones, standard reference particles, particle size measurement by the automatic-type sedimentation balance method, and the recycling process of fly-ash particles. Dr. Yoshida received his Ph.D. degree (1979) in chemical engineering from Kyoto University, Japan.

Editors

Contributors

Motoaki Adachi
Department of Chemical Engineering
Frontier Science Innovation Center
Osaka Prefecture University
Sakai, Osaka, Japan

Masatoshi Chikazawa
Division of Applied Chemistry
Graduate School of Engineering
Tokyo Metropolitan University
Hachioji, Tokyo, Japan

Reg Davies
E. I. Du Pont de Nemours, Inc.
Wilmington, Delaware, USA

Shigehisa Endoh
National Institute of Advanced
  Industrial Science and Technology
Tsukuba, Japan

Yoshiyuki Endo
Process and Production Technology Center
Sumitomo Chemical Co., Ltd.
Osaka, Japan

Richard C. Flagan
Division of Chemistry and Chemical
  Engineering
California Institute of Technology
Pasadena, California, USA

Masayoshi Fuji
Ceramics Research Laboratory
Nagoya Institute of Technology
Tajimi, Gifu, Japan

Toyohisa Fujita
Department of Geosystem Engineering
  Graduate School of Engineering
University of Tokyo
Tokyo, Japan

Yoshinobu Fukumori
Faculty of Pharmaceutical Sciences
Kobe Gakuin University
Kobe, Japan

Mojtaba Ghadiri
Institute of Particle Science and Engineering
School of Process, Environmental and
  Materials Engineering
University of Leeds
Leeds, United Kingdom

Kuniaki Gotoh
Department of Applied Chemistry
Okayama University
Okayama, Japan

Jusuke Hidaka
Department of Chemical Engineering and
  Materials Science
Faculty of Engineering
Doshisha University
Kyotanabe, Kyoto, Japan

Ko Higashitani
Department of Chemical Engineering
Kyoto University
Katsura, Kyoto, Japan

Hideki Ichikawa
Faculty of Pharmaceutical Sciences
Kobe Gakuin University
Kobe, Japan

Hironobu Imakoma
Department of Chemical Science and Engineering
Kobe University
Kobe, Japan

Chikao Kanaoka
Ishikawa National College of Technology
Tsubata, Ishikawa, Japan

Yoshiteru Kanda
Department of Chemistry and Chemical
  Engineering
Yamagata University
Yonezawa, Yamagata, Japan

Yoshiaki Kawashima
Gifu Pharmaceutical University
Mitahora-Higashi, Gifu, Japan

Yasuo Kousaka
Department of Chemical Engineering
Osaka Prefecture University
Sakai, Osaka, Japan

Ryoichi Kurose
Central Research Institute
  of Electric Power Industry
Yokosuka, Kanagawa, Japan

Hisao Makino
Central Research Institute
  of Electric Power Industry
Yokosuka, Kanagawa, Japan

Matsuoka Masakuni
Department of Chemical Engineering
Tokyo University of Agriculture and Technology
Koganei, Tokyo, Japan

Hiroaki Masuda
Department of Chemical Engineering
Kyoto University
Katsura, Kyoto, Japan

Shuji Matsusaka
Department of Chemical Engineering
Kyoto University
Katsura, Kyoto, Japan

Minoru Miyahara
Surface Control Engineering Laboratory
Department of Chemical Engineering
Kyoto University
Katsura, Kyoto, Japan

Yasushige Mori
Department of Chemical Engineering and
  Materials Science
Doshisha University
Kyotanabe, Kyoto, Japan

Yoshio Ohtani
Department of Chemistry and Chemical
  Engineering
Kanazawa University
Kanazawa, Ishikawa, Japan

Kikuo Okuyama
Department of Chemical Engineering
Hiroshima University
Higashi-Hiroshima, Japan

Yasufumi Otsubo
Center of Cooperative Research
Chiba University
Image-ku, Chiba, Japan

Mamoru Senna
Faculty of Science and Technology
Keio University
Yokohama, Kanagawa, Japan

Manabu Shimada
Department of Chemical Engineering
Division of Chemistry and Chemical
  Engineering
Hiroshima University
Higashi-Hiroshima, Japan

Kunio Shinohara
Materials Chemical Engineering, Lab.
Division of Chemical Process
  Engineering
Hokkaido University
Sapporo, Japan

Michitaka Suzuki
Department of Mechanical System
  Engineering
University of Hyogo
Himeji, Hyogou, Japan

Hiroshi Takahashi
Department of Mechanical Systems
  Engineering
Muroran Institute of Technology
Muroran, Hokkaido, Japan

Takashi Takei
Faculty of Urban Environmental Science
Tokyo Metropolitan University
Hachioji, Tokyo, Japan

Ken-ichiro Tanoue
Yamaguchi University
Ube, Yamaguchi, Japan

JunIchiro Tsubaki
Department of Molecular Design and
  Engineering
Nagoya University, Nagoya, Japan

Hirofumi Tsuji
Central Research Institute of Electric
 Power Industry
Yokosuka, Kanagawa, Japan

Hiromoto Usui
Department of Chemical Science and Engineering
Kobe University
Kobe, Japan

Satoru Watano
Department of Chemical Engineering
Osaka Prefecture University
Sakai, Osaka, Japan

Toyokazu Yokoyama
Hosokawa Powder Technology Research
 Institute
Hirakata, Osaka, Japan

Hideto Yoshida
Department of Chemical Engineering
Hiroshima University
Higashi-Hiroshima, Japan

Shinichi Yuu
Ootake R. & D. Consultant Office
Fukuoka, Japan

Contents

PART II   *Fundamental Properties of Particles*

PART IV   *Particle Generation and Fundamentals*

Part I

Particle Characterization and Measurement

1.1 Particle Size

JunIchiro Tsubaki

Nagoya University, Nagoya, Japan

1.1.1 DEFINITION OF PARTICLE DIAMETER

Particle size data are essential to anyone treating powders. Expressing the size of a single particle is not a simple task when the particle is nonspherical. Expressions of individual particle size and a hypothetical equivalent sphere with regard to some properties. Table 1.1 lists the physical meaning of variously defined characteristic diameters.

When a particle is circumscribed by a rectangular prism with length l, width w, and height t, its size is expressed by the diameter, obtained from the three dimensions. l, w, and t are measured with a microscope. Ferct and Martin diameter are statistical diameters, which are affected by the particle orientation or measuring direction. The mean values of them are often defined as characteristic diameters. Unrolled diameter is the mean value of a statistical diameter.

The equivalent diameters are the diameters of spheres having the same geometric or physical properties as those of nonspherical particles.

1.1.2 PARTICLE SIZE DISTRIBUTION

Size Distribution

When a certain characteristic diameter, shown in Table 1.1, is measured for N particles, and the number of particles, dN, having diameters between x and $x + dx$, is counted, the density size distribution $q_0(x)$ is defined as

$$q_o(x) = \frac{dN}{N} \frac{1}{dx}$$
(1.1)

where

$$\int_0^\infty q_0(x)\,dx = 1$$
(1.2)

The cumulative distribution $Q_0(x)$ is given as

$$Q_0(x) = \int_0^x q_0(x)\,dx$$
(1.3)

Therefore,

$$\frac{dQ_0(x)}{dx} = q_o(x)$$
(1.4)

TABLE 1.1   Expression of Particle Size

Definition of characteristic diameters		Physical meaning and corresponding measuring method
Geometric size	Breadth: b Length: l Thickness: t	

$$\frac{b+1}{2}, \frac{b+l+t}{3}, \left(\frac{b}{t}\right)^{1/3}, \frac{3}{1/l+1/b}, \sqrt{lb}\left(\frac{2lb+2bt+2lt}{6}\right)^{1/2}$$

	Feret diameter 	Distance between pairs of parallel tangents to the particle silhouette in some fixed direction
	Martin diameter 	Length of a chord dividing the particle silhouette into two equal areas in some fixed direction
cylinder 	Unrolled diameter 	Chord length through the centroid of the particle silhouette
	Sieve diameter $\frac{1}{2}(a_1+a_2)$ or $\sqrt{a_1 a_2}$	a_1, a_2: Openings of sieves
Volume: v Equivalent diameter	Equivalent projection area diameter (Heywood diameter)	Diameter of the circle having same area as projection area of particle, corresponding to diameter obtained by light extinction
Surface: s	Equivalent surface area diameter (specific surface diameter) $(s/\pi)^{1/2}$	Diameter of the sphere having the same surface as that of a particle
	Equivalent volume diameter $(6v/\pi)^{1/3}$	Diameter of the sphere having the same volume as that of a particle, corresponding to diameter obtained by (electrical sensing zone method)
	Stokes diameter	Diameter of the sphere having the same gravitational settling velocity as that of particle obtained by gravitational or centrifugal sedimentation and impactor
	Aerodynamic diameter	Diameter of the sphere having unity in specific gravity and the same gravitational settling velocity as that of a particle obtained by the same methods as above
	Equivalent light-scattering diameter	Diameter of the sphere giving the same intensity of light scattering as that of a particle, obtained by the light-scattering method

The size distributions thus defined are on a number basis. In the case of mass or volume basis, total mass M and fractional mass dM are used instead of N and dN, respectively, and also the subscript value is changed from 0 to 3, so that $q_3(x)$, $Q_3(x)$ should be used. In general the subscript is described as r, and $r = 0, 1, 2, 3$ corresponds to number, length, area, and mass or volume basis, respectively. Density distribution $q_r(x)$ can be transformed to another basis distribution $q_s(x)$ by

$$q_s(x) = \frac{x^{s-r} q_r(x)\, dx}{\int_0^\infty x^{s-r} q_r(x)\, dx} \tag{1.5}$$

When particle size distributes widely, the distributions are plotted versus $\ln x$ instead of x and defined as

$$Q_r(\ln x) = Q_r(x) \tag{1.6}$$

$$q_r^*(\ln x) = \frac{dQ_r(x)}{d\ln x} = x q_r(x) \tag{1.7}$$

The density distribution in a representation with a logarithmic abscissa is distinguished by superscript *.

The discrete expression, which gives the size distribution histogram, $\bar{q}_{r,i}$, $\bar{q}^*_{r,i,}$ becomes

$$\bar{q}_{r,i} = \frac{\Delta Q_{r,i}}{\Delta x_i} = \frac{Q_r(x_i) - Q_r(x_{i-1})}{x_i - x_{i-1}} \tag{1.8}$$

$$\bar{q}^*_{r,i} = \frac{\Delta Q_{r,i}}{\Delta \ln x_i} = \frac{Q_r(x_i) - Q_r(x_{i-1})}{\ln(x_i / x_{i-1})} \tag{1.9}$$

Normal Distribution

The normal or Gaussian distribution function is defined as

$$q_r(x) = \frac{1}{\sigma \sqrt{2\pi}} \exp\left[-\frac{(x - x_{50})^2}{2\sigma^2} \right] \tag{1.10}$$

where x_{50} is the 50% or median diameter defined as $Q_r(x_{50}) = 0.5$ and σ is the standard deviation expressing the dispersion of the distribution.

$$\sigma = x_{84.13} - x_{50} = x_{50} - x_{15.87} \tag{1.11}$$

Log-Normal Distribution

The log-normal distribution function is given by substituting $\ln x$ and $\ln \sigma_g$, respectively, for x and σ in Equation 1.10 as follows:

$$q_r^*(\ln x) = \frac{1}{\ln \sigma_g \cdot \sqrt{2\pi}} \exp\left[-\frac{(\ln x - \ln x_{50})^2}{2 \ln^2 \sigma_g} \right] \tag{1.12}$$

where σ_g is the geometric standard deviation given as

$$\sigma_g = \frac{x_{84.13}}{x_{50}} = \frac{x_{50}}{x_{15.87}} \tag{1.13}$$

If a size distribution obeys the log-normal form, the other distribution converted by Equation 1.5 also obeys the log-normal form. The geometric standard deviation of any distribution is the same value; meanwhile, the median diameters are different but can convert each other. In the case of the volume and number distribution the median diameters, $x_{50,3}, x_{50,0,}$, can convert each other by Hatch's equation.

$$\ln x_{50,3} = \ln x_{50,0} + 3\ln^2 \sigma_g \tag{1.14}$$

Rosin–Rammler (Weibull) Distribution

The Rosin–Rammler or Weibull distribution function is written as

$$Q_r(x) = 1 - \exp\left(-bx^n\right) \quad \text{or} \quad Q_r(x) = 1 - \exp\left[-\left(\frac{x}{x_e}\right)^n\right] \tag{1.15}$$

where b is a constant equal to x_e^{-n} and x_e is an absolute size constant defined as $x_e = x_{63.2}$, that is, $x_e = x_{63.2}$. n is the distribution constant expressing the dispersion of particle sizes. The density distribution is written as

$$q_r(x) = nbx^{n-1}\exp(-bx^n) \quad \text{or} \quad q_r(x) = \frac{1}{x}\left(\frac{x}{x_e}\right)^n \exp\left[-\left(\frac{x}{x_e}\right)^n\right] \tag{1.16}$$

Graphical Representation

As an example, a set of data obtained by a sieving test is illustrated in Table 1.2, and the size distributions are illustrated in Figure 1.1 with a general abscissa and Figure 1.2 with a logarithmic abscissa.

$Q_{3,i}$ in Table 1.2 is plotted on log-normal and Rosin–Rammler probability paper, as shown in Figure 1.3. Since the plots are on a straight line on a log-normal probability graph, the particle size distribution obeys the log-normal function of which $x_{50} = 1.0$ mm and $x_{84.13} = 2.2$ mm. If we read the two values of $x_{15.87}, x_{50}$ or $x_{84.13}$, σ_g can be calculated. From Figure 1.3 $x_{50} = 1.0$ mm and $x_{84.13} = 2.2$ mm, then $\sigma_g = 2.2$. The dotted line in Figure 1.3 is the number distribution converted from measured volume distribution by Hatch's equation.

ISO 9276-1 (JIS Z 8819-1) standardizes the graphical representation of particle size analysis data.

1.1.3 AVERAGE PARTICLE SIZE

All average particle diameters except the geometric mean diameter are defined by

$$\bar{x}_{k,r} = \sqrt[k]{M_{k,r}} = \sqrt[k]{\frac{M_{k+r,0}}{M_{r,0}}} = \sqrt[k]{\frac{M_{k+r-3,3}}{M_{r-3,3}}} \tag{1.17}$$

TABLE 1.2 Calculation of the Histogram and the Cumulative Distribution

1	2	3	4	5	6	7	8
i	x_i (mm)	ΔM_i ($\times 10^{-3}$kg)	$Q_{3,i}$ (%)	Δx_i (mm)	$\bar{q}_{3,i}$ (%mm^{-1})	$\Delta \ln x_i$	$(\bar{q}_{3,i})$ (%)
0	0.063		0				
1	0.09	0.1	0.1	0.027	3.7	0.357	0.3
2	0.125	0.27	0.37	0.035	7.7	0.329	0.8
3	0.18	1.03	1.4	0.055	18.7	0.365	2.8
4	0.25	2.2	3.6	0.07	31.4	0.329	6.7
5	0.355	5.4	9	0.105	51.4	0.351	15.4
6	0.5	9.3	18.3	0.145	64.1	0.342	27.2
7	0.71	13.7	32	0.21	65.2	0.351	39.1
8	1	17	49	0.29	58.6	0.342	49.6
9	1.4	17	66	0.4	42.5	0.336	50.5
10	2	14.6	80.6	0.6	24.3	0.357	40.9
11	2.8	9.9	90.5	0.8	12.4	0.336	29.4
12	4	5.6	96.1	1.2	4.7	0.357	15.7
13	5.6	2.4	98.5	1.6	1.5	0.336	7.1
14	8	1.1	99.6	2.4	0.5	0.357	3.1
15	11.2	0.3	99.9	3.2	0.1	0.336	0.9
16	16	0.1	100	4.8	0.0	0.357	0.3

Where $M_{k,r}$ is complete kth moment of a $q_r(x)$ -distribution.

$$M_{k,r} = \int_0^\infty x^k q_r(x) dx \tag{1.18}$$

Geometric mean diameter is defined as

$$\ln \bar{x}_{geo,r} = \int_0^\infty \ln x \cdot q_r(x) dx \tag{1.19}$$

Although a lot of average diameters can be defined, the several listed in Table 1.3 are generally used.

If a density distribution is given as a histogram, $M_{k,r}$ is calculated by the following equations. Equation 1.18 is rewritten as follows if $k \neq -1$.

$$
\begin{aligned}
M_{k,r} &= \sum_{i=1}^m \bar{q}_{r,i} \int_{x_{i-1}}^{x_i} x^k dx = \frac{1}{k+1} \sum_{i=1}^m \bar{q}_{r,i} \left(x_i^{k+1} - x_{i-1}^{k+1} \right) \\
&= \frac{1}{k+1} \sum_{i=1}^m \Delta Q_{r,i} \left(\frac{x_i^{k+1} - x_{i-1}^{k+1}}{x_i - x_{i-1}} \right)
\end{aligned} \tag{1.20}
$$

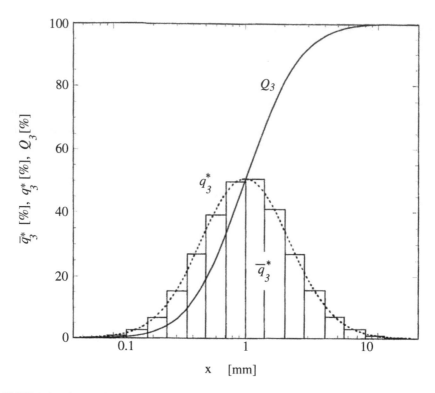

FIGURE 1.1   A graphical representation of particle size distribution with a linear abscissa.

If $k = -1$,

$$M_{-1,r} = \sum_{i=1}^{m} \bar{q}_{r,i} \ln \frac{x_i}{x_{i-1}} = \sum_{i=1}^{m} \Delta Q_{r,i} \frac{\ln \dfrac{x_i}{x_{i-1}}}{x_i - x_{i-1}} \tag{1.21}$$

Equation 1.19 is rewritten as follows:

$$\ln \bar{x}_{geo,r} = \sum_{i=1}^{m} \bar{q}_{r,i} \int_{x_{i-1}}^{x_i} \ln x \, dx = \sum_{i=1}^{m} \frac{\bar{q}_{r,i}}{x_i - x_{i-1}} = \sum_{i=1}^{m} \frac{\Delta Q_{r,i}}{\left(x_i - x_{i-1}\right)^2} \tag{1.22}$$

The values of the average diameters listed in Table 1.3 can be calculated from the data in Table 1.2. The calculation results are illustrated in Table 1.3.

The spread of a size distribution is represented by its variance, which represents the square of the standard deviation, σ. The variance, σ^2, of a $q_r(x)$-distribution is defined as

$$\sigma^2 = \int_0^\infty \left(x - \bar{x}_{1,r}\right)^2 q_r(x) dx = M_{2,r} - \left(M_{1,r}\right)^2 \tag{1.23}$$

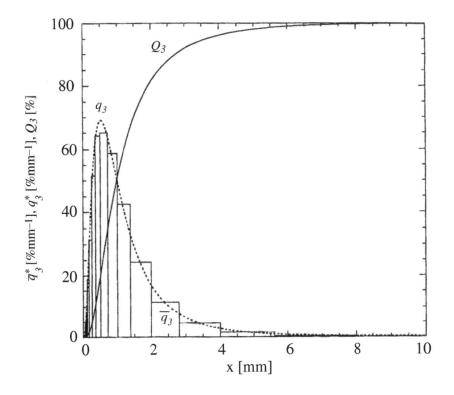

FIGURE 1.2 A graphical representation of particle size distribution with a logarithmic abscissa.

From a histogram,

$$\sigma^2 = \frac{1}{3}\left[\sum_{i=1}^{m}\overline{q}_{r,i}\left(x_i^3 - x_{i-1}^3\right)\right] - \frac{1}{4}\left[\sum_{i=1}^{m}\overline{q}_{r,i}\left(x_i^2 - x_{i-1}^2\right)\right]^2$$

$$= \frac{1}{3}\left[\sum_{i=1}^{m}\Delta Q_{r,i}\left(\frac{x_i^3 - x_{i-1}^3}{x_i - x_{i-1}}\right)\right] - \frac{1}{4}\left[\sum_{i=1}^{m}\Delta Q_{r,i}\left(x_i + x_{i-1}\right)\right]^2$$

(1.24)

The standard deviation of the particle sizes illustrated in Table 1.2 is calculated as σ 1.29 mm by Equation 1.24.

ISO 9276–2 (JIS Z 8819-2) standardizes the calculation of average particle sizes.

Notation

i	number of the size class with upper particle size x_i
k	power of x
m	total number of size classes
$M_{k,r}$	complete kth moment of a $q_r(x)$ -distribution
n	distribution constant

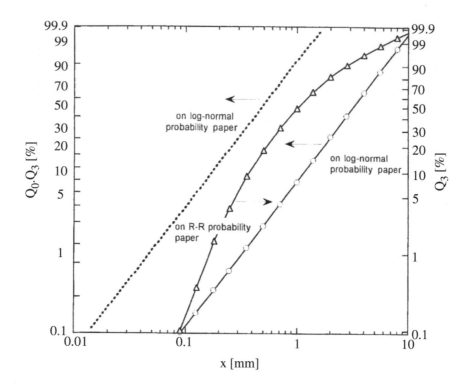

FIGURE 1.3 Cumulative distributions on a log-normal and Rosin–Rammler distribution graph.

TABLE 1.3 Average Diameters

Average diameters	Symbols	$q_0(x)$ from	$q_3(x)$ from	Examples of Table 1.2 [mm]
Arithmetic average length diameter, Weighted average number diameter	$\overline{x}_{1,0}$			0.228
Arithmetic average surface diameter	$\overline{x}_{2,0}$			0.303
Arithmetic average volume diameter	$\overline{x}_{3,0}$			0.410
Weighted average length diameter	$\overline{x}_{1,1}$			0.402
Weighted average surface diameter, Sauter-diameter	$\overline{x}_{1,2}$			0.749
Weighted average volume diameter	$\overline{x}_{1,3}$			1.40
Harmonic mean diameter	$\overline{x}_{1,r}$			0.147 $(r=0)$
				0.749 $(r=3)$
Geometric mean diameter	$\overline{x}_{geo,r}$			3.98 $(r=3)$

N	particle number
$q_r(x)$	density distribution
q_r	$= q_r(x)$
$q_r^*(ln\ x)$	density distribution in a representation with a logarithmic abscissa
q_r^*	$q_r^*(x)$
$\bar{q}_{r,i}$	average density distribution of the class Δx_i, histogram
$\bar{q}_{r,i}$	average density distribution of the class $\Delta\ In\ x_i$, histogram
$Q_r(x)$	cumulative distribution
Q_r	$= Q_r(x)$
$\Delta Q_{r,i}$	$= Q_r(x_i) - Q_r(x_{i-1})$
x	particle size, diameter of a sphere
x_e	absolute size constant
x_i	upper size of a particle size interval
x_{i-1}	lower size of a particle size interval
Δx_i	$= x_i - x_{i-1}$, width of the particle size interval
$x_{15.87}, x_{50}, x_{63.2}, x_{84.13}$	defined as $Q_r(x_{15.87}) = 0.1587, Q_r(x_{50}) = 0.5, Q_r(x_{63.2}) = 0.632\ Q_r(x_{84.13}) = 0.8413$
$x_{50,3}$	median particle size of a cumulative volume distribution
$x_{50,0}$	median particle size of a cumulative number distribution
r, s	type of quantity of a distribution, $r, s = 0, 1, 2, 3$
σ	standard deviation
σ_g	geometrical standard deviation of log-normal distribution

1.2 Size Measurement

Reg Davies

E. I. Du Pont de Nemours, Inc., Wilmington, Delaware, USA

1.2.1 INTRODUCTION

The particle size distribution has been usefully defined as the state of subdivision of a material. For a smooth dense sphere, size (X) is equivalent to diameter (D_p). In most industrial applications, however, size (X) is equivalent to diameter (D) only through the use of an adjustment factor (F), where

$$D_p = FX$$

F can be a shape factor or, in the case of optical methods, a combined shape and extinction coefficient. Unless F is known, accurate particle size measurement of irregular particles is impossible. In these instances, size will be a function of the direction of linear measurement. It will be a function of the physical principle by which it is detected and a function of the instrument by which it is measured. Hence, an average size calculated from data obtained via an envelope—volume diameter of a powder in an electrolyte, a light-scattering measurement of the powder in suspension, or a permeability measurement of dry powder in air—will not be the same. The resulting differences will provide a valuable fingerprint of shape/optical variations and can be used advantageously for this purpose.

Table 2.1 shows many common methods of experiencing the size of irregular particles. The first six of these are obtained from microscopy and image analysis: the sieve diameter from sieving methods, specific surface diameter from permeability measurements, surface diameter from gas adsorption; Stokes' diameter from sedimentation methods, volume diameter from electrical sensing zone methods, and optical diameters from instruments using optical sensing detectors. Again, for most industrial powders, all these diameters will be different. This is not necessarily a disadvantage if one is aware of the reasons for the differences and is able to convert one equivalent diameter to another. It is used to advantage when one wishes to increase the sensitivity of detection of a change in size distribution of a powder during a processing step. For example, if one is concerned with breakage and the reduction of mass, then a volume-sensitive (e.g., electrical sensing zone instrument or X-ray sensing) instrument will be most useful. Whereas in attrition, where volume is conserved but large numbers of very fine particles are abraded from the surfaces of large particles, number counting or surface methods will have the highest sensitivity to the change.

Such considerations lead one to conclude that one should always choose the most suitable method and instrument for the job in hand. No sample should ever be submitted to or analyzed by a physical resting laboratory simply with a request for particle size analysis. Neither should one use an instrument because it is conveniently located near one's workplace. Several hundred instruments are documented in the literature for the measurement of particle size distribution and, their range of application spans six orders of magnitude. The most comprehensive treatise that discusses size measurement and the instruments available is given by Allen.[1] At the time of writing this chapter, Allen's book had just been submitted for its fifth edition. Allen's treatment combines historical size measurement with in-depth theory. Many of the older techniques are described, but many of these are now almost obsolete. In this chapter, theory will be minimal and only modern instruments will be discussed. For information on other methods, theory and detail, simply read the work by Allen.

TABLE 2.1   Several Methods of Expressing the Size of Irregular Particles

Average thickness	The average diameter between the upper and lower surfaces of a particle in its most stable position of rest
Average length	The average diameter of the longest chords measured along the upper surface of a particle in the position of rest
Average breadth	The average diameter at right angles to the diameter of average length along the upper surface of a particle in its position of rest
Feret's diameter	The diameter between the tangents at right angles to the direction of scan, which touch the two extremities of the particle profile in its position of rest
Martin's diameter	The diameter which divides the particles profile into two equal areas measured in the direction of scan when the particle is in a position of rest
Projected area diameter	The diameter of the sphere having the same projected area as the particle profile in the position of rest
Particle sieve diameter	The width of the minimum square aperture through which the particle will pass
Specific surface diameter	The diameter of the sphere having the same ratio of external surface area to volume as the particle
Surface diameter	The diameter of the sphere having the same surface area as the particle
Stokes' diameter	The diameter of the sphere having the same terminal velocity as the particle
Volume diameter	The diameter of the sphere having the same volume as the particle
Optical diameter	The diameter of the sphere having the same optical (e.g., backscattering or forward scattering) cross section as the particle; these will be different.

1.2.2 THE APPROACH

There are several basic steps that might have to be followed from the initial thought that size distribution data are required. Some of these could be the following:

1. Define the particle-processing problem. What is the hypothesis? What are you expecting to find? Why are you measuring size?
2. Define the type and extent of the size distribution data required—if any!
3. Acquire a reproducible sample of the test material.
4. Prepare (disperse) the sample for analysis.
5. Check the approximate upper and lower sizes and the state of dispersion in the sample by microscopy.
6. Select the most relevant (sensitive) physical principle to measure this range of size present. In some instances, this could require more than one principle.
7. Select an instrument that uses this principle. Again, in some instances this could require more than one instrument.
8. Calibrate the instrument(s) against international standards.
9. Conduct the analysis as per international (ISO) or national standards.
10. Select the most relevant data-handling method, calculate the most relevant distribution, and display it as per ISO or national standards.

This chapter will provide information to assist the reader with steps 6 and 7 only—selection of a method/instrument for size measurement. Information on other steps is provided elsewhere.

1.2.3 PARTICLE SIZE ANALYSIS METHODS AND INSTRUMENTATION

For simplicity, the methods and instruments have been loosely classified into six groups:

1. Visual methods (e.g., optical, electron, and scanning electron microscopy and image analysis)
2. Separation methods (e.g., sieving, classification, impaction, electrostatic differential mobility)
3. Stream scanning methods (e.g., electrical resistance zone, and optical sensing zone measurements)
4. Field scanning methods (e.g., laser diffraction, acoustic attenuation, photon correlation spectroscopy)
5. Sedimentation
6. Surface methods (e.g., permeability, adsorption)

Discussion will not be uniform across these classes. Class 6 will have minimum discussion, as the methods are not widely used for measuring distributions; classes 1, 2, and 5 will have moderate discussion, as they receive widespread use in laboratories where high-cost instruments are not common; classes 3 and 4 will receive the most discussion, as many of the instruments in these classes are being used for online, in-process measurement and in today's industrial environment are being implemented into process measurement and control schemes. Use for powders, suspensions, and aerosol will be treated simultaneously; that is, aerosol will not be treated separately, as in previous versions.

Visual Methods: Microscopy

Use of a microscope should always accompany size analysis by any method because it permits an estimate to be made of the range of sizes present and the degree of dispersion. In most instances, particle size measurement rarely follows, as other procedures are faster and less stressful to the operator. Microscope size analysis is used primarily as an absolute method of size analysis, as it is the only one in which individual particles are seen. It is used more widely in aerosol analysis, as airborne particles are more often deposited onto surfaces [e.g., by impactors, thermal precipitators (i.e., in a state more conducive to visual examination)]. It is used perhaps more for shape/morphology analysis than size analysis and for beneficiation studies of minerals when combined with x-ray analysis. It is also used when other methods are not possible (e.g., inclusions in steel, porosity in ceramics).

Clearly, the range of sizes and their degree of dispersion strongly influence the ease and reliability of size measurement. For instances where particles are not already deposited on surfaces, sample preparation is a critical step. This is discussed in detail by Allen.[1] Typically, samples can be extracted from an agitated well-dispersed suspension, but dry, temporary, or permanent mounts also possible. For dry mounts, particles can be dusted onto a surface; for temporary slides, powder can be dispersed-held in viscous media. For permanent slides, a 2% solution of collodion in amylacetate or other similar systems can be used to fix the particles permanently in place for later examination.

In transmission electron microscopy (TEM), particles are deposited on a very thin film that is transparent to the electron beam. This film is supported on metal grids or frames. For scanning electron microscopy (SEM) backscatter measurements, the powder is dispersed onto a metal substrate and made conductive by coating with a thin layer of carbon from a vacuum evaporator.

It cannot be stressed enough that preparation of the sample for visual examination is one critical step and the fact that it receives little space in this chapter does not reflect on its importance.

When a satisfactory dispersion of particles on a relevant substrate has been achieved, particle size analysis can follow. Points to consider are resolution, the total number of particles to be counted, the choice of size distribution (whether number or mass distribution is required), and the effect of material properties. The limit of resolution (i.e., the distance at which two particles in close proximity appear as a single particle) is proportional to the wavelength of the illuminating source and inversely proportional to both the refractive index of the immersion medium and the sine of the angular aperture of the objective. Although absolute limits of resolution exceed the following, the more common ranges of measurement are 3-1000μm for optical microscopy, 2 nm to 1μm for TEM, and 20 nm to 1000μm for SEM.

For manual counting and operator comfort, the total number of particles should be typically below 800, and procedures have been developed to attain acceptable statistical accuracy with this constraint. Counting particles by number is easier than by mass, as statistical reliability does not rely on the omission of one particle, whereas counting by mass (e.g., of a distribution containing 999 particles of diameter 1μm and one 10 μm particle), the omission of the one 10 μm particle removes approximately 50% of the mass. Hence, very accurate assessment of the largest particles in the sample is essential for mass distribution measurement.

Allen[1] shows the expected standard error $S(P_r)$ of the percentage P_r by number in each size class, out of the total number in all size classes, to be

$$S(P_r) = \left(\frac{P_r(100 - P_r)}{Sm_r} \right) \tag{2.1}$$

where Σm_r is the total number of particles of all size classes. The standard error is a maximum when $P_r = 50$; hence, $S(P_r)$ will always be less than 2% — an acceptable error for most instances — if $M_r \geq$ 625 particles.

For mass counting, if $S(M_g)$ is the standard deviation experienced as a percentage of the total by weight, M_g is the percentage by weight in a given size range, and M_r is the number of particles counted in the size range, then

$$S(M_g) = \frac{M_g}{M_r} \tag{2.2}$$

So if 10% by weight of particles lie in the upper or top size range for a similar acceptable accuracy of 2%, it is absolutely necessary to count 25 particles in the top size range. Without special procedures, the total number count to achieve this accuracy would be many thousands and beyond the endurance of an operator. Two approaches can be taken: count the thousands using automatic microscopes/image analyses or maintain the count below 800 by reducing the area examined for the smaller sizes. This procedure is beyond the scope of this chapter, but it is described in British Standard 3406[2] and by Allen.[1]

Furthermore, for most purpose, a choice is made to count on a projected area basis rather than a linear basis. Although this requires an estimate of a circle of equivalent area to the irregular particle, the law of compensating errors generally results in lower errors than by manipulation of any of the possible linear measurements.

Figure 2.1 shows a typical reticule given in British Standard 3406 showing seven circles in a root-2 progression of sizes and five different geometric areas. The importance of a well-dispersed and nonsegregated sample becomes clear when the geometric areas of this reticule are examined. Spatial variations can be minimized by using opposite quadrants.

Materials properties affect the procedure depending on particle strength, wettability, and particle refractive index, to name three factors. Others can be similarly important. Strength and wettability will certainly influence the method of dispersion and the mounting of the sample. The refractive index will influence resolution. With image analysis, gray value discrimination and edge definition is an acute problem for automatic counting. These latter instruments vary widely in cost, and cost generally correlates with the ability to enhance the image electronically. Allen[1] discusses the

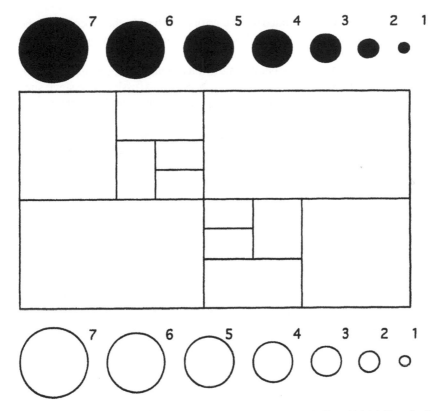

FIGURE 2.1   The British Standard Graticule (Graticules Ltd). From British Standard 3406 (1963).

commercially available instruments today. Typically, manufacturers are sensitive to cost and offer lines of expensive instruments with full image enhancement capability and color representation, moderately priced instruments for general–purpose analysis and lower-cost instruments for applications where resolution, gray value discrimination, and edge detaching are not a problem. Software packages are also appearing whereby images can be analyzed by a personal computer with software interactive capability on a routine basis.

Size analysis by microscopy is performed in our laboratory in less than 2% of size measurement cases, but it is significantly higher when shape information is required.

Separation Methods

These methods are typified by procedures which permit the extraction of a certain size class from the feed sample followed by a measurement of the percentage by number or mass of the removed size class relative to the feed concentration. Sieving is perhaps the oldest of these methods, although classification by winnowing was widely used to separate grain in ancient Egypt. Sieving is very well known, but perhaps at the same time the least well known for potential errors in the analysis. Because it appears to be a simple and low-cost process, errors are commonplace.

Sieving consists of placing a powder on a surface (e.g., a plate with numerous openings of a fixed size) or a mesh of intersecting wires with numerous openings of a fixed size, and agitating the surface (sieve) so that particles of size smaller than the holes pass through. To speed up the analysis, several sieves are placed on top of each other, the coarsest at the top and the rest shaken until the residue on each sieve contains particles which cannot pass through the lower sieve. This is a well-known procedure; what is not always appreciated are the errors involved in the process. For the sake of this discussion, only wire-woven sieves will be discussed, as these generally pose the greatest problems and are the most commonly used.

Separation of sizes on the sieve depends on the maximum breath and maximum thickness of a particle. Length affects the separation only when it becomes excessive. The sieve surface contains apertures with a range of sizes and a range of shapes. They are classified internationally by their mesh sizes and percentage of open area.

The first error arises in the assumption that this classification is accurate for all sieves both new and old. Apertures vary widely and the standard deviation of apertures is smaller (e.g., 3–5% for large aperture sieves of nominal aperture widths of 630 μm and larger, 10% for small aperture widths of 40 μm).[3] Leschonski[4] examined eight 50-μm sieves and the median aperture varied from 47.3 to 63.2 μm. Hence, sieve calibration is necessary at frequent intervals of use if size analysis is to be accurate. Several approaches have been used, including microscopy and sieving with standard powders. The latter is more tedious, but for woven sieves, it gives a better measure of three-dimensional aperture sizes and calibrates the sieve in terms of volume diameter using the actual procedure used for size measurement. Leschonski recommends a counting and weighing technique similar to that of Andreasen[5] applied to the fraction of particles passing the sieve just prior to the completion of the analysis. These are usually of narrow size range and are typical of the cut size of the sieve.

As sieving is a rate process and the passage of a particle through an aperture a statistical procedure, sieving accuracy (i.e., the time of sieving) depends also on the size distribution spread of sizes, the number and mass of particles on the sieve surface, the particle shape, the method of shaking, and the percentage of open area. Agglomeration and strength of particles affect the outcome. For highly agglomerated dry powders, agglomeration effects can be overcome by wet sieving; hence, wet sieving is recommended often for particle sizes less than 200 mesh. Friability strongly influences the choice of the sieving procedure and, again, wet sieving is less aggressive. Sieve wear with time influences aperture size and requires recalibration at frequent intervals. Sieve cleaning is sometimes the primary source of aperture distortion. Sieving machines are numerous and use different methods of agitating the particles on the surface, thus assisting them to pass through the sieve. A Ro-tap machine with horizontal rotary and vertical tapping motion is aggressive, but for strong particles, it is highly effective.

Air-swept sieves (e.g., Hosokawa Alpine and Allan Bradley sonic sifters) can be equally aggressive on the particle, whereas wet sieving is gentler (e.g., Retsch, Hosokawa, Alpine). Automatic versions such as the Gradex Size Analyzer simultaneously reduce analytical times and operator error. The sieving process is simple but probably experiences the most inaccuracy of all sizing methods.

Classifications in liquid and air can be used to separate size classes but is not commonly used for size analysis. Online analyzers have been developed using air classification (e.g., the Humboldt size analyzer), but perhaps the more recent applications have been in classification/fractionation methods using hydrodynamic chromatography (HDC) and field flow fractionation (FFF) for colloidal suspensions.

Figures 2.2 and 2.3 shows the essential differences between the two. In HDC, colloidal particles in a dilute suspension are separated by injecting them into a nonporous packing of large particles in a laminar flow of clean fluid. Particle separation occurs due to the hydrodynamic interaction of the laminar flow profile and the cross section of the particle. Large particles whose diameters interact with the higher-velocity central stream lines of the parabolic laminar flow profile move faster through the column, whereas small particles whose diameters interact only with the slower profiles near the packing surfaces move slower. Size analysis is obtained by measuring the concentration of particles existing the column with time using an ultraviolet detector. Size resolution and size discrimination are not generally as good as FFF. Here, a similar injection of particles is made into a long channel under laminar flow, but this time a field is applied across the channel. In centrifugal fields typified by the DuPont SFFF, shear-induced hydrodynamic lift forces oppose the driving force of the field. Because of diffusional effects, finer particles diffuse nearer the higher velocity central portion of the laminar flow field and are eluted first, whereas larger particles pushed by the

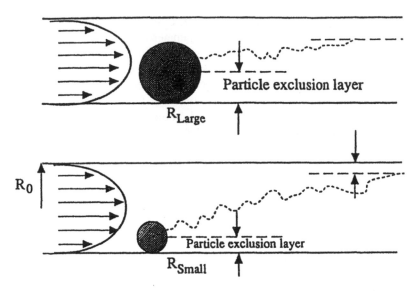

FIGURE 2.2   Capillary model of hydrodynamic chromatography separation.

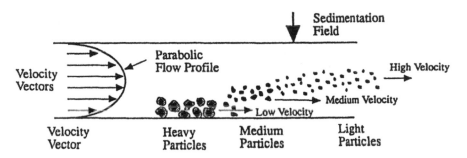

FIGURE 2.3   Centrifugal sedimentation field flow fractionation.

centrifugal field nearer to the outer wall and away from the higher-velocity portions of the flow profile are retarded. Again, concentration exiting the channel is measured by an ultraviolet (UV) detector and size distribution calculated. In the DuPont SFFF, a time-delayed exponential centrifugal field is used. Although a constant force field provides high resolution of particles, an exponential force field in which the centrifugal field is allowed to decay exponentially speeds up the analysis, particularly for wide size distributions. Alternate types of fields have been reported in the literature. These include magnetic, thermal, and steric designs.

For aerosols, typical devices that fractionate feed distributions in gases include impactors and electrostatic differential mobility analyzers.

In impactors, the particles in gas are passed through a nozzle and made to interact normally with a plane surface. The resulting particle trajectories can be calculated using the equation of motion of a particle in a flow field. Above a certain size of particle termed the cut size, particles will strike the surface with the potential for collection. Particles smaller than the cut size will follow the flow of gas, not touch the surface, and so have low potential for collection. The collection efficiency at the impaction surface is expressed in terms of the Stokes number (Stk) where

$$Stk = \frac{\rho_p C_c D_p^2 U_o}{18\mu(W12)} \tag{2.3}$$

Where p_p density of the particle and D_p its diameter, C_c is the Cunningham correction, U_o is the velocity of the gas, μ is the gas viscosity, and W the diameter/width of the nozzle. Impaction surfaces are calibrated by determining the collection effiency as a function of the Stokes number. For a given impactor geometry and operating condition, the Stokes number at the 50% collection efficiency is estimated and used to determine the cut size D_p for those conditions.

Thus, by reducing the value of W, the nozzle diameter for the identical gas flow, the impaction velocity U_p increases and the cut size is reduced. In this way, multiple impaction stages can be designed within an instrument to collect a variety of sizes. These instruments are termed cascade impactors. By weighing or otherwise estimating (microscopy) the weight of particles on each surface, a size distribution can be measured.

Impactors suffer from errors in that there is some interaction between neighboring surfaces, deposition occurs inside the impactor by other mechanisms (e.g., eddy diffusion, thermal, or electrostatic). More importantly, particles rebound from the collection surfaces and are recollected on later stages. To minimize rebound, surfaces can be made tacky. Agglomeration is another concern in that it broadens the size distribution and the collected mass on any impactor stage. The mass of the agglomerate will determine its collection efficiency, and agglomerates will be collected as if they were large particles. Upon impact however, they may mechanically disperse into their smaller components on the collection surface. The mass loading of any stage influences the collection efficiency and impactors are used with low concentrations of particles in gas and are operated for short sampling times. Solids concentrations and sampling times can be increased by redesign of the collection surface (e.g., in the Lundgren impactor, the collection stage is held on a slowly rotating drum). The lower size of collection can be extended by other designs (i.e., high-flow-rate systems, the use of microjets, and low pressure). Impactors are generally used for particles sizes in excess of 1 μm with cyclone or sedimentation pre-stages for sizes greater than 10 μm. They have found widespread use in health and biological studies, including respiratory/lung deposition studies. Typical of these is the Andersen impactor, Figure. 2.4.

For fine particles in gases (e.g., 1 μm and smaller), electrical mobility analyzers have become common. Figure 2.5 shows two types of systems in which the velocity V_e of a charged spherical particle in an electric field is given by

$$V_e = \frac{peC_cE}{3\pi\mu D_p} = B_eE \tag{2.4}$$

where E is the strength of the electric field, e the elementary charge, P the number of elementary charges carried by the particle, C_c the Cunningham correction, μ the gas viscosity, and D_p the particle diameter. B_e, the electrical mobility, is inversely proportional to diameter and can be varied by changing the electrical voltage applied to the central collection rod and other parameter values remaining unchanged. Hence, particles are collected on the center rod according to their mobility. Particles larger than a critical size determined by a critical electrical mobility do not reach the rod and pass through the analyzer. Their concentration is detected by either an electrometer or a condensation nuclei counter.

Thermosystems manufactures both types of instrument: TSI 3071 is a differential mobility analyzer (DMA) fitted with an impactor to preclassify particles coarser than 1 μm, and a condensation nuclei counter to measure number concentration. Two DMA are sometimes used in tandem to study kinetics and growth, where the first acts as a generation and classification stage to present a narrow size distribution to the second. A conditioner between the DMAs can apply humidity or gases (e.g., SO_3, NH_3, etc.) for interaction studies. The TSI 3030 electrical aerosol analyzer uses an electrometer and employs 11 voltage steps which span the range 0.003–1 μm. Automatic and manual modes are available at an aerosol flow rate of 4 L/min.

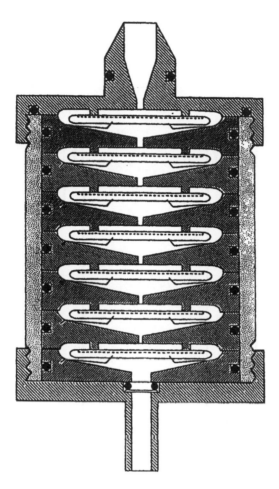

FIGURE 2.4   Anderson impactor.

Stream Scanning Methods

Stream scanning methods are those in which particles are examined one at a time, and their interaction with an external field is taken as a measure of their size. Such methods are usually limited to measurements at low particle concentrations and are most suitable for particle number counting. They have found widespread use in air quality and contamination monitoring. For the determination of mass distributions, many thousands of particles typically have to be counted. They are used for both particles in liquid and particles in gases.

The Coulter principle is used for particles in liquids only. Patented in 1949 and published in 1956 for blood cell counting, the principle has found widespread use in all branches of particle science. Figure 2.6 shows the basic principle. Particles are dispersed in an electrolyte and a tube containing an aperture of known size is inserted into the suspension. By placing an electrode both inside and outside the tube and initiating the flow of an electrical current through the orifice, an electric field of known characteristics becomes a sensing zone. When suspension flow is initiated by applying vacuum on the immersed tube, particles in suspension flow singly through the orifice, and momentary changes in impedance give rise to voltage pulses, the heights of which are proportional to particle volume. Theses pulses are amplified, sized, and counted and expressed as a size distribution. Size ranges from 0.6 to 1200 μm are typical, but several apertures/tubes have to be used to achieve this. This is because

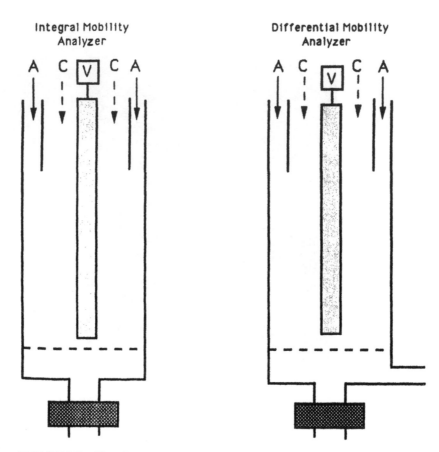

FIGURE 2.5   Electrical mobility analyzer.

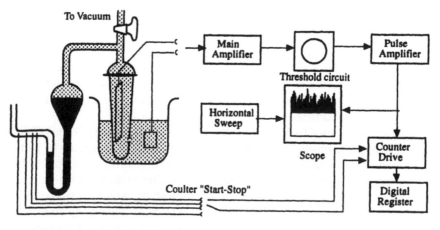

FIGURE 2.6   Coulter counter.

each orifice can sense particles from 2% to 60% of the orifice diameter. Problems can arise due to the necessity to use and match multiple aperture tubes, due to aperture blockage by large particles, by the passage of more than one particle through the sensing zone at a time (coincidence), by the nondetection of particles smaller than the lower detection limit of an aperture, and by the evolution of erroneous pulses due to abnormal interactions with the electric field.

The first problem is easily overcome by trial and error and by calibration. The second is simply overcome by preclassifying the sample. Coincidence errors are routinely overcome by the application of simple formulas supplied by the manufacturer.

Abnormal pulses arise due to the fact that the electric field is not uniform and that the field is dense near the inlet and outlet edges of the orifice. Passage of particles through these regions or interaction of extreme particle shapes with these regions can generate nonuniform electrical pulses. These can be eliminated by the use of mechanical focusing of particles through the orifice center or by electronic editing of abnormal pulses by the electronic data analyzer. The latter is more common today.

The nondetection of particles due to each orifice having a lower detection limit of 2% of its diameter is overcome by the performance of a mass balance. This is discussed at length by Allen.[1]

Accurate particle size analysis is best performed on dilute suspensions in which particles pass through the center of the orifice and mass balances are performed. In addition to Coulter, the principle is also applied in the Elzone electrozone counter manufactured by Particle Data. One special use of the Coulter principle is for the determination of oversize counts in coating slurries. As there are very few, a precision transparent sieve with a ±2% tolerance is used in combination with a Coulter mass balance procedure to prefill the few oversize particles and examine them by microscopy to determine their nature, count them by size, and recombine the data into the total mass balance. Allen describes this procedure on precision sieves manufactured by Collimated Holes Inc.

The primary value of the Coulter principle is that it can be used to measure both mass and population distributions accurately, and as volume diameter is one typical size representation, this agrees closely with most sedimentation analysis and with sieve analyses when calibrated, as discussed earlier in this section.

Of more widespread use are optical sensing zones which operate according to the following principles:

1. Collecting and measuring the light intensity in a forward direction
2. Collecting and measuring the light intensity in other directions (e.g., 90° or backscatter)
3. Light blockage or geometric shadowing
4. The measurement of phase shift

The most common types of instruments fall under class 1. They are available in various degrees of sophistication and wide variations in design. Particle size response is a function of the size, the shape, the orientation of the particle, the flow rate, and the relative refractive index between the particle and its surroundings.

Instrumentation cost and lower measurement limit are determined by the light source, the light collection system, and the efficiency of the detector. White-light sources are cheaper but give rise to a lower detection limit of 0.5 μm. Extension below this limit demands the use of lasers or laser diodes and highly efficient light collection systems. It is a balance among the intensity of the light source, the particle response, and the collection optics.

The relationship between particle size and scattered light intensity at any angle can be obtained for spheres with Mie theory. For a monotonic increase in scattering cross section with size, forward scattering at small angles is used. This has been dependence on particle refractive index, but extraneous light scattered from instrument internals is more of a problem than with large-angle scattering. This is why class 1 instruments are numerous and class 2 are not. Typical manufacturers include HIAC-Royco, Particle Measuring Systems, Climet, Kratel, and Polytec among many others. Each manufacturer has a relatively broad choice of instruments for gas or liquid applications. These are discussed more extensively by Allen.[1.]

Geometric shadowing or light blockage is typified by the HIAC-Royco light-blocking principle. When particles in liquid pass singly through the sensing zone, an amount of light equivalent to the effective cross section/shadow of the particle is cut from the incident beam. Sensors are available for the detection of sizes 1 μm to 3 mm. They are used particularly for sizes larger than 1 μm, and because they are less sensitive to refractive index and to the nature of the surrounding liquid, they have found use in the contamination monitoring of hydraulic oils.

Geometric shadowing has also been used to measure particles in molten polymer. The flow vision analyzer (Figure 2.7) uses fiber optics both to pass light across and detect particles in pipe-lines containing polymer. As particles pass through the beam, strobe-actuated cross sections of the particles are analyzed by image analysis. Particles of size 2–1000 μm are typically counted and Figure 2.8 shows a series of gel events in a polymer line monitered online by the device.[6]

The Partec 200 (Figure 2.9) offers online measurement capabilities at higher concentration in liquids, typically 5–30% volume percent, and so finds applications in the measurement and control of crystallizers, precipitators, and reactors. A light source is focused as a spot in the dispersion by a scanning focusing lens. This spot moves in a circular path. Light intersecting a particle as a chord is backscattered via the identical optics used to focus the light, and the chord length is recorded and scaled. Software to transform chord length distribution into particle diameter is used for some applications. Figure 2.10 shows typical results of a heating and cooling cycle on an organic system. Clearly, nucleation and dissolution are followed as a function of temperature.

Field Scanning

Field scanning methods measure the interaction of an assembly of particles and interpret the signal in terms of the size distribution of the assembly.

Low-angle laser light-scattering instruments (LALLS) collect light scattered from particles in a collimated laser beam by an array of detectors in the focal plane of the collecting lens

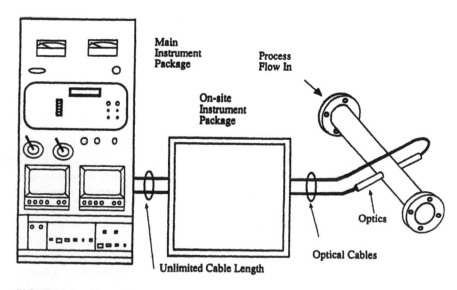

FIGURE 2.7   Flow vision analyzer.

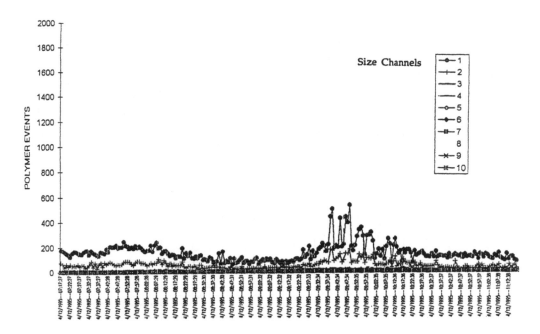

FIGURE 2.8 Output of flow vision analyzer.

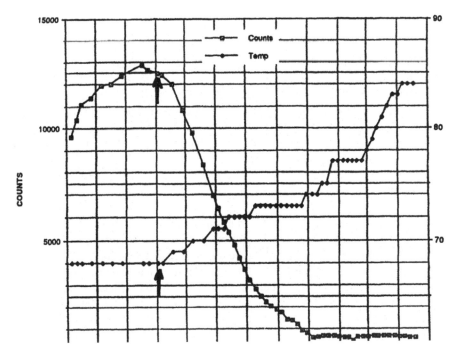

FIGURE 2.9 Partec 200.

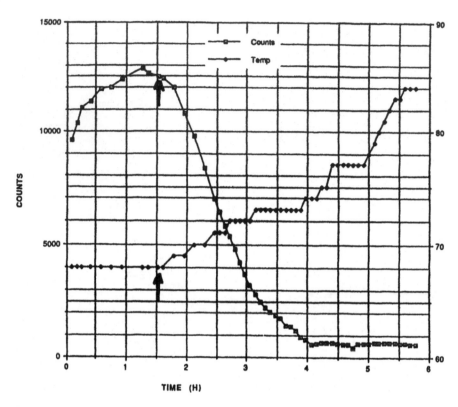

FIGURE 2.10 Typical output of Partec 200 (heating and cooling cycle on a crystal system).

(Figure 2.11). The light-intensity distribution for a single opaque spherical particle as given by the Fraunhofer equation falls off rapidly as particle size is reduced. This results in a diffraction pattern of circular light and dark rings, the intensity due to coarse particles being near the center and fine particles at the periphery. When an array of sizes are illuminated, a similar pattern emerges, but particles contribute to the intensities of more than one ring; hence, the light-scattering pattern is a matrix which has to be deconvoluted in order to determine the size distribution from the scatter pattern measurement.

Many commercial instruments are available that use this principle. They are manufactured by Leeds & Northrup, Cilas, Coulter, Seishin, Shimadzu, Sympatec, Malvern, Fritsch, Insitec, and Horiba & Nitto.

Instruments differ by the way they design the sensor/array to measure the pattern of their scattering algorithm, their mathematical deconvolution routine, their method of extending the lower limit of 0.1 μm, and their method of extending the upper limit to 3000 μm. A full discussion of these commercial instruments, their similarities and their differences are given by Allen.[1]

The devices are limited to a few percent mass concentration and have found extensive use in offline and on-line applications. The instruments are easy to operate and yield highly reproducible data. However, their general tendency is to oversize the coarse end of a distribution and assign an excess of particles to the fine end of the distribution, thus broadening the size distribution. For on-line applications in dense slurries, they require a dilution pre-analysis step, although multiple scattering studies have been made but not widely applied. They are calibrated using standard powders or by photomask reticules.[7]

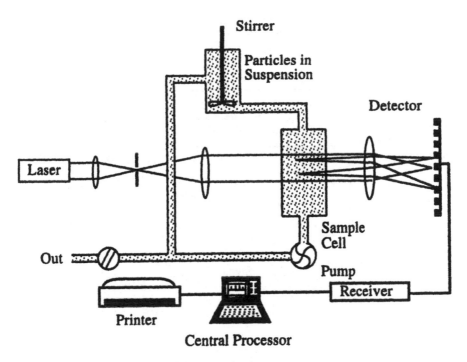

FIGURE 2.11 Low-angle laser light-scattering instrument.

Ultrasonic attenuation systems measure the attenuation of a range of ultrasonic frequencies passed through a concentrated dispersion of particles in liquid. A series of relationships among particle size, mass concentration, and wavelength are obtained, which can be deconvoluted to produce a size distribution. A single-point sizing device manufactured by ARMCO-Autometrics found widespread use in the mining industry in the 1970s and 1980s, whereas later developments by Herbst and Alba[8] developed a multipoint sizing option. This was extended to develop the Alba-DuPont Ultraspec™ technology. This was embodied in an instrument to measure the size distribution of TiO_2 slurries at 70% by weight.[9]

Reibel and Loeffler[10] reported on an online ultrasonic device for particle size distribution that is now available from Sympatec. Figure 2.12 shows the concept. Data on narrow and broad size ranges of glass beads showed good correlation with LALLS for particles greater than 15 μm at volume concentration up to 10%.

Pendse and Sharma[11] developed a laboratory acoustic analyzer for dense slurries up to 70% by weight. This is marketed by Pen-Kem.

Acoustic sensing is becoming more widely used due to the simplicity of the sensor/sample configuration and for its lower sensitivity to volume concentration. The systems are compact and robust, needing no optical benches as with LALLS. Gas bubbles in the suspension create problems however. For very small particles in liquids, photon correlation spectroscopy is used. Figure 2.13 shows the configuration in which a laser beam is passed into a suspension under Brownian motion and the scattered light fluctuations measured at 90° to the incident beam. The autocorrelation function of the scattered light fluctuations is calculated and related to the diffusion coefficient. This then yields an average size together with a polydispersity factor—some measure of the spread of sizes around the average value. Commercial instruments are available from Amtec, Wyatt, Malvern, NiComp, Brookhaven, Coulter, and Munhall. Different designs, some using fiber optics and some multiangle systems, are available. These are discussed more fully by Allen.[1]

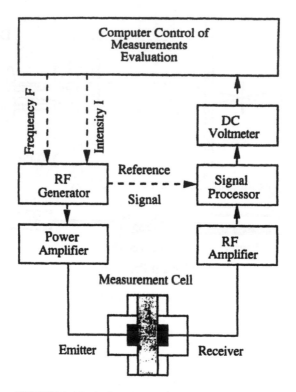

FIGURE 2.12 Online ultrasonic device.

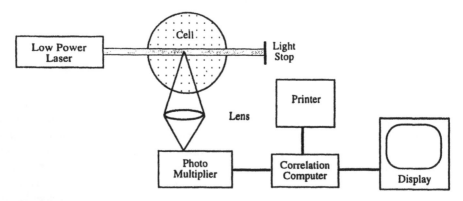

FIGURE 2.13 Photon correlation spectroscopy.

Sedimentation Methods

Sedimentation methods are based on the settling of a single sphere under gravitational or centrifugal fields of infinite extent. Sedimentation techniques have been classified by Allen[1] according to the following principles:

Class	Suspension Type	Measurement Principle	Force Field
1	Homogeneous	Incremental	Gravitational
2	Homogeneous	Incremental	Centrifugal
3	Line start	Incremental	Centrifugal
4	Homogeneous	Cumulative	Gravitational
5	Line Start	Cumulative	Gravitational
6	Homogeneous	Cumulative	Centrifugal
7	Line Start	Incremental	Gravitational
8	Line Start	Cumulative	Centrifugal

Most modern instruments in use today fall into classes 1 and 2. Only examples from these will be discussed here. For all others, see the work by Allen.[1]

In the homogeneous incremental gravitational class of methods, the solids concentration is measured at a known depth below the surface of an initial homogeneous suspension. The correlation will remain constant until the largest particle present has fallen below the measurement zone, when the particle concentration begins to fall. In consequence, the concentration will be that of particles smaller than the Stokes diameter, and a plot of concentration against the Stokes diameter will be the mass undersize distribution.

In class 2, the same principles apply, except that the particles now fall in radial paths, hence, the number of particles entering the measuring zone is less than the number leaving, so the concentration of these particles is less than the initial concentration.

Corrections have to be applied for the following:
- Reynolds number, if that of the particle exceeds 0.2
- Radial dilution
- Extinction coefficient, if optical incremental sensors are employed

An example from class 1 is the Micromeretics Sedigraph 5000 and 5100 (Figure 2.14) in which the measuring principle is x-ray attenuation, which is directly proportional to the atomic mass of the suspended particles (the mass undersize). In this unit, the measuring zone is changed with time by permitting the measurement cell to scan through the x-ray beam, resulting in smaller and smaller sedimentation heights. In this way, analytical measurement is speeded up, and size measurement as low as 0.2 µm is routinely recorded. For other instruments in this class, see the work by Allen.[1]

Examples from class 2 include the following:
- The DuPont/Brookhaven Scanning X-ray Disc Centrifugal Sedimentometer (Figure 2.15), which is used for the accurate measurement of particles less than 0.5 *µm*. In this device, the suspension is measured in a hollow x-ray transparent disk which generally contains 20 ml of suspension at 0.2% by volume. Speed ranges from 750 to 6000 rpm. The source and detector remain stationery for a preset time and then scan toward the surface. The commercial version (i.e., BI-XDC) has a stationary measurement disk and a scanning x-ray source and detector module. Total analysis time is under 8 min for most inorganic pigments.
- The Horiba Cuvet Photocentrifuge is typical of several Japanese-manufactured instruments. Here, the disk is replaced with a rectangular spectrophotometric cell containing a homogeneous suspension. Extinction coefficient and radial dilution corrections

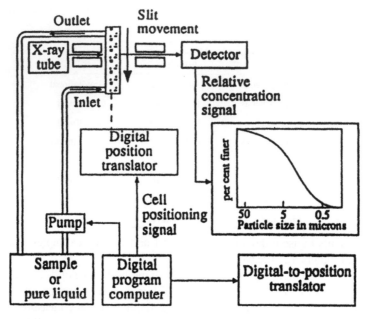

FIGURE 2.14 Sedigraph.

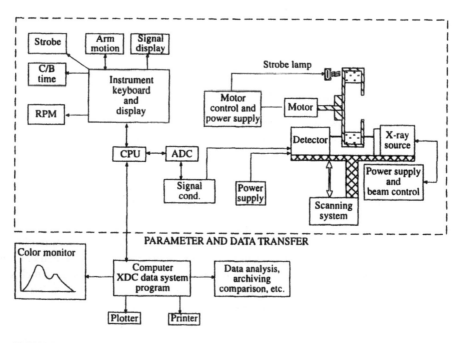

FIGURE 2.15 Scanning X-ray disk centrifugal sedimentometer (DuPont/Brookhaven).

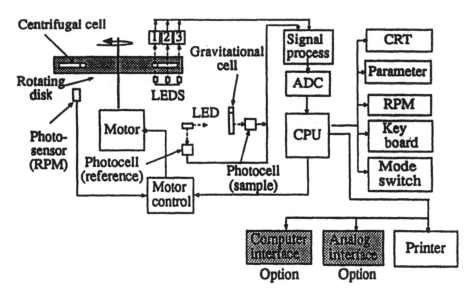

FIGURE 2.16 Centrifugal sedimentometer (Horiba).

must be made. Figure 2.16 shows the Horiba configuration. Clearly, smaller amounts of sample can be used, and as Allen[1] shows, a wide range of operating conditions are available by choice.

Sedimentation methods measure the Stokes diameter which for many particle systems closely resembles the volume diameter of the distribution. Hence, sedimentation and Coulter methods often agree closely, except for extreme shapes.

Surface Methods

For completeness, surface methods can be used to measure the average size of a powder. Generally, this is done dry and by two approaches. For dense, smooth particles of insignificant internal porosity, permeability methods give average specific surface diameters comparable to other methods, providing the lowest size present is greater than 2 μm. Allen[1] describes these. Gas adsorption methods, which measure the internal and external surface area, have been used but significantly undersize powders with internal surface area. These methods are rarely used today.

1.2.4 SUMMARY

This chapter has outlined the more prevalent methods of size analysis in the 1990s. It does not provide detail of each method or the range of choices that are available even with one manufacturer's instrument. Today, most instruments are modular and have options that can be selected for specific needs. Allen[1] provides a detailed list of manufacturers, and they should be contacted in order to understand their options and thereby provide a basis for instrument selection and purchase.

REFERENCES

1. Allen, T., *Particle Size Measurement*, 4th ed., Chapman & Hall, New York, 1990.
2. British Standard 3406, Methods for PSD of Powders, Part 4., Optical Methods, 1963.
3. Ilantzis, M. A., Ann. *Inst. Tech. Taim. Trav. Publ.,* 14 (161):165 and 484, 1961.

4. Leschonski, K. in *Proc. Particle Size Analysis Conference*, M. J. Groves and J. L. Wyatt-Sargent. eds., Society Analytical Chemistry, 165, 1970.
5. Andreasen, A. H. M., *Sprechsaal*, 60:515, 1927.
6. Joche, E. C., *Proc. Control of Particulate Processes*, 1995.
7. Hirrleman et al., *PARTEC 5th European Symp., Part. Characterization*, pp. 412, 655–671, 1992.
8. Herbst, J. A., and Alba, F., *Particle and Multiple Processes, Volume 3, Colloidal and Interfacial Phenomenon*, pp. 297–311, 1989.
9. Alba, F., U.S. Patent 5.121, 629, 416, 1992.
10. Riebel, U. and Loeffler, F., *Eur. Symp. Particle Characterization*, p. 416, 1989.
11. Pendse, H. P., Sharma A., *Proc. PTF Forum*, pp. 136–147, 1994.

1.3 Particle Shape Characterization

Shigehisa Endoh
National Institute of Advanced Industrial Science and Technology,
Tsukuba, Japan

1.3.1 INTRODUCTION

Particles have various shapes depending on their manufacturing method and mechanical properties. Unlike a sphere or a rectangular parallel-piped, which are objects with clear geometrical definitions, powder particles are very complicated objects with no definite form. It is generally very difficult to describe their shapes. They have therefore been conventionally described and classified by the use of various terms.

On the other hand, along with recent rapid advances in computer and information technologies, image processing technology has remarkably progressed in both software and hardware, making it easy to obtain image information and geometrical features of particles. Accordingly, various quantitative methods for expressing particle shape have been proposed and adopted.

The particle shape description or expression is classified by several criteria. In the most primitive manner, the descriptions are classified into quantitative methods and qualitative ones.

In qualitative description, the shape is expressed by several terms such as "spherical," "granular," "blocky," "flaky," "platy," "prismodal," "rodlike," "acicular," "fibrous," "irregular," and so on.[1,2] The verbal description is sometimes convenient to express irregular shape and makes it easy to understand the shape visually. But how are a platy shape and a flaky shape distinguished? Under present conditions, this distinction must depend on human visual judgments in many cases. Therefore, quantitative descriptions of particle shape will be necessary.

Quantitative shape descriptors can be calculated from two- or three-dimensional (2D or 3D) geometrical properties and can be calculated by comparing with physical properties of the reference shape.

The following are required for the shape descriptor[3]:

- Rotation invariance: values of the descriptor should be the same in any orientation.
- Scale invariance: values of the descriptor should be the same for identical shapes of different size.
- Reflection invariance.
- Independence: if the elements of the descriptors are independent, some can be discarded without the need to recalculate the others.
- Uniqueness: one shape always should produce the same set of descriptors, and one set of descriptors should describe only one shape.
- Parsimony: it is desirable that the descriptors are thrifty in the number of terms used to describe a shape.

The form of a particle is essentially a 3D property, but the shape of 2D images is mainly treated in this paper.

33

1.3.2 REPRESENTATIVE SIZE

If a given particle has the k-dimensional geometric property quantity Q_k, perimeter P for one-dimension, surface area S and projected area A for two-dimensions, volume V for three-dimensions, the equivalent diameter x_β of a sphere having the same quantity can generally be expressed by the following equation:

$$x_\beta = \left(Q_k/\phi_\beta\right)^k \tag{3.1}$$

where the coefficient ϕ_β is called the geometrical shape factor. From the mathematical relations for a sphere, the equivalent diameters and the factors are given as follows[2]:

- volume equivalent diameter:

$$x_V = (6V/\pi)^{1/3}, \phi_V = \pi/6 \tag{3.2}$$

- surface equivalent diameter:

$$x_S = \sqrt{S/\pi}, \phi_S = \pi \tag{3.3}$$

- projected are equivalent diameter (Heywood diameter):

$$x_A = \sqrt{4A/\pi}, \phi_A = \pi/4 \tag{3.4}$$

- perimeter equivalent diameter:

$$x_P = P/\pi \tag{3.5}$$

When particles have similar shapes, the averages of volume and surface area are given by the following expressions, respectively:

$$\overline{V} = \phi_\beta M_{3,0} \tag{3.6}$$

$$\overline{S} = \phi_\gamma M_{2,0} \tag{3.7}$$

where $M_{k,0}$ is the complete kth moment of number density distribution $q_0(x_\beta)$ of x_β defined in Equation 8:

$$M_{k,0} = \int x_\beta^k q_0(x_\beta)dx_\beta \tag{3.8}$$

Accordingly, a volume-related specific surface is as follows:

$$S_V = \frac{\overline{S}}{\overline{V}} = \frac{\phi_\gamma M_{2,0}}{\phi_\beta M_{3,0}} = \frac{\phi_{23}}{M_{1,2}} \tag{3.9}$$

where ϕ_{23} is a specific surface area shape factor given by the ratio between ϕ_γ and ϕ_β. $M_{1,2}$, which is a ratio of $M_{3,0}$ to $M_{2,0}$, is called the Sauter diameter $\bar{x}_{vs}(=\bar{x}_{1,2})$ (an average weighted surface diameter):

$$\bar{x}_{vs} = M_{1,2}\left(= M_{3,0}/M_{2,0}\right) = \int x_\beta q_2\left(x_\beta\right)dx_\beta \tag{3.10}$$

where $q_2(x_\beta)$ is area density distribution.

The specific surface area shape factor of spheres is given as $\phi_{23} = 6$ and the specific surface area is $S_V = 6/x_\beta$. By applying a correction coefficient ϕ_c, that is, Carman's shape factor, to a general particle, $S_V = 6/\phi_c x_\beta$ is given.

A 2D image of a particle gives the Feret diameter x_F, which is a distance between parallel tangents as shown in Figure 3.1. Since the Feret diameter changes with the angle of the tangents $x_F(\theta)$, that is, this means statistical diameter, the average value $\bar{x}_F$ can be employed as a representative diameter, given by

$$\bar{x}_F = \frac{1}{\pi} \int_0^\pi x_F\left(\theta\right)d\theta \tag{3.11}$$

The maximum and minimum values of x_F give the "length" and "breadth," respectively.

The unrolled diameter, $x_R(\theta)$, is also defined as a diameter passing through the center of gravity of the image, as shown in Figure 3.1. The average is given by

$$\bar{x}_R = \frac{1}{\pi} \int_0^\pi x_R\left(\theta\right)d\theta \tag{3.12}$$

1.3.3   GEOMETRICAL SHAPE DESCRIPTORS

Quantitative descriptors for the geometry of a particle image can be also classified into three categories based on the scale of the inspection of form:

1. Macroscopic descriptor, expressing overall form and referring to "proportion" or "elongation" of the particle image. The Fourier descriptors of lower-order harmonic number also give overall form features.

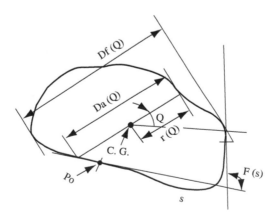

FIGURE 3.1   Statistical diameter and angular function.

2. Mesoscopic descriptor, expressing details of form and profile outline and defining space filling feature (or bulkiness), concavity, and robustness of the particle.
3. Microscopic descriptor, expressing surface structure and texture, giving roundness and fractal dimension of the image. The Fourier descriptors of higher order also express surface properties.

Macroscopic Descriptors

The definition and classification of shape descriptors expressing macroscopic shape properties are calculated from macroscopic geometrical features such as representative diameter, axis lengths, thickness, and so on.

Proportion

This parameter describes the anisotropy or elongation of a particle. It is given by the ratio of lengths of axes (length L, breadth B, thickness T):

$$\text{elongation} = L/B \text{ or } = L'/B' \tag{3.13}$$

$$\text{flatness} = B/T \text{ or } L/T \tag{3.14}$$

where L and B' are maximum and minimum Feret diameters, respectively, and B and L' are Feret diameters perpendicular to L and B', respectively.[4]

The axial ratio of the radius equivalent ellipse of the image that has the same geometrical moments up to the second order as the original particle silhouette (Figure 3.2) is referred to as "anisometry" and also gives the proportional property[4]

$$K = \sqrt{I_1 / I_2} = a / b \, (I_1 > I_2) \tag{3.15}$$

where I_1 and I_2 are the major and minor principle inertia moments of the particle silhouette, and the axis lengths are $a = \sqrt{I_1 / A}$ and $b = \sqrt{I_2 / A}$, respectively.

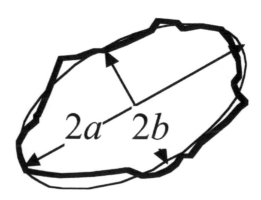

FIGURE 3.2 Radius equivalent inertia ellipse and its principle axis length.

Index Given by the Ratio of Characteristic Particle Diameters

Circularity ψ_{AP} is defined by the ratio of the perimeter P_c of a circle with the same projected area and that of the particle P, and gives the degree of similarity to a circle. The circularity is also given the ratio of the Heywood diameter and the perimeter equivalent diameter, as follows:

$$\psi_{AP} = P_c / P = x_H / x_P \tag{3.16}$$

For 3D inspection, Wadell's sphericity ψ_S defined by the ratio of the surface area S_s of a sphere with the same volume and that of the particle S gives the degree of similarity to a sphere, as follows[5]:

$$\psi_S = S_s / S \tag{3.17}$$

ψ_S is approximately obtained by the ratio x_A/x_{min} where x_{min} is the diameter of circumscribed minimum circle as a projected image (Figure 3.3).

The several kinds of shape descriptors given by the ratios of characteristic particle diameters are[6]

$$\psi_{AF} = x_A / \bar{x}_F, \ \psi_{RA} = \bar{x}_R / x_A, \ x_{RP} = \bar{x}_R / x_P, \dots \tag{3.18}$$

The coefficient of variation of the statistical diameter is thought of as a shape descriptor giving deviation from a circle with mean size, defined as

$$\sigma_i = \frac{1}{\bar{x}_i} \sqrt{\frac{1}{\pi} \int_0^\pi \left(x_i(\theta) - x_i \right)^2 d\theta}, \text{ for } i = F \text{ or } R \tag{3.19}$$

This descriptor is related to the circularity.

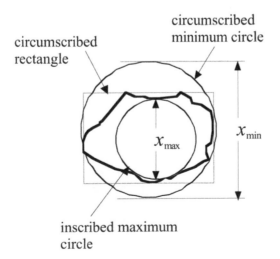

FIGURE 3.3 Circumscribed and inscribed circles and their diameters.

Mesoscopic Descriptors

Space Filling Factor

This factor is called "bulkiness," "bulkiness factor," and so on and expresses the bulky space occupied by a particle. The descriptor also indicates the degree of compactness.

Bulkiness is defined as a ratio between the area of a circumscribed rectangle or circumscribed circle of the image and that of the particle (Figure 3.3), as follows[7]:

$$f_{bR} = A / LB \tag{3.20}$$

$$f_{bC} = 4A / \pi x_{min}^2 \tag{3.21}$$

The bulkiness corresponds to the solidity defined as an area ratio between the circumscribed convex hull A_h and the image, as follows[8]:

$$\text{Solidity} = A / A_h \tag{3.22}$$

Also, the projected area relative to the area of the equivalent ellipse of inertia can be defined as the space filling factor, as follows[9]:

$$f_{BE} = A / \pi ab \tag{3.23}$$

where a and b are the major and minor radii of ellipse of gyration, respectively.

In 3D inspection, a circumscribed polyhedron (rectangular prism) or sphere is employed and the factor is given by the ratio between the volumes.

Irregularity index x_{max}/x_{min}, defined by the ratio of diameters of the inscribed maximum circle to that of the circumscribed minimum circle, gives the bulky property, as shown in Figure 3.3.

For an elongated particle, "curl" is defined by the following equation[10]:

$$Curl = L(= x_{F max}) / L_G \tag{3.24}$$

where L_G is the geodesic length of a fibrous particle, as shown in Figure 3.4.

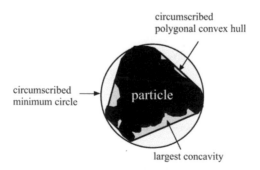

FIGURE 3.4   Polygonal convex hull.

Concavity and Robustness

This descriptor is based on mathematical relation.

The descriptor ψ_{FP} is defined by the ratio of the average Feret diameter and equivalent perimeter one:

$$\psi_{FP} = \bar{x}_F / x_p \tag{3.25}$$

ψ_{FP} for a shape without any convex portions on the surface is unity, thus the descriptor ψ_{FP} can be used to assess the surface concavity.[7]

The descriptors indicating the degree of convexity or concavity on the surface of the image are proposed as follows:

$$Convexity = P / P_h \tag{3.26}$$

$$Concavity = (A_h - A) / A_h \tag{3.27}$$

where P_h is the perimeter of the convex circumscribed polygon.[9]

The image processing technique gives the morphological descriptors. ω_1 and ω_2 are, respectively, the numbers of erosions (shrinking) in image processing necessary to eliminate completely the image and its residual set with respect to the convex hull of area A_h (Figure 3.5). The robustness factor is

$$\Omega_1 = 2\omega_1 / \sqrt{A} \tag{3.28}$$

The largest concavity is

$$\Omega_2 = 2\omega_2 / \sqrt{A} \tag{3.29}$$

Robustness is correlated to the elongation. The plot of Ω_1 versus Ω_2 is used for shape screening.[10,11]

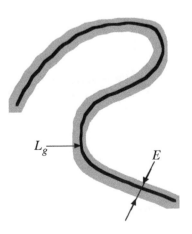

FIGURE 3.5   Elongated concave particle.

Surface Roughness

Hausner's surface index is given by a reciprocal of circularity ψ_{AP}[8]:

$$C = 1 / \psi_{AP}^2 = P^2 / 4\pi A \tag{3.30}$$

If the convex portion of the silhouette outline of particle is an arc, roundness is defined by the ratio of the average value of the curvature radii of the corners and edges to the diameter of the maximum inscribed circle R_{max} (Figure 3.6)[6]:

$$Rn = \frac{\sum \rho_i}{N R_{max}} \tag{3.31}$$

where ρ_i is the curvature radius of the corner i and N is the number of corners.

Application of Fourier Analysis: The Fourier Descriptor

The morphology of a particle can be expressed by the Fourier descriptor. The Fourier descriptor is a set of Fourier coefficient $A_k (= 2|C_k|)$, or pairs of the amplitude A_k and phase angle α_k of the Fourier coefficient that can be obtained by the Fourier transformation of the projected image function $F(\Theta)$:

$$C_k = \frac{a_k - jb_k}{2} = \frac{1}{2\pi} \int_0^{2\pi} F(\Theta)\exp(-jk\Theta)d\Theta, \; A_k = \sqrt{a_k^2 + b_k^2} \tag{3.32}$$

$$\alpha_k = \tan^{-1}(b_k / a_k) \tag{3.33}$$

for the harmonic order $k = 0, \pm 1, \pm 2, ..., j = \sqrt{-1}$.

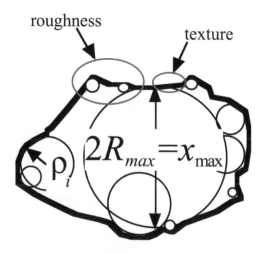

FIGURE 3.6   Surface structure and roundness.

The composite values of the Fourier coefficients and their statistical values are referred to as the Fourier descriptor, also. The values of A_k for lower harmonic orders give the macroscopic form feature,[12-22] that is, A_2 corresponds to the elongation (i.e., aspect ratio), and A_3 and A_4 mean triangularity and squareness, respectively.

The Fourier descriptor is distinguished from other shape descriptors in its representation of the measured image. The representation of original form $F(\Theta)$ can be guaranteed within the number of coordinate points N measured or interpolated for the Fourier transformation, as follows:

$$F(\Theta) = \frac{a_0}{2} + \sum_{k=1}^{N/2-1} A_k \cos(k\Theta - \alpha_k) \tag{3.34}$$

The following methods for various image function of outline are proposed (Figure 3.1 and Figure 3.7).

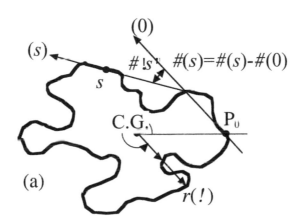

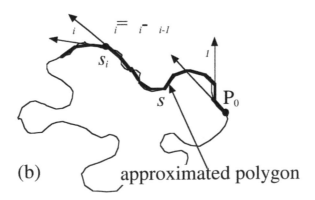

FIGURE 3.7   Angular function.

The Radial Fourier Descriptor: Polar Coordinate Method

In the polar coordinate method, the radius vector $r(\theta)$ from the center of gravity to the boundary of the image as a function of an angle θ of counterclockwise rotation about the center of gravity is transformed into the Fourier space (Figure 3.1).[12–15] The radial length $r(\theta)(=|r(\theta)|)$ is given by the Fourier series, as follows:

$$r(\theta) = r_0 + \sum_{k=1}^{N/2-1} A_k \cos(k\theta - \alpha_k) \tag{3.35}$$

$$r_0 = \frac{1}{2\pi} \int_0^{2\pi} r(\theta)d\theta \tag{3.36}$$

Some of the geometrical quantities, such as area and mean unrolled diameter, are given by the Fourier coefficient in polar coordinates. The mean radial length is $r_0(=\bar{x}_R/2)$, given by Equation 3.35, and the area A is given by

$$A = \pi \left[r_0^2 + \sum_{k=1}^{N/2-1} A_k^2 \right] \tag{3.37}$$

The Fourier descriptors of the polar coordinate method can be defined as follows:

$$IRR = (D_A/2r_0)^2 = \psi_{RA}^{-2} \tag{3.38}$$

$$Randance = \mu_2/r_0^2 = \sum A_k^{*2} \tag{3.39}$$

$$Skewness = \mu_3/r_0^3 \tag{3.40}$$

where A_k^* is the normalized amplitude A_k/A_0, and μ_m is the mth moment around the mean radius r_0, as follows:

$$\mu_m = \int [r(\theta) - r_0]^m d\theta/2\pi \tag{3.41}$$

IRR corresponds to the equivalent dimensionless area diameter and is a measure for irregularity. Dimensionless moments μ_2/r_0^2 and μ_3/r_0^3 are called randance and skewness.

This method, however, cannot be applied unless the radius vector $r(\theta)$ becomes a single-valued function of θ (Figure 3.7a).

The Angular Bend Function Method (ϕ-l Method): ZR Fourier Descriptor

The ZR Fourier descriptors are derived by the Fourier transformation of the angular function $\phi(s)$, which is the direction change (angle of deviation) of the outline tangent on the arc length s from the starting point P_0 (Figure 3.7).[16–19] The normalized angular function $\phi^*(t)$ is meaningful in comparison between different size particles by defining

$$\phi^*(t) = \phi(s) - t = \phi(Pt/2\pi) - t \tag{3.42}$$

where t is a normalized argument with the perimeter P, as follows:

$$t = 2\pi(s/P) \tag{3.43}$$

$\phi^*(t)$ is subjected to the Fourier transformation:

$$\phi^*(t) = \mu_0 + \sum_{k=1}^{\infty} [a_k \cos(kt) + b_k \sin(kt)] \tag{3.44}$$

The profile of the particle image can be approximated by a polygon (Figure 3.7b), then the Fourier descriptors are derived:

$$\mu_0 = \pi + \frac{1}{P} \sum_{j=1}^{m} s_j \Delta\phi_j \tag{3.45}$$

$$a_k = \frac{1}{k\pi} \sum_{j=1}^{m} \Delta\phi_j \cos(kt_j) \tag{3.46}$$

$$b_k = \frac{1}{k\pi} \sum_{j=1}^{m} \Delta\phi_j \sin(kt_j) \tag{3.47}$$

where s_j is the arc length of jth vertex from the starting point, $\Delta\phi_j$ is the change of slope at vertex j, and m is the total number of sides of the polygon.

The angular function method can be applied to every shape, even if the radial function $r(\theta)$ is not unique with θ. In some cases, however, the starting point and the endpoint of the represented profile do not coincide.[18]

The Radius Vector Method

In the case where $r(\theta)$ does not become a single-valued function, the modified function by considering the rotational direction of radius vector is introduced and subjected to the Fourier transformation.[20] The modified radial function is defined as

$$R(\theta') = r(\theta')f_D(\theta') \tag{3.48}$$

where $f_D(\theta')$ indicates the rotational direction of the radius vector. The function gives $f_D(\theta') = +1$ for the counterclockwise rotational direction and $f_D(\theta') = -1$ for the clockwise rotational direction, respectively. θ' is a cumulative value of angle variations at θ from the starting point, such as $\theta' = \Sigma|\Delta\theta|$.

The (x, y) coordinates can be transformed into the Fourier spectrum.[21,22]

Fractal Dimensions

Fractal dimensions[23] are applied to the quantification of particle shape.[24] Generally, the fractal dimension is defined as follows. The geometrical quantity Q_k of a k-dimensional pattern (topological dimension k) measured by a certain scale η is proportional to the number of elements for k-dimensional volume N.

$$Q_k = \eta N \tag{3.49}$$

If the scale η and N is given by the following relation:

$$N = \eta^{-D_f}, D_f \neq k \tag{3.50}$$

the pattern is fractal, which means statistical self-similarity, and D_f is called a fractal dimension.

As shown in Figure 3.8, let us follow the profile by a divider (linear element) of constant width η, and plot the perimeter estimated by Equation 3.48 or number of linear elements N plotted against η on an amphi-logarithmic sheet. The relation shown in Figure 3.9 is generally obtained with changing η. This relation is expressed in Equation 3.49, and the fractal dimension D_f is obtained by its slope:

$$\log(P) = (1 - D_f)\log \eta + c \tag{3.51}$$

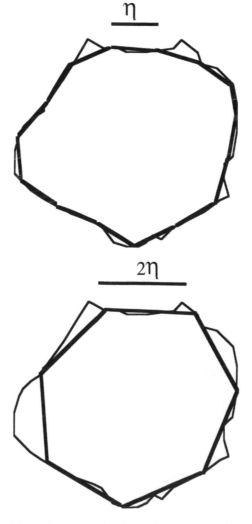

FIGURE 3.8   Estimation of perimeter with changing liner element length η.

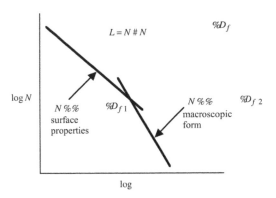

FIGURE 3.9 Log N versus log η plot.

With the influence of the concavity and convexity on the surface, D_f for the perimeter of the outline ($k = 1$) becomes $1 < D_f < 2$. The D_f value of a particle with a smooth surface is close to unity, whereas the fractal dimension is higher for complicated curves.[10,26,27] Therefore, it is believed that the fractal dimension quantitatively expresses the complexity of the surface and structure of the surface.

Usually, the log N-log η plot for the profile with a fractal structure shows the relation as shown in Figure 3.8, that is, the linear relation divided into two regions or more. The region of scale larger than some break point refers to the macroscopic morphology, on the other hand, the smaller region corresponds to the microscopic surface properties, such as surface texture.

When a 2D pattern of a binary image ($k = 2$) seems to be fractal, the correlation function $C(r)$ of density between the patterns with a distance r satisfies the following relation,[28,29]

$$C(r) = < f(r + r') \cdot f(r') > \infty\, r^{-\beta} \tag{3.52}$$

where $<>$ expresses an average operation and β is the slope of the log $C(r)$-log r plot. The fractal dimension is given by the following expression:

$$D_f = k - \beta \tag{3.53}$$

where $f(r')$ is an image function at a point r', which is 0 or 1.

1.3.4 DYNAMIC EQUIVALENT SHAPE

Drag Force Shape Factor

Stokes' fluid drag force F_D, which acts on an irregular-shaped particle moving at a relative velocity v in a fluid with a viscosity μ, is given by the following expression when the characteristic particle diameter is x_β:

$$F_D = 3\pi\mu v x_\beta K_\beta \tag{3.54}$$

where the coefficient K_β is called the drag force shape factor.

The F_D value of an irregular-shaped particle varies depending on the orientation to fluid flow, and thus the drag shape factor K_β depends on the orientation of the particle. Also, the value of K_β depends on the characteristic particle diameter employed in the equation. The value for equivalent volume diameter x_V differs from that for the equivalent surface diameter x_S, respectively.

Stokes diameter x_{st} is generally given using the volume shape factor as follows:

$$x_{st}^2 = \frac{6}{\pi} \frac{\phi_{3,\beta}}{K_\beta} x_\beta^2$$

(3.55)

If the drag shape factors with the axes of an ellipsoid orthogonal and parallel to the flow are $K_{\beta a}$ and $K_{\beta c}$, respectively, the factor for random orientation can be given by the following equation[30]:

$$3/\bar{K}_\beta = (1/K_{\beta a} + 2/K_{\beta c})$$

(3.56)

Dynamic Shape Factor

The dynamic shape factor κ is defined as follows[31]:

$$\kappa = \frac{\text{fluid drag force acting on a particle}}{\text{drag force to a sphere with equal volume}}$$

(3.57)

and is equal to the drag force shape factor of the equivalent volume diameter K_V. It is also given by the ratio of the volume equivalent diameter and the Stokes diameter, as follows:

$$\kappa = (x_V / x_{st})^2$$

(3.58)

Also, according to $(\phi_V/\kappa)x_V^2 = (\phi_A/K_A)x_A^2$, the following relation is obtained:

$$\kappa = \frac{\pi}{6\phi_A}\left(\frac{x_V}{x_A}\right)^2 K_A = \left(\frac{\pi}{6\phi_A}\right)^{1/3} K_A$$

(3.59)

κ of a spheroid is shown in Figure 3.10.[2,32]

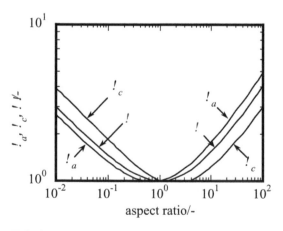

FIGURE 3.10 Dynamic shape factor of spheroid.

The Wadell sphericity is related with the dynamic shape factor in the following expression:

$$\psi_{\mathrm{S}} = (K_{\mathrm{A}} / \kappa)^2 = (6\phi_{\mathrm{A}} / \pi)^{2/3} \tag{3.60}$$

The dynamic shape factor of agglomerates with porosity ε is given by the volume shape factor $\phi_{\mathrm{V,agst}}$ based on the Stokes diameter of aggregate x_{agst}, as follows[33]:

$$\kappa = \frac{(x_{\mathrm{V}} / x_{\mathrm{agst}})^2}{(1-\varepsilon)} = \left(\frac{6}{\pi}\right)^{2/3} \frac{\phi_{\mathrm{V,agst}}^2}{(1-\varepsilon)} \tag{3.61}$$

For the blocky and chainlike aggregate composed by n particles, empirical relationships for κ are obtained, as follows[34]:

$$\begin{aligned} &\kappa \approx 1.232, \quad (\text{blocky}, \ 5 \le n \le 23), \\ &\kappa \approx 0.863 n^{1/3}, \quad (\text{chainlike}, \ n \le 8) \end{aligned} \tag{3.62}$$

1.3.5 CONCLUDING REMARKS

The shape descriptors presented here are mainly classified into two categories based on the scale of inspection. The descriptors derived from geometric quantities represent overall or mesoscopic features of shape. Those descriptors convenient to treat the shapes, however, are not unique with the shapes. This means that the form cannot be reconstructed using the overall or mesoscopic descriptor.

On the other hand, harmonic analysis, such as the Fourier transformation, for several images of a particle is useful. The harmonic spectrum extracted from the image function describes not only the overall shape feature but the microscopic shape. Further, the shape reconstruction can be done using the Fourier coefficients. However, the Fourier descriptor includes too much information on the shape to select appropriate parameters for characterization or distinction of particles.

In this paper, classification and distinction of particles based on shape have not been treated, but these are significant in practical powder processing. The statistical methods and artificial neural networks are effective tools for particle characterization.[35,36,19]

The form of a particle is essentially a 3D property. A few descriptors for 3D shape have been proposed and used in restricted fields. The 3D morphological analysis including measuring techniques in three dimensions should be developed.

REFERENCES

1. Allen, T., in *Particle Measurement,* Chapman and Hall, London, 1981, p. 107.
2. Endoh, S., *J. Powder Technol. Japan,* 29, 854, 1992; 35, 383, 1998.
3. Chan, L. C. L. and Page, N. W., *Part. Part. Syst. Charact.,* 14, 67, 1997.
4. Heywood, H., *Trans. Inst. Chem. Eng.,* 25S, 14, 1947.
5. Wadell, H., *J. Geo.,* 40, 443, 1932; 43, 250, 1935; *J. Franklin Inst.,* 217, 459, 1934.
6. Tsubaki, J. and Jimbo, G., *Powder Technol.,* 22, 161, 1979.
7. Hausner, H. H., *Planseeber. Pulvermetall.,* 14, 75, 1966.
8. Faris, N., Pons, M. N., et al., *Powder Technol.,* 133, 54, 2003.
9. Medaria, A. I., *Powder Technol.,* 4, 117, 1970/71.
10. Pons, M. N., Vivier, H., et al., *Powder Technol.,* 103, 44, 1999.
11. Vocak, M., Pons, M. N., et al., *Powder Technol.,* 97, 1, 1998.
12. Schwarcz, H. P. and Shane, K. C., *Sedimentology,* 13, 213, 1969.

13. Meloy, T. P., *Powder Technol.*, 16, 233, 1977.
14. Luerkins, D. W. et al., *Powder Technol.*, 31, 209, 1982.
15. Beddow, J. K. et al., *Powder Technol.*, 18, 19, 1977.
16. Zahn, C. T. et al., *IEEE Trans. Compt.*, C-21, 269, 1972.
17. Fong, S. T. et al., *Powder Technol.*, 22, 17, 1979.
18. Koga, J. et al., *Report IPCR,* 56, 63, 1980.
19. Hundal, H. S. et al., *Powder Technol.*, 91, 217, 1997.
20. Gotoh, K., *Powder Technol.*, 23, 131, 1979.
21. Shibata, T. et al., *J. Powder Technol. Jpn.*, 24, 217, 1987.
22. Browman, E. T. et al., *Geotechnique,* 51, 545, 2001.
23. Mandelbrot, B. B., *Fractal, Form, Chance and Dimension,* Freeman, San Francisco, 1977.
24. Kaye, B. H., *Powder Technol.*, 21, 1, 1978.
25. Schwarz, H. et al., *Powder Technol.*, 27, 207, 1980.
26. Kaye, B. H., *Part. Charact.*, 1, 14, 1984.
27. Suzuki, M. et al., *J. Powder Technol. Jpn.*, 25, 287, 1988.
28. Meakin, P., *J. Colloid Interface Sci.*, 96, 415, 1983.
29. Meakin, P., *Phys. Rev. A,* 29, 997, 1984.
30. Blumberg, P. N. et al., *AIChE J.*, 14, 331, 1968.
31. Fuchs, N. A., in *The Mechanics of Aerosol,* Pergamon Press, New York, 1964, p. 40.
32. Gans, R., *Sitzber math-physk Klasse Akad. Wiss. München,* 41, 191, 1911.
33. Kosaka, Y. et al., *J. Colloid Interface Sci.*, 84, 91, 1981.
34. Stober, W., in *Fine Particles,* Liu, B. Y. H., Ed., Academic Press, New York, 1970, p. 363.
35. Chien, Y. T., *Interactive Pattern Recognition,* Marcel Dekker, New York, 1978.
36. Endoh, S. et al., *Kagaku Kogaku Ronbunshu,* 8, 476, 1982.

1.4 Particle Density

Yasuo Kousaka
Osaka Prefecture University, Sakai, Osaka, Japan

Yoshiyuki Endo
Sumitomo Chemical Co. Ltd., Osaka, Japan

1.4.1 DEFINITIONS

The density of particles and powder is a physical property as important as the particle size. Although the density is defined as the ratio of mass to volume, the density of the particles does not necessarily agree with that of the particle material, as a particle occasionally includes pores, as shown in Figure 4.1. The density defined for particles and powder is as follows.

True Density, ρ_s

This density is defined as the ratio of the mass of the particle to its actual volume excluding inside pores. If the particles are ground so fine as to make the pores disappear, the true density can be measured by the method described later.

Particle Density, ρ_p

This is defined as the particle mass divided by the particle volume, including the inside closed pores. The volume can be obtained by the method mentioned below.

Bulk Density, ρ_B

The bulk density is defined for powder or particle beds. It is given as the ratio of the mass of the powder bed in a vessel to the volume of the bed, including the voids among primary particles. If the values of ρ_p and ρ_B are measured, the porosity or void fraction ε can be determined by

$$\varepsilon = 1 - \left(\frac{\rho_B}{\rho_p} \right) \tag{4.1}$$

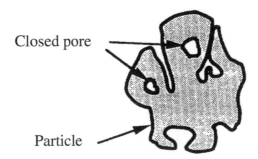

Closed pore

Particle

FIGURE 4.1 Illustration of a cross section of a particle.

49

1.4.2   MEASUREMENT METHOD FOR PARTICLE DENSITY

The measurement of the particle's density consists of measuring its mass and volume. The mass can be accurately measured with an electronic balance. However, it is difficult to measure the volume of particles. As described below, the volume of a particle is determined from the liquid volume increase upon adding the particle to a liquid, or in the case of gaseous medium, it is determined from the increased amount of volume or pressure.

Pycnometer Method

This is the most representative method using a liquid medium. Figure 4.2 shows the Wadon-type pycnometer, whose volume is generally about 2×10^{-5} m^{-3} (20 ml). In this method, the following masses are measured: empty pycnometer m_0, pycnometer containing liquid m_1, pycnometer including sample particles m_{sl}. The particle density ρ_p is given by

$$\rho_p = \frac{\rho_1(m_s - m_0)}{(m_1 - m_0) - (m_{sl} - m_s)} \tag{4.2}$$

where ρ_1 is the liquid density. If ρ_1 is not known, it can be measured with this pycnometer, whose volume is determined by filling it with water. In the present method, one must use a nonvolatile liquid in which the particles are wettable and insoluble. Air bubbles that adhere onto the particle surface can be effectively removed by heating or boiling the liquid under diminished pressure.

Constant Gas Volume Method

In this method, gas such as air or helium is used instead of liquid. The density of almost any particle can be measured because, in contrast with the pycnometer method, there is no problem of particle solubility in the fluid medium.

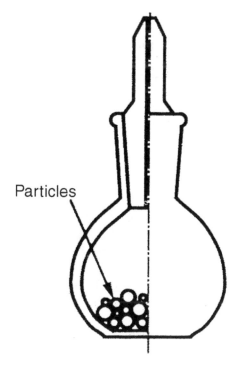

Particles

FIGURE 4.2   Pycnometer.

Figure 4.3 shows the basic principle of this method.[1] The well-known Beckmann method is based on a principle similar to that of the present method. First, without any particle sample in the cylinder, the piston is moved from position A to position B. In the compression process the volume and the pressure of gas in the cylinder change as follows:

$$V + v_0 \rightarrow V$$
$$P_a \rightarrow P_a + \Delta P_1 \tag{4.3}$$

where v_0 is the cylinder volume between planes A and B, and V is the volume between the plane B and the bottom of the cylinder. According to Boyle's law, the following relation holds.

$$P_a(V + v_0) = (P_a + \Delta P_1)V \tag{4.4}$$

This may be rewritten as

$$V = v_0 \left(\frac{P_a}{\Delta P_1} \right) \tag{4.5}$$

Second, after introducing a known weight of particle sample into the cylinder, the above procedure is repeated. In this case, the change in volume and pressure is

$$V + v_0 - V_s \rightarrow V - V_s$$
$$P_a \rightarrow P_a + \Delta P_2 \tag{4.6}$$

Therefore, the following equation is obtained:

$$V - V_s = v_0 \left(\frac{P_a}{\Delta P_2} \right) \tag{4.7}$$

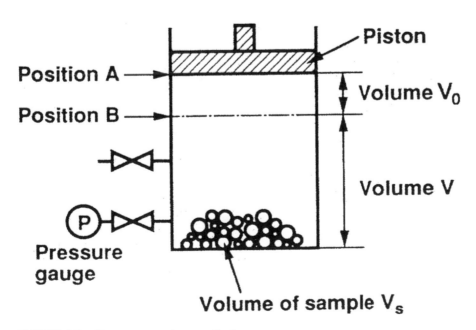

FIGURE 4.3 Constant gas volume method.

From Equation 4.5 and Equation 4.7, the particle volume V_s becomes

$$V_s = v_0 \left(\frac{P_a}{\Delta P_1} - \frac{P_a}{\Delta P_2} \right)$$

(4.8)

Although some other methods are available,[2-4] their principles are almost the same as that described here.

REFERENCES

1. Mular, A. L., Hockings, W. A., and Fuerstenau, D. W., *AIME Trans. Soc. Mining Eng.,* 226, 404–406, 1963.
2. von Neumann, B., Gross, W., Kremser, L., and Schmidt, J., *Brennstoff Chemie,* 15, 161–180, 1934.
3. Poisson, R., *Peintures Pigments Vernis,* 42, 767–769, 1966.
4. von Rennhack, R., *Gas und Wasserfach,* 109, 1209–1213, 1968.

1.5 Hardness, Stiffness, and Toughness of Particles

Mojtaba Ghadiri
University of Leeds, Leeds, United Kingdom

Hardness, stiffness, and toughness represent the resistance of solid materials to plastic and elastic deformations and crack propagation, respectively. They are important in processes such as comminution, tableting, polishing, abrasive and erosive wear, and attrition, as they define the mode and pattern of mechanical failure. Their characterization is highly desirable for product and process improvement and optimization.

1.5.1 INDENTATION HARDNESS

The stress required for plastic deformation can be quantified directly only for a number of well-defined geometric shapes such as spherical, conical, and pyramidal objects[1,2] by, for example, compression of a particle between two platens. For nonspherical particles the indentation hardness is easier to measure (as compared to the deformation stress), from which the resistance to plastic deformation can be inferred. Hardness is related to a number of fundamental properties of solids, such as the yield stress, work-hardening rate, and anisotropy, and is influenced by the geometry of the deformed region.

Indentation hardness is commonly defined as the resistive pressure, H, when a permanent impression is made. It is measured after a certain amount of plastic strain has been induced, typically around 8%,[3] and is therefore affected by anisotropy in the structure and the work-hardening process. However, it is still useful to relate hardness to yield stress, and the ratio of hardness over yield stress is commonly defined as the constraint factor.[4] This factor is usually much greater than unity, because the plastic deformation in a hardness test is constrained by the surrounding region, which is under elastic deformation.

The indentation hardness and constraint factor depend on the properties of the deformed material, the indenter geometry,[5] and the coefficient of friction.[6,7] The material properties of concern here are the elastic modulus, yield stress, rate of work-hardening, and anisotropy. The yield stress is a function of temperature and strain rate.[8] As temperature is decreased, the dislocation mobility is reduced, and hence the yield stress increases. A similar behavior is observed when the strain rate is increased.

The hardness is related to the yield stress and work-hardening by considering the characteristics of the uniaxial stress–strain curve for several types of material failure. Figure 5.1A and B represent materials that do not work-harden; elastic deformation is included in Figure 5.1B, while ideal plastic behavior is shown in Figure 5.1A. The effect of work-hardening is included with and without an elastic deformation in Figure 5.1D and C, respectively. The elastic deformation and rate of work-hardening have a very strong influence on the value of the constraint factor, and this is analyzed by the theoretical models of indentation hardness as outlined below.

Rigid–Perfectly Plastic Indentation

A general treatment of the indentation process for this case has been described by Hill,[9] based on the wedge-cutting mechanism. Here, the deformation process is represented by Figure 5.1A. The deformation for a blunt indenter has also been analyzed by approximating the process to the deformation by a rigid flat punch penetrating a semi-infinite rigid–perfectly plastic medium.[6]

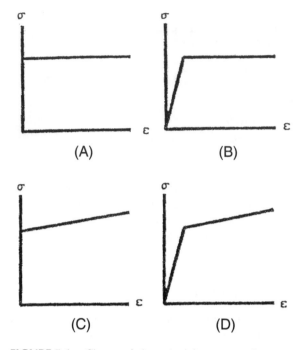

FIGURE 5.1 Characteristic uniaxial stress–strain curves: (A) rigid, perfectly plastic, (B) elastic, perfectly plastic, (C) rigid, plastic with work-hardening, and (D) elastic, plastic with work-hardening.

According to the wedge-cutting mechanism, the indentation is treated as cutting by a smooth wedge, where it is assumed that slip occurs along planes of maximum shear. In this case, a simple relationship exists between the yield stress, Y, and hardness, H:

$$\phi = H/Y \cong 3 \tag{5.1}$$

Tabor[3] tested various ductile metals such as aluminum, copper, and mild steel and found that the experimental results are in good agreement with the above theory. However, the above simple relationship does not apply to anisotropic and work-hardening materials. Furthermore, the experimental observations made by Samuels and Mulhearn[10] have shown that the wedge-cutting mechanism operates only when the wedge angle is less than 30°. For wedge angles greater than 30°, the plastic deformation occurs by radial compression, producing hemispherical surfaces of constant strain, where Hill's theory of expansion of a spherical cavity is more applicable (see below).

Elastic–Perfectly Plastic Indentation

The deformation characteristics are shown in Figure 5.1B. The expansion of a spherical cavity in an infinite elastic–plastic medium has been used to describe the process of indentation, where the mean pressure, p, at which the cavity expands, is given by Hill[9]:

$$\frac{p}{Y} = \frac{2}{3}\left\{1 + \ln\left(\frac{E}{3(1-v)Y}\right)\right\} \tag{5.2}$$

Where E is Young's modulus and v is Poisson's ratio. The contact pressure, p, is equivalent to the hardness of the flat surface, H, and the above model has been widely used to interpret the problems associated with elastic–plastic indentation.

Following Hill's approach, Marsh[11] has proposed a similar expression for the expansion of a hemispherical cavity, where the constraint around the plastic zone is less than that around a spherical cavity. It can be seen from Equation 5.2 that p/Y (or H/Y) is a function of E/Y. Tabor has presented the hardness of a wide range of materials on such a coordinate system.[7] The onset of plastic flow occurs at a ratio of p/Y of about 1, and the rigid–perfectly plastic theory predicts $p/Y = 3$.

Indentation is often made with a conical or pyramidal shape indenter where the angle of indenter affects the indentation pressure. A model of indentation hardness for this case has been presented by Johnson[5] based on the spherical cavity expansion as

$$\frac{H}{Y} = \frac{2}{3}\left(1 + \ln\frac{E\cot\theta}{3Y}\right) \tag{5.3}$$

where θ is the indenter semiangle. For a spherical indenter, $\cot\theta$ may be replaced by a/R, where a is the radius of the impression and R is the indenter radius. This is a simple correlation between the indentation hardness and the indenter angle. However, Equation 5.3 is not valid for very sharp indenters having a narrow apex angle (e.g., wedges[12] and cones[13]).

Work-Hardening Model

For materials which exhibit significant work-hardening, the hardness is much larger than the yield stress. This is because hardness is conventionally measured at a strain of about 8%,[4] where significant work-hardening can take place. It is possible to use the analysis of expansion of a spherical cavity and to incorporate the effect of work-hardening, whose characteristics are shown in Figure 5.1C and D. Here, the stress after yielding can be expressed in the form of

$$\sigma = Y + \Pi \tag{5.4}$$

where Π is the augmented stress associated with work-hardening, and it is a function of strain ε. Π is normally approximated by a linear relationship of the form

$$\Pi = \Pi' \times (\varepsilon - \varepsilon_o), \varepsilon > \varepsilon_o \tag{5.5}$$

where Π' is the rate constant of hardening, and ε_o is the strain corresponding to the onset of work-hardening. For an incompressible material, in other words, no change of volume ($v = 0.5$), a simple relationship between the resistive pressure and the rate of work-hardening has been given by Bishop et al.[14]:

$$H = p = \frac{2}{3}Y\left\{1 + \ln\left(\frac{2E}{3Y}\right)\right\} + \frac{2\pi^2}{27}\Pi' \tag{5.6}$$

An alternative approach to determine the constraint factor for work-hardening materials has been suggested by Tabor,[4] where the constraint factor is described as the ratio of the indentation hardness to the representative uniaxial flow stress, Y_R, at an effective strain ε_R.

$$\phi = H/Y_R \tag{5.7}$$

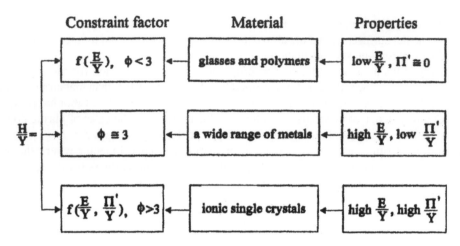

FIGURE 5.2   Relationship between hardness and yield stress.

For a Vickers indenter, the average effective strain produced by the indenter is between 8% and 10%.[3] The flow stress at this value of strain is taken to calculate the constraint factor from Equation 5.7. This approach produces significantly different values for ϕ and does not include the yield stress and work-hardening in an explicit way.

In summary, the relationship between the hardness and yield stress is illustrated in Figure 5.2. It is clear that the material is subjected to more constraint as its morphology changes from the amorphous state to the crystalline state.

1.5.2 MEASUREMENT OF HARDNESS

Measurement and theory of hardness testing have been developed in the field of material testing, where it has been possible to carry out hardness tests on large specimens. Until very recently, it has been difficult to characterize the hardness of small particles. However, with rapid progress in the past two decades in the field of nanotesting and atomic force microscopy, detailed characterization of surfaces of even micrometer-size particles is now possible.

The measurement of hardness started with scratch testing of flat surfaces by various particles having varying degrees of hardness, for example, the Mohs hardness test.[15] This method provided a qualitative indication of resistance to plastic flow, but it was too complex for analysis to obtain quantitative results. The method of indentation hardness testing evolved by the use of well-defined indenter geometry, such as spherical, pyramidal, or conical shapes. Tabor's pioneering work[3] established the basis of understanding of the deformation processes involved in the indentation. A number of common test methods are available using different indenter shapes: Brinell, using a 10 mm sphere of steel or tungsten carbide; Vickers, using a diamond pyramid; Knoop, using also a diamond pyramid; and Rockwell, using a diamond cone with a spherical tip and steel spheres.[7] In all these tests, a load is applied to the surface under testing by slowly penetrating the indenter at the right angle into the specimen. The hardness number is calculated from the applied load, the projected area of the impression, and by taking account of the shape of the indenter. The relevant formula for hardness numbers may be found, for example, from Tabor.[7]

The above techniques cannot be easily applied to the measurement of hardness of a particle unless the particle is large, that is, above a few millimeters. However, the recent development of nanotest methods has enabled a full characterization of mechanical properties as well as topography of fine particulate solids.[16] Because of the small size of impression, which can be submicrometer, almost all nanoindentation methods measure the resistance of material to penetration (i.e., the applied load) as a function of depth of penetration, from which an effective hardness can be inferred.

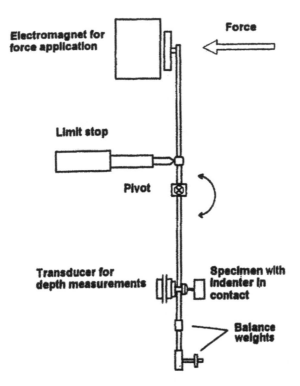

FIGURE 5.3   Schematic diagram of the nanoindenter device.
[From Arteaga, P. A. et al. *Tribol. Int.*, 26 (Suppl 5), 305–310,
1993. With permission.]

A schematic diagram of a typical nanotest device is shown in Figure 5.3. There are several designs.[16-18] The design shown here was originally developed by Pollock et al.[19] and is now manufactured by Micro Materials, Ltd., Wrexham, U.K. The device measures the movement of a calibrated diamond indenter penetrating into a specimen surface at a controlled loading rate. It uses a pendulum pivoted on bearings that are essentially frictionless. A coil is mounted at the top of the pendulum so that when a coil current is present the coil is attracted toward a permanent magnet, producing motion of the indenter toward the specimen and into the specimen surface. The displacement of the diamond is measured by means of a parallel plate capacitor, one plate of which is attached to the diamond holder; when the diamond moves, the capacitance changes, and this change is measured by means of a capacitance bridge. The most commonly used indenter is a Berkovitch diamond, a three-sided pyramid, which is particularly suitable for nanoindentation work because it can be machined down very accurately to a very sharp tip. However, 90° trigonal and spherical indenters have also been used. The above device is capable of measuring load and displacement with typical resolutions around 25 μN and 20 nm, respectively.

In a nanoindentation test, the specimen is first brought in contact with the indenter; then the load is increased at a prescribed rate until the desired maximum load is reached, and then it is decreased back to zero at the same rate. Therefore a continuous recording of the penetration depth and the applied load is made, from which an effective hardness modulus may be inferred at a specified strain rate by invoking a relation between the penetration depth and the indentation area. A typical depth/load cycle is shown in Figure 5.4. During the unloading part of the cycle, the elastic deformation of the sample will recover, and therefore the unloading curve is not horizontal. The maximum depth and characteristics of the unloading curve are used to calculate the hardness. However, it is also possible to evaluate the elastic modulus from the unloading curve. Two distinct regions can be observed here: the initial unloading stage bearing all the features of an elastic unloading, and a

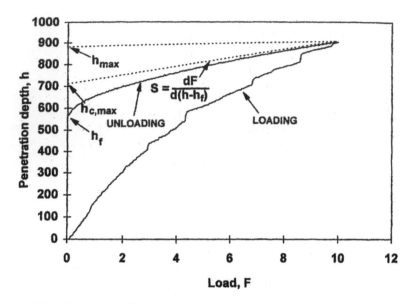

FIGURE 5.4   Typical loading/unloading curve in a nanoindentation test.

highly nonlinear stage toward the end of the unloading cycle. Reasons for this nonlinearity are complex; changes in the contact area arising from an increase in the apical angle of the impression are thought to be responsible.[20-22] There are several methods for the evaluation of hardness from such data.[17,20] The method described below is based on the analysis of Oliver and Pharr[17] and is one of the most widely used methods.

At the initiation of unloading (Figure 5.4), the elastic component of the deformation starts to recover so that contact between the material surface and the indenter is still maintained. As the load is decreased the elastic displacement continues to recover until at zero loads the displacement reaches a final value, h_f. At the peak load, F_{max}, the displacement is h_{max}. A schematic diagram depicting a cross-sectional view of an indentation is shown in Figure 5.5. The total displacement, h, can be described at any time during loading as

$$h = h_c + h_s \tag{5.8}$$

where h_c is the contact depth, that is, the vertical distance along which contact is made, corresponding to the contact size, a, and h_s is the elastic displacement of the surface at the perimeter of the contact. In this analysis, hardness, H, can be determined by

$$H = \frac{F_{max}}{A} \tag{5.9}$$

where A is the projected area of the impression. Now, from the geometry of the indenter

$$A = kh_{c,max}^2 \tag{5.10}$$

with k being the indenter shape factor, and $h_{c,max}$ the contact depth at F_{max}. Now, to determine the contact depth from the experimental data, it can be noted that

$$h_{c,max} = h_{max} - h_{s,max} \tag{5.11}$$

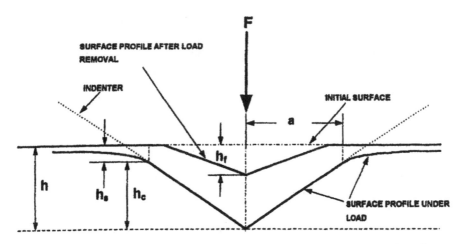

FIGURE 5.5   A schematic diagram depicting a cross-sectional view on an indentation.

which follows directly from Equation 5.8. h_{max} is experimentally measured, and $h_{s,max}$ can be ascertained from the unloading data by the use of an expression derived by Oliver and Pharr[17] from Sneddon's solution[23] for the elastic contact of indenters of different shapes. This expression is given by

$$h_{s,max} = \varepsilon \frac{F_{max}}{S} \tag{5.12}$$

where S is the stiffness at load F_{max} (as shown in Figure 5.4), and ε is a constant that depends upon indenter geometry. The value of ε for a Berkovitch indenter cannot be determined analytically; however, it has been shown that the Berkovitch indenter behaves very much like a paraboloid of revolution for which a solution for the parameter ε exists, that is, $\varepsilon \approx 0.75$.[24] In practice, the stiffness, S, is obtained by fitting the unloading data to a power law relation of the form

$$F = \alpha(h - h_f)^m \tag{5.13}$$

where $(h - h_f)$ is the elastic displacement of the indentation during unloading; α and m are constants. The values of α, m, and h_f are all determined by a nonlinear curve fitting procedure. Equation 5.13 can then be differentiated analytically with respect to $(h - h_f)$ to obtain the value of stiffness

$$S = \frac{dF}{d(h - h_f)} \tag{5.14}$$

which can then be evaluated at the peak load, F_{max}. Then, by first substituting the value of stiffness obtained from Equation 5.14 into Equation 5.12, the value of $h_{s,max}$ is calculated. Equation 5.11, Equation 5.10, and Equation 5.9 are then used to determine the hardness.

1.5.3   MEASUREMENT OF STIFFNESS

Particles subjected to mechanical stresses below the critical level for plastic yielding undergo elastic deformation. The resistance to elastic deformation is represented by the ratio of load over the extent of deformation, termed stiffness. As in the case of plastic deformation, for nonspherical particles the contact geometry is too complex for characterizing the stiffness by standard techniques. The approach outlined above for indentation hardness measurement can also be used to evaluate the stiffness. For materials obeying Hooke's law, the stiffness can be calculated from Young's modulus, and this in turn can be determined from the unloading portion of the depth–load data (as shown in Figure 5.4) by the use of Sneddon's analysis.[23] This analysis is independent of indenter geometry[24] and is therefore applicable to results obtained with the Berkovitch indenter

$$E_r = \frac{\sqrt{\pi}}{2} \frac{S}{\sqrt{A}}$$

(5.15)

where S is calculated from Equation 5.14, A is the contact area given by Equation 5.10, and E_r is the reduced modulus defined by

$$\frac{1}{E_r} = \frac{\left(1 - v^2\right)}{E} + \frac{\left(1 - v_i^2\right)}{E_i}$$

(5.16)

where E and v are Young's modulus and Poisson's ratio of the specimen, and E_i and v_i are the same parameters for the diamond indenter. Since Young's modulus of the indenter is two orders of magnitude greater than that of the test materials, the reduced modulus is effectively Young's modulus of the specimen. In conclusion, the above approach enables the determination of hardness and Young's modulus at the micro- and nanoscale levels. However, great care is needed for the experimental procedure and interpretation of the data because of the scale of operation.

1.5.4 MEASUREMENT OF TOUGHNESS

Fracture toughness represents the resistance of solid materials to crack propagation. The basis of its measurement lies in the principles of linear elastic fracture mechanics and the Griffith energy balance for crack propagation.[25] For large specimens, the measurement of fracture toughness is made using a uniform stress field. A common method is the three-point bend test as shown in Figure 5.6, where a crack of length c is made much shorter than the specimen's depth, d, providing uniform

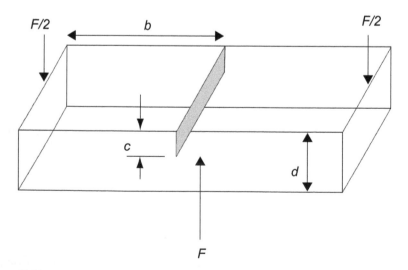

FIGURE 5.6 Three-point bend test.

tension at the crack tip. In this case the fracture toughness is calculated based on the simple beam theory and is given by[25]

$$K = m(3Fb/d^2)(\pi c)^{1/2} \tag{5.17}$$

where m is a factor which depends on c/d and has a limiting value of 1.12. For small particles the indentation method is used to measure the fracture toughness as described below.

Indentation Fracture Toughness

In indentation fracture by a sharp indenter causing plastic deformation, there is a critical size of indentation, r_c, above which fracture is initiated[26,27]:

$$r_c \cong \alpha \frac{E\Gamma}{Y^2} \tag{5.18}$$

where α is a constant, E is Young's modulus, Γ is the fracture surface energy, and Y is the yield stress in a uniaxial compression test. This critical size reflects the influence of the mechanical properties. The yield stress is related to the hardness, as $Y = H/\phi$, where ϕ is the constraint factor. The fracture surface energy and Young's modulus are related to the fracture toughness based on linear elastic fracture mechanics. For the case of plane strain,

$$E\Gamma = K_c^2(1 - v^2) \tag{5.19}$$

where v is Poisson's ratio. Considering that v^2 is much smaller than unity, the critical size can be written approximately in the following form:

$$r_c \approx \alpha'(\frac{K_c}{H})^2 \tag{5.20}$$

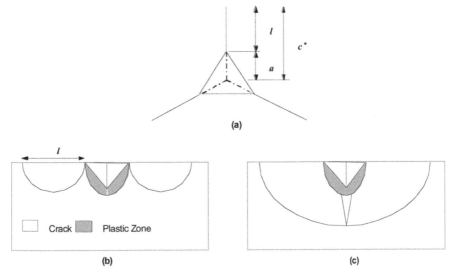

FIGURE 5.7 Indentation crack geometry: (A) Berkovitch impression; cross-section view of (B) radial crack and (C) median crack systems.

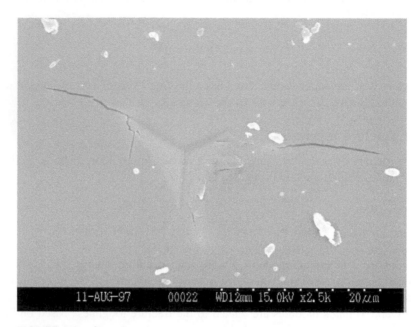

FIGURE 5.8   Scanning electron micrograph of a nanoindentation in the (001) face of lab-produced paracetamol. Applied force: 50 mN. Indenter: Berkovitch pyramid. [From Arteaga, P. A. et al., *Tripartite Programme in Particle Technology,* Final Report, 1997. With permission.]

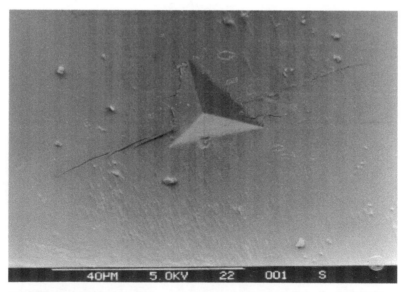

FIGURE 5.9   Scanning electron micrograph of a nanoindentation in the (100) face of lab-produced lactose. Applied force: 100 mN. Indenter: Berkovitch pyramid. [From Arteaga, P. A. et al., *Tripartite Programme in Particle Technology,* Final Report, 1997. With permission.]

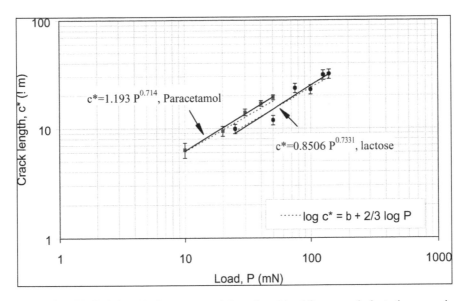

FIGURE 5.10 Relationship between crack length and load from nanoindentation experiments on paracetamol and lactose crystals. [From Arteaga, P. A. et al., *Tripartite Programme in Particle Technology*, Final Report, 1997. With permission.]

When the surface is indented by a sharp indenter, the plastic zone size increases until a critical size is reached, above which cracks propagate. From the indentation fracture mechanics, it can be shown that the size of the stress field scales with $\sqrt{r_c}$. Therefore the crack length is controlled by r_c, which in turn is related to K_c. The fracture toughness can then be determined from the relationship of the crack length with applied load by the use of an expression which has the following general form[28]:

$$K_c = x_v \left(\frac{a}{l}\right)^k \left(\frac{E}{H}\right)^n \frac{P}{c^{*3/2}} \tag{5.21}$$

where x_v is a calibration factor, a/l is the ratio of crack length to impression size as defined in Figure 5.7, c^* ($= a + l$) is the crack size as measured from the center of the impression, P is the load, and k and n are power indices. It is clear that this method relies on the factor $P/c^{*3/2}$ being a constant (for a given material). There are several correlations for fracture toughness cited in the literature, all of which suggest different values for x_v, k, and n. A review of these correlations is given by Ponton and Rawlings.[28]

The main advantage of measuring fracture toughness by indentation is that no special specimen geometry is required. However, since it relies upon the indentations producing a radial or median crack system with well-formed surface traces, its applicability is limited to materials that exhibit brittle and semibrittle fracture during indentation. In a typical test a number of different loads are applied, and the indent size and crack length, as shown in Figure 5.7, are measured. A plot of crack length, c, as a function of $P^{2/3}$ should produce a straight line, the slope of which would give K_c.

As an example, the results of measurement of fracture toughness of paracetamol and lactose crystals by nanoindentation are shown below. Figure 5.8 and Figure 5.9 show indentations on the (001) face of paracetamol and (100) face of lactose crystals, respectively. It can be seen clearly that surface traces of well-defined cracks are produced on nanoindentation of these crystals.

The relationships between crack length and load are shown in Figure 5.10 for indentations in lab-produced crystals of paracetamol and lactose. Error bars of one standard deviation of the mean are also included in the figure. The load–crack length relationship given previously, specifically, $P/c^{*3/2} \approx$ constant, has been added for comparison. The values of the intercept, b, as indicated in the figure, have been chosen to be slightly different from the values of the intercept for the best fits of the experimental data to show the close agreement in the slopes of the lines. It is clear that the load–crack length data for both materials approximate reasonably well the expected relationship. The results of fracture toughness for both paracetamol and lactose are shown in Figure 5.11 as plots of fracture toughness against load. Here, the equation proposed by Laugier[29] has been used to calculate K_c, as this correlation has been developed for materials which are well behaved in their indentation response, and it takes into account the role of plastic deformation in driving the crack growth. This correlation, which is based on the modification of the equation proposed by Anstis et al.[30] to include the effect of the plastic driving force, gives the constants of the equation as $x_v = 0.010$, $k = 0$, and $n = 2/3$.

Care needs to be taken to account for the effect of indenter geometry. Most correlations reported in the literature are based on the Vickers indentation with four associated radial cracks, while with Berkovitch (three-sided pyramid) indentations a smaller number of cracks is produced. The work of Dukino and Swain[31] can be used to account for this difference. These workers compared measurements of fracture toughness obtained from Berkovitch and Vickers indentations and proposed a method to relate the two measurements.

An issue that is very important in predicting the comminution behavior of powders is the determination of the critical particle size for breakage transition. Several models have been proposed.[32–35] All these models, although based on different considerations and assumptions, follow the same general form. Therefore, with the values of fracture toughness and hardness of paracetamol and lactose as measured

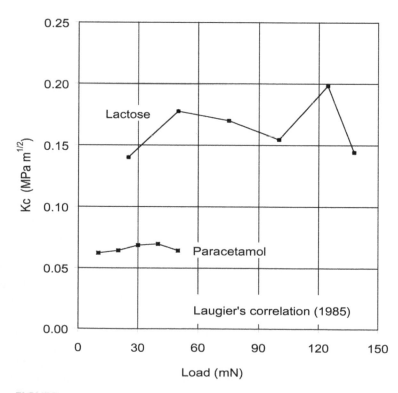

FIGURE 5.11 Plots of fracture toughness against load for paracetamol and lactose crystals.

by nanoindentation, the values of r_c for paracetamol and lactose are calculated as 0.5 μm and 2.4 μm, respectively.[36] These values seem reasonable. However, further work is needed to establish a comparison with the actual sizes of the particles obtained in comminution experiments.

REFERENCES

1. Chaudhri, M. M., *J. Mater. Sci.*, 19, 3028–3042, 1984.
2. Ghadiri, M., Arteaga, P., and Cheung, W., *Proceedings of Second World Congress in Particle Technology*, Kyoto, 1990, pp. 438–446.
3. Tabor, D., *Proc. Roy. Soc.*, A192, 247, 1948.
4. Tabor, D., *Microindentation Techniques in Material Science and Engineering*, ASTM STP889, Blau, P. J. and Lawn, B. R., Eds., ASTM, Philadelphia, 1986, pp. 129–159.
5. Johnson, K. L., *J. Mech. Phys. Solids*, 18, 115–126, 1970.
6. Gilman, J. J., *The Science of Hardness Testing and Its Research Application*, Westbrook, J. H. and Conrad, H., Eds., American Society for Metals, Materials Park, Ohio, 1971, pp. 54–58.
7. Tabor, D., *Rev. Phys. Tech.*, 1, 145–155, 1970.
8. Hull, D. and Bacon, D. J., *Introduction to Dislocations*, Pergamon Press, Oxford, 1984.
9. Hill, R., *The Mathematical Theory of Plasticity*, Clarendon Press, Oxford, 1950.
10. Samuels, L. E. and Mulhearn, T. O., *J. Mech. Phys. Solids*, 5, 125–134, 1957.
11. Marsh, D. W., *Proc. Roy. Soc. A*, 279, 420, 1964.
12. Hirst, W. and Howse, M., *Proc. Roy. Soc. Lond. A*, 311, 429–444, 1969.
13. Atkins, A. G. and Tabor, D., *J. Mech. Phys. Solids*, 13, 149–164, 1965.
14. Bishop, R. F., Hill, R., and Mott, N. F., *Proc. Phys. Soc.*, 57, 147–159, 1945.
15. Tabor, D., *Proc, Phys, Soc. B*, 67, 249–257, 1954.
16. Arteaga, P. A., Ghadiri, M., Lawson, N. S., and Pollock, H. M., *Tribol. Int.*, 26 (Suppl. 5), 305–310, 1993.
17. Oliver, W. C. and Pharr, G. M., *J. Mater. Res.*, 7 (Suppl. 6), 1564–1583, 1992.
18. Field, J. S. and Swain, M. V., *J. Mater. Res.*, 10 (Suppl. 1), 101–112, 1995.
19. Pollock, H. M., Maugis, D., and Barquins, M., *Microindentation Techniques in Materials Science and Engineering*, ASTM STP 889, Blau, P. J. and Lawn, B. R., Eds., ASTM, Philadelphia, 1986, pp. 47–71.
20. Doerner, M. F. and Nix, W. D., *J. Mater. Res.*, 1 (Suppl. 4), 601–609, 1986.
21. Loubet, J. L., Georges, J. M., and Meille, G., *Microindentation Techniques in Materials Science and Engineering*, ASTM STP 889, ASM, Blau, P. J. and Lawn, B. R., Eds., ASTM, Philadelphia, 1986, pp. 72–89.
22. Ion, R. H., Pollock, H. M., and Roques-Carmes, C., *J. Mater. Sci.*, 25, 1444–1454, 1990.
23. Sneddon, I. N., *Int. J. Eng. Sci.*, 3, 47–57, 1965.
24. Pharr, G. M., Oliver, W. C., and Brotzen, F. R., *J. Mater. Res.*, 7 (Suppl. 3), 613–617, 1992.
25. Lawn, B. R. and Wilshaw, T. R., Eds., *Fracture of Brittle Solids*, Cambridge University Press, Cambridge, 1975.
26. Puttick, K. E., *J. Phys. D Appl. Phys.*, 11, 595–604, 1978.
27. Puttick, K. E., *Energy Scaling in Elastic and Plastic–Elastic Fracture*, paper presented at the Griffith Centenary Meeting, U.K., 1993.
28. Ponton, C. B. and Rawlings, R. D., *Mater. Sci. Technol.*, 5, 865–872, 1989.
29. Laugier, M. T., *J. Mater. Sci. Lett.*, 4, 1539–1541, 1987.
30. Anstis, G. R., Chantikul, P., Lawn, B. R., and Marshall, D. B., *J. Am. Ceram. Soc.*, 64, 533–538, 1981.
31. Dukino, R. D. and Swain, M. V., *J. Am. Ceram. Soc.*, 75 (Suppl. 12), 3299–3304, 1992.
32. Lawn, B. R. and Evans, A. G., *J. Mater. Sci.*, 12, 2195–2199, 1977.
33. Kendall, K., *Nature*, 272, 710–711, 1978.
34. Hagan, J. T., *J. Mater. Sci.*, 16, 2909–2911, 1979.
35. Puttick, K. E., *J. Phys. D Appl. Phys.*, 13, 2249–2262, 1980.
36. Arteaga, P. A., Bentham, A. C., and Ghadiri, M., *Tripartite Programme in Particle Technology*, Final Report, University of Surrey, Guildford, U.K., 1997.

1.6 Surface Properties and Analysis

Masayoshi Fuji, Ko Higashitani, and Yoichi Kanda
Nagoya Institute of Technology, Tajimi, Gifu, Japan

1.6.1 SURFACE STRUCTURES AND PROPERTIES

Particle Surface and Character

A material that does not have flowability at temperatures below its melting point can exist as a solid in various shapes. The significant feature is the degree of the unsaturation of the chemical bond to which the internal structure terminates when the surface of the solid is looked at on the microscopic scale. It is not possible to diffuse because the activation energy of the surface diffusivity is generally high, when the adjacent potential energy is different by atoms, ions, or molecules that compose the surface. Therefore, a feature of the solid surface is that it is not able to produce a uniform energy like the liquid surface. Various characteristics of a solid surface are strongly influenced by these two features.

The population of atoms, ions, and molecules that compose the fine particle surface increases in comparison with the particulate inside. In this case, a lot of peculiar characteristics begin to appear in fine particles.[1] Especially, the property concerning handling the powder begins to strongly depend on the character of the particulate surface.[2] Therefore, how one can know the character of the particulate surface and control it are the major technologies in all fields where the powder has an effect. Here, we describe in detail an important matter for understanding the character of the particulate surface.

Surface Relaxation

Various surface relaxations appear for the stabilization because the solid surface has broken chemical bonds and exists in an unstable state. These relaxations are different depending on the kind of chemical bond, such as covalent bond, ionic bond, and metallic bond, and the kind of material. In this section we separately describe chemical relaxation with a chemical change and physical relaxation with a structural change. Moreover, because the solid is usually handled in the atmosphere, we also describe the stabilization of the surface by water adsorption.

Physical Surface Relaxation

An atom, ion, or molecule on the surface receives an unsymmetrical interaction force, because chemical bonds have been cut on the crystal surface. Therefore, these positions change from the position forecast from the internal crystal lattice. In the case of a surface of alkali halides, which are ionic crystals, the anion or the cation is displaced to a position that is positive or negative in the vertical direction against the surface from the position expected from the lattice location. Figure 6.1 shows such an example. In the surface layer of the (100) face of NaCl, Cl⁻, which is the anion, is polarized with the Na⁺ ion on the circumference and, as a result, displaces from the surface to the outside. The cations internally displace.

The electrical double layer is formed as a result and the surface has been stabilized. That is, a center position for both ions does not exist on the same face. Moreover, the distance between the position of the first layer as the average of both ions and the position of the second layer is compressed neater than interlayer distances in the crystal. It is assumed that such displacement is caused from the first layer to the fifth layer in NaCl. A similar phenomenon in other materials is listed in Table 6.1. In the (100) face of MgO, whose bond energy is larger than the alkali halide though it is classified into the same ionic crystal system, the interionic distance of the first surface layer and the second layer is compressed to about 85% of the interlayer distance in the crystal. The physical surface relaxation phenomenon is presumed to be different depending on the difference in the ion species, the coordination number, the crystal face, and the bond energy for the crystal with the same ionic bond. A similar relaxation is also caused on the surface of a metallic crystal. The surface relaxation is high in the crystal face; therefore, the atomic density is small, as shown in Table 6.1. The change region is assumed to be from the surface to about

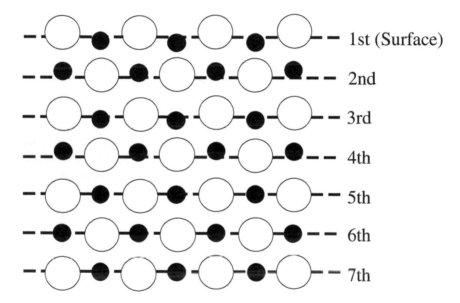

FIGURE 6.1   Physical relaxation model for (100) face of alkali halide alkali. Open circle is anion. Closed circle is cation. Dotted line indicates the position of inner crystal lattice.

TABLE 6.1   Physical Surface Relaxation

	Crystal structure	(110)	(001)	(111)
Cu	face-centered cubic	0.804	0.871	0.944
Al	face-centered cubic	0.900	1.00	1.15
Fe	body-centered cubic	1.00	0.986	0.85
NaF	rock-salt	—	0.972	—
NaCl	rock-salt	—	0.970	—
NaBr	rock-salt	—	0.972	—
MgO	rock-salt	—	0.85	—

* The data indicate the ratio of the length between surface layer and 2nd layer and the length between atoms in bulk.

two layers. The physical relaxation phenomenon also includes the relaxation phenomenon by the two-dimensional rearrangement of the surface atoms. In the (111) face of the Si crystal formed with the covalent bond, one of four sp³ covalent bonds is cut and is unstable. Therefore, the surface atom arrangement is caused by rearrangement in two dimensions, and changes into another regular surface structure (2 × 2). Moreover, the hybrid orbital of the Si atoms, which are adjacent to the surface, changes, and regular relief structures are formed. In the latter case, the bonding orbital of the Si atom alternately changes from the hybrid orbital of sp³ into sp² + p and 3p + s. In the former case, because the sp² orbit is contained, the bond angle in the Si interatomic is large, and the height of the Si atom is less than for the sp³ cases. On the other hand, because the bond angle in the Si interatomic is small when three p³ orbitals are contained, the height of the Si atom is greater than for the sp³ case. Therefore, the height of the Si atom alternately changes, and the formation of a regular relief structure occurs. The (2 × 1) structure in the Si(001) face and the (7 × 7) structure are well known.[6,7]

Chemical Surface Relaxation

Because a chemical bond is unsaturated in the solid surface as previously described, the reactiveness is high. When the solid is exposed to an atmosphere, the surface, where the reactivity is high, chemically reacts with the reactive gas in the atmosphere, that is, oxygen, carbon dioxide, or water vapor, and forms a surface composition different from the chemical composition of the bulk. The metallic surface is oxidized by atmospheric oxygen and water vapor, and corrosion is a daily phenomenon. Moreover, oxygen in the air is chemically adsorbed and an oxide is also formed on the surface of the nitride. This oxide surface further chemically adsorbs the water vapor, and a hydroxyl group is formed on the surface. Examples of the chemisorption of oxygen on a metallic surface and the nitride surface are as follows.

$$2Fe + O_2 \rightarrow 2FeO \tag{6.1}$$

$$4AlN + 3O_2 \rightarrow 2Al_2O_3 + N_2 \tag{6.2}$$

$$Si_3N_4 + 3O_2 \rightarrow 3\ SiO_2 + 4N_2 \tag{6.3}$$

$$2TiB_2 + 6H_2O \rightarrow 2TiO_2 + 2B_2H_6 + O_2 \tag{6.4}$$

The material chemically adsorbs the atmospheric carbon dioxide and forms a carbonic acid salt. For instance, carbonate and a surface hydroxyl group are formed with the carbon dioxide and water vapor in air on MgO, CaO, and the surface of the MgO–ZnO complex oxide. Water vapor is easily chemically adsorbed in general, and the surface becomes covered by hydroxyl groups on the oxide surface. Figure 6.2 shows the chemisorptions of water vapor on the surface of MgO having a simple crystal structure as one example. The chemisorption of water vapor has the ability to absorb only on the surface layer and absorb into a layer in the solid. Not only the first surface layer but even the inside of the solid MgO can absorb. The water vapor is chemically adsorbed, and the oxide is also formed on

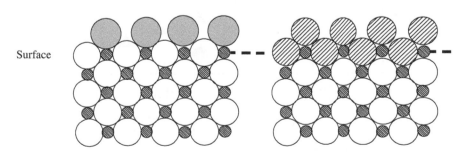

Surface

FIGURE 6.2   Chemical adsorption of water molecules on MgO (100).

the surface of nitride and boride, as shown by the following equations. In addition, the oxide surface changes into a surface of hydroxyl groups by chemisorption of the water vapor.

$$\equiv Si–O–Si\equiv + H_2O \rightarrow 2\equiv Si–OH \tag{6.5}$$

$$Si_3N_4 + 6H_2O \rightarrow 3SiO_2 + 4NH_3$$

$$\equiv Si–O–Si\equiv + H_2O \rightarrow 2\equiv Si–OH \tag{6.6}$$

The surface hydroxyl group is desorbed by heating. The desorption behavior not only differs depending on the oxide type, but also changes depending on synthetic method and the thermal hysteresis of the sample. Figure 6.3 shows the change in the amount of the surface hydroxyl groups by heating various oxides.[8–11] The physical adsorption of the water vapor occurs on the polar surface when the solid has been left in the atmosphere and the surface free energy has decreased. The adsorption layer is formed from about two molecular layers at the relative humidity about 60% on the hydrophilic surface such as glass. Figure 6.4 shows water vapor adsorption isotherms for various glass surfaces. The solid surface is stabilized by the existence of these adsorbed water molecules, and the dynamic, chemical and electrical properties of the material are significantly influenced. Moreover, it is known

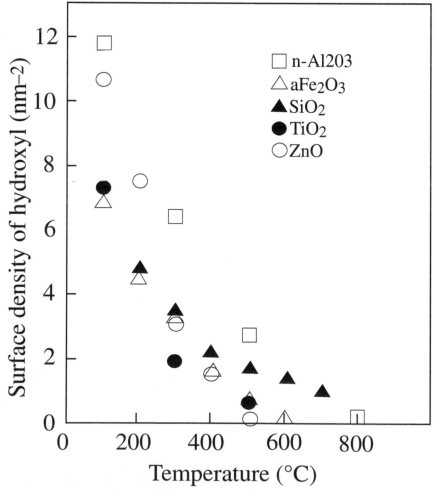

FIGURE 6.3 Changes of surface density of hydroxyl on several oxides with heat-treatment.

that the adhesive force between particles changes and produces a remarkable influence on the aggregation and the solidifying properties of the powder.[12–14]

Surface Acidity and Basicity

Influence of the Element

For the chemical property of the oxide surface, one of the important characteristics for practical use might be the solid acid and base. The acidity and the basicity of the oxide surface are predictable to some degree based on the chemical property of the atom composition. Figure 6.5 shows the relation to electronegativity (χ_i) of the partial charge and the metal ion on the oxide oxygen.[17] It is understood that the negative charge density of oxygen becomes smaller when the electronegativity of the metal ion is high and shows acidity. The electronegativity of the metal ion is shown here by the following equation:

$$\chi_i = (1+2z)\,\chi_0 \tag{6.7}$$

where z is the charge, and χ_0 is the electronegativity of the metal atom. Auroux et al. found a linear relation, as shown in Figure 6.6, from the heat of adsorption measurement of CO_2 by some oxides and the ratios of the charge/metal ion radius. When the ratio of the charge/metal ion radius is low, and the ionicity of chemical bond is high, the basicity is strong.[18]

Influence of Partially Chemical Structure

The strength of the acid and the base changes depending on the chemical structural difference, even if it is a single chemical composition. As for the free surface hydroxyl group observed on the Al_2O_3 surface, two or more structures are possible based on the relation of the coordination number. If the net charge of the hydroxyl of each structure is requested, one can use it becomes as Table 6.2. It is expected that the higher the amount of the positive charge, the stronger the acidity. The complex oxide, which consists of the oxide and contains more than two kinds of metals, occasionally shows an acidity which does not

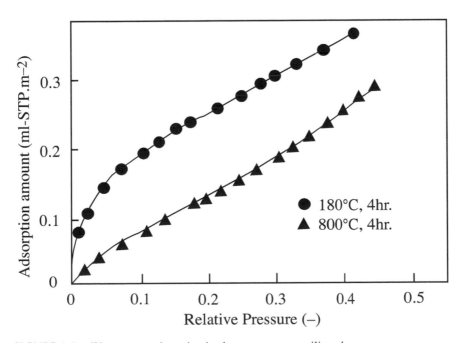

FIGURE 6.4　Water vapor adsorption isotherms on porous silica glass.

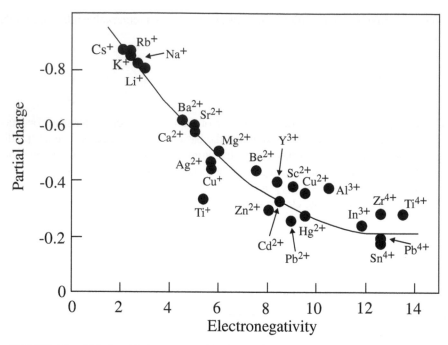

FIGURE 6.5   Relationship between partial charge of oxide surface and electronegativity of metal ion.

appear in a single oxide. Tanabe et al. reported that the acid appearance of the complex oxide can be explained according to the coordination number and each electric charge of the metal ion and oxygen.[19] Moreover, the acid strength of the complex oxide is found to correlate to the average of the electronegativity of the composition metal ion.[20] Recently, the acid and the base on the surface were also explained by the quantum chemical calculation using the cluster model, by which an oxide part is cut out.[21]

Influence of Surface Acidity and Basicity on Dispersion

The surface hydroxyl group receives the proton or dissociates and positively electrifies the particle surface or negatively in water.

$$M–OH + H^+ + OH^- \rightarrow M–OH^{2+} + OH^- \tag{6.8}$$

$$M–OH + H^+ + OH^- \rightarrow M–O^- + H_2O + H^+ \tag{6.9}$$

The charge of the particulate surface strongly influences the dispersibility of the particle in a liquid. The character of the proton receipt or the dissociation of the hydroxyl depends on the acidity and basicity. Therefore, the difference in the positive and negative charges of the surface and the amount of the charge is caused by the oxide type. The typical oxide isoelectric point is shown in Table 6.3.[22] When the oxide powder is dispersed in the medium, it is influenced by dissociation according to which type of the above equation by the character of the H^+ or OH^- acceptor. That is, the state of the interfacial interaction is determined by the relative intensity of an acid, the base character of the powder, and the acid and base characteristics of the medium. The principle of this interfacial phenomenon is shown in Figure 6.7. For instance, when the acidity of the powder surface is strong and the medium is basic, the particle gives the medium H^+ and the surface is charged negatively. On the other hand, when the basicity of the particulate surface is strong and the medium is acidic, the

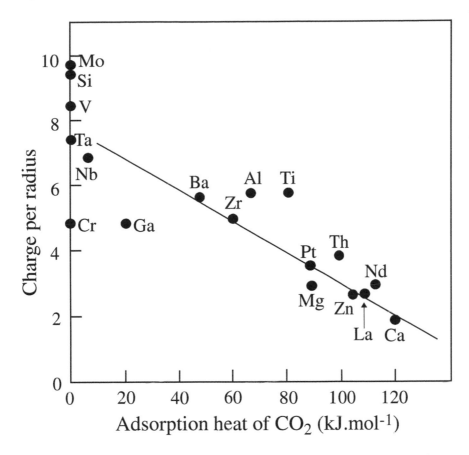

FIGURE 6.6 Relationship between adsorption heat of CO$_2$ and ratio of surface charge and radius of ion.

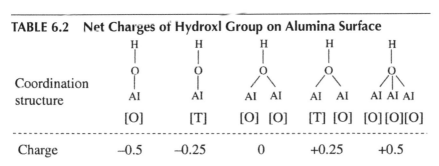

TABLE 6.2 Net Charges of Hydroxl Group on Alumina Surface

Coordination structure	H | O | Al [O]	H | O | Al [T]	H | O ╱╲ Al Al [O] [O]	H | O ╱╲ Al Al [T] [O]	H | O ╱|╲ Al Al Al [O][O][O]
Charge	−0.5	−0.25	0	+0.25	+0.5

Note: O, octahedral coordination; T, tetrahedral coordination.

particulate surface receives H$^+$ from the medium, and the surface is charged positively. When acidity and the basicity of the particulate surface and the dispersion medium are the same strength, the particle is not charged.

Surface Roughness and Porosity

The surface geometry, which influences the particle interaction predicted from the size of the particle when the particle size is submicron, is a nanoscale structure, because a bigger surface geometry than

TABLE 6.3   Isoelectric Point of Various Powders

Material	Isoelectric point	Material	Isoelectric point
WO_3	>0.5	γ-Al_2O_3	7.4–8.6
SiO_2	1.8	Y_2O_3	9.0
SiC	3.4	α-Fe_2O_3	9.04
Au	4.3	α-Al_2O_3	9.1
Al(OH)2	5.0–5.2	ZnO	9.3
SnO_2	6.6	CuO	9.4
α-FeO(OH)	6.7	BeO	10.2
TiO_2	6.7	La_2O_3	10.4
CeO_2	6.75	ZrO_2	10–11
Cr_2O_3	7.0	$Ni(OH)_2$	11.1
γ-FeO(OH)	7.4	$Co(OH)_2$	11.4
$Zn(OH)_2$	7.8	MgO	12.4 ± 0.3

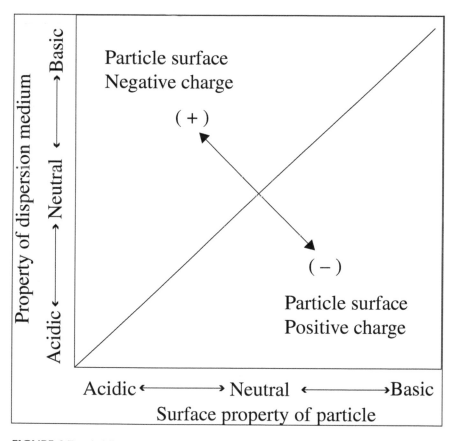

FIGURE 6.7   Acidic and basic properties of particle surface in nonaqueous dispersion medium.

this is roughly thought to be a category of particle shapes including macropores (above 50 nm). The surface geometry of this large scale can also be observed using an electron microscope, and it allows one to observe the relation between the particle shape and the particle interaction. Surface geometry structure can be roughly divided into the four classes shown in Figure 6.8, according to these ideas and the relative sizes of the adsorbed molecule and the particle. The first structure is a smooth surface on an atomic scale, as shown in Figure 6.8A. The second is a surface structure with roughness on an atomic level, as shown in Figure 6.8B. The third structure is a surface structure, shown in Figure 6.8C, which has micropores. The fourth structure, Figure 6.8D, is a surface structure that has mesopores. The studies concerning the quantitative relationship of these surface structures and powders and their various physical properties are few. An example concerning the surface structure, the physical adsorption of water, and the particle–particle adhesive force is introduced here.[23,24] The adhesive force between particles in the presence of water vapor is influenced by the state of adsorption of the water molecule on the particulate surface and the geometric structure of the particulate surface. The main cause of the adhesive force due to the humidity before the liquid bridge forms among particles is a hydrogen bond between particles by mediation of the water molecule. The main cause of the adhesive force in the humidity after the particle–particle capillary condensation occurs is a liquid bridge force. When the particle has pores, these basic mechanisms are the same. Therefore, the adhesive force because of the humidity before the liquid bridge forms among particles is the force of the hydrogen bond among particles by the mediation of the water molecule. However, the adhesive forces become small for a nonporous particle, according to the decrease in the interfacial area. Moreover, only a partial liquid bridge force is formed in the humidity before the capillary condensation in the pore is completed, and the adhesive forces are smaller than for when the capillary condensation in the pore is completed. On the other hand, the liquid bridge force is equal for the nonporous particle that is caused when neither the capillary condensation among particles nor the capillary condensation in the pore is completed. For instance, when the granulation operation is done in humidity, the capillary condensation in the pore is not completed, and the granulation will not be able to be done enough. On the other hand, to cause aggregation, one only has to set the opposite condition. As mentioned above, controlling various physical properties of powder by understanding the surface structure will become possible.

Actual Surface

Various surface relaxations and the physical adsorption of water occur, and the stabilization of the surface energy is achieved as on an actual solid surface. The surface inhomogeneity may exist as a step, a kink, a defect, or a dislocation, plus the adsorption impurities; therefore, the state of the actual surface is more complex. The exposure ratio of the crystal face and the number of heterogeneity sites on the surface differ and depend on the preparation conditions of the solid. The surface

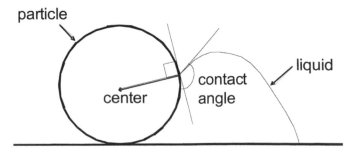

FIGURE 6.8 Conceptual classification of surface structure: (A) smooth surface on molecular level, (B) surface with roughness on molecular level, (C) surface with micropore, (D) surface with mesopore.

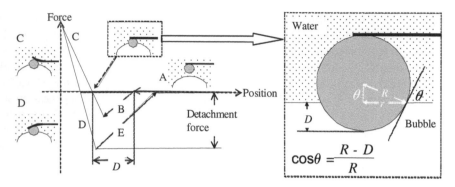

FIGURE 6.9   Ratio of edge length of cubic sodium chloride crystal and number of atoms.

structures on the top and the edge increase when the particle becomes a fine particle. Many of these unsaturated bonds are skillfully used as active sites, as with solid catalysts. The ratio of the surface of the atom to the entire atom that composes the particle increases when the particle of a salt crystal with six coordinations becomes small, as shown in Figure 6.9. For instance, when particle size or atomic diameter becomes 50 or less, it is understood that the ratio of the surface of the atom rapidly increases. Moreover, when the solid is used as a material, it is necessary to process the solid. With respect to processing factors such as generated heat and chemical reaction, the surface layer of the particle is different from the inside; that is, a processing-altered layer is formed on the surface, and defects and dislocations are generated. On the other hand, it is thought that the amorphous solid of glass, for instance, is solidified without making the supercooled liquid a crystal. Therefore, there is no regular array of atoms, ions, or molecules observed in the crystal. The surface of melted glass is a smooth plane because of the surface tension as well as the liquid. When the cooling solidification is done, the surface smoothness is maintained. Because the mobility of the material is high in the melt, the element by which the surface energy is decreased is concentrated at the surface. Especially, alkaline components easily gather on the surface. Moreover, because the melt is at a high temperature, it seems that the vaporization of the element and the alteration of the element along with the disappearance, by the reaction with the reactive gas in the atmosphere, have occurred on the surface. As mentioned above, the actual surface is complex. However, with an understanding of the major characteristics it is possible to predict how it will react by carefully considering the generation process of the powder and the matters explained here.

1.6.2   SURFACE CHARACTERIZATION

Classification of Surface Characterization

The characterization methods for particulate surfaces are shown in Table 6.4. The characterization methods, separate for the dry and the wet processes, are roughly classified; a strict classification is difficult. Many solid surface analyses based on the spectrum can be used to determine composition and chemical conditions. The methods as shown in Table 6.4 are a part of all measurements. For a powder sample, various physical properties are generally measured as the powder sample undergoes compression molding as a disk. When the powder sample is measured using these methods, it is necessary to note the relative size difference between the particle size and the equipment, such as the beam diameter of the probe, penetration depth of the probe, escape depth of the response signal, and diffusion effects of the response signal. It cannot be judged whether the information is for the powder surface or the entire powder. Moreover, the surface roughness of the powder sample that undergoes compression molding influences the detection efficiency. The level of qualitative

TABLE 6.4   Classification of Particle Surface Analysis

	For Dry Powder and/or Dry Process	For Wet Powder and/or Wet Process
Composition chemical condition	Magic angle spinning nuclear magnetic resonance: MAS-NMR Fluorescent X-ray photoelectron spectroscopy: AES Secondary ion mass spectrometry: SIMS Electron spin resonance: ESR	Inductively couple plasma: CP Atomic absorption Atomic absorption spectrometry: AAS
Functional groups	Chemical reaction method Infrared Spectroscopy: IR Thermogravimetric and Differential thermal analysis: TG/DTA	Chemical reaction method Zeta potential measurement
Acidity and basicity	Gas adsorption, adsorption heat: IR Temperature-programmed desorption: TPD Thermal desorption spectroscopy: TDS	Adsorption (CV, titration etc.), ultraviolet and visible spectrophotometry: UV, titration, heat of immersion
Wettability	Contact angle measurement by direct observation of droplet, wet velocity, or AFM, gas adsorption, heat of adsorption Inverted gas chromatography	Heat of immersion, preferential dispersion test
Roughness and porosity	Gas adsorption Scanning electron microscope: SEM Transmission electron microscope: TEM Atomic force microscope: AFM X-ray Computerized tomography: X-CT	Adsorption AFM Magnetic resonance imaging: MRI

measurement is generally appropriate to this method. It is easy for the composition analysis in the wet process to analyze the chemical composition of the entire particle. On the other hand, various instruments are necessary, as only the surface is dissolved to determine information on the surface. However, this method has enough ability to qualitatively analyze the ion, which undergoes segregation to the particulate surface. The analysis of surface functional groups, surface acidity and basicity, wettability, and roughness and porosity, and their measurements are described below.

Surface Functional Groups

The analysis method of a surface functional group is chiefly accomplished with a method based on a chemical reaction, spectroscopy methods such as infrared (IR) and nuclear magnetic resonance (NMR) spectroscopy, and methods that combine both approaches. Here, the measurement method of the surface hydroxyl group is explained. A similar method can be applied for other functional groups. The ignition loss method, Grignard reagent method,[25] alkyl silane reagent method,[16,26-28] and so forth, are chemical reaction methods for surface hydroxyl group determination. The D_2O displacement method, which uses IR spectroscopy, is a kind of chemical reaction method.[29,30] The chemical reaction method and the analysis methods that use IR spectroscopy are discussed below.

The basis of the ignition loss method for an oxide powder is a chemical reaction due to dehydration between the surface hydroxyl groups by heating. The chemical reactions for silica are as follows:

$$\equiv\text{Si–OH} + \text{HO–Si}\equiv \rightarrow \equiv\text{Si–O–Si}\equiv + H_2O \tag{6.10}$$

Therefore, if the water molecule generated along with heating is measured by a suitable method, the amount of the hydroxyl can be estimated. The weight change according to heating or the water capacity generated along with heating is then measured. In the method of applying gas chromatography, the generated water is measured with a thermal conductivity detector. It is possible to measure not only the hydroxyl, which exists in the powder surface, but also the water, which exists inside the particle, and this method is very simple and easy.[16]

It explains the characterization of the surface hydroxyl group on the oxide that uses the chemical reaction method. The chemical reaction method is classified based on the name of the molecule used for the evaluation, such as the Grignard reagent method and the silane method. The Grignard reagent method measures CH_4 generated by the reaction shown in Equation 6.11.

$$\equiv Si{-}OH + CH_3MgI \rightarrow \equiv Si{-}O{-}MgI + CH_4 \tag{6.11}$$

The DDS silane method can be used to distinguish the iso-type hydroxyl, which does not form the hydrogen bond, and the hydroxyl of other types according to the difference in the reaction, as shown in the following equations.[16]

$$\equiv Si-OH \;+\; \underset{Cl}{\overset{Cl}{\diagdown}}\!Si\!\underset{Me}{\overset{Me}{\diagup}} \;\rightarrow\; \equiv Si-O\!\underset{Cl}{\overset{Me}{\diagdown}}\!Si\!\underset{Me}{\overset{Me}{\diagup}} \;+\; HCl{\uparrow} \tag{6.12}$$

$$= Si\!\underset{OH}{\overset{OH}{\diagup}} \;+\; \underset{Cl}{\overset{Cl}{\diagdown}}\!Si\!\underset{Me}{\overset{Me}{\diagup}} \;\rightarrow\; \underset{\equiv Si-O}{\overset{\equiv Si-O}{\diagup}}\!Si\!\underset{Me}{\overset{Me}{\diagup}} \;+\; 2HCl{\uparrow} \tag{6.13}$$

$$\underset{\equiv Si-OH}{\overset{\equiv Si-OH}{}} \;\;\underset{Cl}{\overset{Cl}{\diagdown}}\!Si\!\underset{Me}{\overset{Me}{\diagup}} \;\rightarrow\; \underset{\equiv Si-O}{\overset{\equiv Si-O}{}}\!Si\!\underset{Me}{\overset{Me}{\diagup}} \;+\; 2HCl{\uparrow} \tag{6.14}$$

However, distinction between the gem-type hydroxyl and the H-bonded-type hydroxyl is difficult. The iso-type hydroxyl discharges one molecule of HCl reacting with one molecule of silane. On the other hand, the gem-type hydroxyl or the H-bond-type hydroxyl discharges two molecules of HCl reacting with one DDS molecule of silane. Therefore, the hydroxyl of each type can be separately measured by measuring the amount of consumption of the DDS silane and the amount of HCl generated. The reagents that can be used for the chemical reaction method are shown in Table 6.5.

The surface hydroxyl group gives clear IR spectroscopy results. Therefore, IR spectroscopy analysis is an extremely effective evaluation method of the surface hydroxyl group to observe the O–H stretching vibration (3000 to 3800 cm^{-1}). The O–H stretching vibration of the physisorbed water exists in this wave-number region. Therefore, a cell that can do the degassing operation and the heating operation in the system, as shown in Figure 6.10, is used.[31,32] The type of surface hydroxyl group that can be identified by IR analysis, as shown in Figure 6.11A, is the H-bonded hydroxyl, which forms the hydrogen bond and the free-type hydroxyl, which does not form the hydrogen bond. There is much research about surface hydroxyl groups. It is believed that three types of hydroxyls with different chemical structures, as shown in Figure 6.11B, exist on the surface of SiO_2.[33] The iso-type hydroxyl and the gem-type hydroxyl have a narrow absorption at about 3750 cm^{-1}. It is difficult to distinguish them by IR spectroscopy.[34,35] These can be distinguished using[29] Si solid-state NMR spectroscopy.[36–38] Reports concerning the existence of the tri-type hydroxyl are very few.[39,40] As for the H-bonded hydroxyl, wide peak, which centers on about 3600 cm^{-1}, has been observed. In addition, the

TABLE 6.5   Reagents Used for Determination of Surface Hydroxyls

Reagent	Molecular formula	Detected molecule
Diazomethane	Ch_3N_2	N_2
Lithium aluminium hydride	$LiAlH_4$	H_2
Trimethylchlorosilane	CH_3ClSi	HCl

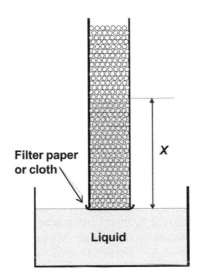

Filter paper or cloth

X

Liquid

FIGURE 6.10 An optical cell of *in situ* measurement for IR spectra.

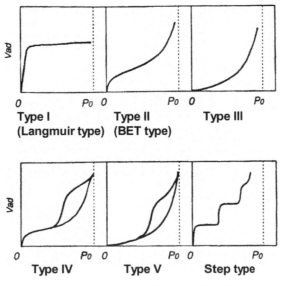

Type I
(Langmuir type)

Type II
(BET type)

Type III

Type IV

Type V

Step type

FIGURE 6.11 Classification of surface silanol (A) according to chemical state, (B) according to chemical structure.

H-bonded hydroxyl marked by the arrow in Figure 6.11A is called the terminal hydroxyl and has an absorption at about 3720 cm^{-1}.[41] The wave number of the O–H stretching vibration changes when the kind of atom that bonds the hydroxyls changes. The relation between the electronegativity of the atom that unites the hydroxyls and the OH stretching vibration is shown in Figure 6.12.[42] In addition, two or more infrared absorption spectra appear when the oxygen coordination number is different, even if the chemical composition is the same. Figure 6.13 shows the structure and absorption spectrum of the surface hydroxyl group observed on the surface of Al_2O_3.[43,44] Moreover, the dehydration behavior caused by heating of the surface hydroxyl group differs depending not only on the oxide type but also on the preparation method and the hysteresis of the sample.

IR spectroscopy is used to measure of strength of the adsorption site and to identify the adsorption site. The history of the research on the observations of the adsorbed molecular state, which uses the infrared spectroscopy and interaction with the surface, is long. Especially, many of the interactions of the surface hydroxyl group and the adsorbed molecule have been determined for the oxide.[45] Kiseleve et al.[46] found the relationship between the heat of formation of the hydrogen bond between the organic molecule and the hydroxyl, and wave number shifts of the O–H stretching vibration by adsorption from the heat of adsorption measurement of the organic molecule and the IR spectroscopy measurement for silica. These results are shown in Figure 6.14. Because the solvent molecule, in addition to the adsorbed molecule, exists in the liquid-phase adsorption, this adsorption behavior is more complex than gas-phase adsorption. However, the interaction between the surface hydroxyl group and the adsorbed molecule can be examined by IR spectroscopy and calorimetry, as well as by gas-phase adsorption.[47,48]

The cleavage reaction of (M–O–M) on the oxide surface in addition to the reaction of the surface hydroxyl group is an important characteristic of the powder surface. Because the reactiveness to water is the same as the mechanism of surface hydroxyl group generation, it is considered an index of the chemical stability of the water molecule on the surface. Moreover, some reactivities with other molecules have been reported. Morrow et al.[49] observed the dissociation of NH_3 to the siloxane linkage ($\equiv$Si–O–Si$\equiv$) and adsorption from the change in the IR absorption spectrum of the NH_3 adsorption on silica gel. Bunker et al.[50] found the structure of reactivity siloxane that tetrahedron SiO_4 share the edge from the analysis of the IR absorption spectrum. SiO_2 is a coupled structure of the SiO_4 tetrahedron. The ring structure with three types of distortion shown in Figure 6.15 is reported in IR spectroscopy[50,51] and Raman spectroscopy.[52] Especially important, the structure that the SiO_4 tetrahedron shares the edge and the D2 structure has a big distortion. Therefore, it is confirmed that various molecules easily

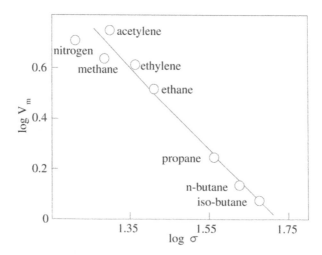

FIGURE 6.12 Relation between electronegativity of atom with hydroxyl and stretching vibration of free hydroxyl.

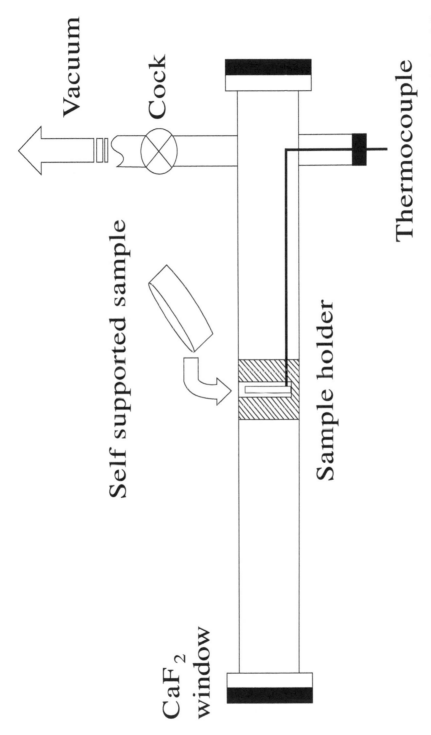

FIGURE 6.13 Structures and stretching frequencies of hydroxyl observed in alumina surface: (A) free hydroxyl, (B) hydrogen-bonded hydroxyl. O, octahedral coordination; T, tetrahedral coordination.

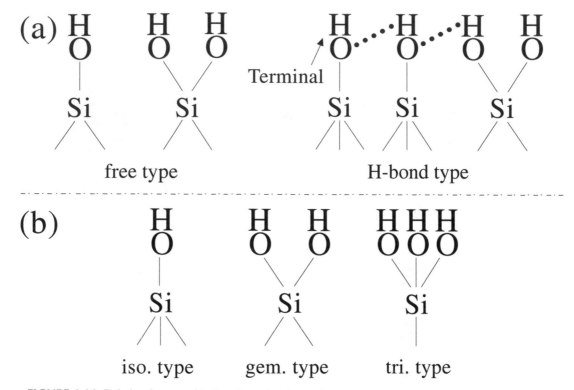

FIGURE 6.14 Relation between binding formation heat of various molecules and silanols and peak shift values of surface hydroxyl by adsorption.

have a dissociation adsorption on the surface with these structures.[53-55] The wave number corresponding to these structures and the calculation value of the strain energy are shown in Table 6.6.[56] It is expected that the amount of siloxane of such reactiveness differs according to the production method and the hysteresis of the silica.

Surface Acidity and Basicity

The strength of the acid–base of the solid surface applies to the definition of the H_0 function and H^- function, which shows the strength of the acidity/basicity of a uniform solvent. That is, the acid strength of the solid surface is defined by H_0, and the base strength is defined by H^-. For instance, when indicator B of basicity is adsorbed on the acid site of the solid surface, a part of the indicator is made a proton with the acid site. The strength H_0 of these acid sites is shown in Equation 6.15 by the value of pK_{BH+} of BH^+, which is the conjugate acid of indicator B when the concentration C_{BH+} of indicator BH^+ made a proton is equal to the concentration C_B of indicator B not made a proton.

$$H_0 = pK_{BH+} - \log(C_{BH+}/C_B) \qquad (6.15)$$

Therefore, the acid strength indicated by the H_0 value shows the ability to change into the conjugate acid by the acid sites which give the protons half of the base indicator B of the neutral which the acid site on the surface adsorbs. In Lewis' definition, the H_0 value shows the ability that the electron pair can be received from half of the adsorbed base indicator B. The investigation of base strength is similar to the investigation of acid strength. When indicator BH, which defines base strength,

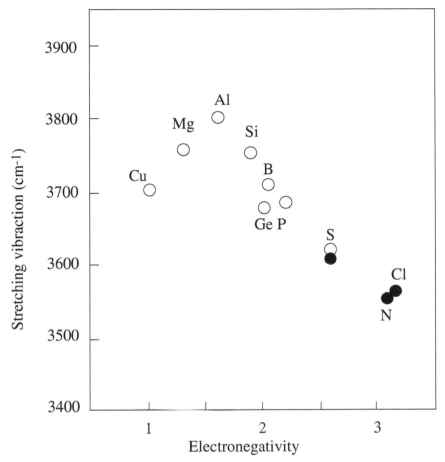

FIGURE 6.15 Distortion ring structures into silica.

TABLE 6.6 Infrared and Raman Absorption and Strain Energy of the Structure of [SiO$_4$] in Silica

Ring structure	Absorption Spectrum	Spectroscopy	Strain energy
4-membered	888/908 cm^{-1}	Infrared	58 kJ/mol (Si–O)
6-membered	605 cm^{-1} (D$_2$)	Raman	17 kJ/mol (Si–O)
8-membered	490 cm^{-1} (D1)	Raman	—

adsorbs on the solid surface, H$^+$ is pulled out by the basic site, and a part of indicator BH becomes the state of B$^-$. The strength H$^-$ of the basic site is shown by the pK$_{BH}$ value of indicator BH when concentration C$_{B-}$ of B$^-$ is equal to the concentration C$_{BH}$ of BH from which H$^+$ is not pulled out. However, because the measurement of the ratio of the concentration (C$_{BH+}$/C$_B$) and (C$_{BH}$/C$_{B-}$) of the indicator on the solid surface is difficult, the following techniques are actually used. Because it is possible to apply it to both the acid and the base, the evaluation method of the acid site is described here. Many indicators of basicity in which the charge neutrality becomes a conjugate acid by receiving H^+ are known as Hammett indicators. Here, each pK_a value of these conjugate acids is assumed to be b, c, d, and e ($b < c < d < e$). The indicator $pK_a = b$ is also assumed not to be changed by the acid structure at the acid site, nor is the indicator $pK_a = c$ assumed to have been changed by the acid structure. At this time, it is understood that the acid strength H_0 value of the surface acid site is between b and c. The indicator causes a discoloration, and the change occurs in the UV spectra when the indicator is changed into the acid structure by the solid acid site. Therefore, the strength and the amount of the acid site are measurable by using these phenomena. An example of the determination of the acid site by the discoloration of the indicator is shown as follows. p-dimethylaminoazobenzene (yellow), which is the indicator, is adsorbed on the solid acid site and becomes red with the following acid structures:

$$\text{yellow} \qquad\qquad\qquad\qquad \text{discolor}$$

$$\langle \; \rangle\text{-N=N-}\langle \; \rangle\text{-N(CH}_3)_2\text{+Acid} \;\rightleftharpoons\; \langle \; \rangle\text{-N-N-}\langle \; \rangle\text{-N}^+\text{(CH}_3)_2 \qquad (6.16)$$
$$\underset{\text{Acid}}{|}$$

p-dimethylaminoazobenzene

n-butylamine, which is a stronger basicity molecule than the indicator, is dropped by the following procedure. Then, n-butylamine causes the exchange adsorption as the indicator of the acid structure which has been adsorbed according to the next equation, and the indicator returns to the former yellow color.

$$\text{discolor} \qquad\qquad\qquad\qquad \text{yellow}$$

$$\langle \; \rangle\text{-N-N-}\langle \; \rangle\text{-N}^+\text{(CH}_3)_2 + C_4H_9NH_2 \;\rightleftharpoons\; \langle \; \rangle\text{-N=N-}\langle \; \rangle\text{-N(CH}_3)_2 + C_4H_9NH_2\cdot\text{Acid} \qquad (6.17)$$
$$\underset{\text{Acid}}{|} \qquad\quad \underset{n\text{-butylamine}}{}$$

Acid strength is understood from the difference in the base strength of the indicator used. Moreover, the amount of the acid site can be obtained according to the amount of n-butylamine necessary to erase the color of the acid structure. The reacted indicator is made the acid structure, and discolor, even if the acid sites are a Brønsted acid site or a Lewis acid site. It is possible to do the same for the evaluation of basicity. Indicators to examine acidity and basicity are shown in Tables 6.7 and 6.8, respectively.[57]

In this procedure, it is necessary to leave it for a long time at the time of each n-butylamine dropping to achieve an adsorption equilibrium and confirm that the color of the acid structure disappears. Therefore, a long time is required for the measurement. The Benesi method in which this fault is improved is widely used. In the Benesi method, many sample solutions, which have different dosages of n-butylamine, are prepared. After the sorption equilibrium isotherm of n-butylamine, various indicators are added for each sample. The discoloration of these indicators to the acid structure is examined, and the acid strength distribution is measured. The intensity distribution of the basic site in the solid surface is obtained by the same principle. That is, one only has to use an acid molecule such as benzoic acid as a titration reagent instead of n-butylamine, and the indicator for the H^- measurement is used.

Adsorption of the basic gaseous molecule is caused on the acid site, and the absorbed amount corresponds to the acidity. The adsorptive interaction of the basicity molecule that adsorbs on a

TABLE 6.7 Indicators Used for Acid Strength Measurements

Indicator	Basic Color	Acid Color	pK_{3H+}	H_2SO_4 wt. %
Neutral red	Colorless	Red	+6.8	8×10^{-8}
Methyl red	Yellow	Red	+4.8	—
Phenylazonaphthylamine	Yellow	Red	+4.0	5×10^{-5}
p-Dimethlaminoazobenzene	Yellow	Red	+3.3	3×10^{-4}
2-Amino-5-azotoluene	Yellow	Red	+2.0	0.005
Benzeneazodiphenylamine	Yellow	Purple	+1.5	0.02
4-Dimethylaminoazo-1-naphthalene	Yellow	Red	+1.2	0.03
Crystal violet	Blue	Purple	+0.8	0.1
p-Nitrobenzeneazo-(p'-nitro)-diphenylamine	Yellow	Purple	+0.473	—
Dicinnamalacetone	Yellow	Red	–3.0	48
Benzalacetophenone	Colorless	Yellow	–5.6	71
Anthraquinone	Colorless	Yellow	–8.2	90
p–Nitrotoluene	290 μm	350 μm	–10.5	99.9
2,4-Dinitrotoluene	255 μm	320 μm	–12.8	106

TABLE 6.8 Indicators Used for Base Strength Measurements

Indicator	Acid Color	Basic Color	pK_{BH}
Bromothymol blue	Yellow	Blue	7.2
Phenolphthalein	Colorless	Pink	9.3
2,4,6-Trinitroaniline	Yellow	Tango	12.2
2,4-Dinitroaniline	Yellow	Purple	15.0
4-Chloro-2-nitroaniline	Yellow	Orange	17.2
4-Nitroaniline	Yellow	Orange	18.4
4-Chloroaniline	Colorless	Pink	26.5

strong acid site is strong, and desorption is difficult at low temperature. Therefore, the desorption temperature of the adsorbed base molecule is closely related to the strength of the acid site. Several measurement examples are given as follows. The base molecules that adsorb onto the surface, and then the physisorbed and hydrogen-bonded molecules are removed at a suitable temperature when basicity gases such as ammonia, pyridine, and alkylamines are used. The amount of the base molecules, which chemically adsorb on the acid site, is measured as follows. The temperature gradually rises while removing the desorption gas, and the amount of chemisorption at each temperature is measured; or the temperature continuously rises while removing the desorption gas, and the intensity distribution of the acid site is obtained as the desorption spectrum of the base molecule. The amount of the desorption base as the amount of chemisorptions is measured by the weight method with a balance or by gas chromatography. That is, the kind of site where the adsorption activity is different is distinguished from the number of desorption peaks in the temperature-programmed desorption method. Moreover, the chemical bonding state and strength of the acid site are obtained from the difference in the desorption temperature, and the amount of the acid site is obtained from the amount of each desorption.

The activation energy of desorption assumes it is almost equal to the heat of adsorption, and it is used as an index of the strength of the acid. The next equation, for the temperature-programmed desorption method, shows a relation between the programming rate (V_t) and desorption peak temperature (T_m), by which the maximum elimination kinetics is given when the gas, in which the strength of the acid site is uniform, and when detached is not adsorbed again.

$$2\log T_m - \log V_t = E_d/(2.303RT_m) + \log(E_d/AR) \qquad (6.18)$$

A is a frequency factor, E_d is the activation energy, and R is the gas constant. The sample with the same absorbed amount of the base probe molecule is prepared, the programming rate is changed, and the desorption peak temperature T_m corresponding to the changes is obtained. The activation energy, E_d, is obtained from the slope of the straight line of the plot of $2\log T_m - \log V_t$ to $1/T_m$. As for the features of the gas absorption method, the solvent has no influence, and the surface acidity at various temperatures can be measured. Moreover, as another advantage, the amount of the initial adsorption can be easily controlled, and a catalytic reaction mechanism and kinetics data might be able to be comparatively facilitated. On the other hand, strong physisorption and chemisorption on powder surfaces cannot be distinguished, which is a disadvantage. In addition, because the diffusion process of the gas cannot be disregarded and re-adsorption of desorption gas occurs, obtaining an accurate kinetic parameter might be difficult. Moreover, when the amount of the acid site is measured using ammonia and n-butylamine is considered, as the adsorption mechanism to the acid site is the same, the amount obtained with ammonia might be large. It is considered that the pore distribution and the distance between acid sites are closely related to this difference.

The evaluation of the acid–base site where IR spectroscopy was used is subsequently introduced. The amounts of the Brønsted acid site and the Lewis acid site are measured from the strength of IR spectroscopy of the chemically adsorbed base molecules. The IR absorption spectrum of pyridine, which adsorbs on the Brønsted acid site and the Lewis acid site, is shown in Figure 6.16.[58] The infrared absorption spectrum in the 1400 to 1700 cm^{-1} region is derived from the in-plane vibration of the pyridine ring. The wave number is remarkably different according to whether the adsorbed pyridine forms a hydrogen bond, a coordinate bond, or an ionic bond. The infrared absorption spectrum features of the adsorbed pyridine are shown in Table 6.9. Moreover, the acid site type can be identified according to the IR absorption spectrum of the adsorption ammonia. The reason for this is that the IR absorption of the deformation vibration of H–N–H changes depending on the kind of adsorption site. As for the ammonium ion (NH_4^+) and ammonia, which formed the coordinate bond, absorption appears at about 1400 cm^{-1} and at about 1620cm^{-1}, respectively. The former is a Brønsted acid site, and the latter is a Lewis acid site.[59,60]

There is a method of measuring the thermal transformation when the probe molecule directly adsorbs on the acid–base site. Because a suitable basic or acidic molecule can be used, the adsorbate and the temperature of the heat of adsorption measurement can be arbitrarily selected according to the calorimeter; acid strength can be obtained under various conditions. For the heat of adsorption measurement, a suitable amount of sample is first placed in the cell, and the heating exhaust processing is done under optimum conditions. The sample is placed in the calorimeter with the heating exhaust processing done. The adsorbate is introduced after it reaches the thermal equilibrium state and the generated heat of adsorption is measured. The absorbed amount is obtained at the same time. In general, because the change in the heat of adsorption is large in the region where the absorbed amount is low, a little probe gas is introduced. The heat of adsorption is converted into the amount of 1 mol adsorbate, and a graph of absorbed amount (V) versus heat of adsorption (q) is obtained. If $\Delta V/\Delta q$ is obtained from the graph based on the V–q curve, and $\Delta V/\Delta q$ is plotted for q, the intensity distribution curve of the acid–base is obtained.

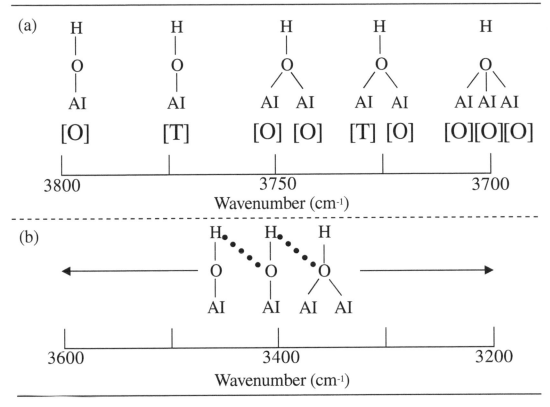

FIGURE 6.16 Infrared absorption spectrum of pyridine: (A) chloroform solution of pyridine, (B) chloroform solution of pyridine: BH3, (C) chloroform solution of (pyridine:H) + Cl⁻. The absorption band at 1520 is wide and is an absorption peak of the chloroform solvent.

TABLE 6.9   IR Absorption Band Assigned Pyridine Molecules Adsorbed on Acidic Surface

Hydrogen Bonded	Coordinately Bonded	Ionically Bonded
1400–1447 (VS)	14.7–1465 (VS)	
1480–1490 (W)	1471–1503 (W-S)	1478–1500(VS)
	1525–1542 (S)	1562–1580 (W)
1580–1600 (S)	1592–1633 (S)	1600–1620 (S)
		1634–1640 (S)

Note: Absorbance strength: VS, very strong; S, strong; W, weak.

Wettability

For the large particle shown in Figure 6.17, the contact angle can be directly measured with a microscope.[61] Moreover, the contact angle is directly obtained by attaching the particle to the AFM probe, as shown in Figure 6.18.[62] However, it is difficult to directly measure the contact angle in a general powder. It assumes as follows, and the contact angle is calculated. It is considered that a minute space in the packed bed of the powder shown in Figure 6.19 is formed with the capillary having a

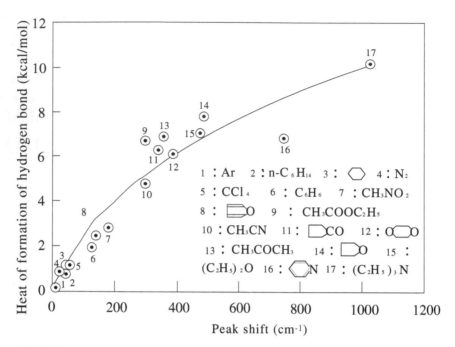

FIGURE 6.17 Contact angle measurement for large particle.

uniform cylindrical shape. The contact angle can be measured according to the velocity of the liquid as it infiltrates into this capillary. The following equation holds when the radius of the cylindrical capillary in the powder-packed bed is assumed to be r.

$$\frac{dx}{dt} = \frac{r\gamma\cos\theta}{4\eta x} - \frac{r^2\rho g}{8\eta}$$ (6.19)

Here, x is the height to which the capillary rises, t is the time, γ is the liquid surface tension, η is the viscosity of the liquid, ρ is the density, g is gravitational acceleration, and θ is the contact angle. The first term on the right in Equation 6.19 is the contribution of the capillary force. The second term is the contribution of gravity. When the second term can be neglected, the following equation is obtained by integrating Equation 6.19.

$$x^2 = \frac{r\gamma\cos\theta}{2\eta}t$$ (6.20)

Therefore, the capillary radius r in the powder-packed bed is obtained from the straight line relation between x^2 and t beforehand by using the liquid which has contact angle $\theta = 0$ where the powder is wet. The contact angle θ is then obtained by measuring the relation between x^2 and t for the powder with a target liquid. Moreover, the wettability evaluation of the powder is determined from the preference dispersing test of the powder in various liquids or mixed solvents. Additionally, the wettability can be judged according to the analysis of the adsorption isotherm, heat of adsorption, and heat of immersion.[63]

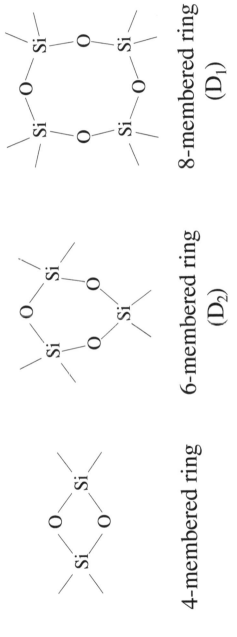

4-membered ring 6-membered ring 8-membered ring

(D_2) (D_1)

FIGURE 6.18 Schematic of a force–position curve between a spherical particle and a bubble. The position is the height position of the bubble. It was assumed that no long-range forces are active between the particle and bubble before contact and that the particle surface is not completely wetted by the liquid. At large distances the cantilever is not deflected (A). When the particle comes into contact with the air–water interface, it jumps into the bubble and a TPC line is formed (B). The reason for the jump-in is the capillary force. Moving the particle further toward the bubble shifts the TPC (three points contact) line over the particle surface (C). The important factor is the receding contact angle θ. When retracing the particle again (D), at some point the force is high enough to draw the air–water interface (E). The drawing of right side is a particle at equilibrium in the air–water interface. θ is the equilibrium contact angle.

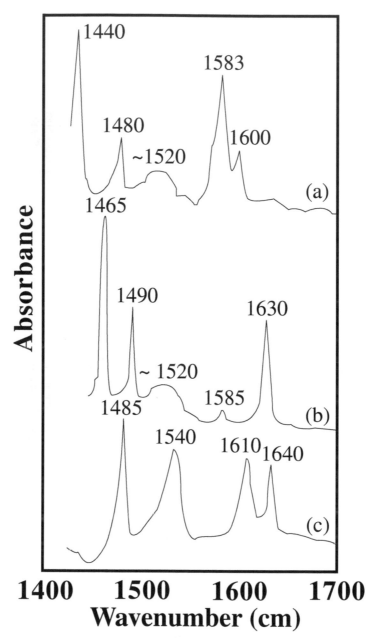

FIGURE 6.19 Contact angle estimated from wet velocity of powder bed.

Surface Roughness and Porosity

In gas absorption, the absorbed amount changes according to the adsorbing gas pressure change observed at a constant temperature. A lot of information can be obtained by analyzing this isothermal line. For instance, the specific surface area, which uses the BET theory, is the most popular analytical result. The adsorption isotherm can be classified according to type. Here, the classification by which the isothermal line of the step is added to the BDDT classification is shown in Figure 6.20.[64] Type I is called the Langmuir type, and much chemisorption and adsorption into the micropores is observed. Type II is adsorption of the multilayers in the isothermal line, which is called the BET type

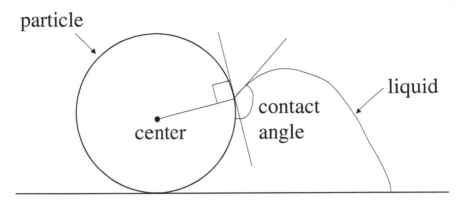

FIGURE 6.20 Classification of adsorption isotherms.

and is often observed because of nitrogen adsorption on a solid without micropores and mesopores. Type III is observed when the interaction of the solid surface and the adsorbed molecule is weak; it is often observed in water vapor adsorption on a hydrophobic surface.[65] Type IV and type V are shapes when hysteresis is observed in the isothermal line. When the solid, which has the adsorption isotherm of type II or type III, has mesopores, the adsorption isotherm becomes type IV and type V, respectively. However, the porous silica with mesopores from which hysteresis in the adsorption isotherm is not observed has been synthesized in recent years,[66] and the disappearance of hysteresis has been discussed.[67] When the adsorption isotherm is on an extremely uniform surface and the phase transition occurs in the adsorption phase, a step is observed. For instance, there is K_r adsorption on a graphite surface.[68] The classification of such an isothermal line is extremely qualitative information. However, it is useful as a means to know the outline of the structure and the characteristics of the powder surface.

The roughness factor, r, is used as an index which shows the complexity of the solid surface.

$$r = \frac{S_{N_2}}{S_d} \qquad (6.21)$$

S_{N2} is the surface area measured by N_2 adsorption, and S_d is the apparent surface area estimated from particle diameter. When r is larger than 1, it is presumed that roughness on a molecular order exists in the surface.

Recently, an attempt was made to describe the complex structure of the solid surface by fractal dimensions.[69] Some methods of obtaining the fractal dimension of the solid surface have been suggested.[70–72] Here, we show an example for obtaining it from the adsorption of the molecule with a different size. The following equation occurs when the surface fractal dimension, D_a, occupation area, σ, of the molecule used for adsorption, and n, a monomolecular absorbed amount are used.

$$\log n = -(D_a/2)\log\sigma + \text{const.} \qquad (6.22)$$

Therefore, D_a is obtained from the plot of log n and log σ. Figure 6.21 shows the result of obtaining the fractal dimension of the activated carbon surface.[73] As a result, D_a of the activated carbon is 3.03. In general, the fractal dimension of the solid surface indicates the value of $2 < D_a < 3$. $D_a = 2$ is a surface as smooth as a graphite surface. On the other hand, D_a approaches 3 as the surface structure becomes complex. Silica gels and active carbon are examples of D_a approaching 3. It is only recently that a solid surface was described by fractal dimensions. Therefore, the relation between a

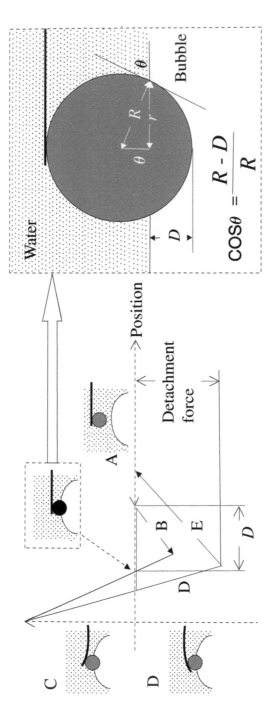

FIGURE 6.21 Monolayer volume, V_m, as a function of the cross-sectional area, σ_x.

fractal dimension and various surface physical properties has not yet been clarified. However, there are some advance reports concerning the relation between catalytic kinetics and the fractal dimensions,[74] the relations between the fractal surface and wettability,[75] and structure predictions of the surface modification group.[76] Future developments are expected in the form of a new index which shows the complexity of the surface structure of a solid.

1.6.3 ATOMIC FORCE MICROSCOPY

These days, various advanced materials have been produced from nanoscale functional particles. This implies that the molecular-scale information of particle properties, such as the shape and surface property of particles and the interaction forces between them in the medium, is extremely important to control the quality of particulate products. The microscopic shape and size of particles have been evaluated, using electron microscopes (SEM and TEM), but they are observed only under the high vacuum state, so the information obtained there is not necessarily appropriate for the real production processes of materials.

In 1982, the Scanning Tunneling Microscope (STM) was invented to observe the atomic image of surfaces by measuring the tunneling current between the cantilever tip and conductive surfaces. This invention of STM stimulated the further invention of the atomic force microscope (AFM) in 1985, which enables us to image the atomic-scale properties of nonconductive surfaces either in vacuum or in any medium. For engineers who want to know what is happening in the real processes, the information given by AFM has been extremely valuable, because of the following features of AFM:

1. Molecular-scale information of local surface properties and interaction forces between surfaces in any medium is obtainable.
2. *In situ* measurements are possible.
3. The apparatus is easy to handle.

The detailed explanation of AFM is given elsewhere,[77,78] but the main principle of AFM and examples of the data obtained are given below.

Apparatus of Atomic Force Microscope

The AFM apparatus is composed of the measuring, controller and data-treatment units. As illustrated in Figure 6.22, the measuring unit is composed of a laser beam generator and a photo detector, the fluid cell with a cantilever and a sealing o-ring and the piezo system whose atomic-scale movement of the horizontal x/y direction and vertical z-direction is controlled by the controller unit. The probe popularly used is made of Si or Si_3N_4. The tip radius of curvature is $10{\sim}60$ nm, the cantilever length is $100{\sim}500$ nm, and the nominal spring constant is $10^{-2}{\sim}10^2$ N/m.

The sample plate glued on the metal plate is attached magnetically on the piezoelectric crystal, which can move to the vertical direction with a constant speed by controlling the voltage of the piezo scanner. The cantilever deflection due to the interaction between the tip and sample surfaces is detected as the voltage change of the photo detector onto which a laser beam reflected from the rear of the cantilever is focused. This deflection is converted into the interaction force, applying Hooke's law to the cantilever. The zero separation is determined from the onset of the "constant compliance" region, where the cantilever deflection is linear with the surface displacement. The separation distance between surfaces is evaluated by the displacement of the sample plate from the constant compliance region. This principle of AFM can be used for two different purposes.

Evaluation of Surface Property by AFM

The AFM was originally constructed to evaluate the surface roughness on the atomic scale by using a so-called contact-mode procedure. In this case, the surface is scanned to x/y direction, keeping the separation between the tip and sample surfaces constant by the feedback control to the piezo scan-

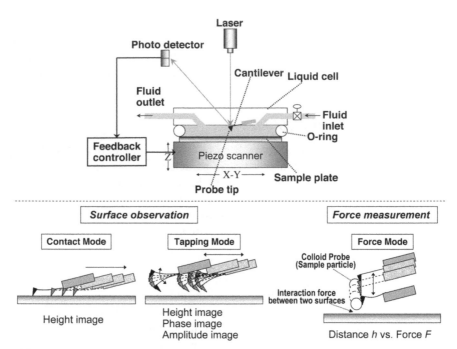

FIGURE 6.22 Schematic illustration of the principle of Atomic Force Microscope (AFM) and the measuring modes.

ner. Then the feedback data are converted to the so-called height image of the surface. It is known that the height of the surface roughness obtained by this method is accurate enough on the molecular scale, but the width depends on the radius of curvature of the tip used. The similar image of roughness is also obtainable by using the data of cantilever deflection accumulated during the scanning without the feedback adjustment. This image is called the deflection image, which is not as accurate as the height image, because the contacting force varies during the scan. Hence, the height mode is popularly employed. These contact-mode procedures can not be used with soft surfaces, because the tip may scratch off the part of the soft surface.

For soft surfaces, the so-called tapping mode is usually employed. In this mode the cantilever piezo system oscillates the cantilever at a given frequency. When the tip interacts with the sample surface, the amplitude of the cantilever damps. Hence, if the sample plate is scanned, keeping the amplitude of the cantilever constant by the feedback control, the height image of the surface can be obtained. Because the possibility of scratching off the surface is very small in this case, the tapping-mode procedure is widely used to gain the image of soft surfaces, especially in the case of the surface with adsorbed materials, such as surfactants and polymers. Figure 6.23 shows an example of AFM image of the polymer chains adsorbed on the mica surface, which was measured *in situ* in water.[79,80] This success to take the AFM image of single polymers was followed by the AFM observation of the nuclei formation of latex particles.[81]

If the phase difference exists between the input given by the cantilever piezo system and the detected output during scanning, which is caused by the interaction between the tip and sample surfaces, the phase image of the surface is obtainable. This image is often used to gain the measure of the rigidity of surfaces. If the data of amplitude are taken by fixing the height constant, the amplitude image can be obtained.

Usage of AFM for Other Surface Properties

In the case of usual AFM measurements, the surface properties are measured by using the intermolecular forces which exist commonly between surfaces. If there exist some other forces between

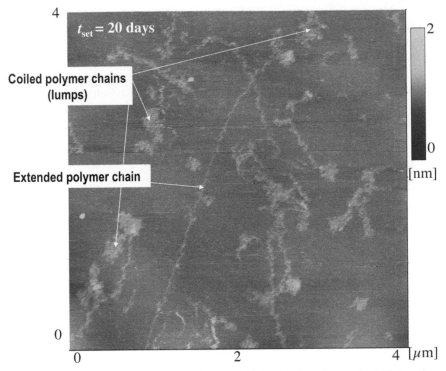

FIGURE 6.23 An example of a height image of high cationic polymers (poly[2-(acryloy loxy)ethyl(trimethyl)ammoniumchloride], molecular weight $= 1.25 \times 10^7$ g/mol) adsorbed on a mica surface measured *in situ* in water.

surfaces, the AFM can be used to detect the force by employing the probe tip, which interacts with surface:

1. Magnetic Force Microscope (MFM)
   If the probe tip is made of magnetic material, the magnetic interaction between the tip and sample surfaces is detected, so that the magnetic map of the surface can be obtained even if the surface is completely flat.

2. Lateral (Friction) Force Microscope (LFM)
   When a probe (usually colloidal probe) is pressed against the substrate at a constant applied force and slides horizontally, the friction force between surfaces can be detected from the torsion of the cantilever. The magnitude of lateral force is determined by using a four divided photo detector. It is found that the lateral force is much more sensitive to the roughness of the sample surface, compared with the normal force measurements.[82]

Evaluation of Interaction Forces between Surfaces in Media

As explained above, the essential principle of AFM is to measure the interaction force between surfaces. By using this principle without scanning, one can evaluate the interaction forces between surfaces in any medium. The interaction between the probe tip and the sample surface is usually very small because of the small radius of curvature of the tip. This difficulty was overcome by gluing a micron-size colloidal particle on the top to the cantilever, shown as a colloid probe in Figure 6.22.[83] A typical force curve given by AFM force measurements is schematically illustrated in Figure 6.24.

From the one cycle of measurement, one can measure the long-range interaction force, the short-range interaction force, and the adhesive force between surfaces simultaneously. The long-range interaction force is usually consistent with the force curve predicted by the DLVO theory described in 2.8.2, but the short-range interaction force depends on the adsorbed layer of water molecules, ions, and so on.[84-86] An interesting example of the short-range interaction force is given in Figure 6.25, where a step-wise force curve was obtained for a 98 wt% alcohol solution. This indicates that the

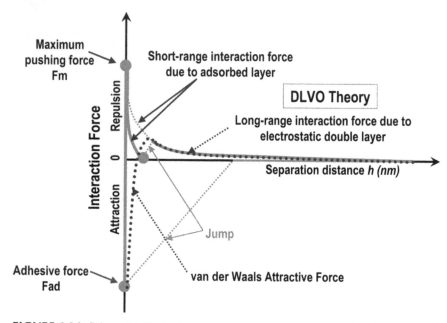

FIGURE 6.24 Schematic illustration of the force curve (the interaction forces vs. the surface separation) obtained by AFM.

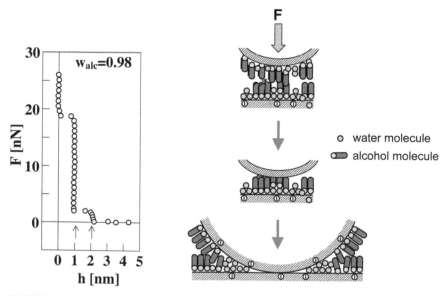

FIGURE 6.25 An interesting example of the short-range interaction, the step-wise force curve found for the interaction between a probe tip and a mica surfaces in an alcohol solution with 2 wt% water, and the corresponding mechanism.

alcohol molecules are adsorbed on the mica surface standing, and the step-wise force curves arise when the structured alcohol layers on both surfaces are broken, as illustrated in the figure.[87,88] Another important contribution by AFM force measurements is to clarify the fact that the long-range attractive force between hydrophobic surfaces in aqueous solutions, as shown in Figure 6.26, which was a debated issue for a long time, originated from the coalescence of nanobubbles attached to the surfaces.[89,90]

As described above, the AFM is a very powerful tool for investigating the surface microstructure and interactions between surfaces in media, mainly because the molecular-scale information is obtainable for any kind of the combination of surfaces and media without any high vacuum as in the case of SEM and TEM. Hence, now a huge number of data obtained by AFM have been reported in the wide range of research fields. Some AFM data overturned mechanisms predicted by macroscopic experiments that have been believed for a long time. However, there exists an inevitable problem of AFM; one cannot know the genuine separation distance between surfaces, because the zero separation is defined at the point where the piezo movement coincides with that of the cantilever. If there exists an adsorbed layer into which the probe tip can not penetrate, the separation between the tip surface and the outer surface of the adsorbed layer is regarded as the zero separation.

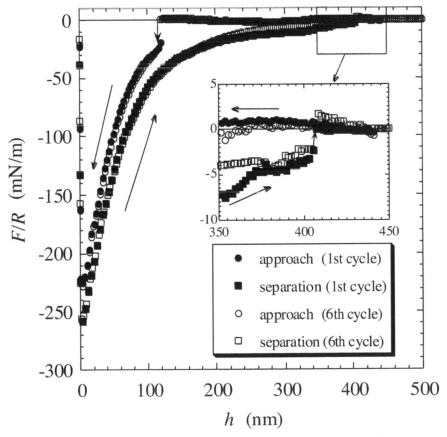

FIGURE 6.26 An interesting example of the long-range attractive force due to nano-bubbles on hydrophobic surfaces in water.

REFERENCES

1. Kubo, R., *Phys. Lett.,* 1, 49–50, 1962.
2. Chikazawa, M. and Takei, T., *Gypsum Lime,* 228, 255–267, 1990.
3. Benson, G. C. and Claxton, T. A., *J. Chem. Phys.,* 48, 1356–1364, 1968.
4. Taroni, A. and Haneman, D., *Surf. Sci.,* 10, 215–221, 1968.
5. Nakamatsu, H., Sudo, A., and Kawai, S., *Surf. Sci.,* 194, 265–271, 1988.
6. Schlier, R. E. and Farnsword, H. E., *J. Chem. Phys.,* 30, 917–925, 1959.
7. Pandey, K. C., *Phys. Rev. Lett.,* 49, 223–229, 1982.
8. Morimoto, T., Nagao, M., and Tokuda, F., *Bull. Chem. Soc. Jpn.,* 41, 1533–1539, 1968.
9. Morimoto, T. and Naono, H., *Bull. Chem. Soc. Jpn.,* 46, 2000–2006, 1973.
10. Naono, H., Kodama, T., and Morimoto, T., *Bull. Chem. Soc. Jpn.,* 48, 1123–1129, 1975.
11. Zhuravlve, L. T., *Langmuir,* 3, 316–322, 1987.
12. Chikazawa, M. and Kanazawa, T., *J. Soc. Powder Technol. Jpn.,* 14, 18–25, 1977.
13. Harnb, N., Hawkins, A. E., and Opalinski, I., *Trans. IchemE,* 74, 605–626, 1996.
14. Chikazawa, M., Kanazawa, T., and Yamaguchi, T., *KONA (Powder Particle),* 2, 54–61, 1984.
15. Chikazawa, M. and Tajima, K., *Kaimenkagaku,* Maruzen, Tokyo, 2001, pp. 14–14.
16. Kanazawa, T., Chikazawa, M., Takei, T., and Mukasa, K., *J. Ceram. Soc. Jpn.,* 92, 11, 654–660, 1984.
17. Ozaki, A., *Shokubai-kougaku-kouza 10,* Chijinsyoten, Tokyo, 1972, p. 778.
18. Auroux, A., Gervasini, A., *J. Phys. Chem.,* 94, 6371–6376, 1990.
19. Tanabe, T., Sumiyoshi, T., Shibata, K., Kiyoura, T., and Kitagawa, J., *Bull. Chem. Soc. Jpn.,* 47, 1064–1070, 1974.
20. Shibata, K., Kiyoura, T., Kitagawa, J., Sumiyoshi, T., and Tanabe, T., *Bull. Chem. Soc. Jpn.,* 46, 2985–2991, 1973.
21. Yoshida, S. and Kawakami, H., *Hyomen,* 21, 737–743, 1983.
22. Yoon, R. H., Salman, T., and Donnay, G., *J. Colloid Interface Sci.,* 70, 483–489, 1979.
23. Fuji, M., Machida, K., Takei, T., Watanabe, T., and Chikazawa, M., *J. Phys. Chem. B,* 102, 8782–8787, 1998.
24. Fuji, M., Machida, K., Takei, T., Watanabe, T., and Chikazawa, M., *Langmuir,* 15, 4584–4589, 1999.
25. Fripiat, J. J. and Uytterhoeven, J., *J. Phys. Chem.,* 66, 800–806, 1962.
26. Hair, M. L. and Hertl, W., *J. Phys. Chem.,* 73, 2372–2378, 1969.
27. Van Der Voort, P., Gillis-D'hamers, I., and Vansant, E. F., *J. Chem. Soc. Faraday Trans.,* 86, 3751–2757, 1990.
28. Yoshinaga, K., Yoshida, H., Yamamoto, Y., Takakura, K., and Komatsu, M., *J. Colloid Interface Sci.,* 153, 207–213, 1992.
29. Madeley, J. D., Richmond, R. C., and Anorg, Z., *Allg. Chem.,* 389, 92–98, 1972.
30. Davydov, V. Ya., Kiserev, A. V., Kiserev, S. A., and Polotnyuk, V. O. V., *J. Colloid Interface Sci.,* 74, 378–386, 1980.
31. Yates, Jr., J. T. and Madey, T. E., *Vibrational Spectroscopy of Molecules on Surfaces.* Plenum, New York, 1987, p. 116.
32. Kondo, J. and Domen, K., *Shokubai (Catalysts Catalysis),* 32, 206–212, 1990.
33. Iler, R. K., *The Chemistry of Silica,* John Wiley & Sons, New York, 1979, p. 622.
34. Hoffmann, P. and Knozinger, E., *Surf. Sci.,* 188, 181–187, 1987.
35. Ferrari, A. M., Ugliengo, P., and Garrone, E., *J. Phys. Chem.,* 97, 2671–2677, 1993.
36. Sindrof, D. W. and Maciel, G. E., *J. Am. Chem. Soc.,* 105, 1487–1493, 1983.
37. Leonardelli, S., Facchini, L., Fretigny, C., Tougne, P., and Legrand, A. P., *J. Am. Chem. Soc.,* 114, 6412–6418, 1992.
38. Hayashi, S., *J. Soc. Powder Technol. Jpn.,* 32, 573–579, 1995.
39. Takei, T., *J. Jpn. Soc. Colour Mater.,* 69, 623–631, 1996.
40. Hauuka, S. and Root, A., *Langmuir,* 98, 1695–1701, 1994.
41. Morrow, B. A., Cody, I. A., and Lee, L. S. M., *J. Phys. Chem.,* 80, 2761–2767, 1976.
42. Smirnov, E. P. and Tsyganenko, A. A., *React. Kinet. Catal. Lett.,* 7, 425–431, 1977.
43. Knozinger, H. and Ratnasamy, P., *Catal. Rev. Sci. Eng.,* 17, 31–37, 1978.
44. Ballinger, T. H. and Yates, Jr., J. T., *Langmuir,* 7, 3041–3047, 1991.
45. Little, L. H., in *Infrared Spectra of Adsorbed Species,* Kagaku-doujin, Kyoto, 1971, pp. 246–308.

46. Curthoys, G., Davydov, V. Y., Kiselev, A. V., and Kiselev, S. A., *J. Colloid Interface Sci.,* 48, 58–64, 1974.
47. Korn, M., Killmann, E., and Eisenlauer, J., *J. Colloid Interface Sci.,* 76, 7–13, 1980.
48. Zhao, Z., Zhang, L., and Lin, Y., *J. Colloid Interface Sci.,* 166, 23–29, 1994.
49. Morrow, B. A. and Cody, I. A., *J. Phys. Chem.,* 80, 1998–2004, 1976.
50. Bunker, B. C., Haaland, D. M., Ward, K. J., Michalske, T. A., Smith, W. L., and Balfe, C. A., *Surf. Sci.,* 210, 406–421, 1989.
51. Chiang, C., Zegarski, B. R., and Dubois, L. H., *J. Phys. Chem.,* 97, 6948–6956, 1993.
52. Krol, D. M. and van Lierop, J. G., *J. Non-Cryst. Solids,* 63, 131–137, 1984.
53. Bunker, B. C., Haaland, D. M., Michalske, T. A., and Smith, W. L., *Surf. Sci.,* 222, 95–101, 1989.
54. Grabbe, A., Michalske, T. A., and Smith, W. L., *J. Phys. Chem.,* 99, 4648–4656, 1995.
55. Wallance, S., West, J. K., and Hench, L. L., *J. Non-Cryst. Solids,* 152, 101–107, 1993.
56. O'Keeffe, M. and Gibbs, G. V., *J. Chem. Phys.,* 81, 876–884, 1984.
57.
58. Little, L. H., in *Infrared Spectra of Adsorbed Species,* Kagaku-doujin, Kyoto, 1971, pp. 202–203.
59. Mapes, J. E. and Eischens, R. P., *J. Phys. Chem.,* 58, 809–815, 1954.
60. Pliskin, W. A., Eischens, R. P., *J. Phys. Chem.,* 59, 1156–1162, 1955.
61. Fuji, M., Fujimori, H., Takei, T., Watanabe, T., and Chikazawa, M., *J. Phys. Chem. B,* 102, 10498–10504, 1998.
62. Preuss, M. and Butt, H.-J., *J. Colloid Interface Sci.,* 208, 468–477, 1998.
63. Fuji, M., Takei, T., Watanabe, T., and Chikazawa, M., *Colloid Surf. A,* 154, 13–24, 1999.
64. Burnauer, S., Deming, L. S., Deming, W. E., and Teller, E., *J. Am. Chem. Soc.,* 62, 1723–1729, 1940.
65. Iwaki, T. and Jellinek, H. H. G., *J. Colloid Interface Sci.,* 69, 17–23, 1979.
66. Kresge, C. T., Leonowicz, M. E., Roth, W. J., Vartuli, J. C., and Beck, J. S., *Nature,* 359, 710–716, 1992.
67. Ravikovitch, P. I., Domohnaill, S. C. O., Neimark, A. V., Schuth, F., and Unger, K. K. *Langmuir,* 11, 4765–4771, 1995.
68. Amberg, C. H., Spencer, W. B., and Beebe, R. A., *Can. J. Chem.,* 33, 305–311, 1995.
69. Pfeifer, P. and Avnir, D., *J. Chem. Phys.,* 79, 3558–3566, 1983.
70. Avnir, D., Farin, D., and Pfeifer, P., *J. Colloid Interface Sci.,* 103, 112–118, 1985.
71. Williams, J. M. and Beebe, Jr., T. P., *J. Phys. Chem.,* 97, 6249, 6255–6261, 1993.
72. Rojanski, D., Huppert, D., Bale, H. D., Xie Dacai, Schmidt, P. W., Farin, D., Seri-Levy, A., and Avnir, D., *Phys. Rev. Lett.,* 56, 2505–2511, 1986.
73. Avnir, D., Farin, D., and Pfeifer, P., *J. Chem. Phys.,* 79, 3566–2572, 1983.
74. Farin, D. and Avnir, D., *J. Phys. Chem.,* 91, 5517–5523, 1987.
75. Hazlett, R. D., *J. Colloid Interface Sci.,* 137, 527–533, 1987.
76. Fuji, M., Ueno, S., Takei, T., Watanabe, T., and Chikazawa, M., *Colloid Poly. Sci.,* 278, 30–36, 2000.
77. Sarid, D., *Scanning Force Microscopy,* Oxford Univ. Press, New York, 1991.
78. Bhushan, B., Fuchs, H., Hosaka, S., Applied *Scanning Probe Methods,* Springer, Berlin, 2004.
79. Arita, T., Kanda, Y., Hamabe, H., Ueno, T., Watanabe, Y., Higashitani, K., *Langmuir,* 19, 6723–6729, 2003.
80. Arita, T., Kanda, Y., Higashitani, K., J. *Colloid Interface Sci.,* 273, 102–105, 2004.
81. Yamamoto, T., Kanda, Y., Higashitani, K., *Langmuir,* 20, 4400–4405, 2004.
82. Donose, B. C., Vakarelski, I. U., Higashitani, K., *Langmuir,* 21, 1834–1839, 2005.
83. Ducker, W.A., Senden, T. J., and Pasheley, R. M., *Langmuir* 8, 1831–1836, 1992
84. Israelachvili, J.N., *Intermolecular and Surface Forces,* 2nd ed. Academic Press, New York, 1992.
85. Vakarelski, I. U., Ishimura, K., Higashitani, K., J. *Colloid Interface Sci.,* 227, 111–118, 2000.
86. Vakarelski, I. U., Higashitani, K., J. *Colloid Interface Sci.,* 242, 110–120, 2001.
87. Kanda, Y., Nakamura, T., Higashitani, *K., Colloids Surf., A,* 139, 55–62, 1998.
88. Kanda, Y., Iwasaki, S., Higashitani, *K., J. Colloid Interface Sci.,* 216, 394–400, 1999.
89. Ishida, N., Sakamoto, M., Miyahara, M., Higashitani, K., *Langmuir,* 16, 5681–5687, 2000.
90. Sakamoto, M., Kanda, Y., Miyahara, M., Higashitani, K., *Langmuir,* 18, 5713–5719, 2002.

Part II

Fundamental Properties of Particles

2.1 Diffusion of Particles

Kikuo Okuyama
Hiroshima University, Higashi-Hiroshima, Japan

Shinichi Yuu
Kyusyhu Institute of Technology, Kitakyushu, Fukuoka, Japan

2.1.1 THERMAL DIFFUSION

When a particle is small, Brownian motion is caused by random variations in the incessant bombardment of molecules against the particle. As the result of Brownian motion, aerosol particles appear to diffuse in a manner analogous to the diffusion of gas molecules.

The Brownian motion of a particle having mass m_p is expressed by the equation of motion of a single particle:

$$m_p \frac{dv}{dt} = -\frac{3\pi\mu D_p}{C_c}\mathbf{v} + m_p a(t) \tag{1.1}$$

where v is the velocity of the particle, $a(t)$ the random acceleration resulting from the thermal motion of the background molecules and Stokes drag, and C_c the Cunningham correction factor accounting for noncontinuum effects. C_c is equal to unity in liquid, and in air it is given by

$$C_c = 1 + 2.514\left(\frac{l}{D_p}\right) + 0.80\left(\frac{l}{D_p}\right)\exp\left[-0.55\left(\frac{D_p}{l}\right)\right] \tag{1.2}$$

where l is the mean free path of gas molecules; $l = 6.5 \times 10^{-6}$ cm at 25°C under atmospheric pressure.

Figure 1.1 shows a particle trajectory in a stationary fluid. In this case, the average square of the particle displacement $\overline{\Delta x^2}$ in a time interval t and the average displacement $|\overline{\Delta x}|$, by Brownian motion, are given as follows:

$$\overline{\Delta x^2} = 2Dt, \quad |\overline{\Delta x}| = \sqrt{\frac{4Dt}{\pi}} \tag{1.3}$$

In the above equations, D (cm²/s) is the Brownian diffusion coefficient of particles with diameter D_p given by

$$D = \frac{C_c kT}{3\pi\mu D_p} \tag{1.4}$$

where k is the Boltzmann constant (1.38×10^{-23} J/K) and T is the temperature (K). Figure 1.2 shows the Brownian diffusion coefficient of particles in air and water at 20°C.

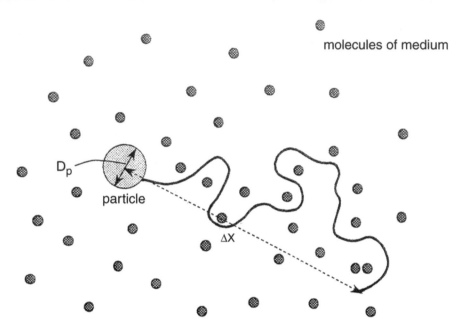

FIGURE 1.1   Particle movement by Brownian motion.

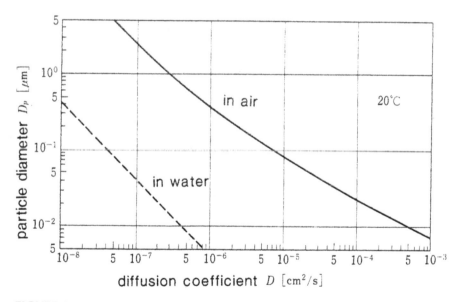

FIGURE 1.2   Diffusion coefficient of particle in air and water.

For the agglomerate particles, the Brownian diffusion coefficient can be expressed by replacing the solid particle diameter by the collision diameter, D_c, of agglomerate in the Fuchs interpolation expression for Brownian coagulation in the free molecule and continuum regimes[1,2]

$$D = \frac{kT}{3\pi\mu D_c}\left(\frac{5 + 4Kn + 6Kn^2 + 18Kn^3}{5 - Kn + (8 + \pi)Kn^2}\right) \tag{1.5}$$

where Kn $(=2l/D_c)$ is the Knudsen number of the agglomerate. The collision diameter of an agglomerate is proposed by[2,3]

$$D_c = d_p \left(\frac{v}{v_p} \right)^{1/D_f} = d_p \left(n_p \right)^{1/D_f} \tag{1.6}$$

where d_p and v_p are the diameter and the volume of the primary particle, respectively, n_p is the number of primary particles in an agglomerate, and D_f is the mass fractal dimension.

The number concentration of particles of a uniform size undergoing Brownian diffusion in a flow with gas velocity u can be determined by solving the following equation of convective diffusion of particles with an allowance for external forces:

$$\frac{\partial n}{\partial t} + \nabla \cdot \mathbf{u} n = D\nabla^2 n - \nabla \cdot \mathbf{v} n, \quad \mathbf{v} = \frac{t\sum F}{m_p} \tag{1.7}$$

where n is the particle number concentration, $\mathbf{v}$ is the particle velocity caused solely by external forces F, $\mathbf{u}$ is the velocity of the surrounding gas or liquid, and τ is the relaxation time of the particle ($\rho_p C_c D_p^2/18\mu$, ρ_p = particle density, μ = viscosity of fluid).

If particles are polydisperse with a number distribution function $n(D_p, t)$ at time t, the corresponding diffusion equation can be given by substituting $n(D_p, t)$ into n in Equation 1.7.

The local deposition rate of particles by Brownian diffusion onto a unit surface area, the deposition flux j (number of deposited particles per unit time and surface area), is given by

$$j = - D\nabla n + vn + \mathbf{u} n \tag{1.8}$$

If the flow is turbulent, the value of the deposition flux depends on the strength of the flow and the Brownian diffusion coefficient.

Table 1.1 indicates the representative solutions for steady-state particle deposition in laminar flow. The solutions in cases (b) and (c) are applicable to size analysis of small aerosol particles based on Brownian diffusion. Figure 1.3 shows the change in number concentration of aerosols in laminar flow through a horizontal tube as a function of the dimensionless parameter μ $[Dx/(u_{xav} R^2); u_{xav} =$ average velocity of fluid, x = pipe length, R = pipe radius].[8] n_0 and n are the inlet and outlet particle number concentrations, respectively. The value of σ ($u_t R/D$; u_t = gravitational settling velocity) indicates the relative effect of gravitational sedimentation to the Brownian diffusion, and $\sigma = 0$ means the case of Brownian diffusion only, case (b) in Table 1.1. With increasing σ, the effect of gravitational sedimentation increases, and the curve of $\sigma = 50$ agrees with that of the gravitational sedimentation without any Brownian diffusion[9]:

$$\frac{\bar{n}}{n_0} = 1 - \frac{2}{\pi} \left(2\alpha\beta - \alpha^{1/3}\beta + \arcsin \alpha^{1/3} \right) \tag{1.9}$$

where $\alpha = (3/8)\sigma\mu$ and $\beta = \sqrt{1 - \alpha^{2/3}}$.

When small particles are enclosed in a chamber and its number concentration is uniform except in the vicinity of the chamber walls, the change in the number concentration of particles with time is given by

$$n = n_0 \exp(- \beta t) \tag{1.10}$$

where n_0 is the initial particle number concentration, β the deposition rate coefficient, and t the elapsed time. In the case of a cylindrical chamber having an inner surface area S and volume V_T,

Table 1.1 Particle Diffusion at Steady State in a Laminar Flow

	Situation	Basic equation	B.C.	Solution or flux j	Reference
(a)	wall	$u_x \dfrac{\partial n}{\partial x} + u_z \dfrac{\partial n}{\partial z} = D \dfrac{\partial^2 n}{\partial z^2}$	$n(x,0)=0$, $n(x,\infty)=n_0$	$n = \dfrac{(0.22Sc)^{1/3} n_0}{0.89}$ $\times \displaystyle\int^{1/2(u_0 z^2/vx)} \exp\left(-0.22Scv^3\right)dv$	Levich (1962)
(b)	pipe	$u_x \dfrac{\partial n}{\partial x} = D\left(\dfrac{\partial n}{r\partial r} + \dfrac{\partial^2 n}{\partial z^2}\right)$, $u_x = 2u_{av}\left(1 - \dfrac{r^2}{R^2}\right)$	$n(0,r)=n_0$, $n(x,R)=0$	$j = 0.34Dn_0\left(\dfrac{u_0}{vx}\right)^{1/2}$ Sc$^{1/2}$, Sc$=v/D$ $\dfrac{n_{av}}{n_0} = 0.819e^{-3.657\mu} + 0.0976e^{-22.3\mu}$ $+ 0.032e^{-57\mu} + \cdots \mu \geq 0.0312$ $\dfrac{n_{av}}{n_0} = 1 - 2.56\mu^{2/3} + 1.2\mu$ $+ 0.177\mu^{4/3} + \cdots \quad \mu < 0.0312$ $\mu = Dx/u_{av}R^2$	Gormely and Kennedy (1948)
(c)	duct	$u_x \dfrac{\partial n}{\partial x} = D\dfrac{\partial^2 n}{\partial z^2}$, $u_x = \dfrac{3u_{av}}{2h^2}(h^2 - z^2)$, $2h \ll 2b \ll$ length	$n(0,z)=n_0$, $n(x,h)=0$, $n(x,-h)=0$	$\dfrac{n_{av}}{n_0} = 0.9149e^{-7.54\mu} + 0.0592e^{-89.2\mu}$ $+ 0.0258e^{-607\mu} + \cdots$ $\mu = Dx/4h^2u_{av}$	De Marcus and Thomas (1952)
(d)	cylinder	$u_r\dfrac{\partial n}{\partial r} + \dfrac{u_\theta}{r}\dfrac{\partial n}{\partial \theta} = D\left(\dfrac{\partial^2 n}{\partial r^2} + \dfrac{1}{r}\dfrac{\partial n}{\partial r} + \dfrac{1}{r^2}\dfrac{\partial^2 n}{\partial \theta^2}\right)$	$n(R,0)=0$, $n(\infty,\theta)=n_0$, Re<1, $D_p/R \approx 0$	$\eta_D = \dfrac{2DR\displaystyle\int_0^\pi \dfrac{\partial n}{\partial r}\Big|_{r=R} d\theta}{2u_0Rn_0}$ $= 2.9k^{-1/3}Pe^{-2/3} + 0.62Pe^{-1}$ η_D: target efficiency, Pe$=2u_0R/D$ parallel cylinders: $k = -\dfrac{1}{2}\ln\alpha - \dfrac{3}{4}$ $+ \alpha - \dfrac{\alpha^2}{4}$ isolated cylinder: $k = 2 - \ln Re$ α: packing fraction	Stechkina and Fuchs (1966)
(e)	sphere	$u_r\dfrac{\partial n}{\partial r} + \dfrac{u_\theta}{r}\dfrac{\partial n}{\partial \theta} = D\left(\dfrac{1}{r^2}\dfrac{\partial}{\partial r}\left(r^2\dfrac{\partial n}{\partial r}\right) + \dfrac{1}{r^2\sin\theta}\dfrac{\partial}{\partial\theta}\left(\sin\theta\dfrac{\partial n}{\partial\theta}\right)\right)$	$n(R,0)=0$, $n(\infty,\theta)=n_0$, Re<1, $D_p/R \approx 0$	$\eta_D = \dfrac{2\pi DR^2\displaystyle\int_0^\pi\left(\dfrac{\partial n}{\partial r}\right)_{r=R}\sin\theta\, d\theta}{\pi R^2 u_0 n_0}$ $= 4.04Pe^{-2/3}$ $Pe=\dfrac{2u_0R}{D}$ η_D: target efficiency	

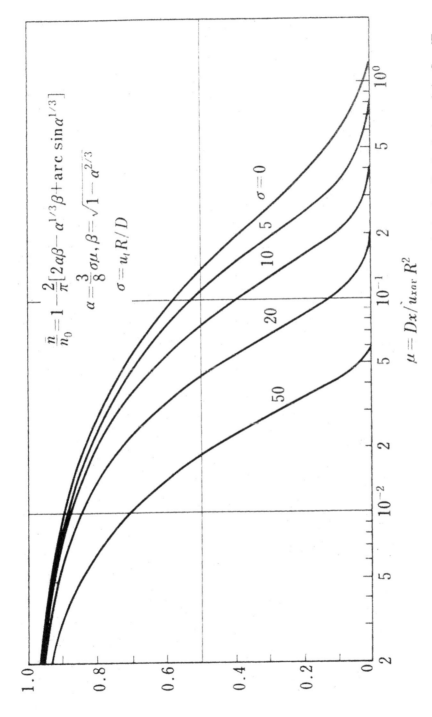

$$\frac{\bar{n}}{n_0} = 1 - \frac{2}{\pi}\left[2\alpha\beta - \alpha^{1/3}\beta + \text{arc}\sin\alpha^{1/3}\right]$$

$$\alpha = \frac{3}{8}\sigma\mu, \quad \beta = \sqrt{1 - \alpha^{2/3}}$$

$$\sigma = u_t R / D$$

$$\mu = Dx/u_{xav} R^2$$

FIGURE 1.3 Decrease in number concentration of aerosol due to Brownian diffusion and gravitational settling in a horizontal pipe flow. [From Taulbee, D. B., *J. Aerosol Sci.*, 9, 17–23, 1978. Reprinted with permission from *J. Aerosol Sci.*, copyright (1978), Pergamon Press, Ltd.]

β corresponds to $DS/\delta V_T$, and the thickness δ of the concentration boundary layer can be given as $2.884D^{-1/3}$ for natural convection flow.[10]

If the flow is turbulent, β becomes a function of the strength of the flow field and the Brownian diffusion coefficient.[10,11] The transport rate of particles suspended in turbulent flow onto a wall surface over which the fluid stream is flowing as shown in Figure 1.4 can be described by

$$N = (D + D_E)\frac{dn}{dy} - nu_t \cos\theta \qquad (1.11)$$

where D and D_E are the Brownian and turbulent diffusion coefficients, u_t the gravitational settling velocity, y the distance from the surface, and θ the angle from the direction of y to the direction of the gravity.

By solving Equation 1.11 analytically, Crump and Seinfeld[11] derived an expression for the deposition velocity $V(\theta)$ [(deposition flux)/(particle number concentration above the surface)] for a wall with the angle θ approximating the turbulent diffusion coefficient D_E to $K_e y^m$ as follows:

$$V(\theta) = u_t \cos\theta \left[\exp\left(\frac{\pi u_t \cos\theta}{m \sin(\pi/m)\sqrt[m]{K_e D^{m-1}}} \right) - 1 \right]^{-1} \qquad (1.12)$$

In the turbulent diffusion coefficient D_E, K_e is considered to be proportional to the velocity gradient of fluid as $k|du/dx|$, where k is a constant depending on the flow condition of fluid. For stirred turbulent flow field, $k = 0.4$ and $m = 2.7$ could explain the experimental values reasonably. When the

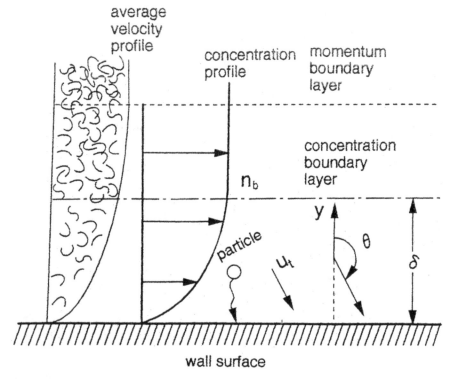

FIGURE 1.4   Schematic diagram of particle deposition onto an inclined flat surface.

surface area of walls of the chamber with the angle θ is $S(\theta)$, the deposition rate coefficient β can be related to the deposition velocity $V(\theta)$ as

$$\beta = \left(\sum_{\theta=0}^{2\pi} V(\theta) S(\theta) \right) V_{T}^{-1} \tag{1.13}$$

where V_{T} is a total volume of the chamber.

When external forces such as electrostatic force have influence on particle transport, the deposition velocity can be predicted by considering the particle velocity induced by the forces in Equation 1.12.[12] In such a case, $u_t \cos \theta$ in Equation 1.12 is replaced by $u_t \cos \theta + v_{ext}$, where v_{ext} is the component of this velocity in the direction normal to the wall. Deposition velocity is also influenced by wall roughness. Because the concentration boundary layer of gas-borne particles is generally much thinner than the momentum boundary layer, even a wall with very small roughness that does not affect the momentum boundary layer (i.e., a hydraulically smooth wall) can enhance particle deposition rate if the height of the roughness is comparable to the thickness of the concentration boundary layer.[13]

2.1.2 TURBULENT DIFFUSION

Turbulent diffusion is defined as the transportation of physical properties such as mass, heat, and momentum by the turbulent random motion of a fluid. Because knowledge of statistical functions describing turbulent motions is limited compared with that of molecular motions, any solution of turbulent transport problems is incomplete and only approximate.

In comparing turbulent mass flux with the corresponding flux caused by molecular motion, it is usually assumed that the turbulent mass flux is similar in description to molecular diffusion; that is, the flux is directly proportional to the concentration gradient (gradient transport). Although the solution based on the gradient transport cannot be correct in all details, the phenomenological theory (the gradient transport theory) is most useful for practical engineering problems. The present section is devoted to a description of the turbulent diffusion, particularly the particle turbulent diffusion based on the gradient transport.

The treatment of transport phenomena in the various flow fields and these applications are omitted in this section, and fundamental descriptions about the turbulent diffusion (i.e., turbulent diffusion equation, turbulence properties, and diffusion coefficient) are presented.

Turbulent Diffusion Equation

The law of conservation of particle mass gives

$$\frac{\partial n}{\partial t} + \nabla \cdot n\mathbf{v} = D\nabla^2 n + \Gamma \tag{1.14}$$

where n, $\mathbf{v}$, D, and T are particle concentration, particle velocity vector, Brownian diffusion coefficient, and source, respectively. ∇ is the nabla operator. The instantaneous values can be written as

$$n = \bar{n} + n', \qquad \mathbf{v} = \bar{\mathbf{v}} + \mathbf{v}' \tag{1.15}$$

where the overbar and prime denote the average and the fluctuating values, respectively. Some rearrangement, after substituting Equation 1.15 into Equation 1.14 gives

$$\frac{\partial \bar{n}}{\partial t} + \nabla \cdot \bar{n}\bar{\mathbf{v}} + \nabla \cdot \overline{n'\mathbf{v}'} = D\nabla^2 \bar{n} + \Gamma \tag{1.16}$$

The first term on the left-hand side represents change with time, the second is a convection term, and the third is a turbulent transport term. The first term on the right-hand side represents mass flux due to Brownian motion and the second is a source or reaction term. Only the third term on the left-hand side has turbulence fluctuating values. If the turbulent transport term $\overline{n'v'}$ is expressed by the time-averaged particle concentration $\overline{n}$, Equation 1.16 can be solved because then the only unknown in Equation 1.16 is $\overline{n}$. The introduction of the gradient transport, which shows an analogy to the Brownian diffusion, gives

$$\overline{n'v_i'} = -\varepsilon_p \frac{\partial \overline{n}}{\partial x_i} \tag{1.17}$$

where ε_p is the particle turbulence diffusion coefficient. If the diffusion coefficient is a tensor of the second order, the relation between $\overline{n'v'}$ and $\partial \overline{n} / \partial x_j$ should be

$$\overline{n'v_i'} = -\varepsilon_{pij} \frac{\partial \overline{n}}{\partial x_j} \tag{1.18}$$

In this section the summation convention with respect to repeated indices is used. Substitution of Equation 1.18 into Equation 1.16 gives

$$\frac{\partial \overline{n}}{\partial t} + \nabla \cdot \overline{n}\,\overline{\mathbf{v}} = \frac{\partial}{\partial x_i} \left(\varepsilon_{pij} \frac{\partial \overline{n}}{\partial x_j} \right) + D\nabla^2 \overline{n} + \Gamma \tag{1.19}$$

Equation 1.19 is a general form of a particle diffusion equation. If we consider the transport by the turbulent motions without molecular diffusion and source, Equation 1.19 reduces to

$$\frac{\partial \overline{n}}{\partial t} + \nabla \cdot \overline{n\mathbf{v}} = \frac{\partial}{\partial x_i} \left(\varepsilon_{pij} \frac{\partial \overline{n}}{\partial x_j} \right) \tag{1.20}$$

When the time-averaged particle velocity $\overline{v}$ and the turbulent diffusion tensor ε_{pij} are known, Equation 1.20 can be solved and the solution with the boundary conditions gives a particle concentration distribution. If the time-averaged particle velocity is equal to that of fluid (i.e., the particle inertia is negligibly small), substitution of the fluid continuity equation into Equation 1.20 gives

$$\frac{\partial \overline{n}}{\partial t} + \overline{\mathbf{v}} \cdot \nabla \overline{n} = \frac{\partial}{\partial x_i} \left(\varepsilon_{pij} \frac{\partial \overline{n}}{\partial x_j} \right) \tag{1.21}$$

where the fluid is assumed to be incompressible.

Turbulent Diffusion Coefficient

The basic problem now is to derive an expression for ε_{pij}. Batchelor[14] showed that when the probability distribution of the displacement of particles has a Gaussian form, the diffusion tensor can be interpreted as

$$\varepsilon_{ij} + \varepsilon_{ji} = \frac{d}{dt} \overline{\left(X_i - \overline{X}_i \right) \left(X_j - \overline{X}_j \right)} \tag{1.22}$$

where X_i and X_j are i and j components of the Lagrangian position of the particle. When $i = j = 2$, Equation 1.22 reduces to

$$\varepsilon_{22} = \frac{1}{2}\frac{d}{dt}\overline{\left(X_2 - \overline{X}_2\right)^2} \tag{1.23}$$

The mean square value $\overline{\left(X_2 - \overline{X}_2\right)^2}$ of the Lagrangian displacement can be obtained by

$$\overline{\left(X_2 - \overline{X}_2\right)^2} = 2\int_0^t \int_0^{t'} \overline{v'_2(t')v'_2(t'-\tau)}d\tau dt' = 2\overline{v'^2_2}\int_0^t (t-\tau)R_L(\tau)d\tau \tag{1.24}$$

The Lagrangian correlation coefficient $R_L(\tau)$ is defined as

$$R_L(\tau) = \frac{\overline{v'_2(t)v'_2(t+\tau)}}{\left[\overline{v'_2(t)^2}\ \overline{v'_2(t+\tau)^2}\right]^{\frac{1}{2}}} \tag{1.25}$$

Now consider some simple cases. The Lagrangian correlation $R_L(\tau)$ is nearly equal to unity for a very small t (i.e., for a very short time diffusion). Substitution of $R_L(\tau) = 1$ into Equation 1.24 gives

$$\overline{\left(X_2 - \overline{X}_2\right)^2} = \overline{v'^2_2}t^2 \tag{1.26}$$

Substituting Equation 1.26 into Equation 1.23, we obtain

$$\varepsilon_{22} = \overline{v'^2_2}t \tag{1.27}$$

Hence, the diffusivity for very short time diffusion depends linearly on the time and the mean square value of fluctuating velocity (i.e., turbulent intensity).

On the other hand, we now consider a long diffusion time. When diffusion time t is much longer than the time t_0 for which $R_L(t_0) \approx 0$, Equation 1.24 is reduced to

$$\begin{aligned} 2\overline{v'^2_2}\int_0^t (t-\tau)R_L(\tau)d\tau &= 2\overline{v'^2_2}t\int_0^{t_0} R_L(\tau)d\tau \\ &= 2\overline{v'^2_2}\int_0^{\infty} R_L(\tau)d\tau \\ &= 2\overline{v'^2_2}tT_L \end{aligned} \tag{1.28}$$

where T_L, defined as $T_L = \int_0^{\infty} R_L(\tau)d\tau$ is the Lagrangian integral timescale. Substitution of Equation 1.28 into Equation 1.23 gives

$$\varepsilon_{22} = \overline{v'^2_2}T_L \tag{1.29}$$

Hence, the diffusivity for a long diffusion time is proportional to the integral timescale and the turbulent intensity.

Solid particles, mist, and small bubbles cannot follow the fluid motion completely due to their inertia. Hinze[15] derived the relation between ε_p and ε_f based on Equation 3.14 as follows:

$$\frac{\varepsilon_p}{\varepsilon_f} = 1 + \frac{1-B^2}{A^2 T_{fL}^2 - 1} \frac{\exp(-At) - \exp(-t/T_{fL})}{1 - \exp(-t/T_{fL})} \tag{1.30}$$

Equation 1.30 shows that $\varepsilon_p \approx \varepsilon_f$ for a long diffusion time. This is reasonable for homogeneous turbulence, even though for a long diffusion time ε_p is not equal to ε_f in an inhomogeneous turbulence in which the turbulence characteristics change spatially. The assumption that the same fluid will surround the particle as it moves (no overshooting) causes an unreasonable result. However, it is difficult to take the effect of the overshooting rigorously into account. Hence, the Lagrangian turbulent trajectories of particles and the turbulent diffusivity are calculated by a simple model.

Yuu et al.,[16] Brown and Hutchinson,[17] and Gosman and Ioannides[18] present models. The model of Yuu et al.[16] is explained here. The particle moves, surrounded by an eddy that has a lifetime T_L defined as

$$T_L = \int_0^\infty R_L(\tau) d\tau \tag{1.31}$$

where $R_L(\tau)$ is the Lagrangian time correlation coefficient and T_L is the averaged longest time during which a fluid particle persists in moving in a given direction. In the model, R_L is 1 when the diffusion time $t \leqslant T_L$, and R_L becomes zero when $t > T_L$. In other words, an independent eddy surrounds the particle and affects the particle motion during the next T_L. The fluid fluctuating velocity is obtained by using a normal random function in which the average value is zero and the variance is equal to the fluid intensity. Substituting the fluid fluctuating velocity into the equation of particle motion, one can calculate the Lagrangian particle trajectory. Similarly, the particle trajectory during the next T_L is calculated, and the process is repeated until the diffusion time passes.

Since the early work by Taylor,[19] the relation between the turbulent diffusivity and Lagrangian displacements has been studied by many investigators to obtain values for the turbulent flux of transferable quantities. Most of them, however, did not completely consider the effect of inhomogeneity of the turbulence field nor the effect of mean velocity distribution in an inhomogeneous turbulent flow. The ensemble average of the Eulerian fluctuating velocities at the Lagrangian positions is not zero in an inhomogeneous turbulent flow. For example, Batchelor's description (Equation 1.22) of the turbulent diffusivity for homogeneous turbulence would not be strictly applied to real turbulent flows, as these usually have mean velocity distributions.

Yuu[20] derived the relation (Equation 1.32) between turbulent diffusivities and the Lagrangian displacement of particles in a flow field with a mean velocity distribution based on the gradient transport:

$$\langle \varepsilon_{ij} \rangle = \frac{1}{2} \frac{\partial}{\partial t} \langle (x_i - \langle x_i \rangle)(x_j - \langle x_j \rangle) \rangle + \{ \langle x_i \rangle \langle \bar{u}_j \rangle - \langle x_i \bar{u}_j \rangle \} \tag{1.32}$$

in which $\langle \rangle$ indicates an ensemble averaged values. Figure 1.5 shows the calculated results of $\langle \tilde{\varepsilon}_{yy} \rangle$ in two interacting plane parallel jets where the distance between nozzles is $24D$. The experimental data of Yuu et al.[21] indicates that the strong interactions between two jets exist primarily in the region where $20 \leqslant x/D \leqslant 80$ when the distance between the two nozzles is $24D$. As shown in Figure 1.5, the effect of the mean velocity distribution on $\langle \tilde{\varepsilon}_{yy} \rangle$ in this region is very large. If Equation 1.22, which assumes that there is homogeneous turbulence, is used for the calculation, we find that this gives about a five-fold overestimation of the diffusivity in this region. This is the curve shown by the open circles.

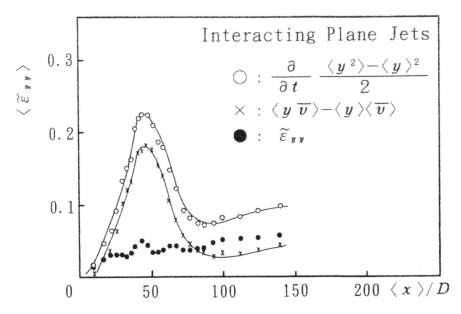

FIGURE 1.5 Calculated results of the dimensionless lateral turbulent diffusivity in two interacting plane parallel jets (the distance between the two nozzle centers is 24D). ●, ○, and × indicates the dimensionless values of the first and second terms of Equation 1.32, respectively. $\langle \tilde{\varepsilon}_{yy} \rangle = \langle \varepsilon_{yy} \rangle / U_0 D$.

REFERENCES

1. Seinfeld, J. H. and Pandis, S. N., *Atmospheric Chemistry and Physics from Air Pollution to Climate Change,* John Wiley & Sons, New York, 1998, pp. 656–664.
2. Kruis, F. E., Kusters, K. A., and Pratsinis, S. E., *Aerosol Sci. Technol.* 19, 514–526, 1993.
3. Matsoukas, T. and Friedlander, S. K., *J. Colloid Interface Sci.* 146, 495–506, 1991.
4. Levich, V. G., *Physicochemical Hydrodynamics,* Prentice-Hall, Englewood Cliffs, N.J., 1962.
5. Gormely, P. G. and Kennedy, M., *Proc. Roy. Irish Acad. A* 52, 163–169, 1948.
6. De Marcus, W. and Thomas, J., U.S. Atomic Energy Commission Report ORN 1–1413, 1952.
7. Stechkina, I. and Fuchs, N. A., *Ann. Occup. Hyg.,* 9, 59–64, 1966.
8. Taulbee, D. B., *J. Aerosol Sci.,* 9, 17–23, 1978.
9. Fuchs, N. A., *The Mechanics of Aerosols,* Dover, New York, 1964, pp. 110–112.
10. Okuyama, K., Kousaka, Y., Yamamoto, S., and Hosokawa, T., *J. Colloid Interface Sci.,* 110, 214–223, 1986.
11. Crump, J. G. and Seinfeld, J. H. *J. Aerosol Sci.* 12, 405–415, 1981.
12. Shimada, M. and Okuyama, K., *J. Colloid Interface Sci.,* 154, 255–263, 1992.
13. Shimada, M., Okuyama, K., Kousaka, Y., and Seinfeld, J. H., *J. Colloid Interface Sci.,* 125, 198–211, 1988.
14. Batchelor, G. K., *Aust. J. Sci. Res.,* A2, 437–451, 1949.
15. Hinze, J. O., *Turbulence,* 2nd Ed., McGraw-Hill, New York, 1975, pp. 460–471.
16. Yuu, S., Yasukouchi, N., Hirosawa, Y., and Jotaki, T., *AIChE J.,* 24, 509–519, 1978.
17. Brown, D.J. and Hutchinson P., *J. Fluids Eng.,* 101, 265–269, 1979.
18. Gosman, A. D. and Ioannides, E., AIAA Paper 81–0323, 1981.
19. Taylor, G. I., *Proc. London Math. Soc. Ser. 2,* 20, 196–215, 1921.
20. Yuu, S., *Phys. Fluids,* 28, 466–472, 1985.
21. Yuu, S., Shimoda, F., and Jotaki, T., *AIChE J.,* 25, 676–685, 1979.

2.2 Optical Properties

Yasushige Mori
Doshisha University, Kyotanabe, Kyoto, Japan

2.2.1 DEFINITIONS

Light is electromagnetic radiation in the wavelength range from 3 nm to 30 μm. Visible light is the part of electromagnetic radiation to which the human eye is sensitive, and the wavelength of it ranges from 400 nm to 750 nm. When a light beam illuminates a particle having a dielectric constant different from unity, a part of the light will be reflected on the surface of particle, and the rest of the light will pass into the particle, as shown in Figure 2.1. The part of the light passed into the particle will go out as refraction, and the rest will be absorbed. When the particle size is much larger than the wavelength of the incident light, the light beam is diffracted near the particle. The scattering phenomenon of light includes diffraction, deflection, and refraction of light. The net results of the absorption and scattering caused by the particle are known as the extinction of incident light. In describing such optical phenomena, the size parameter α defined by the following expression may be important:

$$\alpha = \pi D_{\mathrm{p}}/\lambda \tag{2.1}$$

where D_{p} is the particle diameter and λ is the wavelength of the incident light.

Another important definition is the refractive index. The refractive index of the medium is given by the ratio of the velocity of light in a vacuum, c, to that in the medium, v:

$$m_0 = c/v \tag{2.2}$$

This value is always larger than unity.

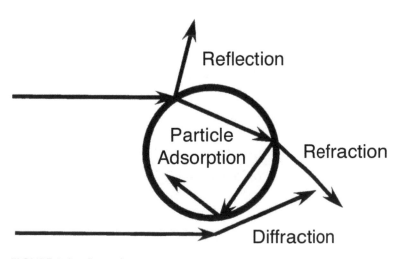

FIGURE 2.1   Scattering geometry.

For an absorbing material having appreciable electrical conductivity, the refractive index is expressed as a complex number:

$$m_m = n_1 - i\,n_2 \quad \left(i = \sqrt{-1}\right) \tag{2.3}$$

The imaginary part n_2 is zero for nonabsorbing particles, as n_2 expresses the property of absorption of light. m is defined as the ratio of the complex refractive index of the medium, m_m, and the refractive index of the absorbing material, m_0, via

$$m = m_m/m_0 \tag{2.4}$$

2.2.2 LIGHT SCATTERING

Mie used electromagnetic theory to obtain the rigorous solution of light scattering by a uniform and spherical particle.[1] When unipolarized light of intensity I_0 illuminates a particle as shown in Figure 2.2, the scattering intensity at a distance r in the direction θ from the particle is given as

$$I = \frac{\lambda^2}{8\pi^2\,r^2}\left(i_1 + i_2\right) I_0 \tag{2.5}$$

where i_1 and i_2 are the Mie intensity parameters for scattered light with perpendicular and parallel polarization against the observed plain, respectively. These parameters are expressed in the following equations as functions of m, α, and θ.[2]

$$i_1 = \left|\sum_{n=1}^{\infty} \frac{2n+1}{n(n+1)}\{a_n\,\pi_n + b_n\,\tau_n\}\right|^2 \tag{2.6}$$

$$i_2 = \left|\sum_{n=1}^{\infty} \frac{2n+1}{n(n+1)}\{b_n\,\pi_n + a_n\,\tau_n\}\right|^2 \tag{2.7}$$

$$a_n = \frac{\psi_n(\alpha)\psi_n'(m\alpha) - m\psi_n'(\alpha)\psi_n(m\alpha)}{\zeta_n(\alpha)\psi_n'(m\alpha) - m\zeta_n'(\alpha)\psi_n(m\alpha)} \tag{2.8}$$

$$b_n = \frac{m\psi_n(\alpha)\psi_n'(m\alpha) - \psi_n'(\alpha)\psi_n(m\alpha)}{m\zeta_n(\alpha)\psi_n'(m\alpha) - \zeta_n'(\alpha)\psi_n(m\alpha)} \tag{2.9}$$

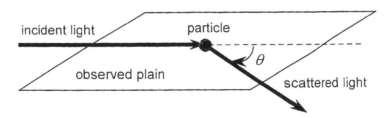

FIGURE 2.2   Scattering and diffraction of light.

where ψ_n' and ζ_n' means the first derivative of ψ_n and ζ_n, respectively. The functions ψ_n and ζ_n are related to the Bessel (J_n) and second-order Hankel ($H_n^{(2)}$) functions via

$$\psi_n(x) = \sqrt{\frac{\pi \alpha}{2}}\, J_{n+\frac{1}{2}}(x) \tag{2.10}$$

$$\zeta_n(x) = \left(\frac{x\pi}{2}\right)^{1/2} H_{n+\frac{1}{2}}^{(2)}(x) \tag{2.11}$$

where x is equal to α or $m\alpha$.

Figure 2.3 shows the angular distribution of scattered intensity calculated for uniform spheres with $m_m = 1.5$.[3] The intensity of scattered light becomes a minimum at a scattering angle between 90° and 150°. Figure 2.4 and Figure 2.5 show the changes in the intensity of scattered light with size parameter α as a function of the scattering angle θ and the refractive index m_m of the particle material.[3] For absorbing particles having the imaginary part of refractive index n_2, the fluctuation of the scattered intensity decreases with the value of n_2.[4]

If a particle is much smaller than the wavelength of light ($m\alpha < 0.5$), the contribution of the nth term rapidly decreases as n increases in Equation 2.6 and Equation 2.7. When the values of a_n and b_n are adopted to a sum of $n = 1$ and 2, and a value only of $n = 1$, respectively, and the value for higher than α^3 is neglected, the Rayleigh scattering equation is obtained:

$$I = \frac{\pi^4 D_p^6}{8 r^2 \lambda^4} \left(\frac{m^2 - 1}{m^2 + 2}\right)^2 (1 + \cos^2 \theta)\, I_0 \tag{2.12}$$

In the Rayleigh scattering regime ($\alpha < 2$), the scattered intensity is proportional to the sixth power of the particle size, as shown as Equation 2.12. In the Mie scattering regime ($\alpha = 2\text{–}10$), the scattered intensity increases with particle size while fluctuating. In the geometrical optics regime ($\alpha > 10$), the intensity increases as the square of particle size, and the intensity fluctuation is more drastic than in the Mie scattering regime.

When the particle size is much larger than the wavelength ($\alpha > 300$), the light beam is diffracted near the particle, as shown in Figure 2.1 or Figure 2.6. For a low diffraction angle ($\theta < 10°$), the angular distribution of the intensity of diffracted beam $I(\theta)$ is given by the Fraunhofer formula:

$$I(\theta) = \frac{\pi^2 D_p^4}{4 \lambda^2} \left\{ \frac{J_1(\alpha \sin \theta)}{\alpha \sin \theta} \right\}^2 I_0 \tag{2.13}$$

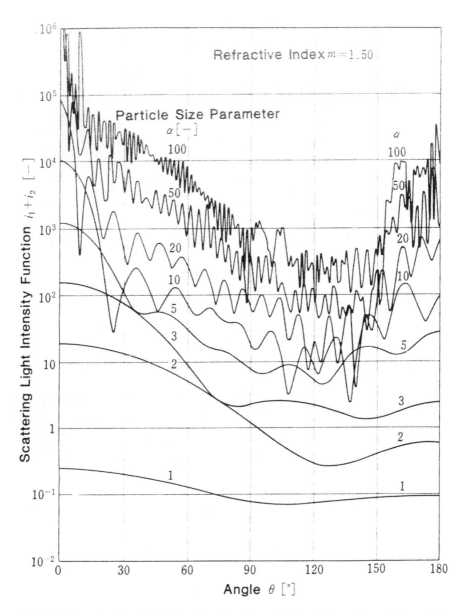

FIGURE 2.3   Angular distribution of scattered intensity for uniform sphere with $m_{\mathrm{m}} = 1.5$.
[From Kanagawa, A., *Kagaku Kogaku*, 34, 521, 1970. With permission.]

where $J_1(\alpha \sin \theta)$ is the first-order Bessel function. As the intensity of the Fraunhofer diffraction depends on the particle size and is independent of the refractive index of the particle material, this fact is often useful in particle-sizing applications.

The light scattering technique is nowadays one of the typical particle size analysis methods.[5] However it is difficult to apply to nonuniform and nonspherical particles for the above theory. The analysis for nonuniform[6] and nonspherical particles[7] was made. Mishchenko et al. recently published a book on light scattering by nonspherical particles in which theoretical and numerical techniques are also included.[8]

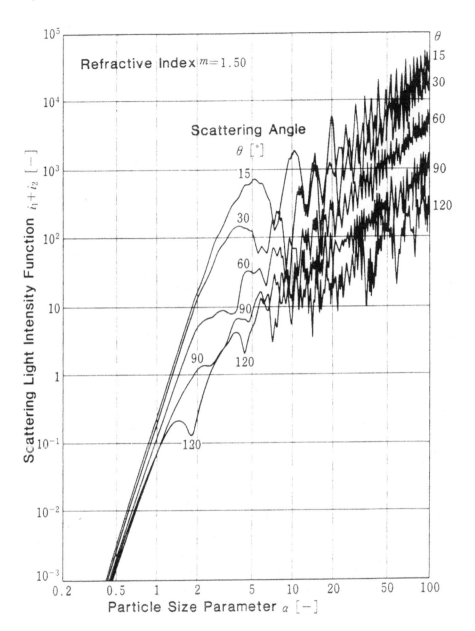

FIGURE 2.4   Changes in the intensity of scattered light with size parameter α as the parameter of the scattering angle θ. [From Kanagawa, A., *Kagaku Kogaku,* 34, 521, 1970. With permission.]

2.2.3   LIGHT EXTINCTION

Aerosol particles or colloidal particles illuminated by a light beam absorb some light, as well as scatter. Thereby, the intensity of the light beam diminishes, as shown in Figure 2.6. According to the Lambert–Beer law, the light extinction can be expressed by

$$I = I_0 \exp(-K L) \tag{2.14}$$

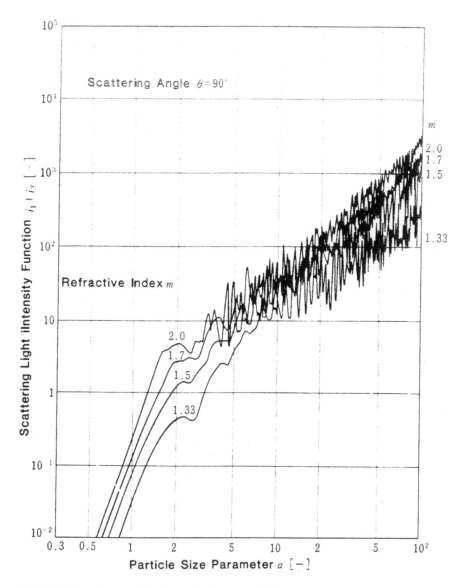

FIGURE 2.5   Changes in the intensity of scattered light with size parameter α as the refractive index m_m. [From Kanagawa, A., *Kagaku Kogaku*, 34, 521, 1970. With permission.]

where I_0 and I are respectively the intensity of incident light and the intensity at penetrated distance L. K is the turbidity given as

$$K = N Q_{\text{ext}}(D_p) \tag{2.15}$$

where N is the particle number concentration, and $Q_{\text{ext}}(D_p)$ is the extinction cross section of one particle, whose projected area equivalent diameter is D_p. As the extinction cross section of a particle is the sum of its scattering cross section and its absorption cross section, the following relation for monodisperse particles holds:

$$Q_{\text{ext}} = Q_{\text{scat}} + Q_{\text{abs}} \tag{2.16}$$

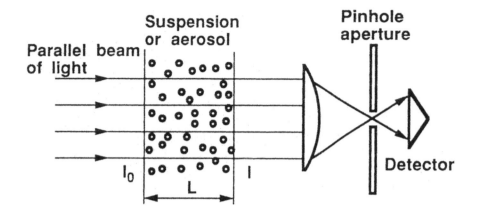

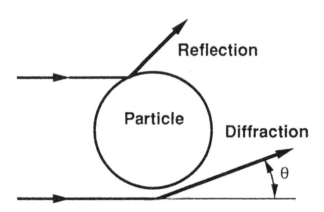

FIGURE 2.6   Setup for light-extinction measurements.

where Q_{scat} and Q_{abs} are respectively the scattering and absorption cross sections. For a spherical particle, Q_{ext} and Q_{scat} can be calculated when Equation 2.5 is integrated on all solid angle.[9]

$$Q_{scat} = \frac{\lambda^2}{2\pi} \sum_{n=1}^{\infty} (2n+1) \left(|a_n|^2 + |b_n|^2 \right)$$ (2.17)

$$Q_{ext} = \frac{\lambda^2}{2\pi} \sum_{n=1}^{\infty} (2n+1) \left(Re\{a_n + b_n\} \right)$$ (2.18)

where $Re\{\ \}$ means the calculation only of the real part. As Q_{scat}, Q_{abs}, and Q_{ext} have the dimension of area, the nondimensional values are obtained from the division by the projected area of a particle and named the scattering, absorption, and extinction efficiencies, respectively. When the sample of particles has a size distribution, the turbidity can be expressed as follows:

$$K = N_T \int_0^{\infty} Q_{ext}(D_p)\, q_0(D_p)\, dD_p$$ (2.19)

where N_T is the total particle number concentration, and $q_0(D_p)$ is the density distribution of particles on a number basis.

If the particle number concentration is sufficiently high, the light scattered by each particle illuminates other particles. When such multiscattering occurs, the coefficient Q_{ext} should be corrected.[10]

2.2.4 DYNAMIC LIGHT SCATTERING

Dynamic light scattering has been used for the size measurement of particles between about 3 nm and 5 μm in diameter.[11] When fine particles are illuminated by monochromatic light, a fluctuating scattering intensity occurs due to the Brownian motion of the particles. The fluctuation is correlated to the diffusion coefficient D of the particles.[12]

$$D = \frac{kT C_c}{3\pi \mu D_p} \tag{2.20}$$

where k is Boltzmann's constant, T is temperature, C_c is the slip correction factor, and μ is viscosity of the medium. From this relationship, the particle size D_p can be determined. The advantage of this measurement is that it is independent of the particle refractive index.

Several techniques have been developed for dynamic light scattering. These techniques can be classified in two ways: by the difference in data analysis (photon correlation spectroscopy and frequency analysis method) and by the difference in optical setup (homodyne and heterodyne detection optics). The scattered light from each particle is mixed with either light scattered from the rest of the particles (homodyne) or with light from the incident beam (heterodyne). The detector signal has two components: a constant level, representing the average intensity of the collected light, and a time-varying component, representing the dynamic light scattering effect. The time-dependent component is analyzed by following two methods: time-based correlation function (photon correlation spectroscopy) or frequency-based power spectrum (frequency analysis method).

The normalized intensity autocorrelation function for photon correlation spectroscopy is defined by

$$g^{(2)}(\tau) = \frac{\langle I(t) I(t+\tau) \rangle}{\langle I(t) \rangle^2} \tag{2.21}$$

where $I(t)$ and $I(t+\tau)$ are the scattered intensities at time t and $t+\tau$, respectively. The autocorrelation function of the scattered intensity is related to the electric field correlation function, $g_1(\tau)$:

$$g^{(2)}(\tau) = 1 + B|g_1(\tau)|^2 \tag{2.22}$$

where B is an instrumental factor. $g_1(\tau)$ is also expressed as follows:

$$g_1(\tau) = \exp(-\Gamma\tau) \tag{2.23}$$

$$\Gamma = D q^2 \qquad (2.24)$$

$$q = \frac{4\pi n \sin(\theta/2)}{\lambda_0} \qquad (2.25)$$

where n is the refractive index of the dispersion medium, λ_0 is the wavelength of the laser in vacuum, and θ is the scattering angle.

In the frequency analysis method, the power spectrum for homodyne and heterodyne detection optics are given by

$$p(\omega) = \frac{2\Gamma}{\omega^2 + (2\Gamma)^2} \text{ for homodyne detection optic} \qquad (2.26)$$

$$p(\omega) = \frac{2\Gamma}{\omega^2 + \Gamma^2} \text{ for heterodyne detection optic} \qquad (2.27)$$

For a sample that has the size distribution, $g_1(\tau)$ and $p(\omega)$ are related to the normalized distribution function of decay rates $G(\Gamma)$, which is a function of particle diameter as shown as Equation 2.20 and Equation 2.24.

$$g_1(\tau) = \int_0^\infty G(\Gamma) \exp(-\Gamma \tau) \, d\Gamma \qquad (2.28)$$

$$p(\omega) = \int_0^\infty G(\Gamma) \frac{2\Gamma}{\omega^2 + (2\Gamma)^2} \, d\Gamma \text{ for homodyne detection optic} \qquad (2.29)$$

$$p(\omega) = \int_0^\infty G(\Gamma) \frac{2\Gamma}{\omega^2 + \Gamma^2} \, d\Gamma \text{ for heterodyne detection optic} \qquad (2.30)$$

The average diameter of sample is calculated by

$$\frac{1}{D_p} = \int_0^\infty \frac{1}{x} G(\Gamma) \, d\left(\frac{1}{x}\right) \qquad (2.31)$$

2.2.5 PHOTOPHORESIS

The mechanism of photophoresis is similar to that of thermophoresis. When a very small particle is illuminated from one side, a temperature gradient is induced in the particle. The difference of strength of the gas molecule impact on the particle causes the particle to move in the direction opposite to the illuminated side. Because photophoresis is a result of an interaction between the particle and the surrounding gas molecules, it cannot occur in the cases of nonabsorbing particles or in a vacuum. The photophoresis force depends on the light intensity and wavelength, the size, shape, and material of the particle, and the gas pressure. Although the evaluation of the force is rather difficult, some analysis has been proposed,[13] and measurement data have been obtained.[14]

REFERENCES

1. Mie, G., *Ann. Phys.*, 25, 377, 1908.
2. Kerker, M., *The Scattering of Light and Other Electromagnetic Radiation.* Academic Press, New York, 1969.
3. Kanagawa, A., *Kagaku Kogaku,* 34, 521, 1970.
4. Kanagawa. A., *Kagaku Kogaku Ronbunshu,* 2, 325, 1976.
5. Xu, R., *Particle Characterization: Light Scattering Methods.* Kluwer Academic, Dordrecht, 2000.
6. Druger, S. D., Kerker, M., Wang, D. S., Cooke, and D. D., *Appl. Opt.,* 18, 3888, 1979.
7. Schuerman, D. W., *Light Scattering by Irregularly Shaped Particles.* Plenum Press, New York, 1980.
8. Mishchenko, M. I., Hovenier, J. W., and Travis, L. D., Eds. *Light Scattering by Nonspherical Particles: Theory, Measurements, and Applications.* Academic Press, San Diego, 2000.
9. Hodkinson, J. R., In *Aerosol Science,* Davies, C. N., Ed. Academic Press, New York, 1966, pp. 287, 324.
10. Born, M. and Wolf. E., *Principles of Optics,* Pergamon Press, Elmsford, NY, 1970.
11. Russel, W. B., Saville, D. A., and Schowalter, W. R., *Colloidal Dispersions,* Cambridge University Press, Cambridge, 1989, p. 441.
12. Berne, B. and Pecora, R.. *Dynamic Light Scattering.* John Wiley and Sons, New York, 1976.
13. Kerker, M. and Cooke, D. D., *Opt. Soc. Am.,* 72, 1267, 1982.
14. Lin, H. B. and Campillo, A. J. *Appl. Opt.,* 24, 422, 1985.

2.3 Particle Motion in Fluid

Shinichi Yuu
Kyushu Institute of Technology, Kitakyushu, Fukuoka, Japan

Yoshio Ohtani
Kanazawa University, Kanazawa, Ishikawa, Japan

2.3.1 INTRODUCTION

Motion of particles in fluid is divided into two classes: rectilinear or curvilinear motion, and random Brownian motion. For particles with diameters larger than about 0.5 μm, inertia or external force acting on the particles and the fluid resistance dominate the motion of particles. In such a case, tracing the motion of single particles is applied to picture the motion of the whole dispersed system of particles in fluid.

Motion of single particles can be further divided into three classes according to the extent of particle–fluid interaction. The simplest case is one in which the particles have the same density as the fluid at a low particle concentration so that the particles exactly follow the motion of fluid. The second case is such that the particle concentration is fairly low and the fluid induces the relative motion of particles but the particle motion does not affect the motion of fluid. The third case is such that the particle motion and fluid motion interact with each other by exchanging momentum. In this section, motions of single particles in the second case without particle–particle interaction are described.

2.3.2 MOTION OF A SINGLE PARTICLE

Resistance Force on a Spherical Particle

Particle motion relative to a fluid is dominated by resistance forces acting on the particle and external forces, such as gravitational, centrifugal, and electrostatic forces. Therefore, we should first know the resistance force on a particle for the description of particle motion.

Because F_D depends on D_p, u, ρ, and μ, F_D would be written as

$$F_D = k D_p^a u^b \rho^c \mu^e \tag{3.1}$$

where F_D, D_p, u, and ρ are resistance force, particle diameter, fluid velocity, and fluid density, respectively, and μ is fluid viscosity. Dimensional analysis of Equation 3.1 gives

$$F_D = k D_p^2 u^2 \rho \left(\frac{D_p u \rho}{\mu} \right)^{-e} \tag{3.2}$$

Substituting the cross-sectional area of spherical particle $A = \frac{1}{4}\pi D_p^2$ into Equation 3.2 gives

$$F_D = \frac{1}{2}\rho u^2 A k'\left(\frac{D_p u \rho}{\mu}\right)^{-e} = \frac{1}{2}\rho u^2 A C_D \qquad (3.3)$$

where C_D is a resistance coefficient. Equation 3.3 shows that C_D depends on only the particle Reynolds number; that is,

$$C_D = f(\mathrm{Re}_p) = k'\,\mathrm{Re}_p^{-e} \qquad (3.4)$$

Experimental results indicate that the exponent e in Equation 3.4 also depends only on Re_p. In other words, the value of e is changed by the fluid moving conditions (i.e., whether the fluid flow is viscous, transitional, or turbulent). As the particle size in powder technology is usually small, in most cases the particle Reynolds numbers are less than unity. When $\mathrm{Re}_p < 1$ (from the engineering point of view, $\mathrm{Re}_p < 2$), the fluid flow is dominated by the viscous force, and then the Navier–Stokes equation of fluid motion reduces to the linear equation, which is the creeping motion equation, by neglecting the inertia terms. Stokes[1] first solved the creeping motion equation and obtained the fluid velocity and stress distributions around a spherical particle settling in a uniform flow. He further obtained the resistance force acting on a spherical particle in a uniform viscous flow by integrating the stress distribution over the spherical surface of the particle. The resistance force is

$$F_D = 3\pi \mu D_p u \qquad (3.5)$$

Equation 3.5 is the well-known Stokes resistance law. Substitution of Equation 3.5 into Equation 3.3 gives the resistance coefficient C_D for the Stokes law:

$$C_D = \frac{24}{\mathrm{Re}_p} \qquad (3.6)$$

As mentioned earlier, the Stokes law applies to the creeping flow only. As the inertia effect on flow surrounding a particle increases with increasing Re_p, the Stokes law cannot apply to the flow field where $\mathrm{Re}_p > 2$. Resistance forces for flow fields where $\mathrm{Re}_p > 2$ are given by the following experimental formulas:

$$2 \leqq \mathrm{Re}_p < 500 \qquad \text{(Allen's region)}$$

$$F_D = \frac{5}{4}\pi\sqrt{\mu\rho}\,(D_p u)^{1.5} \qquad (3.7)$$

$$C_D = \frac{10}{\sqrt{\mathrm{Re}_p}}$$

$$500 \leqq \mathrm{Re}_p < 10^5 \quad \text{(Newton's region)}$$

$$F_D = 0.055\pi\rho\left(D_p u\right)^2 \tag{3.8}$$

$$C_D \cong 0.44$$

Recently, the Schiller–Naumann experimental formula [2] has become more popular:

$$0 < \mathrm{Re_p} \leq 800$$

$$C_D = \frac{24}{\mathrm{Re_p}}\left(1 + 0.15\,\mathrm{Re_p^{0.687}}\right) \tag{3.9}$$

Figure 3.1 shows C_D of these equations, where the bold line indicates the experimental result. When the particle diameter is comparable to or less than the mean free path of the air molecule, the effect of the slip around the particle should be considered. Thus, F_D is

$$F_D = \frac{3\pi\mu u D_p}{C_c} \tag{3.10}$$

where C_c is the slip correction factor. Kennard[3] formulized the slip correction factor using Millikan's experimental data[4]:

$$C_c = 1 + \left[2.46 + 0.82\exp\left(-\frac{0.44 D_p}{l}\right)\right]\frac{l}{D_p} \tag{3.11}$$

Neglecting the exponential term in Equation 3.11 gives the following approximate equation for a particle suspended in air of normal conditions

$$C_c = 1 + \frac{0.165}{D_p} \tag{3.12}$$

where the unit of D_p is in μm.

Equation of Particle Motion

Tchen[5] derived the Lagrange equation of motion for a particle suspended in an unsteady velocity field by extending the Basset–Buossinesq–Oseen equation[6]:

$$\frac{\pi}{6}D_p^3\rho_p\frac{dn}{dt} = 3\pi\mu D_p\left(u - v\right)$$

$$+ \frac{\pi}{6}D_p^3\rho\frac{du}{dt} + \frac{\pi}{12}D_p^3\rho\left(\frac{du}{dt} - \frac{dv}{dt}\right) \tag{3.13}$$

$$+ \frac{3}{2}D_p^2\sqrt{\pi\rho\mu}\int_{t_0}^{t}\frac{du/dt' - dv/dt'}{\sqrt{t - t'}}dt' + F_e$$

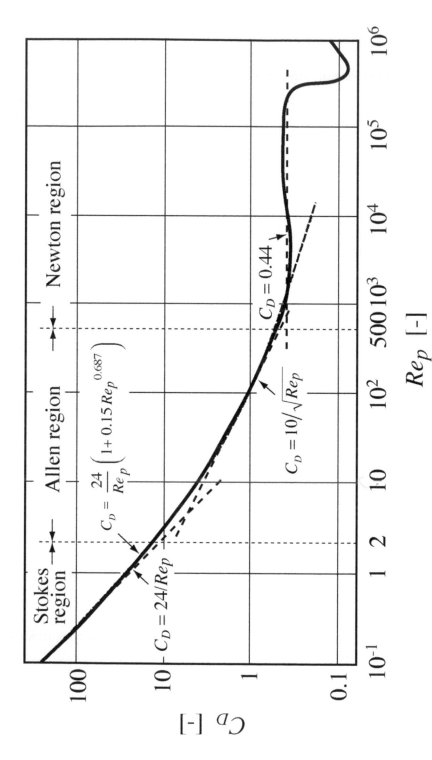

FIGURE 3.1   Resistance coefficient versus particle Reynolds number.

The term on the left-hand side of Equation 3.13 is the force required to accelerate the particle. The first term on the right-hand side of Equation 3.13 is the Stokes resistance force. The second term is the force due to the pressure gradient of the fluid surrounding the particle. The third term is the force to accelerate the virtual added mass of the particle relative to the fluid. The fourth term is the Basset term describing the effect of the deviation in flow pattern from steady state. F_e is the external force.

The condition $|(u-v)/u| \ll 1$ reduces Equation 3.13 to

$$\frac{dv}{dt} + A(v-u) - B\frac{du}{dt} - C\int_{t_0}^{t} \frac{du/dt - dv/dt}{\sqrt{t-t'}}dt'$$
$$- \frac{6F_e}{\pi D_p^3 \rho_p} = 0 \qquad (3.14)$$

where

$$A = \frac{36}{(2\rho_p + \rho)D_p^2}, \quad B = \frac{3r}{2\rho_p + \rho}, \quad C = \frac{18}{(2\rho_p + \rho)D_p}\sqrt{\frac{\rho\mu}{\pi}}$$

As $\rho_p \gg \rho$ and $\mu \ll 1$ in an airflow, B and C in Equation 3.14 are much smaller than A. Therefore, the third and fourth terms on the left-hand side of Equation 3.14 can be neglected in an airflow.

When the resistance force is expressed with the resistance coefficient C_D, the equation of particle motion is

$$m\frac{dv}{dt} = -\frac{C_d A_c \rho}{2C_c}|v-u|(v-u) + m\left(1 - \frac{\rho}{\rho_p}\right)g \qquad (3.15)$$

where the third and fourth terms in Equation 3.14 are neglected, and the particle mass $m = (\pi/6)D_p^3\rho_p$ and the cross-sectional area $A_c = (\pi/4)D_p^2$ for the spherical particle. Two-dimensional normalized equations of Equation 3.15 are

$$\frac{C_c \rho_p u_0 D_p^2}{18\mu D}\frac{24}{C_D \operatorname{Re}_p}\frac{d^2\overline{x}}{d\overline{t}^2} + |\overline{v}-\overline{u}|\left(\frac{d\overline{x}}{d\overline{t}} - \overline{u}_x\right) = 0$$

$$\frac{C_c \rho_p u_0 D_p^2}{18\mu D}\frac{24}{C_D \operatorname{Re}_p}\frac{d^2\overline{y}}{d\overline{t}^2} + |\overline{v}-\overline{u}|\left(\frac{d\overline{y}}{d\overline{t}} - \overline{u}_y\right)$$

$$- \frac{24C_c}{C_D \operatorname{Re}_p}\frac{\rho_p D_p^2}{18\mu u_0}\left(1 - \frac{\rho}{\rho_p}\right)g = 0 \qquad (3.16)$$

where

$$\overline{x} = \frac{x}{D}, \ \ \overline{y} = \frac{y}{D}, \ \ \overline{t} = \frac{u_0 t}{D}, \ \ \overline{u}_x = \frac{u_x}{u_0}, \ \ \overline{u}_y = \frac{u_y}{u_0}$$

(3.17)

$C_D = 24\mu / \left(D_v |\overline{v} - \overline{u}| u_0 \rho \right)$ for the Stokes law reduces Equation 3.16 to the following equations:

$$\psi \frac{d^2\overline{x}}{d\overline{t}^2} + \frac{d\overline{x}}{d\overline{t}} - \overline{u}_x = 0, \quad \psi \frac{d^2\overline{y}}{d\overline{t}^2} + \frac{d\overline{y}}{d\overline{t}} - \overline{u}_y - G = 0$$

(3.18)

where ψ and G are the inertia parameter and normalized settling velocity, respectively, defined as

$$\psi = \frac{C_c \rho_p D_p^2 u_0}{18\mu D}, \quad G = \frac{C_c (\rho_p - \rho) D_p^2 g}{18\mu u_0}$$

(3.19)

The solution of Equation 3.18 based on the initial conditions gives the particle trajectory in the airflow with the velocity distribution (u_x, u_y).

2.3.3   PARTICLE MOTION IN SHEAR FIELDS

When a particle moves in a shear field, lift force (lateral force) is exerted on a particle due to the local velocity gradient of fluid flow and the rotation of particle. The lift force plays an important role in the motion of particles in pipe flow, boundary layers, and other shear fields. Saffman[7] theoretically derived the lift force for spherical particle at a low Reynolds number:

$$F_L = 1.62 u D_p^2 \left(\rho\mu \frac{du}{dy} \right)^{1/2}$$

(3.20)

where u is the local relative velocity between the particle and the fluid. The lift force exerts on particle in the direction of increasing the fluid velocity. In the shear field, the particle rotates due to the viscous force. At a steady state, the angular velocity is given by

$$\omega = \frac{1}{2} \frac{du}{dy}$$

(3.21)

The lift force occurs even in uniform flow when the particle is rotating. At a low Re the lift force due to the rotation of particle is given by

$$F_L = \frac{\pi}{8} u D_p^3 \rho \omega$$

(3.22)

The above expression was theoretically derived by Rubinow and Keller.[8]

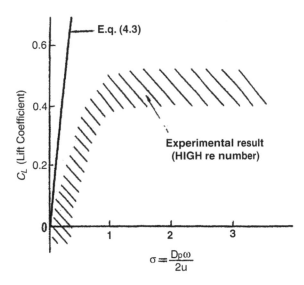

FIGURE 3.2 Lift coefficient as a function of dimension-less angular velocity. [From Tsuji et al., *J. Fluids Eng.*, 107, 484–491, 1985. With permission.]

In general, particles are not spherical in shape. Even a small deviation of particle shape from sphere causes a large lift force. Since the lift force always acts on particles in the direction perpendicular to the fluid flow, it makes the particles concentration nonuniform in the direction perpendicular to the fluid flow.

Equation 3.20 and Equation 3.21 are applicable only at a low *Re* less than unity. At high *Re* the formulae for the lift force are only empirical. The lift force at high *Re* is expressed in a similar manner to the drag force by introducing the lift coefficient:

$$F_L = C_L \left(\frac{\pi}{4} D_p^2\right) \frac{\rho_f v_r^2}{2} \tag{3.23}$$

Unfortunately, experimental data are still limited for the lift force due to the shear at high *Re*. Some data of the lift force were measured for a particle near the wall so that a high shear layer was developed and the wall effects were included in the data. Data by Yamamoto[9] even show that the lift force exerts in the direction of decreasing the velocity, which is completely opposite to the lift force direction predicted by Equation 3.20.

Experimental data of the lift due to the rotation are also very much limited, and the relation between the lift coefficient and the dimensionless angular velocity shown in Figure 3.2 is often used to estimate the lift due the particle rotation at high *Re*.

REFERENCES

1. Stokes, G. G., *Trans. Cambridge Philos. Soc.,* 8, 287–318, 1845.
2. Schiller, L. and Naumann, A., *Z. Ver. Dtsch. Ing.,* 77: 318–321, 1933.
3. Kennard, E. H., *Kinetic Theory of Gases,* McGraw-Hill, New York, 1938, p. 83.
4. Millikan, R. A., *Phys. Rev.,* 22: 1–18, 1923.
5. Tchen, C. M., Dissertation, Delft, Martinus Nijihof, The Hague, 1947.

6. Basset, A. B., *Philos. Trans. Roy. Soc.*, 179, 43–71, 1888.
7. Saffman, P. G., *J. Fluid Mech.* 22, 385–403, 1965.
8. Rubinow, S. and Keller, J. B., *J. Fluid Mech.*, 11, 447–457, 1961.
9. Yamamoto, F., *Trans. JSME*, 57, 3414–3421, 1992.
10. Tsuji, Y., Morikawa, Y., and Mizono, O., *J. Fluids Eng.*, 107, 484–491, 1985.

2.4 Particle Sedimentation

Shinichi Yuu

Kyushu Institute of Technology, Kitakyushu, Fukuoka, Japan

2.4.1 INTRODUCTION

Sedimentation is the most important phenomenon not only in measuring fundamental characteristics of particles but also in understanding mechanisms of unit operations such as dust and mist collection, classification, and pneumatic transport. In this chapter, we describe the resistance force on a spherical particle moving through a fluid, the Lagrangian equation of a particle in motion, and the sedimentation velocities of particles in various situations. Particles moving through a viscous fluid interact with each other; thus, interactions between two or more spherical particles falling through a viscous liquid are also discussed.

2.4.2 TERMINAL SETTLING VELOCITY

Substituting $u = 0$ into Equation 3.15 from 2.3 gives the equation of particle motion in a quiescent fluid:

$$\frac{dv}{dt} = -\frac{3}{4}\frac{C_D v^2 \rho}{\rho_p D_p} + \frac{(\rho_p - \rho)g}{\rho_p} \tag{4.1}$$

When the first term (the hydrodynamic resistance force) on the right-hand side of Equation 3.15 from 2.3 is counterbalanced by the second term (the hydrodynamic resistance force), the spherical particle settling under the gravity attains a uniform velocity. It is called the terminal settling velocity.

Substituting $dv/dt = 0$ and Equation 3.6, Equation 3.7, or Equation 3.8 from 2.3 into Equation 4.1 gives the equation of terminal settling velocity for each region:

$$u_t = \frac{(\rho_p - \rho)D_p^2 g}{18\mu} \quad , \quad \text{Re}_p \leqq 2 \tag{4.2}$$

$$u_t = \left(\frac{4(\rho_p - \rho)^2 g^2}{225\mu\rho}\right)^{1/3} D_p \quad , \quad 2 < \text{Re}_p \leqq 500 \tag{4.3}$$

$$u_t = \left(\frac{3g(\rho_p - \rho)D_p}{\rho}\right)^{1/2} \quad , \quad 500 < \text{Re}_p \leqq 10^5 \tag{4.4}$$

Resistance Force Acting on a Spherical Particle Settling Relative to Cylindrical and Plane Walls

Lorentz[1] derived the theoretical equation of the resistance force acting on a spherical particle settling near a wall, as shown in Figure 4.1:

$$F_D = 3\pi\mu D_p u\left[1 + \frac{9}{16}\left(\frac{D_p}{2H}\right) + \cdots\right] \quad , \quad \text{Re}_p \ll 1 \tag{4.5}$$

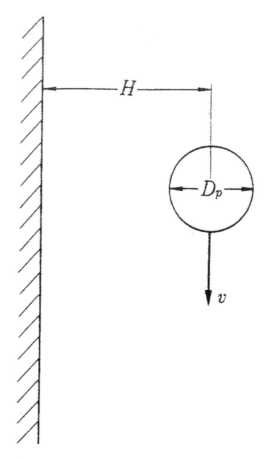

FIGURE 4.1   Spherical particle settling near a wall.

Faxen[2] derived theoretically the following equation, which describes the torque acting on a particle settling near a wall:

$$T = 2\pi\mu D_p^2 u \left[\frac{3}{32}\left(\frac{D_p}{2H}\right)^4 - \frac{9}{256}\left(\frac{D_p}{2H}\right)^5 + \cdots \right] , \quad \mathrm{Re}_p \ll 1 \qquad (4.6)$$

Both F_D and T increase with increasing $D_p/2H$. This means that the settling and rotational velocities decrease with increasing $D_p/2H$.

Bohlin[3] derived the theoretical equation of the resistance force acting on a spherical particle settling on a centerline of a cylindrical pipe as shown in Figure 4.2:

$$F_D = 3\pi\mu D_p u K_1 \qquad (4.7)$$

$$K_1 = \left[1 - 2.10443\left(\frac{D_p}{2R_0}\right) + 2.08877\left(\frac{D_p}{2R_0}\right)^3 - 6.94813\left(\frac{D_p}{2R_0}\right)^5 \right.$$
$$\left. -1.372\left(\frac{D_p}{2R_0}\right)^6 + 3.87\left(\frac{D_p}{2R_0}\right)^8 - 4.19\left(\frac{D_p}{2R_0}\right)^{10} + \cdots \right]^{-1} , \quad \mathrm{Re}_p \ll 1$$
$$(4.8)$$

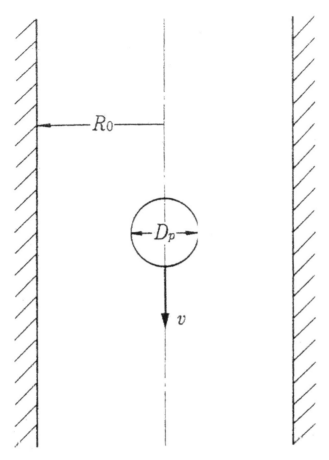

FIGURE 4.2 Spherical particle settling along a centerline of
a cylindrical pipe.

Bohlin[3] also presented the theoretical equation of the resistance force acting on a spherical particle
moving with velocity u in an axisymmetrical Poiseuille flow whose centerline velocity is u_0.

$$F_D = 3\pi\mu D_p\left(uK_1 - u_0K_2\right) \tag{4.9}$$

$$K_2 = K_1\left[1 - \frac{2}{3}\left(\frac{D_p}{2R_0}\right) - 0.1628\left(\frac{D_p}{2R_0}\right)^3 - 0.4059\left(\frac{D_p}{2R_0}\right)^7\right.$$
$$\left. +0.5236\left(\frac{D_p}{2R_0}\right)^9 + 1.51\left(\frac{D_p}{2R_0}\right)^{10} + \cdots\right], \qquad \text{Re}_p \ll 1 \tag{4.10}$$

Fayon and Happel[4] presented the following equation, which represents the resistance force for flows
of higher Reynolds number than the creeping flow:

$$F_D = 3\pi\mu D_p u\left[\frac{1}{1 - 2.104\left(D_p/2R_0\right) + 2.088\left(D_p/2R_0\right)^8}\right.$$
$$\left. +\left(\frac{C_A}{C_S} - 1\right)\right], \qquad \text{Re}_p < 100 \tag{4.11}$$

where C_A is a real resistance coefficient of a spherical particle settling in an unbounded fluid, and C_S is the Stokes resistance coefficient (i.e., $C_S = 24/Re_p$). Therefore, $C_A/C_S - 1$ represents a deviation from Stokes law.

Brenner[5] obtained solution of the creeping motion equation for the steady motion of a spherical particle toward or away from a plane surface of infinite extent, as shown in Figure 4.3. His equation of the resistance force is

$$F_D = 3\pi\mu D_p u\beta \tag{4.12}$$

$$\beta = \frac{4}{3}\sinh\alpha \sum_{n=1}^{\infty} \frac{n(n+1)}{(2n-1)(2n+3)}$$

$$\times \left(\frac{2\sinh(2n+1)\alpha + (2n+1)\sinh^2\alpha}{4\sinh^2\left(n+\frac{1}{2}\right)\alpha - (2n+1)^2\sinh^2\alpha} - 1 \right) \tag{4.13}$$

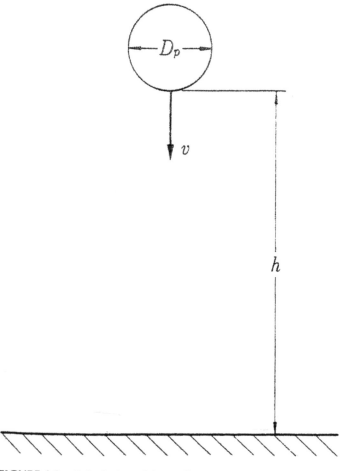

FIGURE 4.3 Spherical particle settling toward a plane surface of infinite extent.

where $\alpha = \cosh^{-1}(2h/D_p)$. The approximate equation of Equation 4.13 presented by Honig et al.[6] is

$$\beta = \frac{6\bar{h}^2 + 13\bar{h} + 2}{6\bar{h}^2 + 4\bar{h}} \quad , \quad \bar{h} = \frac{2h}{D_p} \tag{4.14}$$

Equation 4.12, Equation 4.13, and Equation 4.14 show that in a continuum, the resistance force of a spherical particle moving toward a plane surface (or another particle) increases with decreasing separation distance between them and finally becomes infinite just before touching. However, there are effects of discontinuity in discontinuous media. The atmospheric air in the minimum separation between a particle and a plane surface or two particles whose diameters are less than several tens of microns creates the discontinuum. Therefore, a fine particle collides with another particle or a plane surface more easily. Umekage and Yuu[7] measured resistance forces of a 6.3-mm foam styrene ball under reduced air pressure by using a high-speed camera. At 10 Pa, the mean free path of air is 670 μm, and the Knudsen number of a 6.3-mm foam styrene ball is equivalent to that of a 0.6-μm particle in atmospheric air. The experimental data in Figure 4.4 show that the increment of the resistance force acting on a 6.3-mm foam styrene ball near a plane surface under reduced pressure decreases with decreasing air pressure. Therefore, it is concluded that the resistance force of a fine particle moving toward a plane surface or another particle in the air is much less than that acting in a continuum.

Settling Velocity of a Nonspherical Particle

The dynamic shape factor K defined by Equation 4.15 is used to estimate the difference between the settling velocities of nonspherical and spherical particles:

$$K = \frac{\text{Transport velocity of equispherical particle}}{\text{Transport velocity of nonspherical particle}} \tag{4.15}$$

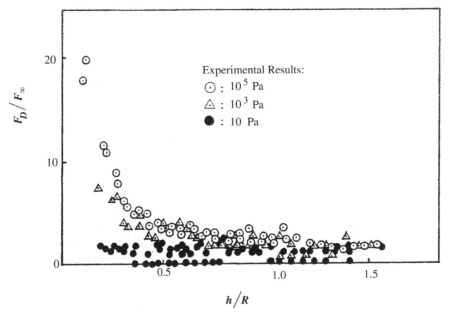

FIGURE 4.4 Resistance force acting on a 6.3-mm foam styrene ball near a plane surface under reduced pressure.

where the transport velocity usually corresponds to the settling velocity. The Stokes resistance law reduces Equation 4.15 to

$$K = \frac{\text{Equispherical diameter}}{\text{Stokes diameter of nonspherical particle}} \qquad (4.16)$$

Stober[8] presented the equations of K for ellipsoids:

$$K_{\perp} = \frac{(8/3)(q^2 - 1)q^{-1/3}}{\left[(2q^2 - 3)/\sqrt{q^2 - 1}\right]\ln\left(q + \sqrt{q^2 - 1}\right) + q} \quad \text{for } q > 1 \qquad (4.17)$$

$$K_{\parallel} = \frac{(4/3)(q^2 - 1)q^{-1/3}}{\left[(2q^2 - 1)/\sqrt{q^2 - 1}\right]\ln\left(q + \sqrt{q^2 - 1}\right) - q} \quad \text{for } q > 1 \qquad (4.18)$$

$$K_{\perp} = \frac{(8/3)(q^2 - 1)q^{-1/3}}{\left[(2q^2 - 3)/\sqrt{1 - q^2}\right]\cos^{-1} q - q} \quad \text{for } q < 1 \qquad (4.19)$$

$$K_{\parallel} = \frac{(4/3)(q^2 - 1)q^{-1/3}}{\left[(2q^2 - 1)/\sqrt{q^2 - 1}\right]\cos^{-1} q - q} \quad \text{for } q < 1 \qquad (4.20)$$

where q = polar diameter of ellipsoid/equator diameter of ellipsoid, and $K_{\perp}$ and $K_{\parallel}$ are dynamic shape factors of nonspherical particles whose settling directions are perpendicular and horizontal to the polar axis, respectively. Stober[8] measured the settling velocities of polystyrene latex aggregates and obtained the following experimental formulas for dynamic shape factors of aggregate particles:

$$K \approx 1.233 \qquad \text{for massive aggregates} \qquad (4.21)$$
$$K \approx 0.862 n^{1/3} \qquad \text{for chainlike aggregates} \qquad (4.22)$$

where n is the number of single particles composing the aggregate.

2.4.3   SETTLING OF TWO SPHERICAL PARTICLES

The exact solution of Stokes equation for the motion of a viscous fluid around two spherical particles settling with equal small constant velocities along their line of centers was presented by Stimson and Jeffery.[9] The resistance force for one of two equal spherical particles is

$$F_{D} = 3\pi\mu D_{p}\beta$$
$$\beta = \frac{4}{3}\sinh\alpha \sum_{n=1}^{\infty} \frac{n(n+1)}{(2n-1)(2n+3)}$$
$$\times \left(1 - \frac{4\sinh^2\left(n + \frac{1}{2}\right)\alpha - 2(n+1)^2\sinh^2\alpha}{2\sinh(2n+1)\alpha + (2n+1)\sinh 2\alpha}\right) \qquad (4.23)$$

where

$$\alpha = \ln \left\{ \frac{l_0}{D_p} + \left[\left(\frac{l_0}{D_p} \right)^2 - 1 \right]^{1/2} \right\} \tag{4.24}$$

When the distance between the centers of two spherical particles l_0 is 1.54 D_p, β is 0.7025; that is, the interaction of two particles gives about a 30% reduction in resistance.

Wakiya[10] gave the solutions of the Stokes equation for the motion of two spherical particles moving in an arbitrary direction:

$$
\begin{aligned}
F_{ax} = 6\pi\mu a \Bigg\{ & u_{ax} \left[1 + \frac{9}{4} \frac{ab}{l_0^2} + \frac{3}{4} \left(-2 \frac{a^3 b}{l_0^4} + \frac{27}{4} \frac{a^2 b^2}{l_0^4} + 3 \frac{ab^3}{l_0^4} \right) + \cdots \right] \\
& - u_{bx} \left[\frac{3}{2} \frac{b}{l_0} - \frac{1}{2} \left(\frac{a^2 b}{l_0^3} - \frac{27}{4} \frac{ab^2}{l_0^3} + \frac{b^3}{l_0^3} \right) \right. \\
& \left. + \frac{9}{4} \left(\frac{a^3 b^2}{l_0^5} + \frac{27}{8} \frac{a^2 b^3}{l_0^5} + \frac{ab^4}{l_0^5} \right) + \cdots \right] \Bigg\}
\end{aligned}
\tag{4.25}
$$

$$
\begin{aligned}
F_{ay} = 6\pi\mu a \Bigg\{ & u_{ay} \left[1 + \frac{9}{16} \frac{ab}{l_0^2} + \frac{3}{8} \left(\frac{a^3 b}{l_0^4} + \frac{27}{32} \frac{a^2 b^2}{l_0^4} + \frac{ab^3}{l_0^4} \right) + \cdots \right] \\
& - u_{by} \left[\frac{3}{4} \frac{b}{l_0} + \frac{1}{4} \left(\frac{a^2 b}{l_0^3} + \frac{27}{16} \frac{ab^2}{l_0^3} + \frac{b^3}{l_0^3} \right) \right. \\
& \left. + \frac{27}{64} \left(\frac{a^3 b^2}{l_0^5} + \frac{9}{16} \frac{a^2 b^3}{l_0^5} + \frac{ab^4}{l_0^5} \right) + \cdots \right] \Bigg\}
\end{aligned}
\tag{4.26}
$$

$$
\begin{aligned}
\omega_a = \frac{3}{4} \frac{b x_0}{l_0^3} \Bigg\{ & u_{ay} \left[1 + \frac{9}{16} \frac{ab}{l_0^2} + \frac{3}{8} \left(\frac{a^3 b}{l_0^4} + \frac{27}{32} \frac{a^2 b^2}{l_0^4} + \frac{ab^3}{l_0^4} \right) + \cdots \right] \\
& - u_{by} \left[\frac{3}{4} \frac{a}{l_0} + \frac{1}{4} \left(\frac{a^3}{l_0^3} + \frac{27}{16} \frac{a^2 b}{l_0^3} \right) + \cdots \right] \Bigg\}
\end{aligned}
\tag{4.27}
$$

where $|x_0| = l_0$.

2.4.4 RATE OF SEDIMENTATION IN CONCENTRATED SUSPENSION

The rate of sedimentation of the particles in a suspension which are well distributed throughout the fluid in a vessel is slower than the velocity given by the Stokes resistance law. Steinour[11] considered that the resistance force on a particle in suspension can be represented by the Stokes resistance law with a correction term that is a function of the concentration of particles only, as follows:

$$F_\mathrm{D} = \frac{3\pi\mu D_\mathrm{p} u}{\phi(\varepsilon)}$$

(4.28)

where u is the relative velocity between a particle and a fluid. ε is the function of the volume occupied by liquid, which is analogous to the porosity in powder beds. When $\varepsilon = 1$ at infinite dilution, $\phi(\varepsilon)$ is equal to unity. The force balance between F_D given by Equation 4.28, and the gravitational force of the particle given by $\frac{1}{6}\pi(\rho_\mathrm{p} - \rho_\mathrm{m})D_\mathrm{p}^3 g$ provides the terminal relative velocity:

$$u = \frac{\pi(\rho_\mathrm{p} - \rho)D_\mathrm{p}^2\,\varepsilon\phi(\varepsilon)}{18\mu} = u_\mathrm{t}\varepsilon\phi(\varepsilon)$$

(4.29)

where the density difference between the particle and suspension is

$$\rho_\mathrm{p} - \rho_\mathrm{m} = \rho_\mathrm{p} - \left[(1-\varepsilon)\rho_\mathrm{p} + \varepsilon\rho\right] = (\rho_\mathrm{p} - \rho)\varepsilon$$

(4.30)

The measured rate of sedimentation u_c is not a relative velocity between a particle and a fluid in suspension, but a settling velocity of the particle relative to a fixed cross section. The continuity condition states that the volume of settling particles is equal to the ascending fluid volume; that is, $(1-\varepsilon)u_\mathrm{c} = \varepsilon(u - u_\mathrm{c})$. Hence,

$$u_\mathrm{c} = \varepsilon u$$

(4.31)

Substituting Equation 4.29 into Equation 4.31, both from 2.3, gives

$$u_\mathrm{c} = u_\mathrm{t}\varepsilon^2\,\phi(\varepsilon)$$

(4.32)

Steinour[11] measured the rate of sedimentation for tapioca particles and glass beads and obtained the experimental formula of $\phi(\varepsilon)$:

$$\phi(\varepsilon) = 10^{-1.82(1-\varepsilon)}\ \text{for}\ \varepsilon > 0.6$$

(4.33)

Notation

C_c	Slip correction factor
C_D	Resistance coefficient
D_p	Particle diameter (μm)
F_D	Resistance force (N)
F_e	External force (N)
$F\infty$	Resistance force of a single particle in an infinite medium (N)
g	Gravitational acceleration (m/s²)
H	Distance between particle center and wall (m)
h	Minimum distance between particle surface and wall (m)
K	Dynamic shape factor
l	Mean free path (μm)

R_0	Radius of pipe (m)
r	Radial coordinate (m)
Re, Re_p	Reynolds number and particle Reynolds number
t	Time (s)
$u\ v$	Fluid and particle velocity (m/s)
u_t	Terminal settling velocity of single particle (m/s)
ε	Fraction of volume occupied by liquid
ρ, ρ_p	Fluid and particle density (kg/m^3)
ψ	Inertia parameter

REFERENCES

1. Lorentz, H. A., *Abh. Theor. Phys. Leipzig,* 1, 23–36, 1907.
2. Faxen, H., *Ann. Phys. IV Folge,* 68, 89–98, 1922.
3. Bohlin, T., *Trans. Roy. Inst. Technol. Stockholm,* No. 155, 1960.
4. Fayon, A. M. and Happel, J., *AIChE J.,* 6, 55–63, 1960.
5. Brenner, H., *Chem. Eng. Sci.,* 16, 242–248, 1961.
6. Honig, E. P., Roebersen, G. J., and Wiersema, P. H., *J. Colloid Interface Sci.,* 36, 97–103, 1971.
7. Umekage, T. and Yuu, S., *Nihon Kikai Gatsukai Ronbunshu Ser. B,* 59, 1559–1565, 1993.
8. Stober, W., in *Fine Particles,* Liu, B. Y. H., Ed., Academic Press, New York, 1976.
9. Stimson, M. and Jeffery, G. B., *Proc. Roy. Soc. London,* A111, 110–122, 1926
10. Wakiya, S., *Niigata Univ. Kogakubu Kenkyu Hokoku,* 5, 155–160, 1957.
11. Steinour, H. H., *Ind. Eng. Chem.* 36, 618–626, 1944.

2.5 Particle Electrification and Electrophoresis

Hiroaki Masuda, Shuji Matsusaka, and Ko Higashitani
Kyoto University, Katsura, Kyoto, Japan

2.5.1 IN GASEOUS STATE

Contact Electrification

When two different metallic materials come into contact, electrons in the metal having the higher Fermi level move into the other metal so as to equalize the Fermi levels of the two materials. Then the number of electrons in the material having a lower Fermi level increases, resulting in electrification to negative charge. As a matter of course, the other material is electrified to positive charge. Contact electrification of semiconductive materials is explained in a similar way. The contact region acts as an electrical capacitor. The amount of charge stored in the capacitor can be obtained by solving Poisson's equation. Electrification of a nonconductive material is, however, more complicated. For metal–insulator contact, the charge density is estimated by the following equation[1]:

$$\sigma = \frac{\phi_s - \phi_m}{\dfrac{1}{eN_s + \sqrt{\varepsilon N_b}} + \dfrac{ez_0}{\varepsilon_0}} \tag{5.1}$$

where

σ = charge density (C/m²)
ϕ_s = work function of nonconductive solid material (eV = 1.602×10^{-19} J)
ϕ_m = work function of metal (eV)
N_s = density of surface states (l/(m² eV))
N_b = density of traps of acceptor type (l/(m³ eV))
e = elementary charge (1.602×10^{-19} C)
ϵ = dielectric constant of solid (F/m)
ϵ_0 = dielectric constant of vacuum (8.85×10^{-12} F/m)
z_0 = gap between solid and metal (4Å = 0.4 nm)

A similar equation is derived for insulator–insulator contact by assuming surface states for both materials.[2]

The difference of work functions divided by elemental charge e is called the contact potential difference. If the surface states have negligible effects ($z_0 = 0$, $N_s = 0$), Equation 5.1 is simplified to

$$\sigma = \frac{\varepsilon}{z} V_c \tag{5.2a}$$

where V_c is the contact potential difference and z is the effective Debye length ($\sqrt{\varepsilon/N_b}/e$). If the density of traps is negligible ($N_b = 0$) and the density of surface states is sufficiently high ($N_s \gg \varepsilon_0/e^2 z_0$), Equation 5.1 is simplified to

$$\sigma = \frac{\varepsilon_0}{z_0} V_c \qquad (5.2b)$$

When the contact bodies are separated, charges are partly neutralized or discharged through the surrounding medium. The process is called charge relaxation.[3]

The maximum charge of a spherical particle is controlled by discharge from the surface of the particle. The breakdown condition in air is given by an electric field strength of 3 MV/m for particles larger than 200 μm and a surface potential of 300 V for those less than 200 μm.[4]

Electrification of a Particle by Impact

Charge transfer between impacted materials proceeds during the very short time of collision. Such a high-speed electrification will be represented by[5]:

$$\Delta q = CV_c \left[1 - \exp\left(-\frac{\Delta t}{\tau} \right) \right] \qquad (5.3)$$

where C is the capacitance of the contact region, τ the time constant for electrification, and Δt the effective duration time of contact. The capacitance C is given by

$$C = \frac{\varepsilon S}{z} \text{ or } C = \frac{\varepsilon_0 S}{z_0} \qquad (5.4)$$

corresponding to Equation 5.2a or 5.2b, where S is the contact area. Equation 5.3 shows that metallic particles ($\tau \ll \Delta t$) may be highly electrified by impact as long as the back discharge is negligible.[6] For nonconductive particles, the time constant τ is given by

$$\tau = \varepsilon \rho_d \qquad (5.5)$$

where ρ_d is the specific resistance of the particle. The charge transferred is given by

$$\Delta q = \frac{SV_c}{z\rho_d} \Delta t \qquad (5.6)$$

Equation 5.6 shows that the perfect insulator ($\rho_d \to \infty$) does not acquire charge, which coincides with the experimental results obtained by the use of solid rare gases.[7]

The contact area S depends on the mode of impact. It is nearly proportional to impact velocity v for inelastic collision and proportional to $v^{0.8}$ for elastic collision. In both cases it is proportional to the square of the particle size.[5]

If particles collide with the wall several times, they are all electrified and produce a strong electric field. Also, the contact potential difference V_c is affected by both the electric field and the image charge of the particle. The electrification process, including the electric-field effect and the image-charge effect, is represented by the following equation[8,9]:

$$\Delta q = q_0 \exp\left(-\frac{N}{N_0} \right) + q_\infty \left[1 - \exp\left(-\frac{N}{N_0} \right) \right] \qquad (5.7)$$

where N is the number of collisions of a particle, N_0 the relaxation number of collisions for charge transfer, q_0 the initial charge of a particle, and q the final charge attainable in the process.

In gas–solids pipe flow, particle charging depends on other factors such as the number of particle collisions, initial charge on particles, and the state of impact charging. The value of each factor is spread, and therefore, the particles have a distribution after passing through the pipe. The charge distribution on particles can be estimated with the distribution of the factors.[10]

Electrification of a Particle through Breakage

Electric charge of particles after breakage is represented by a normal distribution with zero mean. The cause of electrification through breakage is the unequal partition of positive or negative charge. Electrification of particles in a crusher is caused by both the contact between different materials (particles and the wall) and the breakage. Therefore, the mean value of charge is no longer zero. Further, smaller particles produced through breakage are apt to take a negative charge because the work function will be higher.[11]

The charge distribution of mist produced through atomization of water solution is symmetrical. The absolute mean is given by[12]

$$\left|\overline{n_e}\right| = \sqrt{\frac{2}{3} N_i D_p^{1.5}} \tag{5.8}$$

where $\overline{n}_e$ is the average number of elementary charge, N_i the number density of ions, and D_p the particle diameter.

Motion of a Charged Particle

When a charged particle is in an electric field, the following force (Coulomb force) will act on it:

$$\mathbf{F} = q\mathbf{E} \tag{5.9}$$

where $\mathbf{F}$ is the force (N), q the charge (C), and $\mathbf{E}$ the strength of the electric field (V/m). The movement of charged particles produces an electric current. This is called the convective current. The convective current is affected by a magnetic field, and the force acting on a charged particle is given by

$$\mathbf{F} = q\mathbf{E} + q\mathbf{v} \times \mathbf{B} \tag{5.10}$$

where $\mathbf{v}$ is the particle velocity (m/s), and $\mathbf{B}$ is the magnetic flux density (T).

The electrostatic force between the particle and a body having electric charge dq is given by

$$dF = \frac{q}{4\pi\varepsilon_0 r^2} dq \tag{5.11}$$

where r is the distance between the particle and the charged body. When the electric charges are distributed on various surrounding bodies, integration of Equation 5.11 will give the total force acting on the particle. Velocity and trajectory of a charged particle in a uniform electric field can be obtained by solving an equation of motion, including the Coulomb force term. For example, the terminal velocity v in an electric field is obtained as

$$v = \frac{qEC_c}{3\pi\mu D_p} \tag{5.12}$$

where C_c is Cunningham slip correction factor and μ is the viscosity of fluid. The particle velocity under unit strength of the electric field v/E is called the electric mobility.

Electrically charged particles produce an electric field, which is calculated by the following Poisson's equation:

$$\text{div } \mathbf{E} = \frac{\rho}{\varepsilon_0} \tag{5.13}$$

where ρ is the space-charge density (C/m³). Table 5.1 lists the strength of electric fields for one-dimensional cases. Unipolar charged particles repulse each other and gradually disperse. This phenomenon is called electrostatic dispersion or diffusion. Particle deposition on a wall is enhanced by the electrostatic effect as estimated approximately by the following Wilson equation[13] both for a circular tube and a parallel-plate channel[14]:

$$\eta = \frac{t/t_0}{1+t/t_0} \tag{5.14}$$

where η is the deposition fraction, t the residence time, and t_0 the half-life time. The half-life time is given by

$$t_0 = \frac{3\pi\mu D_p \varepsilon_0}{n_0 q^2 C_c} \tag{5.15}$$

where n_0 is the number concentration (m⁻³) and D_p is the equivalent volume diameter.

A charged particle near the equipment wall will be affected by the so-called image force, which is always attractive. If the particle is assumed to be a point charge, the force is given by

$$F = -\frac{q^2}{16\pi\varepsilon_0 r^2} \tag{5.16}$$

where r is the distance from the wall. For a nonconductive wall of specific dielectric constant $\bar{\varepsilon}$, the force is given by

$$F = -\frac{q^2}{16\pi\varepsilon_0 r^2}\frac{\bar{\varepsilon}-1}{\bar{\varepsilon}+1} \tag{5.17}$$

TABLE 5.1   Electric Field Produced by Electrically Charged Particles

	In Cylindrical Tube [E_r (V/m)]	Between Parallel Plates [E_x (V/m)]	In Spherical Vessel [E_r (V/m)]
Inside	$\dfrac{1}{2}\dfrac{\rho}{\varepsilon_0}r$	$\dfrac{\rho}{\varepsilon_0}x$	$\dfrac{1}{3}\dfrac{\rho}{\varepsilon_0}r$
Outside	$\dfrac{1}{2}\dfrac{\rho}{\varepsilon_0}\dfrac{r_0^2}{r}$	$\dfrac{\rho}{\varepsilon_0}x_0$	$\dfrac{1}{3}\dfrac{\rho}{\varepsilon_0}\dfrac{r_0^3}{r^2}$

If the assumption of point charge does not hold, infinite set of images should be adopted.[15]

A particle in an electric field attracts other particles even if the particle is not charged. This is because the electric field is distorted by the particle. The force acting on a particle in a distorted electric field is given by[16]

$$F = \frac{1}{4}\pi D_p^3 \varepsilon_0 \frac{\bar{\varepsilon}-1}{\bar{\varepsilon}+2} \text{grad} |\mathbf{E}|^2 \tag{5.18}$$

2.5.2 IN LIQUID STATE

Particles in solutions are more or less charged. This charge generates the electrostatic interaction between particles when their surfaces are sufficiently close to each other, and also their migration toward the electrode of the opposite sign when particles are exposed to an external electric field. In this section, the electrical characteristics of particles in electrolyte solutions, which are understood at present, are discussed.

Surface Charge[17]

When an interface is formed between solid and liquid, the solid surface is usually charged because of the difference of the affinity of electrons to the surfaces. The charging mechanism depends on both the properties of the solid and the medium. The following mechanisms of charging are known for particles in electrolyte solutions.

Nernst-Type Charging

When crystal particles such as AgI crystals are dispersed in water, an equilibrium will be established between Ag^+ and I^- ions adsorbed on the particle surface and those in the bulk solution. If the concentration differs between adsorbed ions of Ag^+ and I^-, the excess amount of either Ag^+ or I^- results in the corresponding charge of the particle surface. These ions are called potential-determining ions. In this case the surface potential ψ_0 is able to be calculated by the so-called Nernst equation:

$$\psi_0 = -\left(\frac{2.3kT}{e}\right)(\text{pAg} - \text{pAg}_0) \tag{5.19}$$

where k is the Boltzmann constant, T is the temperature, e is the elementary charge, $\text{pAg} = -\log a_{Ag}^+$, and a_{Ag}^+ is the activity of silver ions. The point of zero charge (p.z.c.), pAg_0, is given in Table 5.2. A few crystal particles, such as AgBr, AgCl, Ag_2S, AgCNS, and $BaSO_4$, are known to be charged by the same mechanism.

Charging by Dissociable Groups

Particles with groups, such as $-OH$, $-COOH$, and $-NH_3$, will be charged by their dissociation in an aqueous solution, and the degree of dissociation depends on the pH of the solution. For example, $-COOH$ dissociates as $-COOH \leftrightarrows -COO^- + H^+$, and so the surface potential is always negative but the magnitude increases with the pH, as illustrated by curve I in Figure 5.1. When a surface is zwitterionic because of $-COOH$ and $-NH_3$ groups on the surface, the particle will be charged positively or negatively depending on the pH as known from Equation 5.20, and the p.z.c. exists in between. The potential in this case varies with the pH, as shown schematically by curve II in Figure 5.1.

$$NH_4^+ - R - COOH \xrightarrow{\quad H^+ \quad} NH_3 - R - COOH \xrightarrow{\quad OH^- \quad} NH_3 - R - COO^- \tag{5.20}$$

TABLE 5.2   Points of Zero Charge

Oxides	pH_0
SiO_2 (precipitated)	2–3
SiO_2 (quartz)	3.7
δ-MnO_2	2.8
β-MnO_2	7.2
SnO_2 (cassiterite)	5–6
α-$Al(OH)_3$	5.0
α-Al_2O_3	9.1
γ-$Al(OH)_3$ (gibbsite)	8–9
γ-$AlOOH$	8.2
RuO_2	5.7
TiO_2 (rutile)	5.7–5.8
TiO_2 (antase)	6.2
α-$FeOOH$ (goethite)	8.4–9.4
α-Fe_2O_3 (haematite)	8.5–9.5
ZnO	8.5–9.5
CuO	9.5
MgO	12.4
Other materials	
$AgBr$	$pAg_0 = 4.9$–5.3
AgI	$pAg_0 = 5.64$–5.66
Montmorillonite	$pH_0 = 2.5$
Kaolinite	$pH_0 = 4.6$
Clay platelets, sides	$pH_0 = 6$–7

When oxide particles, such as SiO_2, TiO_2, and Al_2O_3, are dispersed in water, water molecules adsorb and form $-OH$ groups on the particle surface. Then the surface will be charged by the adsorption or desorption of H^+, as illustrated by Equation 5.21; the surface potential of particles varies with the pH in the similar way as that of zwitterionic particles:

$$M - OH_2^+ \xrightleftharpoons{H^+} M - OH \xrightleftharpoons{OH^-} M - O^- \tag{5.21}$$

where M indicates a metal atom. The values of p.z.c. for various oxide particles are given in Table 5.2.

Charging by Isomorphic Substitution

This charging mechanism is found particularly in clay minerals. When there are defects in the crystal lattice of particles in which Si^{4+} is substituted by Al^{3+}, the deficit of positive charge results in charging particles. This charge is not affected by the pH value.

Charging by Other Mechanisms

When ions, ionic surfactants, or ionic polymers are adsorbed on the particle surface by the covalent force, the van der Waals force, or electrostatic force, the particle will be charged because of the

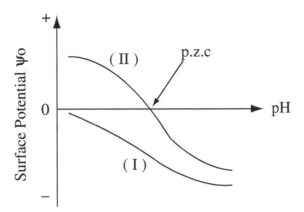

FIGURE 5.1 Typical variation of surface potential with pH.

charge of adsorbed molecules. The adsorption of impurities with charge in the solution also results in the charge of particles. In these cases, the charging mechanism is generally hard to predict.

Electrical Double Layer around Spherical Particles and Their Charge

As described earlier, particles in solutions are usually charged. Then, because of the electroneutrality principle, ions which are of the same amount but different sign are attracted toward the particle surface. A part of the ions is adsorbed directly on the particle surface to form the so called Stern layer. The rest of the ions are in thermal motion, balancing the electrical attractive force to form a diffuse layer. The potential at the outside of the Stern layer is called the Stern potential, ψ_d. Because the structure within the Stern layer is not well known, the Stern potential is often regarded as the surface potential of particles for many engineering purposes.

The potential, ψ, in the diffuse layer around a particle of radius a is expressed by the Poisson–Boltzmann equation, which cannot be solved analytically. However, when the Debye–Hückel approximation that ψ is small everywhere (i.e., $ze\psi \ll kT$) holds, and the electrolyte is symmetric with valency z, ψ is given explicitly as follows:

$$y = \left(\frac{\psi_d a}{r} \right) \exp[-\kappa(r-a)] \tag{5.22}$$

$$\kappa = \left(\frac{2n_0 z^2 e^2}{\varepsilon \kappa T} \right)^{0.5} \tag{5.23}$$

where r is the distance from the center of the particle, n_0 is the ionic concentration of the bulk solution, and ε is the permittivity of the medium. $1/\kappa$ is a measure of the thickness of the diffuse layer. In the case of particles in an aqueous solution at 25°C, the thickness of the diffuse layer is given by

$$\frac{1}{\kappa} = 3 \times 10^{-10} \left(z\sqrt{C} \right)^{-1} \ (m) \tag{5.24}$$

where C is the electrolyte concentration (mol/dm³). It is clear that the diffuse layer becomes thinner as values of z and C increase. At $z = 1$ and $C = 10^{-3}$ mol/dm³, $1/\kappa \sim 10^{-8}$ m ($= 10$ nm).

The total charge, Q, of the particle is given by

$$Q = 4\pi\varepsilon a\,(1 + \kappa a)\psi_{\mathrm{d}} \qquad (5.25)$$

Electrical Interaction between Spherical Particles[18]

When two charged particles are so close that their double layers overlap each other, the ionic concentration between the particle surfaces becomes much higher than that in the bulk solution. This difference in ionic concentration generates an osmotic pressure between the bulk solution and the solution between the surfaces and, therefore, the corresponding interaction force between the particles. The interaction is normally expressed in terms of the potential energy, V_{R}, which is correlated with the interaction force, F, by $F = -dV_{\mathrm{R}}/dr$. The electrostatic interaction potential between dissimilar spherical particles is given by

$$V_{\mathrm{R}} = \frac{\pi\varepsilon a_1 a_2 \left(\psi_{\mathrm{d1}}^2 + \psi_{\mathrm{d2}}^2\right)}{(a_1 + a_2)}\left[\frac{2\psi_{\mathrm{d1}}\psi_{\mathrm{d2}}}{\left(\psi_{\mathrm{d1}}^2 + \psi_{\mathrm{d2}}^2\right)}\ln\left(\frac{1+\exp(-\kappa h)}{1-\exp(-\kappa h)}\right)\right. \\ \left. + j\ln\left[1-\exp(-2kh)\right]\right] \qquad (5.26)$$

where the subscripts 1 and 2 indicate particles 1 and 2, respectively, and h is the distance between the particle surfaces. $j = 1$ when the surface potential is regarded as constant, and $j = -1$ when the surface charge density is regarded as constant. This equation is valid only when the Debye–Hückel approximation holds and the value of κa is reasonably large (say $\kappa a > 10$). Equation 5.26 indicates that the interaction is not only repulsive but also attractive, depending on the values of ψ_{d1} and ψ_{d2}. For particles of equal size and potential, the equation is simplified as

$$V_{\mathrm{R}} = j2\pi\varepsilon a\,\psi_{\mathrm{d}}^2\ln[1 + j\exp(-\kappa h)] \qquad (5.27)$$

It is clear that the interaction is always repulsive between similar particles. The values of V_{R} for $j = \pm1$ coincide well when h is reasonably large (say $\kappa h > 3$), but the value of V_{R} for $j = -1$ becomes greater than that for $j = 1$ when h is small (say $\kappa h < 1$). It is known that the potential curve of real colloids is often found between the potential curves of $j = \pm1$.

Electrophoresis

When an external electric field is applied to a suspension, charged particles will migrate toward the electrode of the opposite sign, as illustrated in Figure 5.2. This phenomenon is called electrophoresis. Measurements of the electrophoretic velocity of particles give information about their surface potential.

When a particle undergoes electrophoresis, a thin layer of liquid around the Stern layer moves with the particles as if it is fixed on the particle surface. The outer surface of this layer is called the slipping plane, and the potential there is called the zeta potential. Hence, the zeta potential can be correlated with the electrophoretic velocity of particles. According to Henry,[19] the electrophoretic velocity, u_{E}, of spherical and cylindrical particles is given as a function of the zeta potential ζ as follows:

$$u_{\mathrm{E}} = \frac{u_\infty}{E} = \frac{2\varepsilon\zeta f(\kappa a, K)}{3\mu} \qquad (5.28)$$

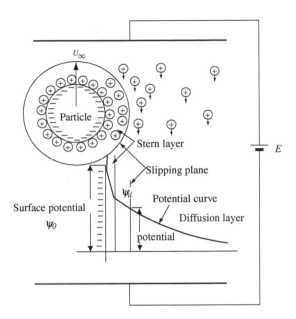

FIGURE 5.2 Electrochemical double layer around particle surface in an applied electric field.

where u_∞ is the velocity of particles, E is the intensity of the external electric field, μ is the viscosity of the medium, a is the particle or cylinder radius, K is λ_p/λ_0, where λ_p and λ_0 are the electric conductivities of the particle and the medium, respectively, and $f(\kappa a, K)$ is the Henry function drawn by the solid lines in Figure 5.3.

When the particle is a nonconducting sphere ($K = 0$) and $\kappa a \gg 1$, $f(\kappa a, K) = 3/2$ and Equation 5.28 results in the so called Smoluchowski equation.[20] This equation is applicable to nonspherical particles also:

$$u_E = \frac{\varepsilon\zeta}{\mu} \qquad (5.29)$$

If $K = 0$ and $\kappa a \ll 1$, Equation 5.28 coincides with the Huckel equation,[21] where $f(\kappa a, K) = 1$,

$$u_E = \frac{2\varepsilon\zeta}{3\mu} \qquad (5.30)$$

When the double layer around a particle is relatively thick (i.e., $0.1 < \kappa a < 100$) and the surface potential is high, the double layer cannot catch up to the particle migration, and the double layer will be distorted. This distortion will decelerate the migration velocity of particles. This is called the relaxation effect. The expression of u_E, in which the relaxation effect is taken into account, is given by[22,23]

$$u_E = \frac{2\varepsilon\zeta f(\kappa a, \zeta)}{3\mu} \qquad (5.31)$$

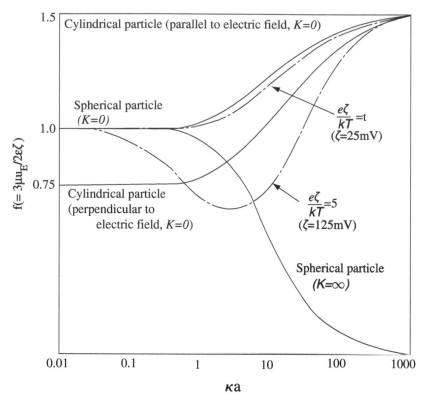

FIGURE 5.3 Henry and Overbeek functions.

where $f(\kappa a, \zeta)$ is given elsewhere,[24] and two examples are drawn by the dashed lines in Figure 5.3. Equation 5.31 is valid only at relatively low zeta potential. A more widely applicable method to estimate u_E was developed by O'Brien and White.[25] u_E for a spherical particle in a KCl solution is shown in Figure 5.4.

The electrophoretic velocity for particles of arbitrary shape is given by the Smoluchowski equation if $\kappa a \gg 1$, and by the following equation in the case of $\kappa a \ll 1$:

$$u_E = \frac{Q}{f_h}$$

(5.32)

where f_h is the hydrodynamic friction factor. Stigter[26,27] calculated the electrophoretic velocity of cylindrical particles in which the relaxation effect is taken into account.

Measuring Methods of Electrophoretic Velocity and Zeta Potential

The measuring methods of electrophoretic velocity have been developed extensively because the zeta potential can be evaluated by measuring the value of u_E. The methods of microelectrophoresis and electrophoretic light scattering are widely employed.

Microelectrophoresis

This method is employed when the movement of particles is detectable by microscope or ultramicroscope. The electrophoretic velocity of colloidal particles is determined by direct observation

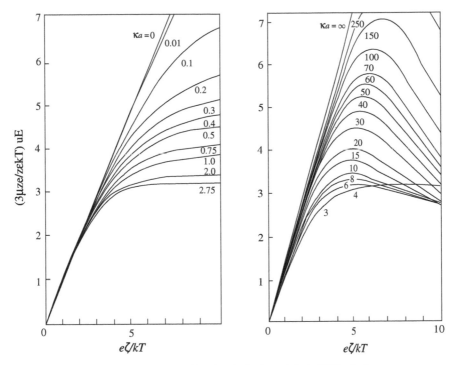

FIGURE 5.4 Electrophoretic velocity of spherical particle in KCl solution.

of particle movement in a closed capillary of circular or rectangular cross section. The medium flows with respect to the cell wall in an electric field because of the electroosmosis, and at the same time the Poisuille flow occurs toward the opposite direction because the cell is closed at both ends. Hence, there exist stationary levels where the net velocity of the medium is zero. The velocity of particles at these levels is their electrophoretic velocity. Measurement at an accurate stationary level is required to minimize the error because the velocity gradient of the medium is quite large around the stationary levels.

Convenient methods have been developed to detect the electrophoretic velocity of particles in the capillary tube. In the rotary prism method, the migrating particles look stationary when their velocity synchronizes with the rotational velocity of the prism; thus, the density distribution of u_E can be evaluated by changing the velocity of the prism. In the rotary grating method, the image of particles is detected by photomultiplier tubes through a rotary grating plate, and the frequency of the detected light is analyzed by a spectrum analyzer with a computer to determine the entire frequency of mobility.

Laser Doppler Method[28]

When colloidal particles in the electrophoretic cell are in Brownian motion, a power spectrum of the laser scattered from the particles is obtained, because of the Doppler broadening of the frequency. Measurement of a small shift of the power spectrum due to the particle displacement in an electric field offers a method to evaluate the electrophoretic velocity. The method is computerized and a prompt measurement for nanosize particles is possible.

Ultrasonic Method

When colloidal particles are placed under the ultrasonic field, both the particles and counter ions oscillate, but their mobility is different. Hence, when electrodes are placed at the distance of half of the sonic wavelength, the potential difference proportional to the zeta potential is detectable. The advantage of this method is that it is applicable to suspensions of high particle concentration, and a prompt measurement for nanosize particles is possible.

Electrokinetic Sonic Amplitude Method

By applying the electric field of high frequency to colloidal suspensions, the ultrasound will be generated from the movement of particles. The zeta potential can be calculated from the strength of the ultrasound. The advantage of this method is that it is applicable to suspensions of high concentration and to a wide range of particle size. It is even applicable to flowing suspensions.

Moving Boundary Method

When a clear boundary between a suspension and a particle-free medium is able to be formed, measurement of the boundary displacement in an electric field offers a method to determine the electrophoretic velocity of particles in the suspension. The Tiselius method[24] and Ottewill method[29] are applications of this principle.

Tracer Electrophoresis[30]

The electrophoretic cell is composed of a central horizontal tube filled with a colloidal solution and has electrodes at both ends. Concentration of particles in the central tube varies with time in an electric field. The electrophoretic velocity is determined by measuring the concentration change of particles with water-insoluble dyes.

Mass Transport Method[31]

The electrophoretic cell is composed of a reservoir and a small collection chamber, which are joined together. When an electric field is applied between the electrodes in the reservoir and in the chamber, particles migrate between them. The electrophoretic velocity is determined by measuring the difference in the net weight of solid in the chamber at the beginning and end of an experiment. This method is suited for highly concentrated suspensions.

REFERENCES

1. Kasai, A., *Proc. Inst. Electrostat. Jpn.,* 1, 46–51, 1977.
2. Gutman, E. J. and Hartmann, G. C., *J. Imaging Sci. Technol.,* 36, 335–349, 1992.
3. Matsuyama, T. and Yamamoto H., *J. Phys. D Appl. Phys.* 28, 2418–2423, 1995.
4. Crowley, J. M., *Fundamentals of Applied Electrostatics,* John Wiley & Sons, New York, 1986, pp. 26–30.
5. Masuda, H. and Iinoya, K., *AIChE J.,* 24, 950–956, 1978.
6. John, W., Reischl, G., and Devor, W., *J. Aerosol Sci.,* 11, 115 138, 1980.
7. Cottrell, G. A., Reed, C., and Rose-Innes, A. C., *Inst. Phys. Conf. Ser.,* 48, 249–356, 1979.
8. Masuda, H., Komatsu, T., and Iinoya, K., *AIChE J.,* 22, 558–564, 1976.
9. Matsusaka, S. and Masuda, H., *Adv. Powder Technol.,* 14, 143–166, 2003.
10. Matsusaka, S., Umemoto, H., Nishitani, M., and Masuda, H., *J. Electrostat.,* 55, 81–96, 2002.
11. Gallo, C. F. and Lama, W. L., *J. Electrostat.,* 2, 145–150, 1976.
12. Smoluchowski, M., *Phys. Z.,* 13, 1069–1080, 1912.
13. Wilson, I. B., *J. Colloid Sci.,* 2, 271–276, 1947.
14. Masuda, H., Ikumi, S., and Ito, T., *KONA,* 3, 17–25, 1985.
15. Smythe, W. R., *Static and Dynamic Electricity,* McGraw-Hill, New York, 1950, pp. 118–122.
16. Pohl, H. A., *Dielectrophoresis,* Cambridge Univ. Press, Cambridge, U.K., 1978, chap. 4.

17. Hunter, R. J., *Foundations of Colloid Science,* Vol. 1, Clarendon Press, Oxford, U.K., 1987.
18. Usui, S., *J. Colloid Interface Sci.,* 44, 107, 1973.
19. Henry, D. C., *Proc. Roy. Soc. London,* A133, 106, 1931.
20. Smoluchowski, M., *Phys. Z.,* 5, 529, 1905.
21. Hückel, E., *Phys. Z.,* 25, 204, 1924.
22. Overbeek, J. T. G., *Kolloid-Beih.,* 54, 287, 1943.
23. Wiersema, P. H., Loeb, A. L., and Overbeek, J. T. G., *J. Colloid Sci.,* 22, 78, 1966.
24. Kruyt, H. R., *Colloid Science,* Elsevier, Amsterdam, 1952, p. 215.
25. O'Brien, R. W. and White, L. R., *J. Chem. Soc. Faraday Trans.,* 74, 1607, 1978.
26. Stigter, D., *J. Phys. Chem.,* 82, 1417, 1978.
27. Stigter, D., *J. Phys. Chem.,* 82, 1670, 1979.
28. Uzgiris, E. E., *Rev. Sci. Instrum.,* 45, 117, 1974.
29. Ottewill, R. H. and Shaw, J. N., *Kolloid-Z.,* 218, 34, 1967.
30. Stigter, D. and Mysels, K. J., *J. Phys. Chem.,* 59, 45, 1955.
31. Oliver, J. P. and Sennett, P., paper presented at the *Third Annual Meeting of the Clay Mineral Society,* 1966.

2.6 Adhesive Force of a Single Particle

Hiroaki Masuda
Kyoto University, Katsura, Kyoto, Japan

Kuniaki Gotoh
Okayama University, Okayama, Japan

Ko Higashitani and Shuji Matsusaka
Kyoto University, Katsura, Kyoto, Japan

The adhesive force between particles or a particle and a solid surface plays an important role in powder-handling processes such as dry dispersion, transportation, and classification of particles. The van der Waals force, the electrostatic force, and the liquid bridge are the main sources of the adhesive force.

2.6.1 VAN DER WAALS FORCE

The van der Waals force is a short-range electromagnetic force interacting between two molecules (or atoms). However, the force also acts between two macroscopic bodies such as particle–particle and particle–wall.

The van der Waals force can be determined by London–van der Waals theory (microscopic theory) or Lifshitz–van der Waals theory (macroscopic theory).[1,2] In London–van der Waals theory, the so-called dispersion force acts between two symmetrical and electrically neutral molecules (or atoms). The potential energy is approximated by

$$E_A = -\beta_{11}/r^6 \tag{6.1}$$

where r is the distance between center points of two molecules (or atoms) and β_{11} is a constant that depends on the molecular (or atom) characteristics. Hamaker[3] assumed that the interaction between molecules is expressed by the London–van der Waals potential energy (Equation 6.1), and the interaction between two macrobodies can be calculated by the integration of all interactions of molecules that exist in the bodies. Therefore, the fundamental equation for the calculation of van der Waals potential energy is

$$E = -\int_{v1} \int_{v2} \frac{q_1^2 \beta_{11}}{r^6} dv_1 dv_2 \tag{6.2}$$

where q_1 is a number of molecules per unit volume in the body and v_1 and v_2 are the volumes of the two bodies. Representative expressions of van der Waals potential energy based on the above equation are summarized in Table 6.1.[4-7] In the table, the equations in which the retardation effect is taken into consideration are also listed.[8,9] The retardation effect should be taken into account when the distance between surfaces of two bodies is beyond 100 nm (= $\lambda/2\pi$, where λ is the wavelength of light). The van der Waals force can be obtained through differentiation of the potential energy; (dE/dr), also

Table 6.1 Representative expression of van der Waals potential energy and van der Waals force

van der Waals Potential Energy (E)	van der Waals Force (F)	
Without retardation effect: $-\dfrac{AR_1R_2}{6(R_1+R_2)z} = E_{sphere}$, $z \ll R_1$	$-\dfrac{AR_1R_2}{6(R_1+R_2)z^2} = F_{sphere}$	(1)
With retardation effect: $-\dfrac{AR_1R_2}{6(R_1+R_2)z}\left[1 - \dfrac{zb}{\lambda}\ln\left(1+\dfrac{\lambda}{zb}\right)\right]$	$-\dfrac{AR_1R_2}{6(R_1+R_2)}\left[\dfrac{R}{z^2} - \dfrac{b}{\lambda}\left(\dfrac{b}{zb+\lambda} - \dfrac{1}{z}\right)\right]$	(2)
Without retardation effect: $-\dfrac{AR}{6z}$, $z \ll R_1$	$-\dfrac{AR}{6z^2}$	(3)
With retardation effect: $-\dfrac{AR}{6z}\left[1 - \dfrac{zb}{\lambda}\ln\left(1+\dfrac{\lambda}{zb}\right)\right]$	$-\dfrac{A}{6}\left[\dfrac{R}{z^2} - \dfrac{b}{\lambda}\left(\dfrac{b}{zb+\lambda} - \dfrac{1}{z}\right)\right]$	(4)
Without retardation effect: $\dfrac{z}{z+0.5(B_1+B_2)}E_{sphere}$	$-\dfrac{z(B_1+B_2)}{[z+0.5(B_1+B_2)]^2}E_{sphere} + \dfrac{z}{z+0.5(B_1+B_2)}F_{sphere}$	(5)

summarized in Table 6.1. The constant A in these equations is called the Hamaker constant, which is given by

$$A = \pi^2 q_1^2 / \beta_{11} \tag{6.3}$$

The Hamaker constant can be related to the Lifshitz–van der Waals constant $\hbar\bar{\omega}$,[10,11] and the Hamaker constant between two different materials in air can be approximately expressed by[12]

$$A_{12} = \sqrt{A_{11} A_{22}} \tag{6.4}$$

where A_{11} and A_{22} are the Hamaker constants for material 1 and material 2, respectively, in a free space. In the presence of a third material between the two bodies, the Hamaker constant is given by

$$A_{132} = \left(A_{11} - A_{33}\right)\left(A_{22} - A_{33}\right) \tag{6.5}$$

where A_{33} is the Hamaker constant of the third material. The values of the Hamaker's constants in air for carbon hydrate, oxides or halides, and metals are $(4 \sim 10) \times 10^{-20}$ J, $(6 \sim 15) \times 10^{-20}$ J, and $(15 \sim 50) \times 10^{-20}$ J, respectively.[13] The separation distance z in the equations in Table 6.1 is determined by Born's repulsion force[2] and is usually taken as 0.4 nm in air.

On the other hand, if the particle or wall is soft, the elastic deformation occurs at the contact point. In Johnson–Kendall–Roberts (JKR) theory,[14] the van der Waals force for the soft material can be expressed by a linear function of the particle diameter, as well as for the hard material. According to Dahneke,[15] the van der Waals force for a soft and large particle can be approximated by a quadratic function of the particle diameter. Derjaguin et al.[16] proposed an equation (Derjaguin–Muller–Toporov [DMT] theory) that is a linear function of the particle diameter. The value calculated by DMT theory is 4/3 times the value calculated by JKR theory. Tabor[17] and Muller et al.[18,19] studied the difference between JKR theory and DMT theory, showing JKR theory and DMT theory are the special case of their model. Tsai et al.[20] proposed a new model introducing a new adhesive parameter that is similar to the model of Muller et al. In their model, when particle deformation is negligible, the van der Waals force is proportional to the diameter, and the value is 0.4 times that calculated by JKR theory, and the force is proportional to the 4/3-th power of the diameter when deformation is large enough.

2.6.2 IN GASEOUS STATE

Electrostatic Force

Electrostatic force in the gas phase arises from (1) particle–charge interaction, (2) image–charge effect, and (3) electrostatic contact potential difference. Between two charged particles, coulombic force F_{ec} acts, and the force is approximately calculated by the following equation (cf. 2.5.1):

$$F_{ec} = \frac{1}{4\pi\varepsilon_0} \frac{q_1 q_2}{r^2} \tag{6.6}$$

where q_1 and q_2 are the charge of the particle, r is the distance between centers of the particles, and ε_0 is a dielectric constant of the medium. If two particles are in contact and the gap between particle surfaces is extremely smaller than the diameter, Equation 6.6 can be expressed as follows:

$$F_{ec} = \frac{\pi \sigma_1 \sigma_2}{\varepsilon_0} D_p^2 \tag{6.7}$$

where σ_1 and σ_2 are the surface charge densities of particles.

The image force F_{ei} acts between a charged particle and a neutral surface:

$$F_{ec} = \frac{1}{4\pi\varepsilon_0} \frac{\varepsilon - \varepsilon_0}{\varepsilon + \varepsilon_0} \frac{q^2}{(2r)^2}$$

(6.8)

where q is the particle charge and ε is the dielectric constant of the wall material.

When two different materials are brought into contact, both of them are electrified because of the contact potential difference (cf. 2.5.1). The induced electrostatic force F_{ed} between a particle and a wall is estimated by the following equation through adopting Dahneke's deformation equation[15] as a first approximation:

$$F_{ed} = \frac{1}{2}\pi\varepsilon_0 \frac{V_c^2}{z^2}\left\{\frac{AkD_p^2}{32z^2}\left(1 + \frac{A^2k^2D_p}{108z^7}\right)\right\}$$

(6.9)

$$k = \frac{1 - v_1^2}{E_1} + \frac{1 - v_2^2}{E_2}$$

where, E_1, E_2 are Young's modulus and v_1, v_2 are Poisson ratio for particle and wall, respectively.

Liquid Bridge Force

When the relative humidity of atmosphere is relatively high (>65%[21]), the liquid bridge is formed at the contact point of two particles, as shown in Figure 6.1. For the completely wettable surface of particles, adhesive force caused by the liquid bridge can be obtained as the sum of the capillary force and the force caused by the surface tension of the liquid as follows:

$$F_L = \pi r_2^2 P_L + 2\pi\sigma r_2$$

(6.10)

where r_2 is a radius of liquid bridge as shown in Figure 6.1, σ is the surface tension of liquid, and P_L is the capillary pressure inside the liquid bridge. If the cross section of the liquid bridge is approximated by a circular arc, capillary pressure P_L is expressed by

$$P_L = \sigma\left(\frac{1}{r_1} - \frac{1}{r_2}\right)$$

(6.11)

Then, the liquid bridge force can be calculated by

$$F_L = \pi r_2^2\sigma\left(\frac{1}{r_1} - \frac{1}{r_2}\right) + 2\pi\sigma r_2$$

(6.12)

If it is assumed that r_1 is very much smaller than r_2, the geometric relations among particle diameter D_p, r_1, and r_2 [$r_1 = \frac{1}{2}D_p(\sec\alpha - 1)$, $r_2 = \frac{1}{2}D_p(1 + \tan\alpha - \sec\alpha)$, $\alpha \to 0$] gives the following equation for contacting spheres of the same size[21]:

$$F_L \cong \pi\sigma D_p$$

(6.13)

For a spherical particle on a plane wall, it becomes

$$F_L \cong 2\pi\sigma D_p$$

(6.14)

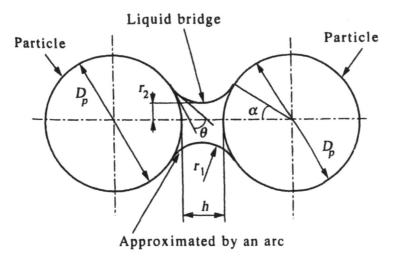

FIGURE 6.1 Liquid bridge formed between two particles.

For a nonwetting surface, the right-hand sides of Equation 6.13 and Equation 6.14 must be multiplied by cos θ, where θ is a contact angle between the liquid and the surface.

On the other hand, r_1 and r_2 can be correlated with a vapor pressure P_d at the vicinity of the surface by Kelvin's equation as follows[22]:

$$\frac{P_d}{P_{s0}} = \exp\left\{-\frac{M\sigma\cos\theta}{RT\rho_L}\left(\frac{1}{r_1} - \frac{1}{r_2}\right)\right\}$$ (6.15)

where
P_{s0} = saturation vapor pressure
M = molecular weight
R = gas constant
T = temperature
ρ_L = density of liquid
θ = contact angle

If the liquid contains a solute, vapor pressure P_d decreases with the increase in the number of solute molecules. The pressure can be expressed by[22,23]

$$\frac{P_d}{P_{s0}} = (1-\gamma)P_{kcl}$$ (6.16)

$$\gamma \equiv \frac{in_s}{(n_w + in_s)}$$

where P_{kcl} ($= P_d$ in Equation 6.15) is the vapor pressure without solute, i is the van't Hoff factor, and n_s and n_w are the number of solute and solvent molecules, respectively.

Comparison of Adhesive Force

Adhesive forces calculated by the above equations are shown in Figure 6.2. The liquid bridge force (Equation 6.14) is the dominating adhesive force, as long as the liquid bridge is formed. Without the liquid bridge, the van der Waals force (Equation 1 in Table 6.1) dominates. For a charged particle such as toner particles,[24] the coulombic force becomes important. When the surface charge density σ is assumed to be 26.5 $\mu C/m^2$, which is the maximum value decided by the electric field limit for discharging, the coulombic force (Equation 6.7) dominates for the particles over 200 μm in diameter. For the particles smaller than 200 μm, the coulombic force may be larger than the van der Waals force because the maximum surface charge density is determined by the voltage limit rather than the field limit.[25]

In the above discussion, the effects of the atmospheric conditions are not taken into consideration. The humidity of the atmosphere changes the shape of the liquid bridge, which results in the alternation of the adhesive force.[26,27] The humidity also changes the adsorbed water layer thickness,[28] and it also affects the adhesive force.[29,30] The effect of the humidity strongly depends on the roughness of a surface.[31] The adhesive force will also be affected by temperature.[32,33]

2.6.3   IN LIQUID STATE

The adhesive force between particles is often assumed to be given by the van der Waals force, even in the case of particles in liquids, without checking the adequacy. It is known that two or three layers of water molecules, ions, and hydrated ions are adsorbed on the solid–liquid interface.[34] Recent measurements of adhesive forces between a plate and a particle with an atomic force microscope indicate that the strength of adhesive force depends greatly on the microstructure of the adsorbed layer, even in

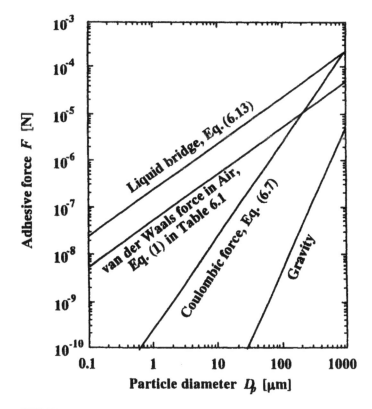

FIGURE 6.2   Comparison of adhesive forces. $A = 10^{-19}$ J (in air), $z = 0.4$, $\sigma = 0.072$ N/m, $\rho_p = 10^3$ Kg/m³, g = 9.8 m/s², $\sigma_1 = \sigma_2 = 26.5$ $\mu C/m^2$, $\varepsilon_0 = 8.8 \ 10^{-12}$ F/m.

simple solutions like electrolyte solutions of monovalent cations. The characteristics are summarized as follows, and the details are given elsewhere.[35,36]

(1) The adhesive force F_{ad} depends on how many particles can break the adsorbed layer to contact directly during the contact time t_c.

(2) The magnitude of F_{ad} depends on the hydration enthalpy of ions ΔH, as well as the electrolyte concentration C_e; F_{ad} decreases with increasing C_e and decreasing value of ΔH. It is especially important to know that the dependence of F_{ad} on t_c varies greatly with ΔH in concentrated solutions, as shown in Figure 6.3. This is explained as follows. Since highly hydrated cations like Li^+ form a thick but weak adsorbed layer, surfaces can contact directly to have a strong adhesion by destroying the adsorbed layer. On the other hand, because poorly hydrated cations like Cs^+ form a thin but strong adsorbed layer, the gap between surfaces at contact reduces the strength of adhesive force greatly. The mechanism is schematically drawn in Figure 6.4.

2.6.4 MEASUREMENT OF ADHESIVE FORCE

When a solid particle comes into contact with a wall or another particle, interaction forces occur between the surfaces. The forces depend on the physical and chemical properties, particle shape, roughness, electrostatic charge, and interaction media. In addition, the forces vary according to the distance between the surfaces. The term "adhesive force" is defined as the maximum attractive interaction force during which the particle is removed in the normal direction; thus, it is called the "pull-off force." In this section, basic techniques for the measurement of the adhesive force are summarized, and the analysis based on the force balance and moment balance is explained. Furthermore, atomic force microscopy, which became a powerful tool for the measurement of a force–distance curve, is described.

Methods of Measurement

The techniques for the measurement of the adhesive force are classified into several categories.[37–39] The principles of the techniques are schematically shown in Figure 6.5. They are applicable to the measurement of particle–particle interaction as well as particle–wall interaction.

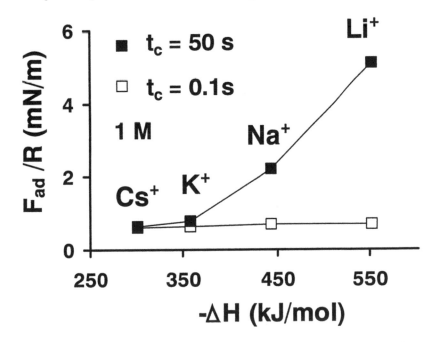

FIGURE 6.3   Difference between the adhesive forces at t_c = 0.1 and 50 s in various 1 M electrolyte solutions.

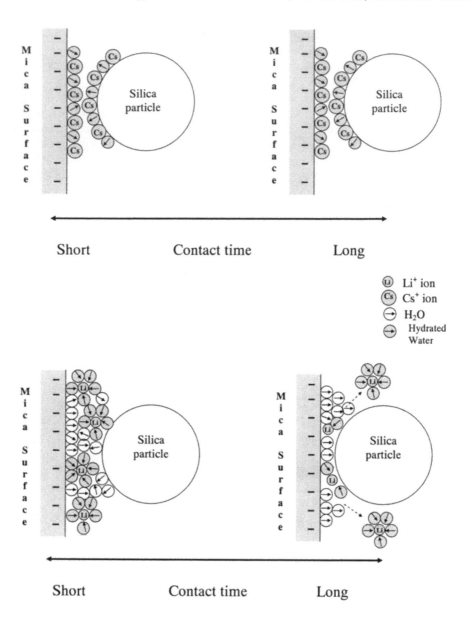

FIGURE 6.4   Mechanism for the difference of adhesive force between cations of low and high hydration enthalpies.

Gravity method (pendulum type): A vertical plate is brought up to a freely hanging particle until contact is made, and then shifted in a direction perpendicular to the contact area. The gravity component that tends to separate the particle is increased with the inclination angle of the plate. From the force balance, the adhesive force F_a can be calculated using the following equation:

$$F_a = M_p g \sin \alpha \qquad (6.17)$$

where M_p is the mass of the particle, g is the gravitational acceleration, and α is the inclination angle. To separate the particle, the gravity component must be larger than the adhesive force. Therefore, this method is used for comparatively large particles of the order of 1 mm in diameter.

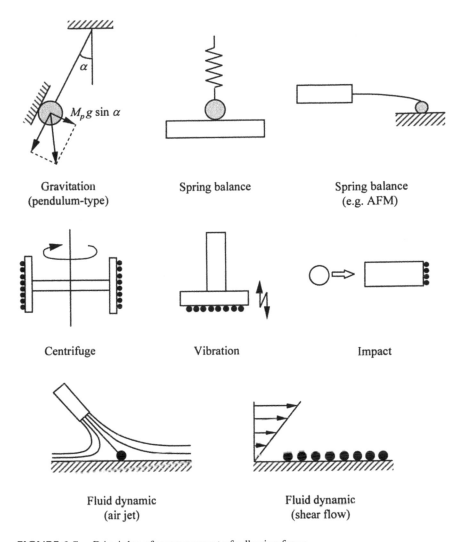

Gravitation
(pendulum-type)

Spring balance

Spring balance
(e.g. AFM)

Centrifuge

Vibration

Impact

Fluid dynamic
(air jet)

Fluid dynamic
(shear flow)

FIGURE 6.5   Principles of measurement of adhesive force.

Spring balance method: The separation force is supplied by a spring. Since the spring force can be made larger than the gravitational force, this method is applicable for micro-sized particles. There are several types of springs, including helical and cantilever springs. Atomic force microscopy is an application using the cantilever spring, which is described later.

Centrifuge method[40–42]: This method is widely used for the measurement of the adhesive forces of micron-sized particles. The important advantages consist in the sensitivity and statistical accuracy. Since the adhesive forces of many particles on a surface can be measured in a single experiment, the adhesive-force distribution is also obtained. The centrifugal force F_{cen} is calculated using the following equation:

$$F_{cen} = M_p \, \alpha_{cen} \tag{6.18}$$

where α_{cen} is the centrifugal acceleration ($= l\omega^2$), in which l is the distance of the object from the axis, and ω is the angular velocity. To prevent the effect of the aerodynamic drag caused by the rotation, the object should be in a closed box or under vacuum. If the particles are sensitive to temperature changes, the frictional heat generated by rotation should be removed.

Vibration method[43]: The apparatus is rather small and simple since there is no mechanical rotation unit. The separation force can be estimated in the way similar to the centrifugal force, namely,

$$F_{vib} = M_p \, \alpha_{vib} \tag{6.19}$$

where α_{vib} is the vibration acceleration ($= y\omega_f^2$), in which y is the amplitude, and ω_f is the angular frequency. If the vibration waveform is not sinusoidal or contains noise, the maximum acceleration must be measured. To separate fine particles, high vibration acceleration is needed. Since such vibration raises the temperature of the vibrating plate, a cooling system is required.

Impact method[44]: Particles are attached on one side of a wall, and an impact is applied to the reverse side. The impact force does not act directly on the particles, but the wall pushes the particles forward. This method is very simple, and the separation force can be calculated from the product of the mass of the particle and the effective acceleration. However, since the accurate determination of the acceleration is not so easy, this method is often used for a qualitative evaluation of the adhesive force.

Fluid dynamic method[45–48]: Particles can be removed by an air jet or a shear flow. This method is suitable for the evaluation of the particle adhesion to a wall in a particle–fluid system. The adhesive strength distribution can also be obtained from the relationship between the particle reentrainment efficiency and the fluid dynamic separation force.

Analysis of Forces Exerted on a Single Particle

Adhesive force is greatly affected by the contact states such as deformation and multiple-point contact, which also depend on the viscoelasticity, geometry, and surface roughness. Figure 6.6 illustrates several contact states between a particle and a wall. In addition to the contact states, the points of application of force and their directions are important to determine the adhesive force.

Figure 6.7 shows two cases of particle–wall contacts in a centrifugal field. The moment balance for the particle adhering to a horizontal wall is expressed as[42]

$$\left(F_a + M_p \, g\right) a \approx M_p \, l\omega_I^2 \frac{D_p}{2} \tag{6.20}$$

where a is the distance from the point of application to the fulcrum. The adhesive force F_a becomes

$$F_a \approx M_p \, (l\omega_I^2 \frac{D_p}{2a} - g) \tag{6.21}$$

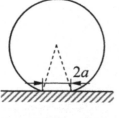
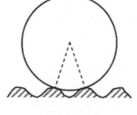
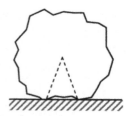

Contact deformation   Rough surface   Irregular shape

FIGURE 6.6   Contact states between a particle and a wall.

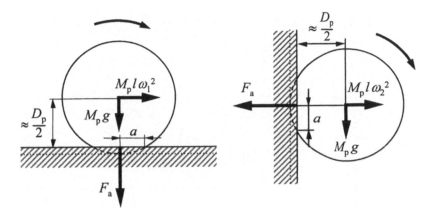

FIGURE 6.7   Particle-wall contacts in a centrifugal field.

For the particle adhering to a vertical wall, the adhesive force F_a can be expressed as

$$F_a \approx M_p \left(l\omega_2^2 + g \frac{D_p}{2a} \right) \tag{6.22}$$

If $l\omega_2^2 \gg gD_p/(2a)$, F_a is simply equal to $M_p l\omega_2^2$. When the gravitational component cannot be neglected, the value of a is important to estimate the adhesive force. However, from Equation 6.21 and Equation 6.22, the term of $D_p/(2a)$ can be eliminated, and the adhesive force is calculated using the following equation:

$$F_a = M_p \frac{(l\,\omega_1\omega_2)^2 + g^2}{l\,\omega_1^2 - g} \tag{6.23}$$

Also, the value of a can be obtained by

$$a \approx \frac{D_p}{2} \frac{l\,\omega_1^2 - g}{l\,\omega_2^2 + g} \tag{6.24}$$

Atomic Force Microscopy

The research on the force–distance curve, as well as the adhesive force, has attracted much attention since the atomic force microscope was invented.[49] In particular, replacing the tip of a cantilever by a spherical particle (i.e., the colloidal probe technique) has accelerated the study of particle–wall interaction.[50–53] Figure 6.8 shows the principle of the atomic force microscope with a colloidal probe. The sample is moved up and down by applying voltage to a piezoelectric translator. When a force acts on the probe, the cantilever bends. The deflection of the cantilever is measured using the optical lever technique. Figure 6.9 shows a force–distance curve. (1) The colloidal probe is approaching but still far from the surface, and the interaction force is negligible. (2) When the probe gets close to the surface, the interaction force will occur; this is supposed to be attractive here. When the attractive force exceeds the spring force of the cantilever, the probe will jump

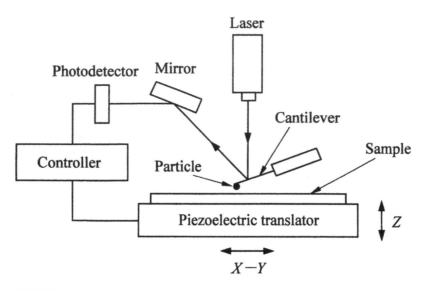

FIGURE 6.8   Atomic force microscope with a colloidal probe.

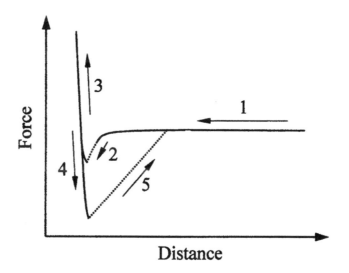

FIGURE 6.9   A force–distance curve.

into contact with the surface. (3) The probe and the surface move in parallel, which is called the "constant compliance region." (4) During retraction, the probe will usually adhere to the surface, causing the cantilever to bend downward. (5) When the spring force of the cantilever becomes larger than the adhesive force, the cantilever will jump out of contact into its equilibrium position. Using this method, interaction forces can be analyzed in detail and compared with theoretical models on van der Waals, liquid bridge, electrostatic forces, and so forth.

REFERENCES

1. Israelachvili, J. N., *Intermolecular and Surface Forces with Applications to Colloidal and Biological Systems*, Academic Press, New York, 1985.
2. Krupp, H., *Adv. Colloid Interface Sci.*, 1, 111–239, 1967.
3. Hamaker, H. C., *Physica*, 4, 1058–1072, 1937.
4. Vold, M. J., *J. Colloid Sci.*, 16, 1–12, 1961.
5. Vincent, B., *J. Colloid Interface Sci.*, 42, 270–285, 1973.
6. Czarnecki, J. and Dabros, T., *J. Colloid Interface Sci.*, 78, 25–30, 1980.
7. Göner, P. and Pich, J., *J. Aerosol Sci.*, 20, 735–747, 1989.
8. Gregory, J., *J. Colloid Interface Sci.*, 83, 138–145, 1981.
9. Clayfield, E. J., Lumb, E. C., and Markley, P. H., *J. Colloid Interface Sci.*, 37, 382–389, 1971.
10. Lifshitz, E. M., *Soviet Phys-JETP*, 2, 73–83, 1956.
11. Langbein, D., *Physical Review B*, 2, 3371–3383, 1970.
12. Israerlachvili, J. N., *Proc. Roy. Soc. Ser. A*, 331, 35–55, 1972.
13. Visser, J., *Adv. Colloid Interface Sci.*, 3, 331–363, 1970.
14. Johnson, K. L., Kendall, K., and Roberts, A. D., *Proc. Roy. Soc. Ser. A*, 324, 301–313, 1971.
15. Dahneke, B., *J. Colloid Interface Sci.*, 40, 1–13, 1972.
16. Derjaguin, B. V., Muller, V. M., and Toporov, Y. P., *J. Colloid Interface Sci.*, 53, 314–326, 1975.
17. Tabor, D., *J. Colloid Interface Sci.*, 58, 2–13, 1977.
18. Muller, V. M., Yushchenko, V. S., and Derjaguin, B. V., *J. Colloid Interface Sci.*, 77, 91–101, 1980.
19. Muller, V. M., Yushchenko, V. S., and Derjaguin, B. V., *J. Colloid Interface Sci.*, 92, 92–101, 1983.
20. Tsai, C. J., Pui, D. Y. H., and Liu, B. Y. H., *Aerosol Sci. Technol.*, 15, 239–255, 1991.
21. Zimon, A. D., *Adhesion of Dust and Powder*, 2nd Ed., Consultant Bureau, New York, 1982, pp. 69, 114.
22. Carman, P. C., *J. Phys. Chem.*, 57, 56–64, 1953.
23. Endo, Y., Kousaka, Y., and Nishie, Y., *Kagaku Kogaku Ronbunshu*, 18, 950–955, 1992.
24. Lee, M. H. and Ayala, J., *J. Imaging Technol.*, 11, 279–284, 1985.
25. Masuda, H. and Matsusaka, S., *J. Soc. Powder Technol. Jpn.*, 30, 713–716, 1993.
26. Endo, Y., Kousaka, Y., and Nishie, Y., *Kagaku Kogaku Ronbunshu*, 19, 55–61, 1993.
27. Chikazawa, M., Nakajima, W., and Kanazawa, T., *J. Res. Assoc. Powder Technol., Japan*, 14, 18–25, 1977.
28. Chikazawa, M., Yamaguchi, T., and Kanazawa, T., *Proceedings of the International Symposium on Powder Technology 1981*, Hemisphere Publ., pp. 202–207, 1981.
29. Danjo, K. and Otsuka, A., *Chem. Pharm. Bull.*, 26, 2705–2709, 1978.
30. Gotoh, K., Takebe, S., Masuda, H., and Banba, Y., *Kagaku Kogaku Ronbunshu*, 20, 205–212, 1994.
31. Gotoh, K., Takebe, S., and Masuda, H., *Kagaku Kogaku Ronbunshu*, 20, 685–692, 1994.
32. Nishino, M., Arakawa, M., and Suito, E., *Zairyo*, 1, 535–540, 1969.
33. Otsuka, A., Iida, K., Danjo, K., and Sunada, H., *Chem. Pharm. Bull.*, 31, 4483–4488, 1983.
34. Israelachvili, J. N., *Intermolecular and Surfaces Forces*, 2nd Ed., Academic Press, New York, 1992.
35. Vakarelski, I. U., Ishimura, K., and Higashitani, K., *J. Colloid Interface Sci.*, 227, 111–118, 2000.
36. Vakarelski, I. U., Ishimura, K., and Higashitani, K., *J. Colloid Interface Sci.*, 242(1), 110–120, 2001.
37. Krupp, H., *Adv. Colloid Interface Sci.*, 1, 79–110, 1967.
38. Zimon, A. D., *Adhesion of Dust and Powder*, 2nd Ed., Consultants Bureau, New York, 1982, pp. 69–91.
39. Israelachvili, J. N., *Intermolec. Surface Forces*, 2nd Ed., Academic Press, London, 1991, pp. 165–175.
40. Asakawa, S. and Jimbo, G., *J. Soc. Mater. Sci., Japan*, 16, 358–363, 1967.
41. Emi, H., Endo, S., Kanaoka, C., and Kawai, S., *Kagaku Kogaku Ronbunshu*, 3, 580–585, 1977.
42. Matsusaka, S., Koumura, M., and Masuda, H., *Kagaku Kogaku Ronbunshu*, 23, 561–568, 1997.
43. Mullins, M. E., Michaels, L. P., Menon, V., Locke, B., and Ranade, M. B., *Aerosol Sci. Technol.*, 17, 105–118, 1992.
44. Otsuka, A., Iida, K., Danjo, K., and Sunada, H., *Chem. Pharm. Bull.*, 31, 4483–4488, 1983.
45. Visser, J., *J. Colloid Interface Sci.*, 34, 26–31, 1970.
46. Masuda, H., Gotoh, K., Fukada, H., and Banba, Y., *Adv. Powder Technol.*, 5, 205–217, 1994.
47. Masuda, H., Matsusaka, S., and Imamura, K., *KONA*, 12, 133–143, 1994.
48. Matsusaka, S., Mizumoto, K., Koumura, M., and Masuda, H., *J. Soc. Powder Technol. Jpn.*, 31, 719–725, 1994.

49. Binnig, G., Quate, C. F., and Gerber, Ch., *Phys. Rev. Lett.,* 56, 930–933, 1986.
50. Ducker ,W. A., Senden, T. J., and Pashley, R. M., *Nature,* 353, 239–241, 1991.
51. Claesson, P. M., Ederth, T., Bergeron, V., and Rutland, M. W., *Adv. Colloid Interface Sci.,* 67, 119 –183, 1996.
52. Cappella, B. and Dietler, G., *Surf. Sci. Rep.,* 34, 1–104, 1999.
53. Kappl, M. and Butt, H.-J., *Part. Part. Syst. Char.,* 19, 129–143, 2002.

2.7 Particle Deposition and Reentrainment

Manabu Shimada
Hiroshima University, Higashi-Hiroshima, Japan

Shuji Matsusaka and Hiroaki Masuda
Kyoto University, Katsura, Kyoto, Japan

Particle deposition is a phenomenon in which particles, suspended in a fluid, are transported to a wall and make permanent or temporary contact with the surface of the wall. Particle deposition is an important phenomenon in many fields: the fouling of channel walls exposed to dusty gas, micro-contamination problems in advanced material processing, the dry deposition of particulate matter in an atmospheric environment, the development of gas filtration devices, the designed deposition of particles for the fabrication of functional devices, and so forth. Particle reentrainment means resuspension of deposited particles. In general, small primary particles are hard to reentrain from a wall; however, aggregate particles are readily reentrained from a particle deposition layer. The aggregate reentrainment affects various engineering applications, such as powder dispersion, particles size classification, dust collection, particle synthesis, and aerosol sampling.

2.7.1 PARTICLE DEPOSITION

To predict and evaluate particle deposition phenomena, mechanisms that govern the transport process of particles must be better understood. Important mechanisms include transport by fluid flow, transport by Brownian diffusion and turbulent eddies, and motion induced by inertial forces and external forces such as gravitational sedimentation, electrostatic migration, thermophoresis, diffusiophoresis, and photophoresis.

Fundamental Concepts

The transport of submicron-sized particles suspended in a laminar flow gas is generally dominated by fluid flow motion, Brownian diffusive motion, and migration by external forces. In such a case, the spatial and temporal change of particle number concentration per unit gas volume, n, is described by the following convection–diffusion equation:

$$\frac{dn}{dt} = \nabla \cdot \left\{ \rho_f D \nabla (n/\rho_f) - (\mathbf{u} + v)n \right\} + Q \tag{7.1}$$

where t is the time, ρ_f the density of the medium gas, D the Brownian diffusion coefficient, $\mathbf{u}$ the medium gas velocity, v the particle velocity induced by the sum of external forces ($\Sigma \mathbf{F}$) and equals $B\Sigma \mathbf{F}$ (mobility $B = C_c/3\pi\mu D_p$; C_c, Cunningham correction factor; μ, viscosity of the medium gas; D_p, particle diameter), and Q the rate of generation or consumption of particles. The gas velocity $\mathbf{u}$

171

can usually be determined by solving the governing equation for fluid flow. The energy equation also needs to be solved in order to incorporate the temperature dependence of ρ_f and D into Equation 7.1 when deposition in a nonisothermal system is under consideration. Sections 2.3–2.5 and literature[1,2] will be referred to for the expression of particle velocity $\mathbf{v}$ induced by various external forces $\mathbf{F}$.

To obtain the rate of deposition, Equation 7.1 can be solved with appropriate boundary conditions that are relevant to the state of the particles on the surface. Strictly speaking, a particle having apparently arrived at a surface is separated from the surface by a very short distance. However, small, submicron particles for which Equation 7.1 holds can be regarded as being held at the surface immediately and permanently, once the surface is reached. Since the separation between a particle and surface is much smaller than the particle diameter, the particle concentration n is usually set to zero at a distance of half the particle diameter from the surface. The number of particles deposited per unit surface area and time, deposition flux j, can then be calculated from

$$j = -D \frac{dn}{dz}\bigg|_{z=D_p/2} \tag{7.2}$$

where z is the distance from the surface. Since j is generally proportional to the particle concentration sufficiently far from the surface, (n_0), $v_d = j / n_0$ gives the deposition flux per unit particle concentration and is denoted as the deposition velocity.

Deposition in Laminar Flow

Analytical or approximate solutions have already been obtained for some geometrically simple systems. As an example, the deposition velocity of particles is approximated as

$$v_d = \frac{1.08 D}{d_d} \left(\frac{\rho_f u_0 d_d}{\mu} \right)^{1/2} \left(\frac{\mu}{\rho_f D} \right)^{1/3} - \frac{2D}{D_p} \mathrm{Ex} \tag{7.3}$$

for particles suspended in a laminar flow impinging against a disc with diameter d_d and, at the same time, experiencing a constant external force in the direction normal to the wall.[3] u_0 and Ex in the above equation are the flow velocity and a parameter that indicates the magnitude of the external force, respectively. Ex equals $EqC_c/(6\pi\mu D)$ (E is electric field strength, q is particle charge) for Coulomb force in an electric field, and $(\rho_p - \rho_f)D_p^3 g C_c/(36 \ \mu D)$ (ρ_p is density of the particle, g is acceleration of gravity) for gravitational force. Figure 7.1 shows examples of some calculated results for v_d for particles flowing above a horizontally oriented disc. Examples of analytical or approximated solutions of v_d for several other systems are also found in Section 2.1 and in a report by Adamczyc et al.[4] Although systems for which such analytical or approximated solutions can be directly applicable are limited, such solutions are often very useful for a rough estimation of deposition with the actual systems being simplified. However, if a detailed distribution of the deposition velocity must be known, a rigorous calculation of the flow field and particle concentration distribution is necessary. Distributions of electric potential and temperature must be analyzed in detail when a nonuniform electrostatic or thermophoretic force influences the deposition.[5,6]

Deposition in Turbulent Flow

In addition to Brownian diffusion, the mixing and transport of particles suspended in a turbulent flow field are enhanced by turbulent eddies in the medium gas. To evaluate the rate of deposition in a turbulent flow, the detailed temporal change (fluctuation) of particle concentration induced by turbulent eddies is not usually analyzed. In most cases, a time-averaged rate of deposition is derived from the basic equations for gas velocity and particle concentration distribution, which have been

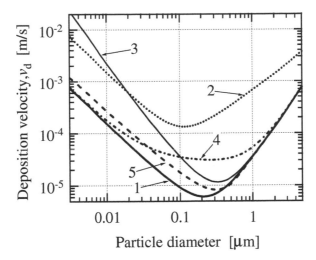

FIGURE 7.1 Deposition velocity of airborne particles on the upper surface of a horizontally oriented disc 125 mm in diameter; (1) for particles in downward flow at a speed of 0.3 m/s; (2) same conditions as (1) except that pressure is reduced to 1.01 kPa; (3) same conditions as (1) except that each particle possesses a unit elementary charge and that uniform electric field of 1 V/m exists; (4) same conditions as (1) except that uniform temperature gradient of 1 K/mm exists; (5) for particles suspended in turbulent flow at a certain intensity.

averaged with respect to time and space ("Reynolds average"). In this case, the transport of particles due to turbulent eddies can be regarded as a diffusion phenomenon, and the magnitude of the diffusion is described by the turbulent diffusion coefficient, D_t. As a result, $\mathbf{u}$ and D in Equation 7.1 are replaced by $\bar{\mathbf{u}}$ and $D + D_t$, respectively.

Several notable experimental results and theoretical models for particle deposition in a turbulent flow have been reviewed by Papavergos and Hedley.[7] Figure 7.2 shows representative experimental results for turbulent deposition in a circular pipe. The ordinate and abscissa of the figure are the dimensionless deposition velocity v_d^+ ($= v_d/u*$, $u*$, friction velocity) and the dimensionless relaxation time of particles τ^+ ($= \rho_f \rho_p D_p^2 u*^2 C_c/(18\,\mu^2)$), respectively. Turbulent deposition is roughly divided into the following three regimes on the basis of the magnitude of τ^+, which indicate the effect of particle inertia.

In the regime in which τ^+ is less than about one and thus particles will follow the fluid motion in turbulent eddies (referred to as "turbulent particle diffusion regime"), the magnitude of D_t is of the same order as the eddy kinematic viscosity of the fluid v_t and has a relationship of $D_t = v_t/Sc_t$ (Sc_t, turbulent Schmidt number (~ 0.7–1.0)). v_t can be determined by measurement, empirical formula, and numerical simulations of flow based on the turbulence models[8–10] When τ^+ is in the range of about 1–10 (referred to as the "eddy diffusion–impact regime"), the deposition velocity depends strongly on τ^+, since the motion of particles tends to deviate from that of fluid near a wall, as the result of inertial force. A number of theoretical or semiempirical models have been proposed to explain the measured deposition velocities: the free-flight model (stopping-distance model), in which particles are assumed to "fly" due to inertia in the viscous sublayer on a wall[11,12]; the effective-diffusivity model, in which D_t is assumed to be enhanced apparently by inertial motion[13,14]; and so on. On the other hand, the deposition velocity for values of τ^+ larger than about 10 (referred to as the "particle inertia moderated regime") is only slightly dependent on τ^+, since the inertia-induced motion of particles tends to prevail over the entire region of the fluid. Despite many existing models,

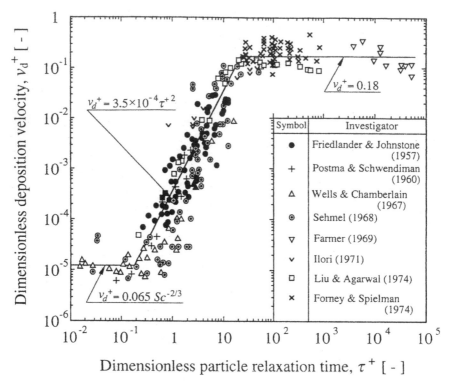

FIGURE 7.2 Summary of experimental deposition data in vertical flow system. [From Papavergos, P. G. and Hedley, A. B., *Chem. Eng. Res. Des.* 62, 275–295, 1984. With permission.]

a method capable of accurately evaluating turbulent deposition has not been fully established. For the present, the following empirical equations will be useful as a rough estimation[7]:

$$v_d^+ = 0.065\{\mu/(\rho_f D)\}^{-2/3} \quad (\tau^+ < 0.2) \tag{7.4}$$

$$v_d^+ = 3.5 \times 10^{-4} \tau^{+2} \quad (0.2 < \tau^+ < 20) \tag{7.5}$$

$$v_d^+ = 0.18 \quad (\tau^+ > 20) \tag{7.6}$$

In addition to the above-mentioned models based on the convection–diffusion equation, recent advances in computer performance have made it possible to more accurately calculate the transport of particles influenced by microscopic fluid motion in turbulent eddies. This type of calculation is based on the equation of motion for each particle and often employs a Monte Carlo simulation. The effects of spatial nonuniformity of turbulent intensity and external force on particle motion and deposition can be analyzed.[15–18]

2.7.2 PARTICLE REENTRAINMENT

When particles are deposited on a surface, adhesive forces act on the particles. However, if aerodynamic forces or other separation forces based on particle collision and so on are sufficiently large, the deposited particles will be reentrained into the flow. The reentrainment phenomenon should be distinguished from bounce or saltation of coarse particles, since the contacting time and the state of the interaction are very different. In general, fine particles immersing deeply within the viscous sublayer in a turbulent boundary layer are hard to reentrain; however, when fine particles accumulate on the surface, aggregate particles are readily reentrained.[19] The aggregate reentrainment affects various engineering applications, such as powder dispersion, particle size classification, dust collection, particle synthesis, and aerosol sampling.

Concept of Reentrainment of Aggregates

The concept of the reentrainment of aggregates from a particle deposition layer is illustrated in Figure 7.3. Since the surface of the particle layer is not smooth, projecting particles experience a relatively large drag; consequently, the aggregate particles can be reentrained. A simple model for the reentrainment of a spherical aggregate is shown in Figure 7.4. When the flow around the aggregate is very slow, namely, creeping flow, the drag on a small segment of the aggregate in the shear flow is approximated by[20,21]

$$dR_D(y) = \frac{24\tau_w}{D_{ag}}(y - y^*)\sqrt{y(D_{ag} - y)}dy \tag{7.7}$$

where τ_w is the wall shear stress. The moment of force M is represented by

$$M = \int_{y^*}^{D_{ag}}(y - y^*)\,dR_D(y) \tag{7.8}$$

The maximum bending stress σ_b in the break point is given by

$$\sigma_b = \frac{M}{Z} \tag{7.9}$$

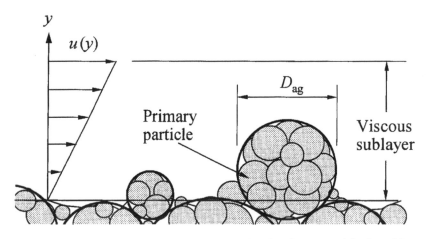

FIGURE 7.3 Concept of reentrainment of aggregates from a particle deposition layer in turbulent pipe flow.

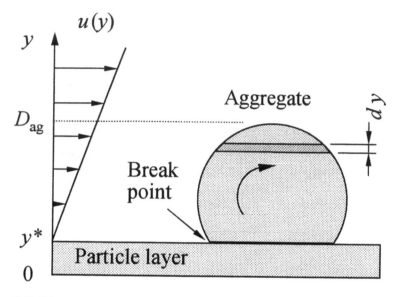

FIGURE 7.4   A reentrainment model of an aggregate particle.

where Z is the section modulus:

$$Z = \frac{\pi\, d^3}{32} = \frac{\pi}{4}\left\{y^{*}(\mathrm{D}_{ag} - y^{*})\right\}^{3/2} \tag{7.10}$$

From Equation 7.7 through Equation 7.10, the following equation is obtained:

$$\sigma_{b} = \frac{96}{\pi}\, \tau_{w} f(Y^{*}) \tag{7.11}$$

where $f(Y^{*})$ is represented by

$$f(Y^{*}) = \left[Y^{*}\left(1 - Y^{*}\right)\right]^{-3/2} \int_{Y*}^{1}\left(Y - Y^{*}\right)^{2} \sqrt{Y(1-Y)}\,dY \tag{7.12}$$

where

$$Y^{*} = \frac{y^{*}}{D_{ag}} \approx 0.5 \tag{7.13}$$

$$Y = \frac{y}{D_{ag}} \tag{7.14}$$

From Equation 7.11 through Equation 7.14, $\sigma_{b} = 3\tau_{w}$ is obtained; the adhesive strength can be calculated from the wall shear stress.

Particle Reentrainment Efficiency

The adhesive strength distribution is related to the reentrainment efficiency. In general, the distribution function is represented by the following log-normal equation[20]:

$$\gamma(\tau_c) = \frac{1}{\sqrt{2\pi}\,\ln\sigma_g}\int_0^{s_c}\exp\left(\frac{-\left(\ln\tau_c - \ln\tau_{c50}\right)^2}{2\ln^2\sigma_g}\right)d(\ln\tau_c) \qquad (7.15)$$

where τ_c is the critical wall shear stress corresponding to the reentrainment. For turbulent pipe flow: $3\times10^3 < Re < 10^5$, $\gamma(\tau_c)$ is rewritten as a function of the average velocity $\bar{u}$:

$$\gamma(\bar{u}) = \frac{1}{\sqrt{2\pi}\,\ln\sigma_g'}\int_0^{\bar{u}}\frac{1}{u}\exp\left(\frac{-\left(\ln\bar{u} - \ln\bar{u}_{50}\right)^2}{2\ln^2\sigma_g'}\right)d\bar{u} \qquad (7.16)$$

where

$$\sigma_g' = \sigma_g^{4/7} \qquad (7.17)$$

When aggregate particles are reentrained from the surface of the particle layer, a new surface of the particle layer is exposed to the flow; the next reentrainment can occur from the newly exposed surface with the same reentrainment efficiency γ. This is a hierarchical phenomenon. After renewal of the surface n times, the total reentrainment efficiency $\Gamma(n)$ is represented by

$$\Gamma(n) = \sum_{s=1}^{n}\gamma^s = \frac{\gamma(1-\gamma^n)}{1-\gamma} \qquad (7.18)$$

The mass of particles reentrained per unit area W/A_p is proportional to $\Gamma(n\to\infty)$:

$$\frac{W}{A_p} = k\Gamma(n\to\infty) = \frac{k\gamma}{1-\gamma} \qquad (7.19)$$

where k is a constant.

Particle Reentrainment Rate

The particle reentrainment is not instantaneous, but takes place over a period of time. This phenomenon cannot be explained solely by the static concept. It should be considered that the reentrainment has a statistical origin associated with flow characteristics such as turbulent bursts. Using both the burst generation probability η_B and the adhesive strength distribution (or reentrainment efficiency γ), the total reentrainment efficiency $\Gamma(m)$ after fluid fluctuation m times is represented by

$$\Gamma(m) = \sum_{n=1}^{m}\sum_{s=1}^{n}{}_{n-1}C_{s-1}(1-\eta_B)^{n-s}(\eta_B\gamma)^s \qquad (7.20)$$

$$= \frac{\gamma[1-(1-\eta_B+\eta_B\gamma)^m]}{1-\gamma}$$

Using the relation $m = t/\Delta t$, where Δt is the interval of the fluctuation, $\Gamma(m)$ is rewritten as a function of elapsed time t:

$$\Gamma(t) = \frac{\gamma}{1-\gamma}\left[1 - \exp\left(\frac{-t}{T_c}\right)\right] \qquad (7.21)$$

where T_c is a time constant, which is given by

$$T_c = \frac{-\Delta t}{\ln(1 - \eta_B + \eta_B\gamma)} \qquad (7.22)$$

As $0 < \eta_B < 1$ and $0 < \gamma < 1$, T_c has a positive value. Also, the normalized total reentrainment efficiency $R(t)$ is represented by

$$R(t) = \frac{\Gamma(t)}{\Gamma(t \to \infty)} = 1 - \exp\left(\frac{-t}{T_c}\right) \qquad (7.23)$$

There are actually two types of reentrainment,[22,23] namely, short-delay and long-delay reentrainment; therefore, the reentrainment flux J is expressed as

$$J = J_S + J_L \qquad (7.24)$$

When average fluid velocity increases from $\bar{u} = 0$ at a constant rate $\alpha\ (= d\bar{u}/dt)$, the amount of reentrained particles increases with elapsed time. Taking into account the time delay (see Equation 7.23), the reentrainment fluxes $J_S(t)$ and $J_L(t)$ can be calculated by the convolution, respectively, as follows[20]:

$$J_S(t) = \frac{ak\alpha}{T_S}\int_0^t \frac{1}{(1-\gamma)^2}\frac{d\gamma}{d\overline{u'}}\exp\left(-\frac{t-t'}{T_S}\right)dt' \qquad (7.25)$$

$$J_L(t) = \frac{(1-a)k\alpha}{T_L}\int_0^t \frac{1}{(1-\gamma)^2}\frac{d\gamma}{d\overline{u'}}\exp\left(-\frac{t-t'}{T_L}\right)dt' \qquad (7.26)$$

where a is the mass ratio of the short-delay reentrainment to total reentrainment. Equation 7.25 and Equation 7.26 can be rewritten as follows:

$$J_S(\bar{u}) = \frac{ak}{T_S}\int_0^{\bar{u}} \frac{1}{(1-\gamma)^2}\frac{d\gamma}{du'}\exp\left(-\frac{\bar{u}-\bar{u}}{\alpha T_S}\right)d\bar{u} \qquad (7.27)$$

$$J_L(\bar{u}) = \frac{(1-a)k}{T_L}\int_0^{\bar{u}} \frac{1}{(1-\gamma)^2}\frac{d\gamma}{du'}\exp\left(-\frac{\bar{u}-\bar{u}}{\alpha T_L}\right)d\bar{u} \qquad (7.28)$$

When the fluid velocity increases at a constant acceleration, the reentrainment flux is calculated from Equation 7.25 and Equation 7.26, or Equation 7.27 and Equation 7.28. When the fluid velocity is maintained at $\bar{u} = \bar{u}_0$ after stopping the fluid acceleration, the reentrainment flux in the steady-state flow is calculated by substituting zero for $d\gamma/du'$.

Figure 7.5 shows the experimental and calculated values for the reentrainment flux in accelerated airflow (α = 0.01, 0.05, 0.1, and 0.6 m/s²). The reentrainment flux increases with the airflow acceleration as well as the velocity. Figure 7.6 shows the reentrainment flux in steady-state flow after

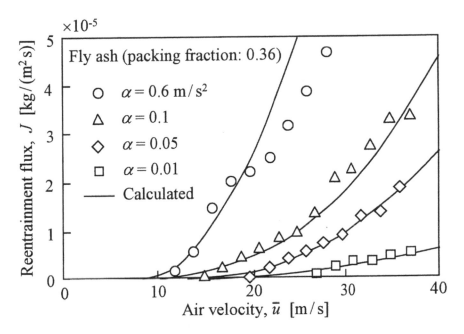

FIGURE 7.5 Reentrainment flux of aggregate particles from a particle deposition layer by accelerated air flow (τ_{c50} = 4.4 Pa, $\bar{u}_{50}$ = 29 m/s, σ_g = 2.0, k = 1.7×10⁻³ kg/m², a = 0.6, T_L = 250 s, T_S = 1/α).

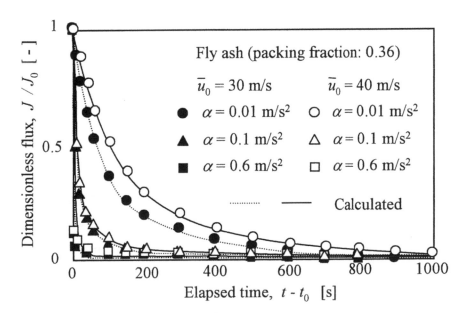

FIGURE 7.6 Dimensionless Reentrainment in steady-state flow after stopping air acceleration (τ_{c50} = 4.4 Pa, $\bar{u}_{50}$ = 29 m/s, σ_g = 2.0, k = 1.7 × 10⁻³ kg/m², a = 0.5~0.7, T_L = 200~300 s, T_S = 1/α).

stopping the air acceleration. The average air velocity is increased at a constant acceleration up to $\bar{u}_0 = 30$ or 40 m/s, and maintained for 1000 s. The reentrainment flux is normalized by the flux J_0 at $t = t_0$. The dimensionless flux J/J_0 decreases rapidly in the early stage and approaches zero gradually. The rate of decrease is larger for higher acceleration, and the time dependence of the reentrainment in the steady-state flow is still affected by the preceding unsteady-state flow (acceleration effect).

Simultaneous Particle Deposition and Reentrainment

In gas–solids pipe flow, particle deposition and reentrainment often occur simultaneously. The amount of particles deposited on the surface increases with elapsed time; however, the increasing rate gradually decreases and finally reaches a certain value, as shown in Figure 7.7. This is because aggregate particles are more readily reentrained as the amount of particles on the wall increases.[19] The variation of the mass of particles per unit area can be expressed as[24]

$$\frac{W}{A} = \left(\frac{W}{A}\right)^* \left\{ 1 - \exp\left(-\frac{t}{\tau_0}\right) \right\} \qquad (7.29)$$

where $(W/A)^*$ is the equilibrium mass of particles per unit area, and τ_0 is a time constant. The calculated lines are also shown in Figure 7.7. The value of $(W/A)^*$ depends on the adhesion and separation strengths. Figure 7.8 shows the photographs of typical particle deposition layers formed on the wall.[24] The form of the deposition layer is classified into two categories: a continuous (filmy) deposition layer covering over the wall and a striped-pattern deposition layer; the state depends on the particle diameter and the air velocity, as shown in Figure 7.9. The submicron particles form only a filmy deposition layer, and micron-sized particles form a striped deposition layer as well as a filmy deposition layer. The formation of the striped deposition layer depends on the inertia of the suspended particles.[25,26] In addition, other factors such as surface roughness[27–30] and external vibration[31] should be taken into consideration.

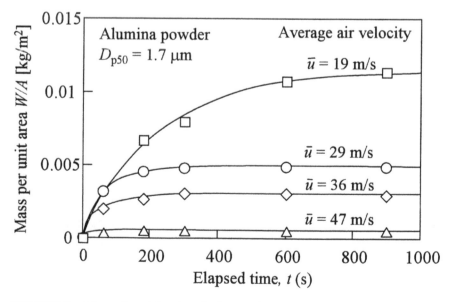

FIGURE 7.7   Mass of particles deposited per unit area as a function of time elapsed (horizontal rectangular channel: 3 mm high, 10 mm wide; particle concentration 0.05kg/m³).

Aerosol flow

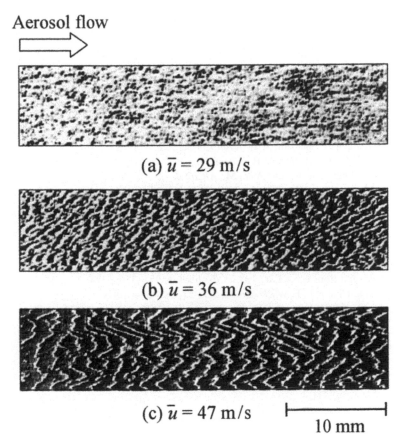

(a) $\bar{u} = 29$ m/s

(b) $\bar{u} = 36$ m/s

(c) $\bar{u} = 47$ m/s   10 mm

FIGURE 7.8   Photographs of particle deposition layers (horizontal rectangular channel 3 mm high, 10 mm wide, Alumina powder $D_{p50} = 1.7$ μm).

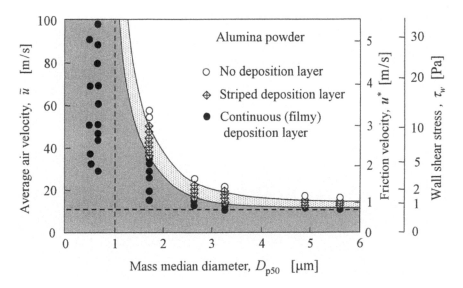

FIGURE 7.9   State diagram of particle deposition layer as a function of average air velocity and particle diameter.

REFERENCES

1. Rosner, D. E., Mackowski, D. W., Tassopoulos, M., and Castillo, J., Garcia-Ybarra, P., *Ind. Eng. Chem. Res.,* 31, 760–769, 1992.
2. Chen, S. H., *Aerosol Sci. Technol.,* 30, 364–382, 1999.
3. Cooper, D. W., Peters, M. H., and Miller, R. J., *Aerosol Sci. Technol.,* 11, 133–143, 1989.
4. Adamczyc, T., Dabros, T., Czarnecki, J., and van de Ven, T. G. M., *Adv. Colloid Interface Sci.,* 19, 183–252, 1982.
5. Shimada, M., Seto, T., and Okuyama, K., *AIChE J.,* 39, 1859–1869, 1993.
6. Tsai, R., Chang, Y. P., and Lin, T. Y., *J. Aerosol Sci.,* 29, 811–825, 1998.
7. Papavergos, P. G. and Hedley, A. B., *Chem. Eng. Res. Des.,* 62, 275–295, 1984.
8. Shimada, M., Okuyama, K., and Asai, M., *AIChE J.,* 39, 17–26, 1993.
9. Shimada, M., Okuyama, K., Okazaki, S., Asai, T., Matsukura, M., and Ishizu, Y., *Aerosol Sci. Technol.,* 25, 242–255, 1996.
10. Schmidt, F., Gartz, K., and Fissan, H., *J. Aerosol Sci.,* 28, 973–984, 1997.
11. Friedlander, S. K. and Johnstone, H. F., *Ind. Eng. Chem.,* 49, 1151–1156, 1957.
12. Sehmel, G. A., *J. Geophys. Res.,* 75, 1766–1781, 1970.
13. Sehmel, G. A., *J. Aerosol Sci.,* 4, 125–138, 1973.
14. Liu, B. Y. H. and Ilori, T. A., *Environ. Sci. Technol.,* 8:351–356, 1974.
15. Lin, C. H. and Chang, L. F. W., *J. Aerosol Sci.,* 27, 681–694, 1996.
16. Brooke, J. W., Kontomaris, K., Hanratty, T. J., and McLaulin, J. B., *Phys. Fluids A,* 4, 825–834, 1992.
17. Wang, L. P. and Maxey, M. R., *J. Fluid Mech.,* 256, 27–68, 1993.
18. He, C. and Ahmadi, G., *J. Aerosol Sci.,* 30, 739–758, 1999.
19. Ikumi, S., Wakayama, H., and Masuda, H. *Kagaku Kogaku Ronbunshu,* 12, 589–594, 1986.
20. Matsusaka, S. and Masuda, H., *Aerosol Sci. Technol.,* 24, 69–84, 1996.
21. Kousaka, Y., Okuyama, K., and Endo, Y., *J. Chem. Eng. Jpn.,* 13, 143–147, 1980.
22. Reeks, M. W., Reed, J., and Hall, D., *J. Phys. D Appl. Phys.,* 21, 574–589, 1988.
23. Wen, H. Y., Kasper, G., and Udischas, R., *J. Aerosol Sci.,* 20, 923–926, 1989.
24. Matsusaka, S., Theerachaisupakij, W., Yoshida, H., and Masuda, H., *Powder Technol.,* 118, 130–135, 2001.
25. Matsusaka, S., Adhiwidjaja, I., Nishio, T., and Masuda, H., *Adv. Power Technol.,* 9, 207–218, 1998.
26. Theerachaisupakij, W., Matsusaka, S., Akashi, Y., and Masuda, H., *J. Aerosol Sci.,* 34, 261–274, 2003.
27. Adhiwidjaja, I., Matsusaka, S., Tanaka, H., and Masuda, H., *Aerosol Sci. Technol.,* 33, 323–333, 2000.
28. Reeks, M. W. and Hall, D., *J. Aerosol Sci.,* 32, 1–31, 2001.
29. Ziskind, G., Fichman, M., and Gutfinger, C., *J. Aerosol Sci.,* 31, 703–719, 2000.
30. Soltani, M. and Ahmadi, G., *J. Adhesion,* 51, 105–123, 1995.
31. Theerachaisupakij, W., Matsusaka, S., Kataoka, M., and Masuda, H., *Adv. Powder Technol.,* 13, 287–300, 2002.

2.8 Agglomeration (Coagulation)

Kikuo Okuyama

Hiroshima University, Higashi-Hiroshima, Japan

Ko Higashitani

Kyoto University, Katsura, Kyoto, Japan

2.8.1 IN GASEOUS STATE

Coagulation of aerosols, which is used to describe the growing process of aerosol particles in contact with each other, causes a continuous change in number, concentration, and size distribution of agglomerates, keeping the total particle volume constant. Of the various coagulations classified by the kinds of force to cause collision, Brownian coagulation (thermal coagulation) may be the most familiar and fundamental. Others are gradient coagulation, turbulent coagulation, electrostatic coagulation, acoustic coagulation, coagulation due to the velocity difference under gravity or centrifugal force, and so on.

When two particles having diameters D_{pi} and D_{pj} collide, the number of collisions per unit time per unit volume can be written as

$$N = K\left(D_{pi}, D_{pj}\right) n_i n_j \tag{8.1}$$

where n_i and n_j are the number concentrations of particles i and j, respectively, and $K(D_{pi}, D_{pj})$ is the coagulation rate function or the collision rate between particles of diameters D_{pi} and D_{pj}.

In the special case of the initial stage of coagulation of a monodisperse aerosol having uniform diameter D_p, the decreasing rate of particle number concentration n can be given from Equation 8.1 as

$$\frac{dn}{dt} = -0.5 K_0 n^2 \tag{8.2}$$

where $K_0 = K(D_p, D_p)$. When the coagulation rate function is not a function of time, the decrease in particle number concentration from n_0 to n can be obtained from the integration of Equation 8.2 over a time period from zero to t:

$$n = \frac{n_0}{1 + 0.5 K_0 n_0 t} \tag{8.3}$$

Because the particle volume is kept constant during the coagulation process, the average particle diameter D_p based on particle volume after the period of time t can be given by

$$D_p = D_{p0}\left(1 + 0.5 K_0 n_0 t\right)^{1/3} \tag{8.4}$$

However, in the case of coagulation of a polydisperse aerosol, the basic equation that describes the time-dependent change in particle number concentration of volume v at time t, $n(v,t)dv$, can be given as

$$\frac{\partial n(v,t)}{\partial t} = \frac{1}{2}\int_0^v K(v', v-v')\, n(v',t)\, n(v-v',t)\, dv'$$
$$-n(v,t)\int_0^\infty K(v,v')\, n(v',t)\, dv' \tag{8.5}$$

The first term on the right-hand side represents the rate of formation of particles of volume v due to coagulation, and the second represents the rate of loss of particles of volume v by coagulation with all other particles.

The time-dependent change in number concentration and size distribution of aerosol particles undergoing coagulation can basically be obtained by solving Equation 8.5 for a given initial particle size distribution. In the derivation of Equation 8.5, however, the following assumptions are made: (1) particles are electrically neutral, and (2) particles are spherical and collide with each other to form another spherical particle whose mass is the same as the combined mass of the two smaller particles. Accordingly, Equation 8.5 cannot be applied to evaluate the coagulation of solid particles as particle fusion takes place slowly after the collision. In order to consider the nonspherical effect of agglomerated particles on coagulation, the population balance equation for collision and sintering has recently been derived.[1] Although their model predicted the average primary particles sizes of agglomerates, the effect of agglomerate structure on the collision rate was neglected by using the collision rate of spherical particles having an equivalent volume of the agglomerate.[2]

Brownian Coagulation

For aerosol particles of submicron diameter, Brownian coagulation by Brownian motion of particles is essential for the characterization of the behavior of aerosols. In the general case, the rate function of Brownian coagulation is characterized by the Knudsen number $Kn = 2l/D_p$, where l is the mean path of the background gas.

In the continuum regime, having very small values of Kn, Smoluchowski[3] applied the theory of Brownian diffusion to the statistical problem of the collision of particles thermally agitated in a continuum.

In the diffusion process where a certain particle (radius r_i and diffusion coefficient D_i) among other particles is fixed, the surrounding particles (radius r_j and diffusion coefficient D_j) collide with the fixed particle due to their Brownian motion, as shown in Figure 8.1. The number of particle r_j diffusing in unit time is

$$N = 4\pi r_{ij}^2 D_j \left(\frac{\partial n}{\partial r}\right)_{r=r_{ij}} \tag{8.6}$$

where $r_{ij} = r_i + r_j$ is the distance between the centers of the two particles at the moment of contact, and n is the concentration of particles r_j at time t as a function of distance r from particle r_i.

The value of the concentration gradient $\partial n/\partial r$ can be determined by solving the diffusion equation in spherical coordinates. For submicron particles, this concentration instantaneously shifts to the stationary one by the following equation:

$$n = n_j \left(1 - \frac{r_{ij}}{r}\right) \tag{8.7}$$

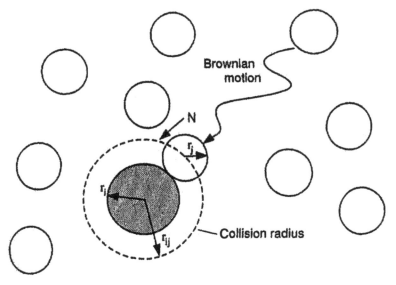

N : Diffusion flux

FIGURE 8.1 Brownian coagulation by Brownian diffusive deposition.

Substitution of this equation to Equation 8.6 gives

$$N = 4\pi D_j r_{ij} n_j \tag{8.8}$$

Because the central fixed particle is also in Brownian movement, Equation 8.8 leads to the next equation by adding the Brownian diffusion coefficient D_i of particle r_i.

$$N = 4\pi \left(D_i + D_j\right)\left(r_i + r_j\right) n_j \tag{8.9}$$

Accordingly, the Brownian coagulation rate function in the continuum regime can be given from its definition as Equation i in Table 8.1, which lists the available coagulation function $K_B(D_{pi}, D_{pj})$ for Brownian coagulation.[4,5] Figure 8.2 shows the coagulation rate functions of monodispersed particles, $0.5K_B(D_p, D_p)$, as a function of particle diameter in air at ambient conditions. There exists a distinct maxima in the size range from 0.01 to 0.1 μm, depending on particle diameter.[6]

At the present stage, for Brownian coagulation in the transition regime, Fuchs' interpolation formula (Equation ii in Table 8.1) is considered to be reasonable due to the better agreement with more rigorous theories. Furthermore, the effect of agglomerate structure on the collision kernel is incorporated by replacing the particle diameter with the collision diameter.[2,7] The collision diameter of agglomerate, $D_{c,i}$ with a primary particle diameter, $d_{p,i}$ and the number of primaries per an agglomerate, $n_{p,i}$ is given by

$$D_{c,i} = d_{p,i} \left(\frac{v_i}{v_{p,i}}\right)^{1/D_f} = d_{p,i} \left(n_{p,i}\right)^{1/D_f} \tag{8.10}$$

TABLE 8.1 Available Expressions for Brownian Coagulation Rate Function

Continum Regime $Kn < 0.1$	Transition Regime $0.1 \leq Kn \leq 10$	Free-Molecular Regime $Kn > 10$
(i) $2\mathrm{p}\left(D_i + D_j\right)\left(D_{pi} + D_{pj}\right)$ $\dots\dots(i)$	(ii) $2\pi\left(D_i + D_j\right)\left(D_{pi} + D_{pj}\right)$ $\times\left[\dfrac{D_{pi} + D_{pj}}{D_{pi} + D_{pj} + 2g_{ij}} + \dfrac{8(D_i + D_j)}{\bar{v}_{ij}(D_{pi} + D_{pj})}\right]^{-1}$ $\dots\dots(ii)$	(iii) $\dfrac{\pi}{4}\left(D_{pi} + D_{pj}\right)^2 \bar{v}_{ij}$ $\dots\dots(iii)$

k = Boltzmann constant ($= 1.38 \times 10^{-23}$ J/K), T = temperature

$$D_i = \frac{kT}{3\pi\mu D_{pi}}$$

$$D_i = \frac{C_c kT}{3\pi\mu D_{pi}}$$

$$C_c = 1 + Kn\left[1.257 + 0.4\exp\left(\frac{-1.1}{Kn}\right)\right]$$

$$g_{ij} = \left(g_i^2 + g_j^2\right)^{0.5}$$

$$g_i = \frac{\left(D_{pi} + l_i\right)^3 - \left(D_{pi}^2 + l_i^2\right)^{1.5}}{3D_{pi}l_i} - D_{pi}$$

$$l_i = \frac{8D_i}{\pi\bar{v}_i}$$

$$\bar{v}_{ij} = \left(\bar{v}_i^2 + \bar{v}_j^2\right)^{0.5}$$

$$\bar{v}_i = \left(\frac{8kT}{\pi m_i}\right)^{0.5}$$

$$m_i = \frac{\pi}{6}D_{pi}^3\rho_p$$

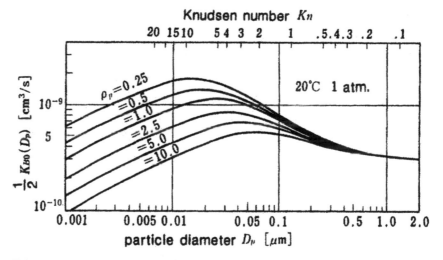

FIGURE 8.2   Brownian coagulation rate against particle diameter.

and

$$d_{p,i} = \frac{6v_i}{a_i} \tag{8.11}$$

where $v_{p,i}$ is the volume of a spherical primary particle in agglomerate size i, and D_f is the mass fractal dimension. However, the enhancement of coagulation rate due to the influence of van der Waals or the dispersion force between colliding particles was quantitatively explained by Marlow's theory.[8]

In order to evaluate the change in particle size distribution of aerosols, the following three types of solutions have been obtained: (1) analytical solution, (2) asymptotic solution, and (3) numerical solution. Table 8.2 shows representative analytical and approximate solutions. In a series of reports, Friedlander[9] developed the similarity theory for the uniformity in shape of observed size spectra of atmospheric aerosols. The asymptotic size distribution obtained is called the self-preserving distribution function (SPDF). This similarity theory evaluates the size spectra only after a sufficiently

---

TABLE 8.2   Representative Analytical and Approximate Solutions for Brownian Coagulation

(1) $Kn < 0.01$, initially monodispersed particles[a]

$$n_k = n_0 \left(\frac{t}{t_B}\right)^{k-1} \left[\left(1+\frac{t}{t_B}\right)^{k+1}\right]^{-1} \quad \text{where } \tau_B = \frac{1}{K_{B0}n_0}, \; K_{B0} = 0.5 K_B \left(D_p, D_p\right)^a$$

(2) $Kn > 10$, initially monodispersed particles[b]

$$n = 0.1624 \alpha^{-6/5} \left(kT\right)^{-3/5} \left(\frac{Mn_0 t^6}{\rho_p^4 N_{av}}\right)^{-1/5}, \; d_v = 2.274 \alpha^{2/5} \left(\frac{kT}{\rho_p^4}\right)^{1/5} \left(\frac{Mn_0 t}{N_{av}}\right)^{2/5}$$

(3) $Kn < 0.1$, initially polydispersed particles[c]

$$\frac{n}{n_0} = \frac{1}{A}, \frac{v_g}{v_{g0}} = A \exp\left(4.5 \ln^2 \sigma_{g0}\right) \left[2 + \frac{\exp\left(9 \ln^2 \sigma_{g0}\right)-2}{A}\right]^{-1/2}, \; \ln^2 \sigma_g = \frac{1}{9} \ln\left(2 + \frac{\exp\left(9\ln^2 \sigma_{g0}\right)-2}{A}\right)$$

where $A = 1 + \left\{1 + \exp\left(\ln^2 \sigma_{g0}\right)\right\} 0.5 K_{B0} n_0 t$, $K_{B0} = 0.5 K_B\left(D_p, D_p\right)$

(4) $Kn > 10$, initially polydispersed particles[d]

$$\frac{n}{n_0} = \frac{1}{A^{6/5}}, \frac{v_g}{v_{g0}} = E^{9/2} A^{6/5} \left(2 + \frac{E^{15/2}-2}{A}\right)^{-3/5}, \; \sigma_g - \exp\left[\frac{2}{15} \ln\left(2 + \frac{E^{15/2}-2}{A}\right)\right]^{1/2}$$

where $A = 1 + \left(\frac{5}{8}\right) H \tau$, $\tau = \left(6kT/\rho_p\right)^{1/2} r_{g0}^{1/2} n_0 t$, $H = E^{1/8} + 2E^{5/8} + E^{25/8}$, $E = \exp\left(z_0\right)$, $z_0 = \ln^2 \sigma_{gc}$

(5) $Kn > 10$, for fractal agglomerates[e]

$$\frac{n}{n_0} = \left[1 + \left(\frac{3D_f - 4}{2D_f}\right) H_0 bK v_{g0}^{\frac{4-D_f}{2D_f}} n_0 t\right]^{-\frac{2D_f}{3D_f-4}},$$

$$\frac{v_g}{v_{g0}} = \exp\left(\frac{9}{2}\ln^2 \sigma_0\right) \left\{1 + \left(\frac{3D_f - 4}{2D_f}\right) H_0 bK v_{g0}^{\frac{4-D_f}{2D_f}} n_0 t\right\}^{\frac{2D_f}{3D_f-4}} \left[2 + \frac{\exp\left\{\frac{9(3D_f-4)}{2D_f}\ln^2 \sigma_0\right\}-2}{1 + \frac{3D_f - 4}{2D_f} H_0 bK v_{g0}^{\frac{4-D_f}{2D_f}} n_0 t}\right]^{-\frac{D_f}{3D_f-4}},$$

$$\ln^2 \sigma = \frac{2D_f}{9(3D_f - 4)} \ln\left[2 + \frac{\exp\left\{\frac{9(3D_f-4)}{2D_f}\ln^2 \sigma_0\right\}-2}{1 + \frac{3D_f - 4}{2D_f} H_0 bK v_{g0}^{\frac{4-D_f}{2D_f}} n_0 t}\right]$$

(continued)

TABLE 8.2   Representative Analytical and Approximate Solutions for Brownian Coagulation (Continued)

where $H = \exp\left\{\frac{9}{2}\left(\frac{4}{D_f^2} - \frac{2}{D_f} + \frac{1}{4}\right)\ln^2\sigma\right\} + 2\exp\left\{\frac{9}{2}\left(\frac{2}{D_f^2} - \frac{1}{D_f} + \frac{1}{4}\right)\ln^2\sigma\right\} + \exp\left\{\frac{9}{2}\left(\frac{4}{D_f^2} + \frac{1}{4}\right)\ln^2\sigma\right\},$

$K = \left(\frac{3}{4\pi}\right)^{\frac{2}{D_f} - \frac{1}{2}}\left(\frac{6kT}{\rho_p}\right)^{1/2} r_p^{2 - 6/D_f}$

---

n, particle number concentration (m^{-3})

r_g, geometric mean radius

r_p, primary particle radius (m)

n_k, number concentration of particle containing k initial particles

v_g, geometric mean volume [$= \pi d_{pg}^3/6$, d_{pg} : geometric mean diameter (m)]

σ_g, geometric standard deviation

k, Boltzmann constant (1.38×10^{-23} J/K)

D_f, mass fractal dimension

T, absolute temperature (K)

ρ_p, particle density (kg/m^3)

t, time (s)

α, sticking probability between collision

N_{av}, Avagadro's number

M, molecular weight

0, initial value

b, values depend on σ_{g0} and D_f^e.

[a] Friedlander, S. K., *Smoke, Dust, and Haze: Fundamentals of Aerosol Dynamics*, 2nd Ed., Oxford University Press, New York, 2000, pp. 188–196.

[b] Ulrich, G. D., *Combust. Sci. Technol.*, 4, 47–58, 1971.

[c] Lee, K. W., *J. Colloid Interface Sci.*, 92, 315–325, 1983.

[d] Lee, K. W., Curtis, L. A., and Chen, H., *Aerosol Sci. Technol.*, 12, 457–462, 1990.

[e] Park, S. H. and Lee, K. W., *J. Colloid Interface Sci*, 246, 85–91, 2002.

long time has passed. Figure 8.3 depicts the time-dependent change of particle size distribution obtained numerically solving Equation 8.5.[10]

Laminar and Turbulent Coagulation

When aerosol particles are suspended in a laminar shear flow, particles tend to coagulate due to their relative motion, and the coagulation rate function can be given as follows[3]:

$$K_L\left(D_{pi}, D_{pj}\right) = 0.17\eta_L\left(D_{pi} + D_{pj}\right)^3\left|\frac{du}{dx}\right| \tag{8.12}$$

where du/dx is the velocity gradient of the flow and η_L is the collision efficiency, which expresses the effect of the aerodynamic interaction between particles.

The collisions between particles suspended in a turbulent flow are generally considered to be caused by the two essentially independent mechanisms. In the first mechanism, particles may collide with each other due to the velocity difference between particles caused by the spatial

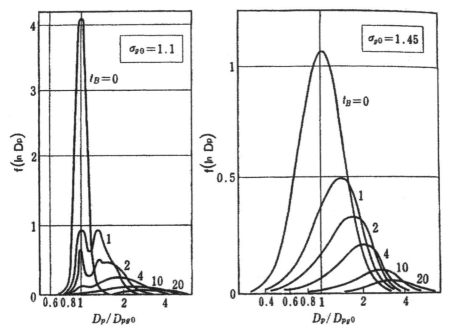

FIGURE 8.3   Time-dependent changes in size distribution of particles undergoing Brownian coagulation.

nonhomogeneity characteristic of turbulent flow. The theoretical rate function by Saffman and Turner[11] is

$$L_{T1}\left(D_{pi}, D_{pj}\right) = 0.16\eta_T \left(D_{pi} + D_{pj}\right)^3 \left(\frac{\varepsilon_0}{v}\right)^{1/2}$$ (8.13)

where ε_0 is the energy-dissipation per unit mass of fluid, v the kinematic viscosity, and η_T the collision efficiency. The second mechanism may be caused by a relative motion of each particle to the local turbulent motion of the fluid, because the inertia of a particle is not the same between unequal-sized particles. Saffman and Turner[11] also obtained the following equation:

$$K_{T2}\left(D_{pi}, D_{pj}\right) = 1.43\eta_T \left(D_{pi} + D_{pj}\right)^2 \left|\tau\left(D_{pi}\right) - \tau\left(D_{pj}\right)\right| \left(1 - \frac{\rho_f}{\rho_p}\right)\left(\frac{\varepsilon_0^3}{v}\right)^{0.25}$$ (8.14)

where $\tau(D_p)$ is the particle relaxation time ($= \rho_p C_c D_p^2/18\mu$). Figure 8.4 shows the turbulent coagulation function. It is seen that the turbulent coagulation plays an important role in the coagulation of aerosol particles larger than a few microns even under the small value of ε_0.[12]

Acoustic Coagulation

In an acoustic field, particles coagulate mainly by the following three mechanisms: (1) orthokinetic coagulation due to relative oscillating motion, (2) coagulation due to hydrodynamic attractive force between particles, and (3) turbulent coagulation due to acoustically induced turbulence in a highly intensive acoustic field. Shaw[13] has published an excellent review on acoustic coagulation.

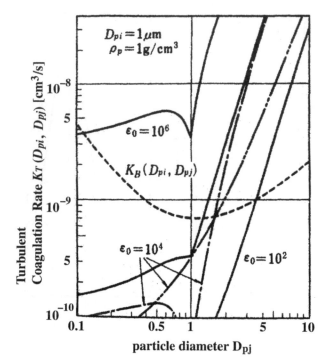

FIGURE 8.4   Turbulent coagulation rate against particle diameter. Solid line, Equation 8.13 and Equation 8.14; dotted–dashed line, Equation 8.14; dashed line, Equation 8.13.

2.8.2   IN LIQUID STATE

The fundamental behavior of particles in liquids is very similar to that in gases, except that (1) the viscosity and permitivity of the medium are much greater, (2) the charge of particles is related closely with their stability, and (3) the effect of particle inertia is usually negligible.

The motion of a particle in a solution is expressed by

$$m\ddot{\mathbf{r}} = f_{T} + f_{B} + f_{F} \tag{8.15}$$

where m and $\ddot{\mathbf{r}}$ are the mass and acceleration of the particle; f_{T}, the total static interaction force with other particles; f_{B}, the force due to the Brownian motion; and f_{F}, the force generated by the flow of the medium. Subtraction of equations between an arbitrary pair of particles gives an equation of the relative motion as follows. Here the effect of particle inertia is neglected.

$$F_{T} + F_{B} + F_{F} = 0 \tag{8.16}$$

where F_{T}, F_{B}, and F_{F} are the corresponding forces acting relatively between particles. F_{T} is the static forces, such as the electrical repulsive force and the van der Waals attractive force. F_{B} and F_{F} are the Brownian and hydrodynamic forces caused by the relative motion of particles respectively, which may be named the dynamic interaction force. If Equation 8.16 is able to be solved, the relative motion of particles will be available to calculate the coagulation rate.

Static Interaction and Coagulation (DLVO Theory)

In order to calculate the coagulation process quantitatively, F_{T}, F_{B} and F_{F} must be known. However, as far as the stability of a suspension is concerned, it may be estimated using the static interaction force $F_T = |F_T|$ only. For example, if F_T is only attractive, one will find that the particles are unstable and, if the maximum value of F_T is very large, the particles must be free from coagulation, even though collisions between particles occur. In this section, the relationship between the static interaction and particle stability is discussed, employing the interparticle potential energy V_T, given by $F_T = -dV_T/dr$, where r is the distance between particle centers.

The charge of particles in aqueous solutions makes them repulsive to each other as explained in Section 2.5.3. The repulsive force between similar particles of radius a is explicitly given by the following equations, when the surface potential, ψ_d, is sufficiently low (say, $\psi_d < 25$ mV)[14]:

$$V_R = \pm 2\pi\varepsilon a\psi_d^2 \ln\{1 \pm \exp[-\kappa a(s-2)]\} \ (\kappa a > 10) \tag{8.17}$$

$$V_R = 4\pi\varepsilon a\psi_d^2 \exp[-\kappa a(s-2)]/s \ (\kappa a < 5) \tag{8.18}$$

where ε is the permitivity of the medium; s, the dimensionless distance between particle centers normalized by a; $\kappa = (2n_0 z^2 e^2/\varepsilon kT)^{0.5}$; n_0, the ionic concentration of bulk solution; z, the ionic valency; e, the elementary charge; k, the Boltzmann constant; and T, the temperature. κ is a measure of the electrolyte concentration. The $+$ sign in Equation 8.17 indicates that the equation is for the constant surface potential, and the $-$ sign indicates that the equation is for the constant surface charge. Equation 8.18 is applicable for both cases.

The instantaneous dipole moment generated by the fluctuation of electrons around an atomic nucleus polarizes the neighboring atoms and generates the attractive force between atoms. This interaction force is called the disperson force. Permanent dipoles also generate the interaction force. The sum of these forces is called the van der Waals force. Because the contribution by the permanent dipole is small in most cases, the dispersion force is considered to be the van der Waals force. Integrating the contribution due to all the atoms in particles, the van der Waals force between spherical particles is given by

$$V_A = \left(\frac{A}{6}\right)\left[\frac{2}{s^2-4} + \frac{2}{s^2}\ln\left(\frac{s^2-4}{s^2}\right)\right] \tag{8.19}$$

where A is the Hamaker constant. The value of A is evaluated either theoretically or experimentally.[15] The value of A between dissimilar particles 1 and 2, in a medium, 3, may be estimated by

$$A_{132} = \left(\sqrt{A_{11}} - \sqrt{A_{33}}\right)\left(\sqrt{A_{22}} - \sqrt{A_{33}}\right) \tag{8.20}$$

where A_{ij} is the Hamaker constant of the material, i, in free space.

When particle surfaces are so close that the electron clouds overlap, a very strong Born repulsion force appears. It is said that the Born potential, V_{Born}, is infinitely large when the separation distance is a few angstroms (say, 4Å) and zero elsewhere.

The total interaction potential, V_T, is given by the sum of these potentials:

$$V_T = V_R + V_A + V_{Born} \tag{8.21}$$

A typical potential curve is illustrated in Figure 8.5a. When the energy given by the Brownian or fluid motion overcomes the maximum potential energy, particles will coagulate. Otherwise, particles will be either dispersed or coagulated loosely in the secondary minimum even if particles collide with each other.

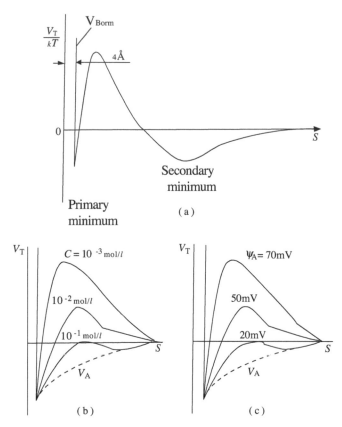

FIGURE 8.5 Interparticle potential and the dependence on electrolyte concentration and surface potential.

Most parameters in Equation 8.21 are not variable for a given suspension, but the values of κ and ψ_d may vary by changing the electrolyte concentration and the pH of the solution, respectively. As illustrated in Figure 8.5b and 8.5c, the maximum value of the potential decreases as the value of κ, that is, the electrolyte concentration, increases and also as the absolute value of ψ_d decreases. At a sufficiently high electrolyte concentration, the peak of the potential curve disappears. This indicates that every collision results in coagulation. This electrolyte concentration is called the critical coagulation concentration (CCC). Coagulation where the electrolyte concentration is higher than the CCC is called rapid coagulation, whereas coagulation where the concentration is lower than the CCC is called slow coagulation. The theoretical value of CCC for particles with a relatively high surface potential in an aqueous solution of 25°C is given by

$$CCC(mol/L) = \frac{87 \times 10^{-40}}{z^6 A^2} \tag{8.22}$$

This indicates that the CCC is very sensitive to the ionic valency and, therefore, so is the stability. The correlation has been confirmed experimentally and is known as "Shultze–Hardy rule."

The stability of a suspension is controllable also by changing the value of ψ_d. The variation of ψ_d depends on the mechanism of surface charge. ψ_d varies with the pH of the medium in the case of oxide particles, and with the concentration of the potential determining ions in the case of the Nernst-type particles, as described in Section 2.5.3.

Various surfactants and polymers are possibly employed to control the stability of particles, but these methods are discussed in Section 4.5.2.

Dynamic Interaction and Coagulation Rate

Particles interact with each other through their Brownian motion and the fluid motion. As explained earlier, the stability of particles can be evaluated roughly by the static interaction, but the dynamic interactions should be taken into account to know the stability quantitatively and to calculate the coagulation rate.

Brownian Coagulation and Stability Ratio

In the case of rapid coagulation, every collision due to the Brownian motion results in coagulation. The rate of the Brownian coagulation between particles i and j in a stationary medium, J_{ij}, is given by[16]

$$J_{ij} = \left[\frac{2KT}{3\mu}\right](a_i + a_j)\left(\frac{1}{a_i} + \frac{1}{a_j}\right)n_i n_j \left(\int_{a_i+a_j}^{\infty} \frac{\exp(V_{Tij}/kT)}{r^2 dr}\right)^{-1} \qquad (8.23)$$

where a_i and n_i are the radius and number concentration of particle i, and V_{Tij} is the interparticle potential between particles i and j. Then the change of particle concentration is calculated by the population balance equation:

$$\frac{dn_k}{dt} = \frac{1}{2}\sum_{\substack{i=1 \\ i+j=k}}^{k-1} J_{ij} - \sum_{i=1}^{\infty} J_{ik} \qquad (8.24)$$

Under the condition that $V_{Tij} = 0$ and the particles are initially monodispersed, substitution of Equation 8.23 into Equation 8.24 gives the following equations for total and i-fold particle concentrations at times t, N_t, and n_i, respectively:

$$\frac{1}{N_t} - \frac{1}{N_0} = \frac{t}{N_0 t_{1/2}} \qquad (8.25)$$

$$n_i = \frac{N_0(t/t_{1/2})^{i-1}}{(1+t/t_{1/2})^{i+1}} \qquad (8.26)$$

where N_0 is the initial total particle concentration and $t_{1/2}$ is $3\mu/4kTN_0$. Equation 8.25 indicates $1/N_t$ varies linearly with time. It is found that this theory overestimates the rapid coagulation rate by about 40%, and this is because no hydrodynamic interaction is taken into account.[17] Because the van der Waals force and the hydrodynamic interaction act even in rapid coagulation, the correction factor representing these effects, α_B (= α_{Bii}), can be calculated by Equation 8.16 as shown in Figure 8.6.[19] Then the change of n_i can be evaluated numerically as shown in Figure 8.7, and N_t is expressed analytically by

$$\frac{dN_t}{dt} = -\left(\frac{4\alpha_B kT}{3\mu}\right)N_t^2 \qquad (8.27)$$

In the case of the slow coagulation ($V_{Tij} \neq 0$), the prediction of the coagulation rate is not easy, but the stability of suspensions is evaluated by the so-called stability ratio W. W is a measure of the

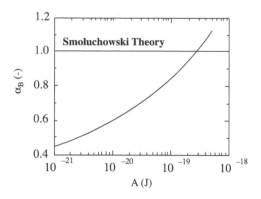

FIGURE 8.6   Dependence of α_B on the Hamaker constant A.

deviation from the rapid coagulation at the initial stage of coagulation where the coagulation between particles of the same size is dominant:

$$W = \frac{J_0^R}{J_0^S} \tag{8.28}$$

where the superscripts R and S indicate the rapid and slow coagulation, the subscript 0 indicates $t = 0$, and $J = J_{11}$. Then, the theoretical value of W is derived from Equation 8.23 as

$$W = 2a \int_{2a}^{\infty} \frac{\exp(V_{T11}/kT)}{r^2} dr \tag{8.29}$$

The experimental value of W is determinable using the following equation from the initial change of the turbidity, τ, of suspensions[19]:

$$W = \left(\frac{d\tau}{dt}\right)_0^R \bigg/ \left(\frac{d\tau}{dt}\right)_0^S \tag{8.30}$$

It is known that the theoretical and experimental values of W agree qualitatively but not quantitatively. This discrepancy is one of the unsolved problems of colloidal phenomena.[16]

Coagulation in Shear Flow

Particles in flow fields collide with each other because of their relative velocity. It depends on the balance between the energy of particles given by the flow and the interparticle potential energy whether the collision results in coagulation or not.

The simplest model for the shear coagulation was proposed by Smoluchowski,[20] in which it is assumed that $F_T = F_B = F_F = 0$, but particles collide because of the geometrical interception. In this case, the collision rate is given by

$$J_{ij} = \frac{4}{3}\gamma R_{ij}^3 n_i n_j \tag{8.31}$$

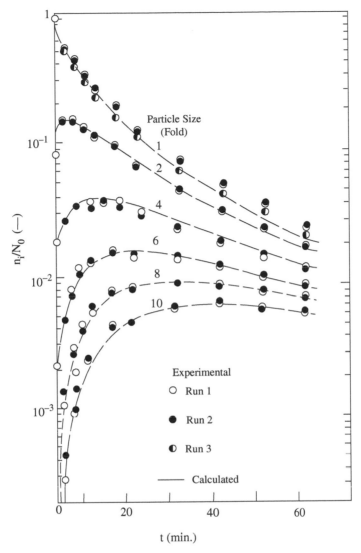

FIGURE 8.7 Comparison of n_i versus t between theoretical prediction and experimental results of Brownian coagulation.

Where γ is the shear rate and R_{ij} ($= a_i + a_j$) is the coagulation radius. Then, the total particle concentration is given by

$$\frac{dN_t}{dt} = -\left(\frac{4\gamma\phi}{\pi}\right)N_t \qquad (8.32)$$

where ϕ is the volume fraction of particles. The trajectory assumed here is valid for particles with a large inertia, such as particles in air. But in the case of particles in liquid, the effect of the inertia is small and the hydrodynamic interaction plays an important role so that the trajectory becomes complicated. The trajectory for this case was solved by Equation 8.16 with $F_B = 0$, and the correction factor, α_s, for the rapid shear coagulation rate was obtained numerically as shown in Figure 8.8.[21] If the contributions of the Brownian and shear coagulations are additive, the change of N_t is given by

$$\frac{dN_t}{dt} = -\left(\frac{4\alpha_B kT}{3\mu}\right)N_t^2 - \left(\frac{4\alpha_s \gamma f}{\pi}\right)N_t \tag{8.33}$$

This equation is easily solved with the proper initial condition, and it is found that the prediction agrees with the experimental results quantitatively.[22]

Taking all the contributions of the hydrodynamic interaction and interaction potentials into account except that of the Brownian motion, the relative trajectory of a pair of particles in the shear flow was calculated by Equation 8.16, and the regions of coagulation and dispersion between equal spherical particles are clarified, as shown in Figure 8.9.[23] This diagram indicates that the stability of suspensions is determined by the balance between the static and dynamic interactions.

Coagulation in Turbulent Flow

It is not easy to estimate the coagulation rate of particles in turbulent flow because of the complicated flow field. Saffman and Turner[24] derived an expression for the rapid coagulation rate in a turbulent

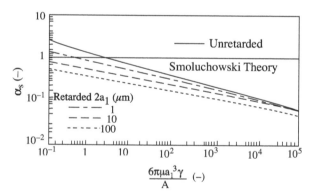

FIGURE 8.8 Dependence of α_s on the dimensionless parameter $(6\pi\mu a_1 3\,\gamma/A)$. The solid curve was obtained by the unretarded potential of Equation 8.19 and the other curves by the retarded potentials 16.

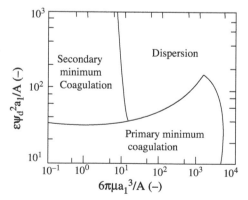

FIGURE 8.9 Regions of coagulation and dispersion.

flow without taking the hydrodynamic interaction into account. Their expression was modified by introducing the correction factor, α_s, for the rapid shear coagulation rate as follows[25]:

$$J_{ij} = 1.429\alpha_s R_{ij}^3 \left(\frac{\Phi}{v} \right) n_i n_j \tag{8.34}$$

where Φ is the energy dissipation per unit mass of fluid and v is the kinetic viscosity of the medium. Substituting Equation 8.34 into Equation 8.24, the change of the particle size distribution can be predicted. It is confirmed that the prediction by Equation 8.34 coincides with experimental results quantitatively.

REFERENCES

1. Koch, W. and Friedlander, S. K., *J. Colloid Interface Sci.*, 140, 419–427, 1990.
2. Tsantilis, S. and Pratsinis, S. E., *AIChE J.*, 46, 407–415, 2000.
3. Smoluchowski, M. W., *Phys. Z.*, 17, 557–571, 1916.
4. Fuchs, N. A., *The Mechanics of Aerosols*, Dover, New York, 1964, pp. 288–302.
5. Sitarski, M. and Seinfeld, J. H., *J. Colloid Interface Sci.*, 61, 261–271, 1977.
6. Okuyama, K., Kousaka, Y., and Hayashi, K., *J. Colloid Interface Sci.*, 101, 98–109, 1984.
7. Kruis, F. E., Kusters, K. A., and Pratsinis, S. E., *Aerosol Sci. Technol.*, 19, 514–526, 1993.
8. Marlow, W. H., *J. Chem. Phys.*, 73, 6284–6287, 1980.
9. Friedlander, S. K., *Smoke, Dust, and Haze: Fundamentals of Aerosol Dynamics*, 2nd Ed., Oxford University Press, New York, 2000, pp. 210–219.
10. Yoshida, T., Okuyama, K., Kousaka, Y., and Kida, Y., *J. Chem. Eng. Jpn.*, 8, 317–322, 1975.
11. Saffman, P. G. and Turner, J. S., *J. Fluid Mech.*, 1, 16–30, 1956.
12. Okuyama, K., Kousaka, Y., and Yoshida, T., *J. Aerosol Sci.*, 9, 399–410, 1978.
13. Shaw, D. T., *Recent Development in Aerosol Science*, Wiley, New York, 1978.
14. Hunter, R. J., *Foundations of Colloid Science*, Vol. 1, Clarendon Press, Oxford, 1987.
15. Israelachvili, N. J., *Intermolecular and Surface Force*, Academic Press, London, 1985.
16. Kruyt, H. R., *Colloid Science*, Elsevier, Amsterdam, 1952.
17. Higashitani, K. and Matsuno, Y., *J. Chem. Eng. Jpn.*, 12, 460, 1979.
18. Higashitani, K., Tanaka, T., and Matsuno, Y., *J. Colloid Interface Sci.*, 63, 551, 1978.
19. Troelstra, S. A. and Kruyt, H. R., *Kolloid-Beihefte*, 54, 225, 1942.
20. Smoluchowski, M., *Z. Phys. Chem.*, 92, 129, 1917.
21. Higashitani, K., Ogawa, R., Hosokawa, G., and Matsuno, Y., *J. Chem. Eng. Jpn.*, 15, 299, 1982.
22. Higashitani, K., *Gyoushu Kougaku*, Nikkankogyo Shinbunsha, Tokyo, 1982, p. 53.
23. Zeichner, G. R. and Showalter, W. R., *AIChE J.*, 23, 243, 1977.
24. Saffman, P. G. and Turner, J. S., *J. Fluid Mech.*, 1, 16, 1956.
25. Higashitani, K., Yamauchi, K., Matsuno, Y., and Hosokawa, G., *J. Chem. Eng. Jpn.*, 16, 299, 1983.

2.9 Viscosity of Slurry

Hiromoto Usui
Kobe University, Nada-ku, Kobe, Japan

2.9.1 INTRODUCTION

The technical term *slurry* contains quite a wide concept of solid–liquid mixtures. If a solid–liquid mixture has some fluidity, we call it a slurry. However, suspended solid particles vary in size from very fine colloidal particles to sedimentable coarse particles. Also, the surface characteristics change according to the combination of solid material and liquid phase. Therefore, the flow characteristics of slurry change widely. The solid concentration greatly affects the viscosity of slurry.

In addition to the effect of solid concentration, the effects of particle size distribution, the shape, and the surface properties of particles on the viscosity of slurry are significant. In this section, viscosity behavior of a wide variety of slurries is described, and a brief description of the measuring technique of slurry viscosity is also given.

2.9.2 BASIC FLOW CHARACTERISTICS

The representative flow models for various kinds of slurries are shown in Figure 9.1, in terms of the plot of shear stress τ versus shear rate (du/dr).[1] Figure 9.1a represents the Newtonian model,

$$\tau = \mu_N \frac{du}{dr} \tag{9.1}$$

where μ_N is a Newtonian viscosity. Figure 9.1b represents the power-law model,

$$\tau = K \left(\frac{du}{dr} \right)^n \tag{9.2}$$

where K and n are material constants. If n is less than unity, this model represents a pseudoplastic fluid, and if n is greater than unity, a dilatant fluid is simulated by this model. When n is equal to unity, this model coincides with the Newtonian fluid model.

Figure 9.1c indicates the Bingham plastic model,

$$\tau - \tau_y = \mu_B \frac{du}{dr} \tag{9.3}$$

where μ_B and τ_y are Bingham plastic viscosity and yield stress, respectively. Figure 9.1d indicates the Herschel–Bulkley model,

$$\tau - \tau_y = K \left(\frac{du}{dr} \right)^n \tag{9.4}$$

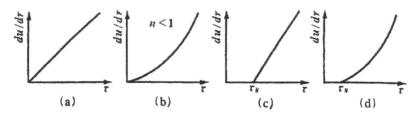

FIGURE 9.1   Classification of fluid models: (a) Newtonian model; (b) power-law model; (c) Bingham model; (d) Herschel–Bulkley model.

More complex models have been proposed for non-Newtonian fluids. However, the foregoing models are simple and widely used for engineering problems.

2.9.3   TIME-DEPENDENT FLOW CHARACTERISTICS

Highly loaded slurries of very fine particles (e.g., plaster, clay, paint pigment, and ceramic powder) show not only the non-Newtonian behavior described above but also time-dependent flow characteristics. A slurry often shows the time-dependent decrease of apparent viscosity with increase in shear rate, but the viscosity is recovered when the slurry is settled at rest. This reversible flow characteristic change under constant temperature is called thixotropy. The basic idea for a phenomenological thixotropy model was proposed by Cheng and Evans,[2] and articles on thixotropy models have been recently reviewed by Galassi et al.[3] and Barnes.[4]

Rheological characteristics of certain kinds of suspension are remarkably changed by applying an external electric field. This phenomenon is called an electrorheology effect. Electrorheological fluids have been extensively investigated[5–7] because they are expected to be applicable as new functional fluids to many engineering purposes.

2.9.4   VISCOSITY EQUATIONS FOR SUSPENSIONS OF SPHERICAL PARTICLES OF NARROW PARTICLE SIZE DISTRIBUTION

Superlative reviews on this subject are given by Thomas[8] and Metzner.[9] Figure 9.2 shows the increase in viscosity with the solid concentration. There appears to be no effect of particle size on fluid viscosity. The volume fraction of solid particle, ϕ, is a unique function for the relative viscosity, μ_r. The classic Einstein equation,[10]

$$\mu_r = 1 + 2.5\phi \tag{9.5}$$

may represent accurately the behavior of the suspension within only a vanishingly small range of solid concentration.

Guth and Simha[11] proposed the following viscosity equation, which is applicable to a higher solid concentration range. They modified the Einstein equation by taking the effect of interaction between solid particles into account.

$$\mu_r = 1 + 2.5\phi + 14.1\phi^2 \tag{9.6}$$

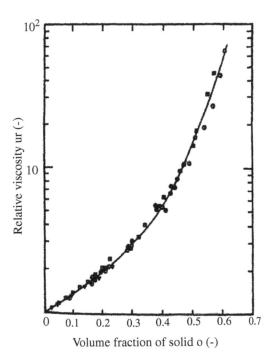

FIGURE 9.2 Relative viscosity versus concentration behavior for suspensions of spheres having narrow particle size distributions. Particle diameters range from 0.1 to 400 μm. The solid line depicts Equation 12.7, and keys for the data are given in the article by Thomas. [From Thomas, D. G., *J. Colloid Sci.*, 20, 267, 1965. With permission.]

Thomas[8] recommended the following empirical equation:

$$\mu_r = 1 + 2.5\phi + 10.05\phi^2 + A\exp(B\phi) \tag{9.7}$$

with $A = 0.00273$ and $B = 16.6$. This equation gives a prediction similar to the oft-quoted Moony equation[12]:

$$\mu_r = \exp\left(\frac{2.5\phi}{1 - k\phi}\right) \tag{9.8}$$

where the parameter k, called the crowding factor, is determined empirically and usually in the range of $1.3 \le k \le 1.9$. The empirical equation proposed by Kitano et al.[13]

$$\mu_r = \left(1 - \frac{\phi}{A}\right)^{-2} \tag{9.9}$$

gives a good prediction for highly loaded slurries. The single empirical constant A has a value of 0.680 for suspensions of smooth spheres. All the equations mentioned above are applicable for completely dispersed suspensions. However, the agglomeration of suspended particles affects significantly the fluidity of slurries. The effect of agglomeration on the non-Newtonian flow characteristics has recently been discussed by Usui et al.[14]

2.9.5 EFFECT OF PARTICLE SIZE DISTRIBUTION ON SLURRY VISCOSITY

Most systems of interest are, of course, more complex than the monodisperse suspensions considered thus far. If particles of a wide size distribution are employed, the maximum possible packing density of particles ($1/k$ in Equation 9.8 and A in Equation 9.9) will change considerably. This change causes a reduction in slurry viscosity. Chong et al.[15] investigated the effect of bimodal particle size distribution on the slurry viscosity. They showed experimentally that the viscosity of a bimodal suspension is reduced to almost 14%, in magnitude, of unimodal suspension viscosity at $\phi = 0.6$ and $d/D = 0.137$, where d and D are the small and large particle diameters, respectively. The volume fraction of maximum packing in bimodal systems was investigated experimentally by McGeary.[16] Using the experimental results of McGeary, Lee[17] proposed a prediction method for the maximum packing volume fraction of multimodal suspensions. Recently, the estimation method of the packing density of particle mixtures by a linear-mixture packing model was proposed by Yu and Standish.[18] Funk[19] modified the theoretical equation proposed by Andreasen and Andersen[20] to give an optimum size distribution for particles in a limited size range. Recently, Usui[21] has developed a non-Newtonian viscosity prediction model, which is applicable for nonspherical particle suspension system with particle size distribution.

2.9.6 MEASUREMENT OF SLURRY VISCOSITY BY A CAPILLARY VISCOMETER

A capillary viscometer is the most simple and the most frequently used viscometer. The pressurized fluid is forced to flow through a capillary of known diameter D, and the volumetric flow rate is measured. The wall shear stress τ_w is calculated by

$$\tau_w = \frac{D \Delta P}{4L} \qquad (9.10)$$

where ΔP and L are the pressure drop and capillary length, respectively. The relationships between the flow rate and the wall shear stress for the fluid models are given as follows:

1. Newtonian model:

$$Q = \frac{\pi D^3}{32 \mu_N} \tau_w \quad \text{(Hagen–Poiseuille equation)} \qquad (9.11)$$

2. Power-law model:

$$Q = \frac{n \pi D^3}{8(3n+1)K^{1/n}} \tau_w \qquad (9.12)$$

3. Bingham model at $\tau_{\mathrm{w}} > \tau_{\mathrm{y}}$:

$$Q = \frac{\pi D^3}{32 \mu_{\mathrm{B}}} \tau_{\mathrm{w}} \left[1 - \frac{1}{3}\left(\frac{\tau_{\mathrm{y}}}{\tau_{\mathrm{w}}}\right) + \frac{1}{3}\left(\frac{\tau_{\mathrm{y}}}{\tau_{\mathrm{w}}}\right)^4 \right] \quad \text{(Buckingham–Reiner equation)} \qquad (9.13)$$

4. Herschel–Bulkley model:

$$Q = \frac{\pi D^3}{32} \frac{4n}{3n+1} \left(\frac{\tau_{\mathrm{w}}}{K}\right)^{1/n} \left(1 - \frac{\tau_{\mathrm{y}}}{\tau_{\mathrm{w}}}\right)$$
$$\left\{ 1 - \frac{1}{2n+1}\left(\frac{\tau_{\mathrm{y}}}{\tau_{\mathrm{w}}}\right) \left[1 + \frac{2n}{n+1}\left(\frac{\tau_{\mathrm{y}}}{\tau_{\mathrm{w}}}\right)\left(1 - n\frac{\tau_{\mathrm{y}}}{\tau_{\mathrm{w}}}\right) \right] \right\} \qquad (9.14)$$

If it is not known what kind of non-Newtonian fluid model is suitable for a given slurry, one cannot use the equations shown above. In this case, a more complex data acquisition technique, called Maron–Krieger's method,[22] should be employed.

2.9.7 MEASUREMENT OF SLURRY VISCOSITY BY A ROTATING VISCOMETER

The rotating viscometers are subdivided into three categories: coaxial rotating cylinders, cone and plate, and plate and plate viscometers. The clearance between cone and plate or between plate and plate is generally so narrow that the use of a coaxial rotating cylinder viscometer is preferred for the measurement of slurry viscosity. For the case of an outer cylinder rotating (Couette type) viscometer, the relationship between the rotating speed Ω and torque T is given as follows:

1. Newtonian model:

$$V = \frac{n}{4\pi L \mu_{\mathrm{N}}}\left(\frac{1}{R_1^2} - \frac{1}{R_2^2}\right) \qquad (9.15)$$

where L, R_1, and R_2 are the length of the inner cylinder, and radii of inner and outer cylinders, respectively.

2. Power-law model:

$$\Omega = \frac{n}{2K^{1/n}}\left(\frac{1}{2\pi R_1^2 L}\right)^{1/n} \frac{R_2^{2/n} - R_1^{2/n}}{R_2^{2/n}} \qquad (9.16)$$

3. Bingham model at $T > 2\pi R_2^2 L \tau_{\mathrm{y}}$:

$$\Omega = \frac{n}{4\pi L \mu_{\mathrm{B}}}\left(\frac{1}{R_1^2} - \frac{1}{R_1^2}\right) - \frac{\tau_{\mathrm{y}}}{\mu_{\mathrm{B}}} \ln \frac{R_2}{R_1} \qquad (9.17)$$

The estimation of shear rate, du/dr, on the surface of the inner cylinder is complex if the fluid is non-Newtonian. The reader should be careful in determining the shear rate using the general description of a Couette viscometer.[22]

REFERENCES

1. Bird, R. B., Stewart, W. E., and Lightfoot, E. N., *Transport Phenomena*, Wiley, New York, 1960, p. 10.
2. Cheng, D. C.-H. and Evans, F. *Br. J. Appl. Phys.*, 16, 1599, 1965.
3. Galassi, C., Rastelli, E., and Laspin, R. *Science, Technology, and Application of Colloidal Suspensions*, American Ceramics Society, 1995, p. 3.
4. Barnes, H. A. *J. Non-Newtonian Fluid Mech.*, 70, 1, 1997.
5. Brooks, D. A., in *Proceedings of the Eleventh International Congress on Rheology*, Vol. 2, Elsevier Science, Amsterdam, 1992, p. 763.
6. Otsubo, Y. and Watanabe, K., *Nihon Rheology Gakkaishi (J. Soc. Rheol. Jpn.)*, 18, 111, 1990.
7. Hasegawa, Y., Isobe, T., Senna, M., and Otsubo, Y., *J. Soc. Rheol. Jpn.*, 27, 131, 1999.
8. Thomas, D. G., *J. Colloid Sci.*, 20, 267, 1965.
9. Metzner, A. B., *J. Rheol.*, 29, 739, 1985.
10. Einstein, A. *Ann. Phys.*, 19, 289, 1906.
11. Guth, E. and Simha, R., *Kolloid Z.*, 74, 266, 1936.
12. Mooney, M., *J. Colloid Sci.*, 6, 162, 1951.
13. Kitano, T., Kataoka, T., and Shirota, T., *Rheol. Acta*, 20, 207, 1981.
14. Usui, H., Kishimoto, K., and Suzuki, H., *Chem. Eng. Sci.*, 56, 2979, 2001.
15. Chong, J. S., Christensen, E. B., and Baer, A. D., *J. Appl. Polym. Sci.*, 15, 2007, 1971.
16. McGreary, R. K., *J. Am. Ceram. Soc.*, 44, 513, 1961.
17. Lee, D. I., *J. Paint Technol.*, 42, 579, 1970.
18. Yu, A. B. and Standish, N., *Ind. Eng. Chem. Res.*, 30, 1372, 1991.
19. Funk, J. E., U.S. Patent 4,282,006, 1979.
20. Andreasen, A. H. M. and Andersen, J., *Kolloid Z.*, 50, 217, 1930.
21. Usui, H., *J. Chem. Eng. Jpn.*, 35, 815, 2002.
22. Middleman, S., *The Flow of High Polymer*, Wiley, New York, 1968, p. 13.

2.10 Particle Impact Breakage

Mojtaba Ghadiri
University of Leeds, Leeds, United Kingdom

In processing of particulate solids, there are circumstances under which interparticle and particle–wall impacts cause damage to the particles. In some cases, such as comminution, this is desirable, while in other cases it is undesirable, as for example in attrition of weak and friable particles in pneumatic conveying. These processes share the same underlying mechanisms, namely, the formation and propagation of various types of crack. It is therefore of great interest to identify and quantify the conditions under which particulate solids are damaged.

2.10.1 IMPACT FORCE

Impact of particles causes a transient stress whose magnitude depends on a number of factors such as impact velocity, particle size, material properties, and contact geometry.

Elastic

If the transient stress does not exceed the plastic yield stress, then the impact is treated as elastic. The maximum impact force, F, is commonly obtained by assuming that the relationship between the impact force and displacement is the same as that of a static elastic contact, given by Hertz analysis.[1] For normal impact between two spheres

$$F = 1.283 \; m^{3/5} \; R^{1/5} \; E^{*2/5} \; V^{6/5} \tag{10.1}$$

where $1/m = 1/m_1 + 1/m_2$, $1/R = 1/R_1 + 1/R_2$, $1/E^* = (1 - v_1^2)/E_1 + (1 - v_2^2)/E_2$, V is the impact velocity, m is the mass, R is the radius, E is Young's modulus, v is Poisson's ratio, and subscripts 1 and 2 refer to the two impacting bodies. Particle damage in this case is by brittle failure.

Plastic

The maximum contact force in this case is obtained by first calculating the onset of yield. The contact stress remains at its yield value and if the material does not strain-harden, the maximum force can be estimated from the contact area. The calculation of contact area is, in turn, based on the impact energy balancing the work of plastic deformation, from which the maximum contact force can be calculated[2]:

$$F = \pi \; R^2 \left(\frac{4\rho Y}{3} \right)^{1/2} V \tag{10.2}$$

where ρ is the particle density, Y is the yield stress, and V is the normal impact velocity. Particle breakdown in this case is by plastic rupture.

Elastic–Plastic

A mathematical treatment of the process has been proposed by Ning and Thornton,[3] where Hertz analysis is used to describe the preyield behavior, and a "coupled" elastic/plastic analysis is used to describe the postyield behavior. The maximum impact force, F, is given by Ning[2] as

$$\frac{F}{F_Y} = \sqrt{\frac{6}{5}(V/V_Y)^2 - \frac{1}{5}} \tag{10.3}$$

where F_Y and V_Y are the impact force and velocity at the onset of yield, respectively, given by

$$F_Y = \pi^3 R^2 Y^3 / (6E^{*2}) \tag{10.4}$$

and

$$V_Y = \left(\frac{\pi}{2E^*}\right)^2 \left(\frac{2}{5\rho}\right)^{1/2} Y^{5/2} \tag{10.5}$$

Particle breakdown in this case is by fracture, which is preceded by limited plastic deformation.

2.10.2 MODE OF BREAKAGE

The current understanding of particle breakage has recently been collated in a special issue of *Powder Technology*, edited by Salman,[4] covering the classic brittle and semibrittle failure modes. Schönert,[5] Salman et al.,[6] and Wu et al.[7] have provided detailed reviews of failure of spheres and discs under both impact and quasi-static loading. Failure of particles is a process that depends on material properties and the mode of loading. The latter is related to the strain rate and contact geometry. Increasing the strain rate from quasi-static to impact can cause a switch in the failure mode from semibrittle to brittle. The geometric effects are in practice related to particle size and shape, as they affect the size of the contact area. If large particles of a material fail by the brittle mode, it is likely that reducing the size can switch the mode of failure to the semibrittle failure and ultimately to the ductile failure. Therefore, the determination of critical transition conditions is important and will be addressed below.

Brittle Failure Mode

This mode of failure is due to the presence of preexisting internal or surface flaws. In their absence, particles are strong and can only fail when shear deformations can generate microcracks. Surface flaws often produce orange-segmented fragments and play a major role in damage when the elastic compliance of the particle or the contacting surface is high, producing large tensile hoop stresses.[8] Failure due to internal flaws is dominant when the elastic compliance of the contact is low.[8] It usually produces diametrical cracks, splitting the particles into fragments, as the diametrical plane is under the greatest tensile stress.[9] When a very large preexisting flaw is present elsewhere, a crack initiating from the flaw follows the tensile stress trajectory. A deterministic analysis of particle breakdown in the case of brittle failure requires a knowledge of the size and position of the flaws. In the absence of this information, the empirical determination of the crushing strength of the particles is the only way to characterize the breakdown. The interpretation of data is commonly carried out by statistical analysis, for example, by the use of the Weibull distribution.[10,11]

Highly localized loading in this mode of failure may lead to the formation of Hertzian cone cracks. Oblique impacts cause tilting of the cone angle, as the tensile stress trajectories are modified by the frictional traction, and this can produce small chips from the particles, which is responsible for the erosive wear of the particles.[12]

Semibrittle Failure Mode

This failure mode is identified by *limited* plastic flow, which is responsible for crack initiation. Plastic flow occurs because the impact stress exceeds the onset of yield, whose characteristics are defined by the critical elastic–plastic transition size, as defined by Puttick.[13] The plastic zone produces compressive radial stresses and tensile hoop stresses. The latter type of stress propagates radial and median cracks, initiated from the plastic zone. When the load is removed, the residual tensile stresses, formed by the elastic unloading component, generate subsurface lateral cracks. Gorham et al.[14] and Chaudhri[15] have investigated the impact breakage of spheres failing in this mode, using polymethyl methacrylate (PMMA) spheres. The most important properties here are the hardness, Young's modulus, and toughness (see Chapter 1.5). Qualitatively "hard" materials undergo less plastic deformation than "soft" materials but can store greater residual stresses, depending on the extent of elastic deformation in the hinterland beyond the plastic zone. Therefore, their tendency for generation of lateral cracks is greater than that of soft materials. Impact damage analysis of particles in this mode of failure has been carried out by indentation fracture mechanics.[16] Particle breakdown in the semibrittle failure mode can be characterized by crack morphology and extension: particle *fragmentation* occurs by the formation of median and radial cracks, and *chipping* occurs by the formation of lateral cracks.

Ductile Failure Mode

Soft materials, such as some polymers, are usually damaged under this failure mode. Ploughing and cutting are the main two mechanisms of material removal.[17] Slip-line field plasticity analysis has been widely used to calculate the deformation pattern.[18] Important factors in the process here are the attack angle, the ratio of Young's modulus to hardness of the surface, the ratio of the hardness values of the two surfaces, and the shear strength of the interface, in other words, the ratio of the shear stress at the interface, and the shear yield stress of the material. In ductile failure, cracking does not readily occur, but instead the plastic ruptures. The breakdown of particulate solids by this mode of failure has not been widely investigated so far.

Analysis of Breakage for the Brittle Failure Mode

Weibull analysis[19] is commonly used to fit experimental observations of breakage in this failure mode, where the probability of breakage, S, is related to the applied stress, σ, using two fitting parameters, z and σ_s, representing a characteristic flaw density and strength, respectively.

$$S = 1 - \exp\left[-z\left(\frac{\sigma}{\sigma_s}\right)^m\right] \qquad (10.6)$$

Vogel and Peukert[11] have recently applied the above analysis to the impact breakage of particles by relating the applied stress to the incident kinetic energy W_k using Weichert's approach[20]:

$$S = 1 - \exp\left[-f_{mat} \ x \ \left(W_k - W_{k,min}\right)\right] \qquad (10.7)$$

where f_{mat} and $W_{k,min}$ are two fitting parameters similar to those of Equation 10.6, and x is particle size. In line with the significance of the parameters of Equation 10.6, Vogel and Peukert[11] suggest f_{mat} and $W_{k,min}$ reflect the material properties and the minimum kinetic energy that cause breakage, respectively. Similar work has also been done by Salman et al.,[21] who related the fraction of unbroken particles to the impact velocity by the use of Weiball analysis.

2.10.3 ANALYSIS OF BREAKAGE FOR THE SEMIBRITTLE FAILURE MODE

To address the impact breakage of particulate solids on a fundamental basis, fracture mechanics is used to define the conditions for propagation of various types of cracks, namely, radial and median cracks for the fragmentation, and lateral cracks for chipping.

Chipping

Figure 10.1 shows a typical example of chipping. A sequence of images from the impact process has been recorded by a high-speed digital video recorder at 27,000 frames per second, where the detachment of a large chip can be seen.[22] The chipping process has been described theoretically by a mechanistic model developed by Ghadiri and Zhang.[16] The formulation of the model is based on the indentation fracture mechanics of lateral cracks. A fractional loss per impact, ξ, is defined as the ratio of the volume of chips removed from a particle to the volume of the original (mother) particle, and it is used as a measure of the breakage propensity break under impact conditions. According to the model, ξ is given by

$$\xi = \alpha\eta = \alpha\frac{\rho\,V^2DH}{K_c^2} \tag{10.8}$$

where η is a dimensionless group describing the attrition propensity, ρ is the particle density, V is the impact velocity, D is a linear dimension of the particle, H is the hardness, K_c is the fracture toughness, and α is a proportionality factor, which depends on particle shape and impact geometry and is determined experimentally.

Fragmentation

Figure 10.2 shows a sequence of digital video images of a particle fragmenting on impact. The recording speed is 27,000 frames per second, as for Figure 10.1.[22,23] Fragmentation occurs when the radial or median cracks extend to the full length of the particles. The formation of two large fragments by the propagation of a median crack can be seen in Figure 10.2. At higher impact energies, a more extensive cracking takes place, producing a larger number of fragments. However, at present there is no theory that can relate the product size distribution to the impact conditions and material properties in a predictive way. The force for fracture of a sphere of diameter D can be estimated based on indentation fracture. Based on the relationship proposed by Ghadiri and Zhang[16] for crack extension, the fragmentation force is given by

$$F_{fr} \propto K_c^{4/3}D^{4/3}H^{-1/3} \tag{10.9}$$

Transition Velocities

The transition velocities from plastic deformation to chipping and from chipping to fragmentation are important features of particle breakage by impact. The dependence of the transition velocities on particle size and material properties has not been widely quantified, and it can only be construed by theoretical considerations at this stage.

Frame no. 1 | 2 | 3 | 4 | 5 | 6

FIGURE 10.1 High-speed video sequence of impact of porous alumina catalyst carrier beads that broke by chipping when impacted at 16 m s^{-1}, recorded at 27,000 fps. [From Couroyer, C., Ghadiri, M., Laval, P., Brunard, N., and Kolenda, F., *Oil and Gas Science and Technology—Revue de l'Institut Français du Pétrole*, 55, 67–85, 2000. With permission.]

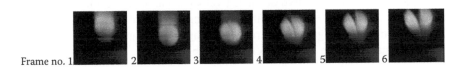

Frame no. 1 | 2 | 3 | 4 | 5 | 6

FIGURE 10.2 High-speed video sequence of impact of porous alumina catalyst carrier beads that broke by fragmentation when impacted at 20 m s^{-1}, recorded at 27,000 fps. [From

Plastic Deformation–Chipping Transition

Marshall et al.[24] have shown that there is a critical load to cause lateral fracture, given by

$$F_{cl} \propto E \left(\frac{K_c}{H} \right)^4 \tag{10.10}$$

Hutchings[25] used this criterion to predict the minimum particle size that can cause erosion of surfaces by impact. The same approach can be followed here to define the critical transition velocities for chipping and fragmentation. In the case of a round (or relatively flat) contact between a particle and target, the critical transition velocity for plastic deformation–chipping is given by

$$V_{ch} \propto \left(\frac{K_c}{H} \right)^4 \frac{E}{H^{1/2}} \rho^{-1/2} D^{-2} \tag{10.11}$$

The above expression indicates that the critical velocity for the onset of chipping is inversely proportional to the square of the particle size.

Chipping–Fragmentation Transition. A similar approach can be followed to specify the threshold conditions for particle fragmentation. Hutchings[25] specified a critical particle size below which no fragmentation occurs, based on the indentation fracture model of Hagan.[26] The critical load for indentation fracture proposed by Hagan[26] is given by

$$F_{cf} \propto \frac{K_c^4}{H^3} \tag{10.12}$$

This equation can be combined with the impact dynamics by making the same assumptions as in the approach used for the chipping case: the contact deformation under impact follows a quasi-static indentation model. This enables the impact force to be described in terms of the hardness times the impression area, to give a threshold velocity for fragmentation:

$$V_{fr} \propto \left(\frac{K_c}{H} \right)^4 H^{1/2} \rho^{-1/2} D^{-2} \qquad (10.13)$$

It is interesting to note that the dependence of V on the particle size D is the same as that of chipping.

Limit of Breakdown. Equation 10.11 has a lower limit of validity for particle size. This limit is given by an ultimate particle size below which the particles can only be deformed plastically and cannot be fractured at all, irrespective of impact velocity. There are several models for this limit, proposed by Kendall,[27] Puttick,[13] and Hagan.[26] These models are all based on energy requirements for crack nucleation and propagation. The model of Hagan[26] appears to provide a closer agreement with the experimental evidence and is therefore given below.

$$D_c \cong 30 \left(\frac{K_c}{H} \right)^2 \qquad (10.14)$$

In conclusion, impact damage depends on the mode of failure, which in turn depends on material properties and contact geometry. The rate of breakage and particle transition size and velocities are all affected by mechanical characteristics such as hardness, toughness, and stiffness. Therefore, the determination of these parameters is important for a better understanding of particle breakage (see Section 1.5).

2.10.4   ANALYSIS OF BREAKAGE OF AGGLOMERATES

The failure mode of agglomerates can cover macroscopically the three modes described in Section 2.10.2 because of their large degree of freedom arising from many factors that can influence agglomerate strength, such as void fraction, primary particle size, interparticle bond characteristics, density, and so forth. Consequently the breakage map of agglomerates is not established yet. The disintegration of weak agglomerates failing macroscopically in a "ductile" mode has been studied by Boerefijn et al.[28] Patterns of failure of large agglomerates of glass ballotini bonded together with a brittle glue have been reported by Subero and Ghadiri[29] and more recently for various dry and wet granules by Salman et al.[6] For the simple case of auto-adhesive primary particles, where the interparticle adhesion follows the JKR model,[30] extensive work has been reported in the literature based on the development of the distinct element analysis of agglomerates by Thornton and his coworkers (see, e.g., Thornton et al.[31] and Thornton and Liu[32]). The effects of interface energy, impact angle, and agglomerate morphology have been investigated by distinct element analysis by Subero et al.,[33] Moreno et al.,[34] and Golchert et al.,[35] respectively. Kafui and Thornton[36] and Moreno[37] have simulated the impact damage of agglomerates on collision with a wall by the distinct element method. The simulation of impact fragmentation of an agglomerate carried out by Moreno[37] is shown in Figure 10.3, where the fragments formed at the end of impact are shown with different (gray) density levels according to the number of particles in each fragment. The agglomerate is made of 10,000 spheres 100 μm in diameter, having the surface energy of 3.5 J m^{-2}, elastic modulus of 31 Gpa, and packing density of 0.55, and it has been impacted at velocity of 2 m s^{-1}.

Kafui and Thornton[36] suggested that the extent of damage described by the damage ratio, Δ, (i.e., the number of broken interparticle bonds divided by the total number of bonds present in the agglomerate) is related to the Weber number, *We*:

$$We = \frac{\rho \, V^2 D}{\Gamma} \qquad (10.15)$$

FIGURE 10.3 Simulation of impact fragmentation of an agglomerate by the distinct element method. [From Moreno, R., Ph.D. Dissertation, University of Surrey, 2003. With permission.]

where ρ, D, and Γ are the primary particle density, diameter, and surface energy, and V is the impact velocity. The breakage propensity parameter η given by Equation 10.8 incorporates the Weber number, because K_c can be related to the surface energy Γ by the use of linear elastic fracture mechanics. For example, for the case of plane strain,

$$\frac{K_c^2}{1-\nu^2} = 2\,E\Gamma \tag{10.16}$$

it then follows that

$$\eta \propto \frac{\rho\,V^2\,D}{\Gamma} \times \frac{H}{E} \tag{10.17}$$

The form H/E is attributed to the elastic–plastic deformation characteristics of the agglomerate. When the simulation results are interpreted in the form of fractional loss per impact, Thornton et al.[37] report that at low impact velocities, corresponding to the chipping regime, the fractional loss per impact varies linearly with the Weber number, which is in agreement with the model of Ghadiri and Zhang.[16]

To model agglomerate breakage, Moreno[38] explored a simple case where the energy required for breakage was linearly related to the incident kinetic energy. Considering the work spent to break a bond, he found that the damage ratio is given by

$$\Delta \propto \frac{\rho\,V^2\,D}{\Gamma} \times \left(\frac{ED}{\Gamma}\right)^{2/3} \tag{10.18}$$

Clearly the Weber number and other dimensionless groups such as ED/Γ influence the breakage of agglomerates, as demonstrated by the numerical simulations of Moreno.[38] More extensive work is required to describe the breakage characteristics of agglomerates with binders.

REFERENCES

1. Johnson, K. L., *Contact Mechanics,* Cambridge University Press, Cambridge, U.K., 1985.
2. Ning, Z., Ph.D. Dissertation, Aston University, U.K., 1995.
3. Ning, Z. and Thornton, C., in *Powders and Grains 93,* Thornton, C., Ed., Balkema, Rotterdam, 1993, pp. 33–35.

4. Salman, A. D., *Powder Technol.,* 1, 143–144, 1, 2004.
5. Schönert, K., *Powder Technol.,* 143–144, 2–18, 2004.
6. Salman, A. D., Reynolds, G. K., Fu, J. S., Cheong, Y. S., Biggs, C. A., Adams, M. J., Gorham, D. A., Lukenics, J., and Hounslow, M. J., *Powder Technol.,* 143–144, 19–30, 2004.
7. Wu, S. Z., Chau, K. T., and Yu, T. X., *Powder Technol.,* 143–144, 41–55, 2004.
8. Shipway, P. H., Hutchings, I. M., *Powder Technol.,* 76, 23–30, 1993.
9. Salman, A. D., Gorham, D. A., Verba, A., *Wear,* 186–187, 92–98, 1995.
10. Van den Born, I. C., Ph.D. Dissertation, University of Groningen, 1992.
11. Vogel, L. and Peukert, W., *Powder Technol.,* 129, 101–110, 2003.
12. Lawn, B. R., in *Fundamentals of Friction: Macroscopic and Microscopic Processes,* Singer, I. L. and Pollock, H. M., Eds., Kluwer Academic Publishers, London, 1991, pp. 137–165.
13. Puttick, K. E., *J. Phys. D: Appl. Phys.,* 13, 2249–2262, 1980.
14. Gorham, D. A., Salman, A. D., and Pitt, M. J., *Powder Technol.,* 138, 229–238, 2003.
15. Chaudhri, M. M., *Powder Technol.,* 143–144, 31–40, 2004.
16. Ghadiri, M. and Zhang, Z., *Chem. Eng. Sci.,* 57, 3659–3669, 2002.
17. Hutchings, I. M., *Powder Technol.,* 76, 3–13, 1993.
18. Childs, T. H. C., in *Fundamentals of Friction: Macroscopic and Microscopic Processes,* Singer, I. L. and Pollock, H. M., Eds., Kluwer Academic Publishers, London, 1991, pp. 209–226.
19. Weiball, W., *J. Appl. Mech.,* 9, 293–297, 1951.
20. Weichert, R., *Zement -Kalk-Gips,* 45 (Suppl. 1), 1–8, 1992.
21. Salman, A. D., Biggs, C. A., Fu, J., Angyal, L., Szabo, M., and Hounslow, M. J., *Powder Technol.,* 128, 36–46, 2002.
22. Couroyer, C., Ghadiri, M., Laval, P., Brunard, N., and Kolenda, F., *Oil and Gas Science and Technology— Revue de l'Institut Français du Pètrole,* 55, 67–85, 2000.
23. Couroyer, C., Ph.D. Dissertation, University of Surrey, 2000.
24. Marshall, D. B., Lawn, B. R., and Evans, A. G., *J. Am. Ceram. Soc.,* 65 (Suppl. 11), 561–566, 1982.
25. Hutchings, I. M., in *Erosion of Ceramic Materials,* Ritter, J. E., Ed., Trans Tech Publications, 1992, pp. 75–92.
26. Hagan, J. T., *J. Mater. Sci.,* 16, 2909–2911, 1981.
27. Kendall, K., *Nature,* 272, 710–711, 1978.
28. Boerefijin, R., Ning, Z., and Ghadiri, M., *Int. J. Pharm.,* 172 (Suppl. 1–2), 199–209, 1998.
29. Subero, J. and Ghadiri, M., *Powder Technol.,* 120 (Suppl. 3), 232–243, 2001.
30. Johnson, K. L., Kendall, K., and Roberts, A. D., *Proceedings of the Royal Society of London A,* 324, 301–313, 1971.
31. Thornton, C., Ciomocos, M. T., and Adams, M. J., *Powder Technol.,* 105, 74–82, 1999.
32. Thornton, C. and Liu, L., *Powder Technol.,* 143–144, 110–116, 2004.
33. Subero, J., Ning, Z., Ghadiri, M., and Thornton, C., *Powder Technol.,* 105 (Suppl. 1–3), 66–73, 1999.
34. Moreno, R., Ghadiri, M., and Antony, S. J., *Powder Technol.,* 130 (Suppl. 1–3), 132–137, 2003.
35. Golchert, D., Moreno, R., Ghadiri, M., and Litster, J., *Powder Technol.,* 143–144, 84–96, 2004.
36. Kafui, K. D. and Thornton, C., in *Powders and Grains 93,* Thornton, C., Ed., Balkema, Rotterdam, 1993, pp. 401–406.
37. Thornton, C., Kafui, D., and Ciomocos, T., Paper presented at the IFPRI Annual Conference, Urbana, Illinois, 1995.
38. Moreno, R., Ph.D. Dissertation, University of Surrey, 2003.

2.11 Sintering

Kikuo Okuyama

Hiroshima University, Higashi-Hiroshima, Japan

Sintering or densification is an irreversible thermodynamic phenomenon to convert unstable packed powder having excess free energy to stable sintered agglomerates. The sintering phenomenon involves the fusion of particles, volume reduction, decrease in porosity, and increase in grain size.

2.11.1 MECHANISMS OF SOLID-PHASE SINTERING

Sintering kinetics have been studied by many investigators from both experimental and theoretical points of view. Coble[1] divided the sintering process into three successive elementary stages, as shown in Figure 11.1. The initial stage corresponds to neck formation and growth, the intermediate stage to the growth of cylindrical vacancies, and the final stage to diffusion and disappearance of the vacancies.

Formation of Necks

Crystalline particles usually contain dislocations at the surface, created while they were produced. These dislocations are moved or recovered at an initial stage of sintering to form a neck at the contact point of particles. The phenomenon usually begins to be observed at the absolute temperature ratio to the melting point $\propto = 0.23$.

Neck Growth

Neck growth is caused by mass transfer such as evaporation–condensation, diffusion, and plastic flow. The principal mass transfer depends on the composition of the material, sintering conditions, and the sintering step.

Evaporation–Condensation

The material evaporates at convex particle points or on the surface of necks with concave curvatures. The process makes round grains.

Diffusion

The equilibrium concentration of atomic or ionic vacancies at the particle surface and necks for crystalline particles varies with the chemical potential at the respective location, which is relatively higher than in the interior of particles. Hence, diffusion occurs from the inside (volumetric diffusion) to the surface (surface diffusion) to the crystalline grain boundary (grain boundary diffusion). Surface diffusion usually occurs at a lower temperature ($\propto = 0.33$ to 0.45) and there is little access of coalesced particles, whereas at a higher temperature ($\propto = 0.42$ to 0.8), volumetric diffusion becomes active and the growth rate of necks is increased due to an increase in the mass transfer rate.

Plastic Flow (Viscous Flow)

Sintering mechanisms of amorphous materials such as glass and resine are usually controlled by plastic flow. Plastic flow also plays a significant role in rapid shrinkage and densification at the initial stage of sintering.

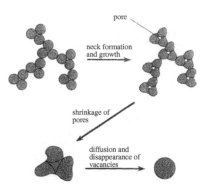

FIGURE 11.1 Sintering of multiple primary particles.

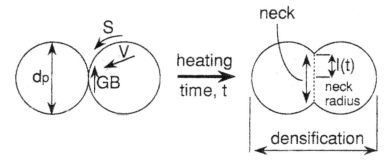

FIGURE 11.2 Two-sphere model of initial-stage sintering.

Mechanism of Shrinkage and Bloating of Pores

Shrinkage and bloating of pores at the final stage of sintering are phenomena due to sink or source of mass transfer, and therefore volumetric diffusion and grain boundary diffusion play major roles regardless of mass transfer mechanisms such as evaporation–condensation and surface diffusion.

2.11.2   MODELING OF SINTERING OF AGGLOMERATES

Coblentz et al.[2] proposed sintering rate equations based on several existing sintering models. Ashby[3] presented a sintering diagram to identify the dominant mechanisms of sintering. Models describing the sintering of crystalline agglomerates have been presented by numerous researchers. One of the most representative and popular modes uses the ideal geometry illustrated in Figure 11.2. The center-to-center approach and the neck growth between two equal spheres are described to evaluate the rate of sintering in the initial stage. Because the neck growth rate depends on the sintering mechanism, it is important to know what mechanism is dominant.

The change in neck radius l with time t is given as[4]

$$\left(\frac{2l}{d_{p0}}\right)^{n} = \frac{Kt}{d_{p}^{m}} \tag{11.1}$$

where d_{p} and d_{p0} are the primary particle diameters at $t = t$ and $t = 0$, respectively. K, m, and n are the constants which depend on the physical properties and dominant mass transport mechanisms for sintering, as shown in Table 11.1.

TABLE 11.1   Values of Constants in Equation 11.1

m	n	Mechanism
1	2	Plastic flow or viscous flow
2	3	Evaporation and condensation
3	5	Volume diffusion
4	6	Grain boundary diffusion
4	7	Surface diffusion

When the neck size reaches a certain value, a channel-like vacancy is formed (intermediate stage). In the final stage of sintering, the vacancy grows to be spherical, and sintering rates are expressed in terms of the diffusion of vacancies. There is no kinetic expression that holds over all the stages of sintering. From many scaling factors available to describe the shrinkage due to sintering, the surface area or the density of the sintered body is usually chosen.

Over all the stages of sintering, the reduction rate of surface area due to sintering is approximately given as[5]

$$\frac{da_s}{dt} = -\frac{1}{\tau}(a_s - a_{sc}) \tag{11.2}$$

where a_{sc} is the surface area of the final single sphere after complete fusion. τ is the rate constant called the sintering time. Assuming that the sintering of two spheres is complete when the ratio of neck radius l to the primary particle radius $d_p/2$ reaches 0.83, τ is given by

$$\tau = \frac{(2l_f/d_{p0})^n d_p^m}{K} = A d_p^m \exp\left(\frac{E}{R_g T}\right) \tag{11.3}$$

where l_f is the neck radius at the equilibrium state, E is the activation energy for self-diffusion, and A is a constant depending on the sintering mechanism. For example, for grain boundary diffusion[2] A is given by

$$A = \frac{\kappa T (2l_f/d_{p0})^6}{12 b D_0 \gamma \Omega} \tag{11.4}$$

where k, D_0, γ, and Ω are respectively the Boltzmann constant, preexponential factor for the diffusion coefficient, surface tension, and atomic volume. Figure 11.3 shows the temperature dependency of τ for ultrafine silver and titania particles as a function of primary particle diameter d_p. τ is a strong function of temperature and changes also with d_p. If τ is constant (i.e., isothermal sintering without grain growth), Equation 11.2 indicates that the surface area of the agglomerates decays exponentially. However, τ does not, in general, remain constant because of the increase in the primary particle diameter during sintering. The primary particle diameter d_p is related to the surface area of an agglomerate by

$$d_p = \frac{6V}{a_s} \tag{11.5}$$

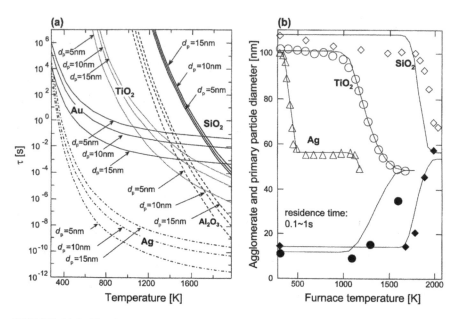

FIGURE 11.3 The change in sintering time, τ, with temperature; d_p is the primary particle diameter.

where V is the volume of the agglomerate that remains constant over the entire sintering process. Equation 11.5 states that d_p is inversely proportional to the surface area—thus, that the growth rate of primary particles can be evaluated from the observed decrease in a_s.

Ultrafine particle synthesis via gas-phase reaction is reported by many researchers. In the synthesis processes, however, many phenomena affect particle morphology (e.g., coagulation, condensation, crystallization, and sintering). Sintering is especially important because it influences particle size, shape, and crystal structure. The sintering of silver and titania agglomerates consisting of nanometer primary particles in a heated gas flow was investigated experimentally. The densification of agglomerates accompanying primary particle growth was explained quantitatively by solving the equation of sintering, Equation 11.2, under the temperature profile.[6,7] The smaller the primary particles are, the shorter the time required for sintering. The sintering rate, therefore, slows as primary particles grow in size. This size dependency in sintering rate will be a problem in sintering of nanophase materials from this point of view, as sintering at low temperature without grain growth is necessary to obtain a compact of high density.[7]

2.11.3 SINTERING PROCESS OF PACKED POWDER

Solid-phase sintering of packed powder of a single composition is conducted at a temperature lower than the melting point. Materials having an inexact melting point begin to sinter near the softening temperature, whereas materials having a clear melting point begin to sinter at three fifths of the melting temperature (in kelvin). Bonding between particles takes place at contact points to form necks that grow with time, resulting in shrinkage of the pores and forming isolated spherical voids as shown in Figure 11.4. The densification of packed powder due to sintering depends on the density of the compact before sintering and, hence, is evaluated by the densification parameter D:

$$D = \frac{\rho_s - \rho_g}{\rho_t - \rho_g}$$

(11.6)

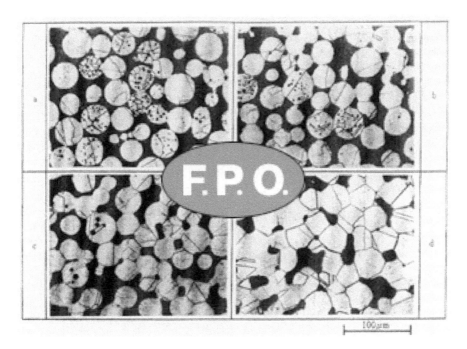

FIGURE 11.4 Structures of sintered grounds of spherical copper particles observed by an optical microscope (sintered in hydrogen after being packed with tapping): (a) 1273 K (1000°C), 5 min; (b) 1273 K, 23 min; (c) 1223 K (950°C), 6 h; (d) 1223 K, 20 h.

Where ρ_s is the density of the sintered body, ρ_g is the density before sintering, and ρ_t is the net density of the particles.

When solid-phase sintering progresses smoothly, the empirical equation $D = Kt^n$ is widely used as the relationship between sintering time t and densification parameter D. The constant K depends on the temperature, and the exponent n depends on the transfer mechanism of materials and is usually less than unity.

The sintering process of mixed materials depends primarily on wettability between particles of different types. Mutual solubility and diffusivity are also significant. A binary mixture of A and B with surface energies γ_A and γ_B, is respectively, and with the interfacial free energy γ_{AB} does not mutually wet if $\gamma_{AB} > \gamma_A = \gamma_B$. Hence, the sintering progresses selectively between particles of the same material. In this case, if a third material reactive to both components is added in small amounts or if the atmosphere is controlled, sintering becomes feasible. Reaction between particles is often applied to sintering (reaction sintering), but the sintered ground is sometimes bloated or broken if the reaction product is of less density.

Mutual diffusion plays a major role in solid-phase sintering. When there is a large difference in the mutual diffusivities, pores remain where particles with larger diffusivity migrate. Grooves and pores are also formed around the neck between particles of different composition.[8] When mass transfer is one-sided, segregation or concentration distribution takes place to form heterogeneous sintering.

REFERENCES

1. Coble, R. L., *J. Appl. Phys.*, 32, 787, 1961.
2. Coblentz, W. S., Dynys, J. M., Cannon, R. M., and Brook, R. J., *Mater. Sci. Res.*, 13, 141, 1980.

3. Ashby, M. F., *Acta Metallurgica,* 22, 275, 1974.
4. German, R. M. and Munir, Z. A., *J. Ann. Ceram. Soc.,* 71, 225, 1976.
5. Koch, W. and Friedlander, K., *J. Colloid Interface Sci.,* 140, 209, 1990.
6. Shimada, M., Seto, T., and Okuyama, K., *J. Chem. Eng. Jpn.,* 27, 795, 1994.
7. Siegel, R. W., Ramasamy, S., Hahn, H., Zongquan, L., Ting, L., and Gronsky, R., *J. Mater. Res.,* 3, 1367, 1988.
8. Kuczynski, G. C., Hooten, N. A., and Gibbon. C. F., *Sintering and Related Phenomena,* Gordon and Breach, New York, 1967.
9. Seto, T., Shimada, M., and Okuyama, K., *Aerosol Sci. Technol.,* 23, 183, 1995.

2.12 Ignition and Combustion Reaction

Hisao Makino, Hirofumi Tsuji, and Ryoichi Kurose

Central Research Institute of Electric Power Industry, Yokosuka, Kanagawa, Japan

2.12.1 COMBUSTION PROFILE

Fuel Properties

There are many kinds of solid fuels, including coal, oil sand, oil shale, refuse fuel, and biomass, and coal is the most abundant among these solid fuels. The most common and useful methods for analyzing these solid fuels are the proximate and ultimate analyses.

Proximate analysis is used to quantify the amounts of moisture, volatile matter, fixed carbon, and ash contents. The amounts of moisture and volatile matter contents of a sample of solid fuel are evaluated by measuring the weight losses in air at $107 \pm 2°C$ for 1 h and in an inert gas at $900 \pm 20°C$ for 7 min., respectively. The amount of fixed carbon content is supposed to correspond to the weight loss after further heating in air at $815 \pm 10°C$ for 1 h, and the residual is regarded as ash. Thus, the dried sample consists of combustible matter and ash, and the combustible matter comprises volatile matter, which easily volatilizes to the gas phase, and fixed carbon, which remains in the char even at high temperature.

On the other hand, the major elements such as carbon, hydrogen, oxygen, nitrogen, and sulfur in solid fuels are evaluated using ultimate analysis. In general, ultimate analysis is roughly related to proximate analysis: for example, coals with high fixed carbon content contain much carbon, and coals with high volatile matter content contain much hydrogen (see Table 12.1[1]).

Combustion Process of Solid Fuel

Figure 12.1[2] shows a schematic of the typical coal combustion processes on pulverized coal, which is one of the most common methods for burning solid fuels. The combustion processes are as follows: (1) moisture in solid fuels is immediately vaporized when solid fuels enter into a high-temperature region in a furnace, (2) evolution of volatile matter (devolatilization) takes place, (3) the volatilized gas is ignited and volatile combustion occurs, (4) char combustion (combustion of fixed carbon) follows the volatile combustion, (5) the combustion is terminated (a piece of the fixed carbon remains in the ash).

2.12.2 DEVOLATILIZATION AND IGNITION

The combustion of solid fuels begins by the ignition of volatilized gas after the evolution of volatile matter takes place in a high-temperature region. Although the devolatilization is believed to finish in about 100 ms, it is very difficult to accurately understand the devolatilization mechanism because devolatilization is an extremely complicated phenomenon. The devolatilization processes are strongly affected by coal properties, temperature, gas compositions, and so on, and these

TABLE 12.1   Analysis Results of Solid Fuels

		Coal A	Coal B	Refuse fuel[*1]	Biomass (Australian Bagasse)
High heating value[*2]	kcal/kg	7,270	7,030	—	—
Low heating value[*2]	kcal/kg	9,980	6,750	—	—
Proximate analysis					
Moisture	wt%	1.1[*2]	2.4[*2]	52.7[*2]	44.0–53.0[*2]
Ash	wt%	12.2[*3]	11.6[*3]	9.5[*2]	4.0–30.0[*3]
Volatile matter	wt%	29.1[*3]	34.1[*3]	—	75.0–87.0[*4]
Fixed carbon	wt%	58.7[*3]	54.3[*3]	—	—
Fuel ratio	—	2.02	1.59	—	—
Ultimate analysis					
Carbon	wt%	74.1[*3]	72.5[*3]	26.45[*2]	43.0–52.05[*4]
Hydrogen	wt%	5.29[*3]	5.57[*3]	3.93[*2]	5.2–6.9[*4]
Nitrogen	wt%	1.54[*3]	1.67[*3]	0.65[*2]	0.2–0.4[*4]
Oxygen	wt%	6.5[*3]	8.4[*3]	16.21[*2]	40.8–52.0[*4]
Total sulfur	wt%	0.34[*3]	0.32[*3]	0.08[*2]	0.02–0.11[*4]
Combustible sulfur	wt%	0.33[*3]	0.31[*3]	—	—

[1] An example of Japanese refuse fuel in Japanese urban areas. *Source:* Makino, H. and Ito, S., *J. Soc. Powder Technol. Japan,* 34, 247–254, 1997.

[2] Equilibrium moisture basis.

[3] Dry basis.

[4] Dry ash-free basis.

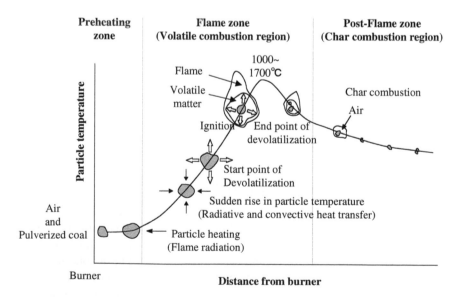

FIGURE 12.1 Combustion profile of a coal particle in pulverized coal combustion. [From Tominaga, H., in *Sangyou Nenshou Gijutsu,* The Energy Conservation Center, Japan, 2000, pp. 93–100. With permission.]

conditions change moment to moment in actual combustion fields. The devolatilization rate of volatile matter is commonly modeled by a single-step Arrhenius reaction scheme,[3] such as

$$\frac{dV}{dt} = K_v(V*-V) \tag{12.1}$$

$$K_v = A_v \exp\left(-\frac{E_v}{RT_p}\right) \tag{12.2}$$

where $V*$ and V indicate the total volatile matter content in a solid fuel particle and volatilized mass released from the particle, T_p the particle temperature, and R the universal gas constant. Kinematic parameters of preexponential factor A_v and activation energy E_v in Equation 12.2 are determined experimentally.[4] Besides, a two-step Arrhenius reaction scheme[5] is also proposed to improve the adaptability.

The definition of ignition temperature has not been explicitly settled yet, since the measurements of ignition temperature of solid fuels with complicated properties are very difficult. Typically, the ignition temperature of brown coals, whose volatile matter content is relatively high, is about 250°C, and that of anthracite, which is a low-volatile coal, is about 500°C. The ignition temperature of bituminous coals, which are generally used in pulverized coal combustion in Japanese coal-fired thermal power plants, is 300–400°C.

For the solid fuels with nonvolatile matter, the ignition temperature is regarded as the temperature at which the heat generated by the combustion on the surface of a solid fuel is beyond the heat loss due to convection and radiation.

2.12.3 GASEOUS COMBUSTION

The combustion of the volatilized gas due to the evolution of volatile matter with combustion air takes place in the gaseous phase. The simple model of gaseous combustion is as follows. The chemical mechanism has two global reactions:

$$C_aH_bO_c + 0.5O_2 \rightarrow \alpha CO + \beta H_2O$$
$$CO + O_2 \rightarrow CO_2 \tag{12.3}$$

where $C_aH_bO_c$ is the volatilized fuel gas, and a, b, c, α, and β are governed by the coal property. The gaseous combustion provided in Equation 12.3 is often calculated using the combination of the kinetics and eddy dissipation models.[6] Regarding the kinetics, the rate of reaction for reactants such as $C_aH_bO_c$ is given as an Arrhenius expression:

$$R_g = A_g \exp\left(-\frac{E_g}{RT_g}\right)[\text{Reactant}]^d[O_2]^e \tag{12.4}$$

where $[\psi]$ means the mol fractions of chemical species ψ. The values of the preexponential factor A_g, activation energy E_g, and orders d and e are determined experimentally.[7,8]

Since the gaseous combustion between volatilized gas and combustion air is strongly affected by the mixing of them, it should be discussed with flow behavior. The reaction rate of volatilized gas is fast, but soot particles are formed under the condition that the mixing of volatilized gases and combustion air is not enough. Figure 12.2[9] shows the emission characteristics of fine particles of less than 1 μm against the excess O_2 concentration in the exhaust gas. As the excess O_2 concentration decreases, the concentration of particles of the order of 0.1 μm, which are thought to be soot, increases.

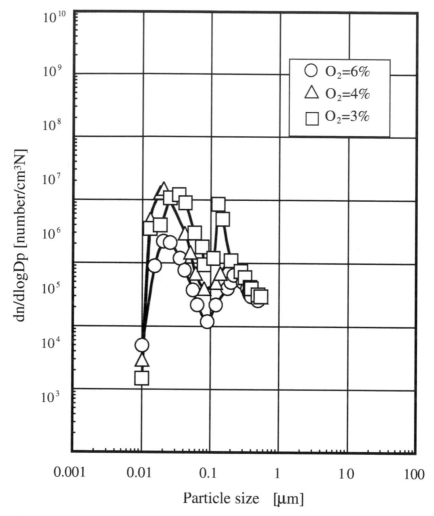

FIGURE 12.2 Emission characteristics of fine particles from pulverized coal combustion. [From Makino, H., *J. Aerosol Research Japan*, 4, 206–210, 1989. With permission.]

2.12.4 SOLID COMBUSTION

The combustion of fixed carbon, which is often referred to as char, is explained. The char burning rate is modeled calculated using Field *et al.*'s model[10]:

$$\frac{dC}{dt} = -\left(\frac{K_c K_d}{K_c + K_d}\right) P_g \pi D_p^{2} \qquad (12.5)$$

$$K_d = \frac{5.06 \times 10^{-7}}{D_p}\left(\frac{T_p + T_g}{2}\right)^{0.75} \qquad (12.6)$$

$$K_c = A_c \exp\left(-\frac{E_c}{RT_p}\right) \qquad (12.7)$$

where C is the char mass, K_c and K_d are the chemical and diffusion rate coefficients, respectively, and P_g is the partial pressure of oxygen in the bulk gas. This model is obtained under the assumption that the char burning rate is controlled by both the chemical reaction rate and the diffusion rate of oxygen to the surface of the char particle. The values of the kinematic parameters of the preexponential factor A_c and activation energy E_c in Equation 12.7 are determined experimentally.[4] It is considered that the char burning rate is dominated by the chemical reaction rate at a temperature less than 1000°C, whereas it is dominated by the diffusion rate at higher temperatures.

The remaining particles consist of ash and char. If the particle temperature is higher than the melting points of the particles, the particles become spherical due to the surface tension and solidify again as the particle temperature decreases. The remaining combustible char is exhausted as pure char particles or contained in ash particles.

REFERENCES

1. Makino, H. and Ito, S., *J. Soc. Powder Technol. Jpn.*, 34, 247–254, 1997.
2. Tominaga, H., in *Sangyou Nenshou Gijutsu,* The Energy Conservation Center, Japan, 2000, pp. 93–100.
3. Van Krevelen, D. W., Van Heerden, C., and Huntjens, F. J., *Fuel,* 30, 253–258, 1951.
4. Kurose, R., Tsuji, H., and Makino, H., *Fuel,* 80, 1457–1465, 2001.
5. Kobayashi, H., Howard, J. B., and Sarofin, A. F., in *Sixteenth Symposium (International) on Combustion,* The Combustion Institute, 1976, pp. 411–425.
6. Magnussen, B. F. and Hjertager, B. W., in *Sixteenth Symposium (International) on Combustion,* The Combustion Institute, 1976, pp. 719–729.
7. Borman, G. L. and Ragland, K. W., in *Combustion Engineering,* McGraw-Hill, 1998, pp. 120–122.
8. Kurose, R., Makino, H., and Suzuki, A., *Fuel,* 83, 693–703, 2004.
9. Makino, H., *J. Aerosol Res. Jpn.,* 4, 206–210, 1989.
10. Field, M. A., Gill, D. W., Morgan, B. B., and Hawksley, P. G. W., *The Combustion of Pulverised Coal,* British Coal Utilisation Research Association, Leatherhead, Surrey, 1967.

2.13 Solubility and Dissolution Rate

Yoshiaki Kawashima

Gifu Pharmaceutical University, Mitahora-Higashi, Gifu, Japan

2.13.1 SOLUBILITY OF FINE PARTICLES

Finely divided particles have a greater solubility than large particles. Therefore, the smaller particles will dissolve and their mass will reprecipitate on the larger particles. But solubility cannot be simply related to gross particle size, because it depends somewhat on the particle (crystal) face exposed, surface roughness, and irregularity. Under ideal conditions, the solubility of spherical particles is expressed by

$$\log \frac{a_2}{a_1} = \frac{2\sigma_m M}{2.303 \rho RT} (\frac{1}{r_2} - \frac{1}{r_1}) \tag{13.1}$$

Assuming that the activity of the solution is proportional to the molar concentration, Equation 13.1 can be rewritten as

$$\log \frac{S_2}{S_1} = \frac{2\sigma_m M}{2.303 \rho RT} \frac{1}{r_2} \tag{13.2}$$

2.13.2 FACTORS TO INCREASE SOLUBILITY

Particle (Crystal) Size

The size and shape of fine particles (diameter, $\cong \mu m$) affect the solubility, which increases with decreasing particle size as predicted by Equation 13.2.

Crystalline Form (Polymorphous and Amorphous Form)

Polymorphism occurs owing to different molecular arrangements in the solid phase, resulting in two or more crystalline forms. The difference in crystal energy of polymorphs generally leads to different physical properties, such as melting point, solubility, density, and so on, although the crystals are chemically identical. Metastable solid polymorphs having higher thermodynamic activity increase solubility, resulting in the improved bioavailability of a poorly soluble drug. Chloramphenicol palmitate is a representative drug. The solubility of polymorph B of chloramphenicol palmitate is roughly two times that of polymorph A.[1] Phenylbutazone has five different polymorphic forms. Among them Form I is thermodynamically most stable, and its equilibrium solubility is the lowest.[2] The polymorphism effect of cimetidine with four different crystalline forms (A, B, C, D) on stress ulceration in the rat was reported.[3] Amorphous forms have clearly the highest free energy, resulting in the largest solubility ratio. Amorphous forms are sometimes used rather than crystalline forms, to increase the solubility and bioavailability of antibiotics (e.g., novobiocin).[4] Solubility ratios of some other amorphous drugs compared to those of crystalline form are listed in Table 13.1.[5,6]

TABLE 13.1   Solubility Ratio for Amorphous to Crystalline Form

Drug (Temp. in °C)	More Soluble Phase / Less Soluble Phase	Solubility Ratio	Ref.
Caffeine (25)	Amorphous/crystalline	6.5	(5)
Diacetylmorphine (25)	Amorphous/crystalline	16	(5)
Theophylline (17)	Amorphous/crystalline	58	(5)
Theobromine (16)	Amorphous/crystalline	50	(5)
Morphine (20)	Amorphous/crystalline	268	(5)
Hydrochlorothiazide (37)	Amorphous/crystalline	1.1	(6)
Bendrofluazide (37)	Amorphous/crystalline	2.8	(6)
Cyclothiazide (37)	Amorphous/crystalline	6.2	(6)
Cyclopenthiazide (37)	Amorphous/crystalline	8.3	(6)

Cosolvents

A solute is frequently more soluble in a mixture of solvents (e.g., water and water-miscible organic solvent) than in one solvent alone (e.g., water). This phenomenon is known as cosolvency, and the mixed solvents are called cosolvents. The solubility of phenobarbital in a water–alcohol–glycerin mixture increases dramatically compared to that in water.[7] A linear relationship between the logarithm of the observed solubility (S_m) of nonpolar nonelectrolyte solutes in a cosolvent–water mixture and the volume fraction of the cosolvent (f) was found, as shown in Figure 13.1.[8]

$$\log S_m = S_w + \sigma f \tag{13.3}$$

Surfactants

Surface active agents in solution form the micelles above the critical micelle concentration at which insoluble or poorly soluble solids are solubilized. This solubilization results in increasing the solubility of particles. Comprehensive reviews of solubilization in surfactant systems were carried out by Swarbrick[9] and Elworthy et al.[10]

Complexation

When poorly soluble substances interact with a second substance added in solution to form soluble complexes, the solubilities of poorly soluble substances can be improved. The additive, called a solubilizing agent, whose optimum concentration in solution increases solubility, should be properly chosen. Representative solubilizing agents are listed in Table 13.2.

2.13.3   THEORIES OF DISSOLUTION

Diffusion in a Liquid Film

When a solid particle is agitated in liquid and allowed to dissolve, Noyes and Whitney,[11] Nernst,[12] and Brünner and Tolloczko[13] assumed a stagnant liquid film, called a diffusion layer, around the particle surface. The concentrations at the solid surface and at the outside of the diffusion layer are assumed to be

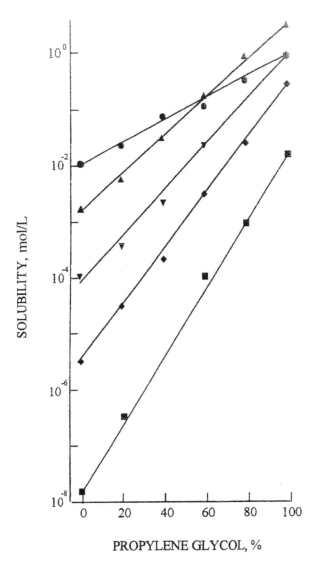

SOLUBILITY, mol/L

PROPYLENE GLYCOL, %

FIGURE 13.1 Dependence of solubility of some alkyl
p-aminobenzoates upon solvent composition. All measurements were performed at 37°C. Key: ●, ethyl; ▲, butyl; ▼,
hexyl; ◆, octyl; ■, dodecyl. [From Yalkowsky, S. H., Flynn,
G. L., and Amidon, G. L., *J. Pharm. Sci.*, 61, 983–984, 1972.
With permission.]

equal to the particle solubility (C_s) and the bulk concentration (C), respectively, as shown in Figure 13.2.
A steady state is assumed and Fick's first law is employed to derive the dissolution rate equation:

$$\frac{dC}{dt} = \frac{DS}{Vd}(C_s - C) = k\frac{S}{V}(C_s - C) \tag{13.4}$$

The thickness of the diffusion layer δ at the liquid–solid interface of the rotating disk at a constant
angular velocity was derived by Levich[14]:

$$\delta = 1.612 \times D^{1/3}V^{1/6}W^{-1/2} \tag{13.5}$$

TABLE 13.2   Solubilizing Agents for Poorly Soluble Substances

Poorly soluble substances	Solubilizing agents
Iodine	Potassium iodide
Caffeine	Sodium benzoate, sodium salicylate
Theobromine	Sodium salicylate
Calcium theobromine	Calcium salicylate
Mephenesin	Salicylic acid
Quinine hydrochloride	Urethane, carbamide
Theophylline	Sodium acetate, ethylenediamine
Riboflavin	Nicotinic amide

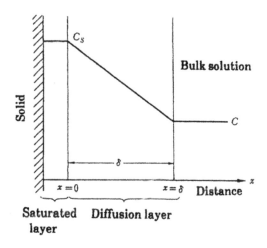

FIGURE 13.2 Concentration profile at solid–liquid interface in Noyes–Nernst model. [From Nernst, W., *Z. Phys. Chem.*, 47, 52–55, 1904; Noyes, A. and Whitney, W. I., *Z. Phys. Chem.*, 23, 689–692, 1897. With permission.]

Hixon and Crowell[15] derived a dissolution rate equation, called the cube root law for spherical, cubic, or cylindrical monodispersed particles, assuming an isotropic dissolution and sink condition ($C_s \gg C$):

$$W_0^{1/3} - W^{1/3} = (\frac{\pi N P}{6})^{1/3} \frac{2DC_s}{\delta P} t \qquad (13.6)$$

Surface Renewal of Solvent Surrounding Particle by Eddy Diffusion

Danckwerts[16] assumed that the surface of a solid particle dispersed in liquid is continuously renewed with fresh liquid by eddy diffusion. According to this model, there is not a stagnant liquid film and the solute concentration at the surface of the particle is not C_s, but a lower limiting concentration,

C_i. The solute dissolves into renewed fresh liquid during its stay at the interface. The solvent with the dissolved solute is continuously replaced by fresh solvent. The model proposed is described by Equation 13.7:

$$\frac{dm}{dt} = S\sqrt{\gamma D}(C_s - C) \tag{13.7}$$

Reaction in a Diffusion Layer

When the dissolved molecules or ions from solid particles react in the diffusion layer, the Nernst–Brunner model exhibited schematically in Figure 13.3 is applicable. Here, the solid particle is an acid, HA, the solute is a base, BOH, and δ is the effective diffusion layer in which the dissolved HA and the BOH diffused from bulk react. Dissolutions of benzoic acid, sulfadiazine, and triethanolamine in a strong aqueous basic solution are explained by this model.

Double-Layer (Reaction and Diffusion Layers) Model

Kawashima and Takenaka[17] proposed a double-layer (i.e., reaction and diffusion layers) model for acid neutralization with antacid (e.g., magnesium carbonate) by considering hydrogen ion transfer from the bulk through the diffusion layer into the neutralizing reaction layer on the solid surface. A steady concentration of hydrogen ions in the reaction layer is established, and the net rate of change in hydrogen ion concentration is set to zero. By assuming the neutralization reaction in the reaction layer is of first order with respect to hydrogen ion concentration, the acidity change (i.e., pH change) is described by Equation 13.8:

$$pH - pH_o - \frac{k_1 k_2 A}{2.303V(k_1 A + k_2)} t \tag{13.8}$$

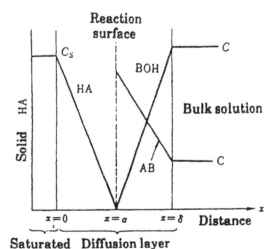

FIGURE 13.3 Concentration profile in Nernst–Brunner Model. [From Brünner, E. and Tolloczko, S. T., *Z. Phys. Chem.*, 47, 56–102, 1904; Nernst, W., *Z. Phys. Chem.*, 47, 52–55, 1904. With permission.]

Chemical Equilibrium in a Diffusion Layer

Higuchi et al.[18] postulated that chemical equilibrium is established between the dissolved molecules and ions from solid particles in the diffusion layer. The dissolution of sulfadiazine particles in dibasic sodium phosphate is explained by this mechanism.

Dissolution of a Metastable Solid

Metastable solid particles such as anhydride dissolve faster than stable solids such as hydride. The concentration of the resultant solution becomes higher than that of the saturated solution of a stable solid (i.e., supersaturated). The concentration of a supersaturated solution gradually decreases to the solubility of a stable solid. Nogami et al.[19] found that the dissolution of p-hydroxybenzoic acid and phenobarbital belongs to the foregoing case and proposed a model modified with the crystallization theory.

Dissolution Controlled by Reaction at the Solid–Liquid Interface

When the dissolution of a solid is controlled by the reaction at the solid–liquid interface, the active energy for dissolution is higher (e.g., >15 to 16 Kcal/mol) than that for dissolution-controlled diffusion (e.g., 5–6 Kcal/mol), although the rate equation of dissolution is identical with Equation 13.4.

Consecutive Process of Reaction and Diffusion

In this mechanism, dissolution is controlled by both the reaction at the solid–liquid interface and the diffusion of solute molecules in the diffusion layer. Therefore, the process is consecutive. Dissolutions of metastable crystals of prednisolone and barbital obey this mechanism.

2.13.4   MEASUREMENT OF DISSOLUTION RATE

The rotating disk method shown in Figure 13.4[20] and the stationary disk method shown in Figure 13.5 are useful to measure intrinsic dissolution rate, as the rate is controlled only by agitation speed in those methods. A typical constant surface area disk, shown in Figure 13.4, is constructed by mounting a tablet on a holder with paraffin wax such that the single surface of the tablet is exposed to the dissolution medium. Wood et al.[21] devised an assembly for compression of the tablet, which can also be used as the tablet holder during dissolution. The apparent dissolution rate can be measured by a simple beaker method.[22] The column method in Figure 13.6[23] can measure the dissolution rate automatically. The flow-through apparatus is operated under either closed mode when the fluid is recirculated or is of fixed volume, or open mode when there is continuous supply of the fluids. Such facilities can resolve some of the problems associated with nonsink conditions. The material under testing is placed in the vertically mounted dissolution cell in Figure 13.6. Yajima et al.[24] have devised a minicolumn running in open mode to evaluate the bitterness of clarithromycin dry syrup and found the threshold of the dissolution rate well correlated to bitterness (Figure 13.7).

2.13.5   METHODS TO INCREASE THE DISSOLUTION RATE

Agitation Speed

The diffusion model indicates that decreasing the thickness of the diffusion layer increases the dissolution rate. The thickness of the diffusion layer is a function of the stirring rate, as described previously. The dissolution rate is correlated experimentally with the stirring rate, as represented by

$$K_d = a N_v^b \qquad (13.9)$$

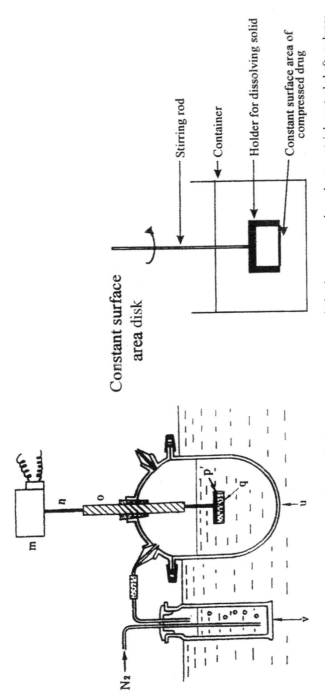

FIGURE 13.4 Rotating disk apparatus for dissolution rate measurement. m, induction motor and gearbox; n, stainless steel shaft; o, brass sleeve; p', disk holder; q, disk; u, flask; v, scrubbing bottle. [From Nogami, H., Nagai, T., and Suzuki, A., *Chem. Pharm. Bull.*, 14, 329–338, 1966. With permission.]

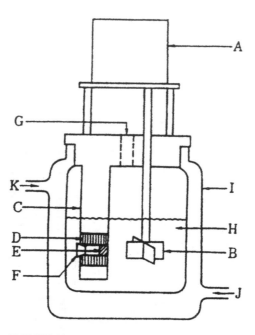

FIGURE 13.5 Stationary disk apparatus for dissolution rate measurement. B, stirrer; C, die holder; D, die; G, sampling port; H, solvent; I, jacked beaker; J, inlet for water. [From Simonelli, A. P., Mehta, S. C., and Higuchi, W. I., *J. Pharm. Sci.*, 58, 538–549, 1969. With permission.]

With the constant surface area disk method, Hersey and Marty[25] relates the intrinsic dissolution rate to the rotation rate of disk by

$$\frac{dW}{Sdt} = K_i R_n^{0.5} \tag{13.10}$$

Diffusion Coefficient

The diffusion coefficient, D, is defined by Einstein as represented by

$$D = \frac{\kappa T}{6\pi \eta \gamma_m} \tag{13.11}$$

Increasing the diffusion coefficient increases the dissolution rate, which can be accomplished by decreasing the viscosity of the dissolution medium.

Surface Area

The increasing effective surface area, which contacts the dissolution medium directly, effectively increases the dissolution rate. Grinding the material is one of the methods. Poorly soluble pharmaceuticals ground with microcrystalline cellulose, methylcellulose, chitin, or chitosan using a ball mill significantly increase the dissolution rate and solubility. In this process, crystalline drugs become disordered frequently, which results in increasing the solubilities.[26] Takahata et al.[27] found that copulverizing

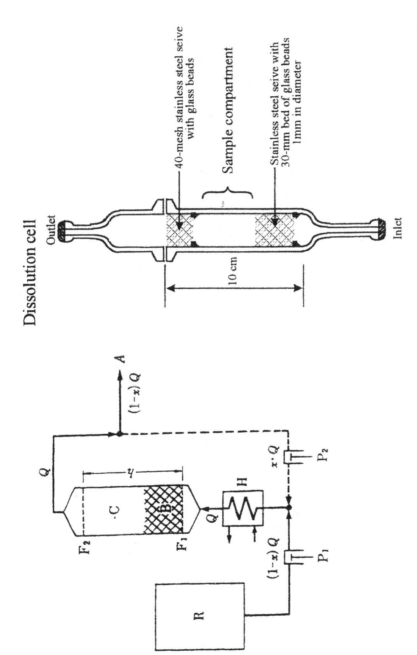

FIGURE 13.6 Apparatus for measurement of dissolution rate by column method. A, synchronous motor; B, particle bed; C, cell; F1, F2, screens; H, heat exchangeer; h, height of cell; P1, P2, volumetric pump; R, liquid reservoir; x, circulation factor; Q, volumetric flow rate. [From Langenbucher, F., *J. Pharm. Sci.*, 58, 1265–1272, 1969. With permission.]

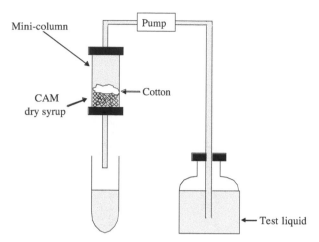

FIGURE 13.7 Minicolumn apparatus for evaluating the bitterness of CAM dry syrup.

poorly soluble pharmaceuticals, such as griseofulvin, oxolinic acid, phenytoin, and so forth, with water-soluble additives (e.g., mannitol, sorbitol, glucose, glycine, etc.) was significantly useful to submicronize such drug particles without changing crystallinities. The submicronized particles dramatically increased their dissolution rates in aqueous medium and bioavailabilities in dogs more than those of conventionally pulverized particles (average diameter, 2–3 μm) without copulverizing additives. Müller et al.[28] prepared nanosuspension (nanoparticle <1 μm suspension) by high-pressure homogenization of a poorly soluble drug microparticle suspension with surfactant to improve its bioavailability by increasing saturation solubility.

Solubility

The dissolution rate can be controlled by the solubility, as indicated by the dissolution model. The solubility varies with the solution pH, salt formation, solubilization by surface-active agents, change in crystal form, complexation, and sufficient reduction in particle size.

Polymorphs

The use of metastable polymorphic forms of a chemical is effective in increasing the solubility and dissolution rate. Many organic chemicals and drugs, such as prednisolone, aspirin, ampicillin, barbital, benzoic acid, sulfathiazole, chloramphenicol, and novobiocin, exhibit polymorphisms. Amorphous forms are also effective in increasing the dissolution rate. Takeuchi et al.[29] created a stable supersaturated system of indomethacin by dissolving its amorphous forms with porous silica prepared by spray drying.

Coprecipitate and Inclusion Compound

Coprecipitation of a poorly soluble drug, such as sulfathiazole[30] and griseofulvin,[31] with polyvinylpyrrolidone, polyethylene glycol, or urea can increase the dissolution rate. An inclusion compound with cyclodextrin can also increase the dissolution rate.[32] The increase in rate is mainly due to the increase in solubility and/or the decrease in crystallinity of the drug by inclusion complexation. The solubilities of cyclodextrin–drug complexes can be improved by utilizing chemically modified cyclodextrin such as dimethyl-β-cyclodextrin. Dimethyl-β-cyclodextrin facilitates particularly the solubilities of steroid hormones, cardiac glycosides, and fat-soluble vitamins.[33] Significant solubility

TABLE 13.3   Solubility Enhancement through the Use of High Concentrations of HP-β-CD

Solute	%HP-β-CD	Solubility Enhancement
Estriol	50	13,666
Estradiol	40	7,000
Progesterone	40	2,266
Spironolactone	40	1,400
Testosterone	40	1,461
Digoxin	50	971
Dexamethasone	50	240
Chlorthalidone	50	87.5
Diphenylhydantoin	50	57
Furosamide	50	24
Nitroglycerin	40	8.3
Acetamidopen	50	6
Apomorphine	50	5.8
Theophylline	50	1.3

enhancement of drug with high concentration of hydroxypropyl-β-cyclodextrin was found by Pitha et al.,[34] as shown in Table 13.3.

Dispersing on Adsorbent or Disintegrant

Poorly soluble drugs can increase their dissolution rate by being adsorbed in fine adsorbents such as fumed silicon dioxide, charcoal, or montomorillonite clay.[35,36] Takeuchi et al.[37] improved the dissolution rate of a poorly water-soluble drug (e.g., tolbutamide) by depositing the fine drug crystals on a disintegrant (e.g., partly pregelatinized cornstarch by a spray-drying solvent deposition method).

Notation

A	Surface area of antacid
a	Constant
a_1	Activity of the solute in a solution produced with larger particle
a_2	Activity of the solute in a solution produced with smaller particle
b	Constant
C	Concentration of solute in bulk liquid
Cs	Solubility of particle
D	Diffusion coefficient of solute
f	Volume fraction of cosolvent
K	Constant
K_d	Dissolution rate constant
K_i	Intrinsic dissolution rate constant
k_1	Mass transfer coefficient
k_2	Reaction rate constant
M	Molecular weight of particle
m	Dissolved mass

N Number of particles
N_v Agitation speed
pH_o Initial pH
R Gas constant
R_n Rotation rate of disk
r_1 Radius of larger particle
r_2 Radius of smaller particle
r_m Radius of molecule
S Surface area
S_w Solubility of drug in water
S_1 Solubility of larger particle
S_2 Solubility of smaller particle
T Absolute temperature
t Time
V Volume of medium
W Weight of particles at time t
W_o Initial weight of particles
X Distance from solid surface
γ Rate of surface renewal
δ Thickness of diffusion layer
η Viscosity of medium
κ Boltzmann constant
ν Kinematic viscosity of medium
ρ Density of particle
σ Slope of a plot of log S_m against f
σ_m Mean interfacial tension
ω Angular velocity

REFERENCES

1. Aguiar, A. J., Krc, J., Kinkel, A. W., and Samyn, J. C., *J. Pharm. Sci.*, 56, 847–853, 1967.
2. Tuladhar, M. D., Carless, J. E., and Summers, M. P., *J. Pharm. Pharmacol.*, 35(5), 269–274, 1983.
3. Kokubo, H., Morimoto, K., Ishida, T., Inoue, M., and Morisaka, K., *Int. J. Pharm.*, 35, 181–183, 1987.
4. Mullins, J. D. and Macek, T. J., *J. Am. Pharm. Assoc. Sci. Ed.*, 49, 245–248, 1960.
5. Toffoli, F., Avico, U., Signoretti, C. E., DiFrancesco, R., and Di Palumbo, V. S., *Ann. Chem.*, 63, 1–4, 1973.
6. Corrigan, O. I., Holohan, E. M., Sabra, K., *Int. J. Pharm.*, 18, 195–200, 1984.
7. Krause, G. M. and Cross, J. M., *J. Am. Pharm. Assoc. Sci. Ed.*, 40, 137–139, 1951.
8. Yalkowsky, S. H., Flynn, G. L., and Amidon, G. L., *J. Pharm. Sci.*, 61, 983–984, 1972.
9. Swarbrick, J., *J. Pharm. Sci.*, 54, 1229–1237, 1965.
10. Elworthy, P. H., Florence, A. T., and Macfarlane, C. B., in *Solubilization by Surface Active Agents*, Chapman & Hall, London, 1968, p. 335.
11. Noyes, A. and Whitney, W. I., *Z. Phys. Chem.*, 23, 689–692, 1897.
12. Nernst, W., *Z. Phys. Chem.*, 47, 52–55, 1904.
13. Brünner, E. and Tolloczko, S. T., *Z. Phys. Chem.*, 47, 56–102, 1904.
14. Levich, V. G., *Acta Physicochim URSS*, 17, 257–307, 1942.
15. Hixon, A. and Crowell, J., *Ind. Eng. Chem.*, 23, 923–931, 1931.
16. Danckwerts, P. V., *Ind. Eng. Chem.*, 43, 1460–1467, 1951.
17. Kawashima, Y. and Takenaka, H., *J. Pharm. Sci.*, 63, 1546–1551, 1974.
18. Higuchi, W. I., Parrott, E. L., Wurster, D. E., and Higuchi, T. J., *Am. Pharm. Assoc. Sci. Ed.*, 47, 376–383, 1958.
19. Nogami, H., Nagai, T., and Yotsuyanagi, T., *Chem. Pharm. Bull.*, 17, 499–509, 1969.

20. Nogami, H., Nagai, T., and Suzuki, A., *Chem. Pharm. Bull.,* 14, 329–338, 1966.
21. Wood, J. H., Syarto, J. E., and Letterman, H., *J. Pharm. Sci.,* 54, 1068, 1965.
22. Levy, G. and Hayes, B. A. N., *Engl. J. Med.,* 262, 1053–1058, 1960.
23. Langenbucher, F., *J. Pharm. Sci.,* 58, 1265–1272, 1969.
24. Yajima, T., Fukushima, Y., Itai, S., and Kawashima, Y., *Chem. Pharm. Bull.,* 50(2), 147–152, 2002.
25. Hersey, J. A., Marty, J., and Man, J., *Chem. Aerosol News,* 46(6), 43*, 1975.
26. Nakai, Y., *Yakugaku Zasshi,* 105, 801–811, 1985.
27. Takahata, H., Nishioka, Y., and Osawa, T., *Funtai to Kogyo (Powder and Industry),* 24, 53*, 1982.
28. Müller, R. H., Becker, R., et al., Int. Patent PCT/EP95/04401, 1996.
29. Takeuchi, H., Nagira, S., Yamamoto, H., and Kawashima, Y., *Powder Technol.,* 141, 187–195, 2004.
30. Simonelli, A. P., Mehta, S. C., and Higuchi, W. I., *J. Pharm. Sci.,* 58, 538–549, 1969.
31. Chiou, W. L. and Riegelman, S., *J. Pharm. Sci.,* 58, 1505–1510, 1969.
32. Hamada, Y., Nambu, N., and Nagai, T., *Chem. Pharm. Bull.,* 23, 1205–1211, 1975.
33. Uekama, K. and Otagiri, M., in *CRC Critical Reviews in Therapeutic Drug Carrier Systems,* Vol. 3, CRC Press, New York, 1987, p. 1.
34. Pitha, J., Milecki, H., Fales, H., Pannell, L., and Uekama, K., *Int. J. Pharm.,* 29, 73–82, 1986.
35. Kreuter, J., *Acta Pharm. Fenn.,* 90, 95–98, 1981.
36. Chiou, W. L. and Riegelman, S., *J. Pharm. Sci.,* 60, 1376–1380, 1971.
37. Takeuchi, H., Handa, T., and Kawashima, Y., *J. Pharm. Pharmacol.,* 39, 769–773, 1987.

2.14 Mechanochemistry

Mamoru Senna

Keio University, Yokohama City, Kanagawa, Japan

2.14.1 TERMINOLOGY AND CONCEPT

Mechanochemistry, first defined by Ostwald[1] at the beginning of the twentieth century, deals with an interplay between mechanical energy and chemical states of matters. Early mechanochemical studies were focused on the chemical transformation induced by gravitational and kinetic energy, as well as energies stored in solids. Mechanochemical phenomena were then broadly divided into two categories: those related with chemical reactions *per se,* or those dealing with the change in the activity or reactivity of solids. These are called mechanochemical reactions and mechanical activation, respectively. The boundary between these subdivisions is diffuse, however, since both of them involve common physicochemical changes of solids under the influence of mechanical energy.

Mechanochemistry was regarded from the beginning as an independent discipline of chemistry, like electro-, photo-, or irradiation chemistry. Biological tissues can also deform as a consequence of chemical changes, as in the origin of every movement of organisms, or in animals through nutrition. However, the latter aspect of the definition is excluded from this chapter, since it is far from powder technology. Tribochemistry, or impact chemistry, came from a different origin but deals with almost the same subjects as mechanochemistry.[2] These different but similar expressions are being unified as mechanochemistry.

There are a number of monographs or books devoted, at least partly, to this field. They are authored, among others, by Boldyrev and Meyer,[3] Kubo,[4] Beke,[5] Heinicke,[2] Butyagin,[6] Tkacova,[7] Juhasz and Opoczky,[8] Gutman,[9,10] Balaz,[11] and Avvakumov et al.[12]

2.14.2 PHENOMENOLOGY OF MECHANOCHEMISTRY

Mechanochemistry originated from industrial phenomenology, in other words, from mineral processing and the related practices of comminution and grinding. Downsizing of solid particles by grinding or milling cannot continue infinitely to individual atoms but is limited generally to a single micron regime or, at most, to a fraction of micrometer, as far as the size of separately available single particles is concerned. The lower limit of comminution comes mainly from two factors: microplasticity[13] and agglomeration. The first concept is understood as an increasing tendency of plastic deformation when the particle size becomes smaller than several micrometers.[14] Below this critical size, it becomes much more difficult to induce fracture via a crack formation and propagation. Increase in the surface free energy prevents powders from limitless size decrease but promotes agglomeration. Fine particles are more than a fragment of solids with a smaller dimension. Excess surface energy, which plays an important role in mechanochemistry, is partly due to a small radius of curvature and surface defects. In addition, the downsizing operation brings about severe defects, which directly combine with surface and bulk chemical properties of solids. Mechanochemistry is, therefore, a very general concept, which all scientists and engineers should understand, as long as they are dealing with fine particulate materials.

Mechanochemical reaction during milling has long been known. Phase transformations of crystalline solids are extensively studied, not only from a metastable to a stable phase, but also many unusual transformations, from an otherwise stable phase to a metastable phase. An apparent equilibrium or mechanochemical stationary state is often attained. Likewise, mechanochemical dissociation of hydroxides and carbonates has been widely studied. Addition reactions, in their broadest definition, can be divided into two types, according to their reaction partner: with gases or liquids existing in a grinding milieu as a continuum, and with solids during milling a solid mixture. In the former category, stress corrosion or tribosorption is also included. A very large number of mechanochemical solid–solid reactions are reported, including mechanical alloying. Concrete examples are given below in their respective sections.

2.14.3 THEORETICAL BACKGROUND

The theoretical approach to mechanochemistry is complicated and manifold. Introduction of grinding limits due to the physicochemical changes of solids in the fine grinding was one of the earliest trials. Khodakov[15] took the following factors into account in his energy density equation for fine grinding: (1) plastic and elastic deformation, (2) surface free energy, and (3) excess work consumed by particle interaction and comminution of particles. All these factors are directly related to excess free energy and, hence, mechanical activation.

A triboplasma model[16] is one of the most famous models, visually explaining triboluminescence and other processes of relaxation. Efforts were mainly paid to verify that mechanochemical processes are fundamentally different from thermal processes, taking place in very limited areas, termed *hot spots,* during mechanical stressing. This is evidenced from different kinetic orders between thermal and mechanical ways, as well as different decomposition products.[17] It was established that the state of excitation under impact stressing is more likely to be that of photochemistry in the sense that electronic energy can be in some of the excited states.[17] Electron excitation is not possible by usual thermal processes.

Factors associated with the periodical nature of practical machines, notably high-intensity mills, where stressing and relaxation take place simultaneously, were introduced.[18,19] Accumulation of energy and related mechanochemical changes, then, are described as a function of frequency of mechanical stressing. There was a trial to explain the apparent stationary state of mechanochemical change as that of higher order, on the basis of thermodynamics of irreversible systems.[20]

Energy transfer from any equipment to the body of fine particles is decisive for the argument of mechanochemical changes. It is important to notice that most of the energy once absorbed is converted into a joule effect.[6] This can be regarded as a useless dissipation of energy, lowering the grinding efficiency, from the viewpoint of comminution.[21] However, a certain amount of energy may be retained in the substance to give an excess energy, which serves as the origin of elevated activity and reactivity of solids. Higher reactivity of solids is beneficial in many industrial aspects. Energy storage in the particles as a result of mechanical treatment, often called as excess energy, or excess free energy when entropy term is also taken into account, is after all the source of all the mechanochemical phenomena.[6]

Mechanochemical processes observed *ex situ* are the consequences of relaxation of the excess energy. Relaxation per reaction can be divided into physical and chemical ones. Butyagin[6] summarized these relations by using a concept of energy yield. The relative importance of the mechanochemical changes is determined by the excess free energy and the relative amount of physical and chemical relaxation times. Computer simulation studies have also been carried out semistatically[22] or dynamically.[6]

Modification of Gibbs thermodynamics was tried by Gutman[9] in order to apply it to the mechanochemical processes. He tried to explain dissolution of mechanically activated solids, notably metals, by using a concept of concentration polarization because of surface heterogeneity. This, together with stress corrosion, is still to be researched in the area of electrochemistry.

2.14.4 STRUCTURAL CHANGE OF SOLIDS UNDER MECHANICAL STRESS

Morphological changes observable under a conventional scanning electron microscope inevitably bring about changes of the internal structure of solids simultaneously. Direct observation of microplasticity was thoroughly made by Hess.[14] A number of model experiments were carried out for the purpose of basic understanding of the topochemical nature of mechanochemistry. Some were compared with laser chemistry or tribology.[23] Silicon single crystals were observed as one of the most easily available nearly ideal solids. Reactivity studies of Si(111) with well-defined stressing by indentation or scratching were carried out in detail for the purpose of studying the topochemical elementary processes,[24] as shown in Figure 14.1.

Every plastic deformation is related to a number of dislocations, which can store elastic energy in the core. Increase in the dislocation density further contributes to structural degradation and ends, in an extreme case, with amorphization. A trend of amorphization depends strongly on the nature of chemical bonds in a crystal. More anisotropic crystals tend to cleave with less extensive structural damage, while crystals with homopolar, isotropic bonds with larger free space tend to amorphization on mechanical treatment.[25]

Thus, formation of lattice imperfection and amorphization of solids during fine grinding are very frequently observed and evaluated.[26] Broadening of X-ray diffraction peaks has been interpreted in detail.[27] Even glasses, which are noncrystalline from the beginning, are further damaged under mechanical stress by loosing connectivity of their structural units, notably SiO_4^{4-}, and the density of dangling bonds increases.[28]

When structural imperfection induces some kind of stacking faults, serving as embryos of other crystalline phases, topotactic polymorphic transformation can take place. These, together with other types of transformations have been observed, for example, for PbO, PbO_2, Sb_2O_3, ZrO_2, ZnS, γ-Fe_2O_3, γ-Al_2O_3, and $CaCO_3$.[29] Phase transformation of organic crystals can be used to improve the rate of dissolution or solubility,[30] which can be utilized for pharmaceutical purposes. It is also to be noted that there is an opposite process of amorphization or decomposition during mechanical treatments, namely, crystallization and grain growth. Grinding of iron oxyhydroxide is one such example.[31]

There is an idea of phase equilibrium under mechanochemical condition.[32,33] Particularly interesting is the formation of the metastable, high-pressure form. Some of the phase transformation systems have been studied with reference to their high-pressure chemistry.[34] While hydrostatic pressure can describe the stability range of the high-pressure form, they can be formed only when kinetic conditions

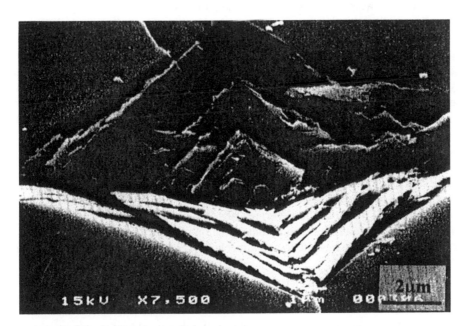

FIGURE 14.1 Scanning electron micrograph of a single crystal of Si(111), indented at 1.96 N and etched by a 1 N aqueous solution of HF. [From Katayama, K. and Senna, M., *Solid State Ionics*, 73, 127, 1994. With permission.]

are fulfilled, and shear stress is inevitable. This is associated with a fundamental profile of mechano-chemical structural change processes, where isotropic hydrostatic pressure plays only a partial role.

Hydroxides, carbonates, chlorates, nitrates, or similar thermally decomposable compounds can also be decomposed under mechanical stresses.[8,35,36] Enhanced reactivity of solids at the moment of structural change, including phase transformation, is known as the Hedvall effect.[37] This concept seems also to work in the course of mechanical treatment of solid mixtures. Carbonates or hydroxides, as well as some crystalline phases, easy to transform under mechanical stress, are therefore often superior to corresponding oxides as a starting material for solid-state reactions, although mechanochemical decomposition takes place via routes different from thermal ones. Organic solids can be caused to react with either organic or inorganic materials by grinding a mixture.[38,39] These phenomena can also be expected to be used in the field of pharmaceutics.

2.14.5   MECHANOCHEMICAL SOLID-STATE REACTION AND MECHANICAL ALLOYING

There have been numerous attempts to synthesize various complex compounds only by mechanochemical process, for example, by grinding a mixture of Ag_2SO_4 with ZnS or CdS.[40] Formation of zinc ferrite has also been reported.[41] For these mechanochemical solid-state reactions, material transport is accomplished via short-range diffusion by repeated close contact of dissimilar particles with the aid of high-density defects.

Mechanical alloying (MA) is one of the simplest reactions, but a very important example of *in situ* reaction. Studies on MA were stimulated for the production of a Ni-based superalloy with finely dispersed oxide particles[42] by high-energy ball milling. Afterward, an enormous number of trials were made to obtain various alloys by using energy-intensive mills. From its birth about a quarter century ago until recently, MA has been developed exclusively at the hands of metallurgists and metal physicists. Nowadays, they are jointly working with materials scientists, including mechanochemists. Metals and oxides have also been ground together to obtain an exchange reaction similar to thermal reactions.[43,44]

Mechanochemical reactions between solids and gases have been extensively studied. One of the most outstanding studies is that between hydrogen or hydrocarbons and silicon carbide.[45] Grinding SiC in a hydrogen atmosphere produced various saturated and unsaturated hydrocarbon compounds. This is strong evidence of elevated reactivity of SiC under the influence of mechanical stress. Reactions with a gaseous reaction partner with more complex systems have been compiled by Juhasz and Opoczky. The formation of metal carbonyls has been extensively studied.[35] Starting from Ni + CO or $NiCO_3$ and H_2S + CO, $Ni(CO)_4$ was obtained. For the latter, the following mechanisms are proposed: (1) $NiCO_3 \rightarrow NiO + CO_2$; (2) $NiO + H_2S \rightarrow NiS + H_2O$; and (3) $NiS + 4CO + H_2O \rightarrow Ni(CO)_4 + H_2S$. A direct synthesis of silicon tetrachloride and gaseous chlorine has also been reported.[47]

A mechanochemical route has an advantage of producing complex or composite materials by the use of apparatus similar to or identical with grinding mills, which are popular in most of industries. At the same time, mechanochemistry has turned out to be a versatile tool, enabling chemical reactions which are quite laborious or not possible via conventional synthetic methods. For solid-state mechanochemical reactions, however, reactions often take hundreds of hours, and hence the products have often suffered from serious contamination from the grinding elements. This is one of the reasons why people often hesitate to apply mechanochemical processes to industrial practices.

2.14.6   SOFT MECHANOCHEMICAL PROCESSES AND THEIR APPLICATION

Mechanochemical synthesis of precursors coupled with subsequent heat treatments can open a new way of synthesizing complex oxide materials,[48] since initial complex formation via a mechanical route can be done quickly enough under gentle conditions when a starting mixture contains ingredients with hydroxyl groups or water.[49] Examples and mechanistic arguments are given below.

Mg(OH)$_2$ remains crystalline even after intensive vibromilling for two days. When grinding with TiO$_2$, however, amorphization takes place in two hours, accompanied by dehydration.[50] One of the predominating mechanisms of this kind of complex formation is a proton transfer from solid acids such as silica or titania to a basic site of the hydroxides, where proton affinity is sufficiently large.[51] As a consequence of the proton transfer, two metallic species, for example, magnesium and titanium, are cross-linked, mediated by an oxygen atom. This is nothing but an incipient chemical reaction between Mg(OH)$_2$ and TiO$_2$, leading to a chemical complex. A complex oxide such as MgTiO$_3$ is conventionally prepared by mixing and firing MgO and TiO$_2$ powders above 1700 K. Grinding a mixture a little improves the fabrication process. However, when we replace MgO with Mg(OH)$_2$, the synthesis becomes much easier. As a matter of fact, MgTiO$_3$ was observed as a single phase on heating a mechanochemically treated mixture, while other mixtures resulted in different phases as well, as shown in Figure 14.2.[52] There are a number of similar examples of the system Sr–O–Ti leading to SrTiO$_3$.[53]

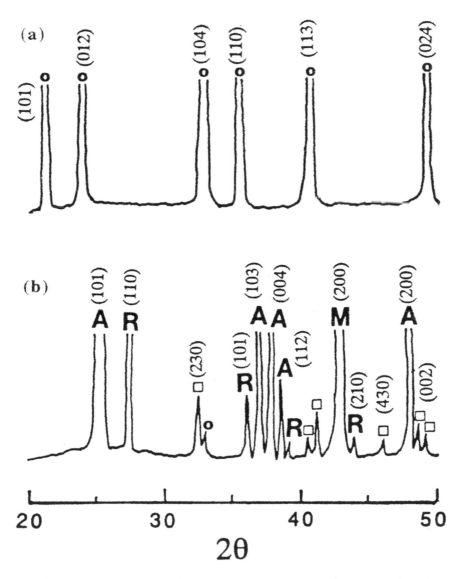

FIGURE 14.2 X-ray diffraction patterns of equimolar Mg(OH)$_2$/TiO$_2$ mixtures after calcining at 1173 K for 2 h with (a) and without (b) preliminary grinding for 1 h. A, anatase; R, rutile; M, MgO; squares, MgTi$_2$O$_5$; circles, MgTiO$_3$. [From Baek et al.[52] With permission.]

Very recently, some complex ferroelectric materials were synthesized on the basis of above-mentioned precursor mechanochemistry.[52] In contrast to many complicated procedures, application of mechanical stresses on precursors comprising PbO, MgO, Nb_2O_5, TiO_2, and $Mg(OH)_2$ in a simple stoichiometric proportion brought about pure perovskite already at the stage of calcination. Firing at increasing temperatures brought about crystal growth, enabling the maximum dielectric constants to reach as high as 30,000, with characteristics of a relaxer, as shown in Figure 14.3.

2.14.7   RECENT DEVELOPMENTS AND FUTURE OUTLOOK

Concepts and applications of mechanochemistry are ever expanding. In recent developments in mechanochemistry, it is particularly noteworthy that successful application is being directed to many high-profile technological fields. In the case of materials for microelectronics, it is essential to obtain phase pure and micrograined material. For that purpose, a soft mechanochemical process is particularly useful, since firing at low temperature suppresses grain growth.

In the field of fast-developing high-frequency telecommunication, hexaferrites are needed, but they are generally not easy to obtain as a pure phase. Phase pure Y-phase hexaferrites were obtained by firing at temperatures as low as 1000°C after appropriate mechanical activation of the starting mixture.[54] Likewise, ferroelectric relaxers such as solid solutions of PMN and PZN, the latter being a similar perovskite where Mg in PMN is substituted by Zn, are available with a PZN fraction up to 0.7 via a soft-mechanochemical route.[55]

In the field of organic chemistry, the concepts and principles of mechanochemistry are also increasingly utilized. In the field of polymeric species, there is a tradition of mechanochemistry for polymers.[2] A similar concept has been extended to degradation of polymers, for environmental purposes.[56]

Much more innovative, however, is the systematic application of mechanochemistry to organic synthesis. A number of peculiar phenomena in a mortar were reported without recognizing that they definitely involved mechanochemistry.[57] The mechanisms of such mechanochemical organic reactions

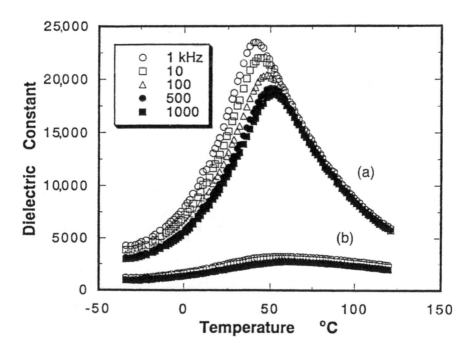

FIGURE 14.3 Change in the dielectric properties of PT-PMN with temperature after firing at 1473 K for 2 h with (a) and without (b) preliminary grinding for 1 h. [From Baek et al.[52] With permission.]

were also elucidated via a computer simulation study.[58] Mechanochromism is also an interesting phenomenon and related topic.[59]

Science and technology associated with mechanochemistry are developing with increasing acceleration. Reference to one of the latest reviews is recommended.[60]

REFERENCES

1. Ostwald, W., *Handbuch der allgemeinen Chemie,* Vol. 1, Akademie-Verlagsanstalt, Leipzig, 1919.
2. Heinicke, G., in *Tribochemistry,* Akademie-Verlag, Berlin, 1984, pp. 335–341.
3. Boldyrev, V. V. and Meyer, K., *Festkoerperchemie,* VEB Deutscher Verlag fuer Grundstoffindustrie, Leipzig, 1973.
4. Kubo, T., *Introduction to Mechanochemistry,* 2nd Ed., Tokyo Kagaku Dojin, Tokyo, 1978.
5. Beke, B., *The Process of Fine Grinding,* Martinus Nijhoff, The Hague, 1981.
6. Butyagin, P. Yu., *Sov. Sci. Rev. B,.* 14(1), 1989.
7. Tkacova, K., *Mechanical Activation of Minerals,* Elsevier, Amsterdam, 1989.
8. Juhasz, A. Z. and Opoczky, L., *Mechanical Activation of Minerals by Grinding,* Akademiai Kiado, Budapest, 1990.
9. Gutman, E. M., *Mechanochemistry of Solid Surfaces,* World Scientific, Singapore, 1994.
10. Gutman, E. M., *Mechanochemistry of Materials,* Cambridge International Science Publishing, 2000.
11. Balaz, P., *Extractive Metallurgy of Activated Minerals,* Elsevier, 2000.
12. Avvakumov, E., Senna, M., and Kosova, N., *Soft Mechanochemical Synthesis. A Basics for New Chemical Technologies,* Kluwer Academic Publishers, 2001.
13. Smekal, A. G., *Anz. Oesterr. Akad. Wiss.,* 92, 733, 1955.
14. Hess, W., Dissertation, Univ. Karlsruhe, 1980.
15. Khodakov, G. S., *Physics of Grinding,* Izd. Nauka, Moscow, 1972.
16. Thiessen, P.A., Meyer, K., Heinicke, G., in *Abh. dtsch. Akad. Wiss. Berlin,* Kl. Chem. Geol., Biol., Berlin, 1966, p. 11.
17. Boldyrev, V. V. and Heinicke, G., *Z. Chem.,* 19, 353, 1979.
18. Lyakhov, N. Z. and Boldyrev, V. V., *Izv. Sib. Otd. Nauk. SSSR. Ser. Thim.,* 3, 1982.
19. Lyakhov, N. Z., in *Proceedings of the Second Japan–Russia Symposium on Mechanochemistry,* Jimbo, G., Senna, M., and Kuwahara, Y., Eds., Soc. Powder Technol., Japan, Kyoto, pp. 59, 291.
20. Thiessen, K. P., *Z. Phys. Chem. Leipzig,* 260, 403, 1979.
21. Rumpf, H., *Aufber.-Tech.,* 2, 59, 1973.
22. Karagedov, G. R., in *Proceedings of the Fourth Japan–Russia Symposium on Mechanochemistry,* Jimbo, G., Kuwahara, Y., and Senna, M., Eds., Soc. Powder Technol., Japan, Kyoto, p. 137.
23. Meyer, K. and Meier, W., *Krist. Tech.,* 3, 399, 1968.
24. Katayama, K. and Senna, M., *Solid State Ionics,* 73, 127, 1994.
25. Steinike, U., Kretzschmer, U., Ebert, I., Henning, H.-P., *Reactivity of Solids,* 4, 1, 1987.
26. Fricke, R. and Gwinner, E., *Z. Phys. Chem.,* A183, 165, 1938.
27. Hall, W. H., *Proc. Phys. Soc.,* 62A, 741, 1949.
28. Zachariasen, W. H., *J. Am. Chem. Soc.,* 54, 3841, 1932.
29. Senna, M., *Cryst. Res. Technol.,* 20, 209, 1985.
30. Otsuka, M. and Kaneniwa, N., *J. Pharm. Sci.,* 75, 506, 1986.
31. Mendelovici, E., Villalba, R., and Sagaraz, A., *Mater. Res. Bull.,* 17, 241, 1982.
32. Schrader, R. and Hoffmann, B., *Z. Anorg. Allgem. Chem.,* 369, 41, 1969.
33. Iguchi, Y. and Senna, M., *Powder Technol.,* 43, 155, 1985.
34. Dachille, F. and Roy, R., *Nature,* 186, 34 and 71, 1960.
35. Heinicke, G. and Harenz, H., *Z. Anorg. Allgem. Chem.,* 329, 185, 1963.
36. Nonat, A. and Mutin, J. C., *Mater. Chem.,* 7, 455 and 479, 1982.
37. Hedvall, J., *Einfuehrung in die Festkoerperchemie,* Friedrich Vieweg u. Sohn Verl., Braunschweig, 1952.
38. Nakai, Y., Yamamoto, K., Terada, K., and Kajiyama, A., *Chem. Pharm. Bull.,* 33, 5110, 1985.
39. Dushkin, A. V., Nagovitsina, E. V., Boldyrev, V. V., and Druganov, A. G., *Siberian J. Chem.,* 5, 75, 1991.
40. Lin, I. J. and Somasundaran, P., *Powder Technol.,* 6, 171, 1972.
41. Lin, I. J. and Nadiv, S., *Mater. Sci. Eng.,* 39, 193, 1979.

42. Benjamin, J. S., *Met. Trans.,* 1, 2943, 1970.
43. Takacs, L., *Mater. Lett.,* 13, 119, 1992.
44. Yang, H. and MacCormick, P. G., *J. Solid State Chem.,* 110, 136, 1994.
45. Heinicke, G. and Hennig, H.-P., *Silikattech,* 14, 86, 1967.
46. Opoczky, L., *Powder Technol.,* 17, 1–7, 1977. See also Ref. 8.
47. Koester, A., *Angew. Chem.,* 69, 563, 1957.
48. Senna, M., *Solid State Ionics,* 3, 63–65, 1993.
49. Avvakumov, E. G., *Chemistry for Sustainable Development,* 2, 1, 1994.
50. Liao, J. and Senna, M., *Solid State Ionics,* 66, 313, 1994.
51. Watanabe, T., Liao, J., and Senna, M., *J. Solid State Chem.,* 115, 390, 1995.
52. Baek, D., Isobe, T., and Senna, M., *Solid State Ionics,* in press, 1996.
53. Kamei, Isobe, T., and Senna, M., *Mater. Sci. Eng., B,* in press, 1996.
54. Temuujina, J., Aoyamaa, M., Senna, M., Masukob, T., Andob, C., and Kishib, H., *J. Solid State Chem.,* in press, 2004.
55. Shinohara, S., Baek, J. G., Isobe, Isobe, Senna, M., *J. Am. Ceram. Soc.,* 83, 208, 2000.
56. Mio, H., Saeki, S., Kano, J., Saito, F., *Environ. Sci. Technol.,* 36 (6), 1344, 2002.
57. Murata, Y., Kato, N., Fujiwara, K., and Komatsu, K., *J. Org. Chem.,* 64, 3483, 1999.
58. Fajar Pradipta, M., Watanabe, H., and Senna, M., *Solid State Ionics,* in press, 2004.
59. Tipikin, D. S., *Russ. J. Phys. Chem.,* 75, 1720, 2001; *Zh. Fiz. Khim.,* 75, 1876–1879, 2001.
60. Zhang, D. L., *Progr. Mater. Sci.,* 49, 537–560, 2004.

Part III

Fundamental Properties of Powder Beds

3.1 Adsorption Characteristics

Masatoshi Chikazawa and Takashi Takei

Tokyo Metropolitan University, Hachioji, Tokyo, Japan

3.1.1 INTRODUCTION

Adsorption has been defined as the phenomenon that occurs when concentration of a component in an interfacial layer of vapor–solid, vapor–liquid, liquid–solid, or liquid–liquid becomes higher or lower than that of the bulk phase. The former case is called a positive adsorption and the latter a negative adsorption. The adsorption phenomenon should be distinguished from absorption, where adsorbate molecules penetrate the interior of a solid. In this section the interactions between vapors and solids, or between liquids and solids, are described. Vapor–solid systems are dealt with in detail.

Generally, adsorption phenomena are classified into two types: physical adsorption and chemical adsorption. In physical adsorption, attractive forces between individual adsorbate molecules and atoms or ions composing a solid surface originate in a van der Waals force that contains dispersion force and orientation forces of permanent and induced dipoles. If the solid surface consists of ions or polar groups, they will produce an electric field that induces dipoles in the adsorbed molecules. Moreover, if the adsorbate molecules possess permanent dipoles, their dipoles interact with the field, and hence, oriented adsorption occurs. On the other hand, chemical adsorption takes place only in special cases of vapor–solid and liquid–solid systems where adsorbate molecules are adsorbed by chemical bonds. Various adsorption phenomena are classified into these two types through detailed discussion of their mechanism. The adsorption mechanism, such as the existence of a chemical bond, can be studied by ultraviolet (UV), infrared (IR), and electron spin resonance (ESR) spectroscopy methods. Other methods (e.g., measurement of adsorption heats and adsorption isotherms) are necessary for further accurate classification.

Generally the heat of physical adsorption is smaller than that of chemical adsorption, and the rate of physical adsorption is faster than that of chemical adsorption. Moreover, physical adsorption and desorption processes are reversible, and a relatively large amount of molecules is adsorbed on a solid surface at temperatures below the boiling point of the adsorptive or under high concentration of adsorptive in solution. Namely, multilayer adsorption occurs. On the other hand, in chemical adsorption, chemical bonds must be formed between a solid surface and adsorbed individual molecule, so the multilayer observed in physical adsorption does not occur. Hence the coverage $\theta = V/V_m$ is always lower than unity, where V is the amount of adsorbed molecules and V_m is that for monolayer completion.

Powder properties, ease of handling, and various problems in powder processes are closely related to surface properties. Hence inspection of the surface properties of powder is very important. In order to estimate the surface properties such as surface area, pore structure, and surface chemical properties, adsorption techniques are widely used.

3.1.2 ADSORPTION MEASUREMENT

Measuring Methods

Vapor–Solid System

Measurements of the amount of adsorbed molecules are carried out by direct and indirect methods. Gravimetric and volumetric methods correspond to the direct method. In volumetric methods the adsorbed amount is calculated from the vapor pressure of adsorptive and known volume (dead volume) of an apparatus. On the other hand, IR and UV spectroscopy and other instrumental methods are classified as indirect methods. Generally, a direct method is more accurate than an indirect method. The adsorbed amount is expressed by ml STP or μmol per unit surface area or per unit weight of a sample powder.

Liquid–Solid System

An adsorbed amount of molecules or ions is determined from the difference in concentration of a component in a solution before and after adsorption. An adsorption isotherm in a liquid–solid system is similar to that in a vapor–solid system. For example, adsorption isotherms of alcohols or fatty acids that have a long hydrocarbon chain belong to the Langmuir-type isotherm. On the other hand, acetic acid adsorption on active carbon belongs to the Freundlich-type isotherm. The particular point of adsorption in a liquid–solid system is the existence of competition adsorption of solvent and adsorptive molecules.

Preparation of Samples

The pretreatment of samples is very important to obtain an accurate adsorbed amount. A clean surface is required for the adsorption measurement. The powder particles are usually handled or reserved under atmosphere. Therefore, water molecules are adsorbed according to their relative vapor pressure surrounding the powder. To obtain a clean surface of powder particles, elimination of the adsorbed water molecules is necessary. Especially in metal oxides, water molecules are chemically adsorbed, and various types of hydroxyl groups are formed on their surfaces. These hydroxyl groups are characteristic for each substance. Since the bonding strength of the surface hydroxyl groups varies with the nature of the metal oxides, the elimination temperatures of the hydroxyl groups are different from each other.

In Figure 1.1 the contents of the surface hydroxyl groups on several oxides are shown as a function of the pretreatment temperature T.[1] In all samples, the hydroxyl groups are markedly eliminated at temperatures above 400°C. When the temperature reaches 800°C, almost all hydroxyl groups are removed from the surfaces. The existence of surface hydroxyl groups has serious effects on the surface properties, such as selective adsorbability, water vapor affinity, catalytic activity, chemical activity, and electric field strength. Moreover, the hydroxyl groups play an important role in the adsorption of polar substances and molecules that can form hydrogen bonds. From these points of view, careful attention should be paid to determine the treatment temperature of sample powders. The surface properties may also be changed for other reasons. For example, the release of CO_2 from $CaCO_3$ powder is found at temperatures above 300°C.[2] In the cases of oxides, oxygen defects are produced by high-temperature treatment, and hence, the samples must be kept under oxygen pressure after pretreatment to remove the defects. A decrease in surface area usually occurs in the high-temperature treatment, which stabilizes the surface condition, causing disappearance of active sites and growing stable crystal planes.

3.1.3   THEORY OF ADSORPTION ISOTHERMS

Types of Isotherms

The adsorption phenomenon gives valuable information on the physical and chemical properties of a solid surface and is represented by an adsorption isotherm. In the case of a vapor–solid system, the isotherm is usually expressed by the relationship between the adsorbed amount V and the equilibrium vapor pressure P of an adsorptive at a fixed temperature T.

$$V = f_T(P) \tag{1.1}$$

On the other hand, in the case of a liquid–solid system, concentration (C) of adsorptive in a solution is used instead of vapor pressure.

$$V = f_T(C) \tag{1.2}$$

The other expressions obtained at a fixed pressure or at a constant amount of adsorbed molecules are given as follows:

$$V = f_P(T) \tag{1.3}$$

$$V = f_V(T) \tag{1.4}$$

Many equations for the adsorption isotherm are introduced by assuming various adsorption mechanisms. The isotherms of physical adsorption are grouped conveniently into eight classes. The typical classification is proposed by IUPAC.[3] Besides these isotherms, types of isotherms are shown in Figure 1.2, which is useful for an understanding of adsorption phenomena.

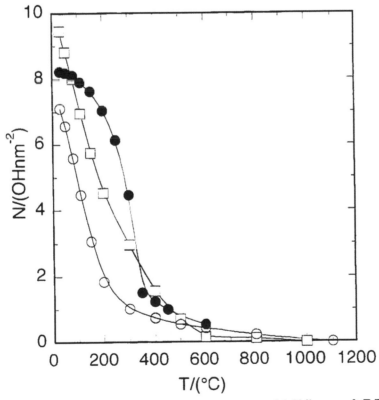

M.Chikazawa & T.Takei
III.1 Adsorption Characteristics

FIGURE 1.1   Change in surface OH groups by heat treatment. ○: α-Fe_2O_3; □: ZnO; ●: TiO_2. [From Morimoto, T., Nagao, M., and Tokuda, F., *Bull. Chem. Soc. Jpn.*, 41, 1533–1537, 1968. With permission.]

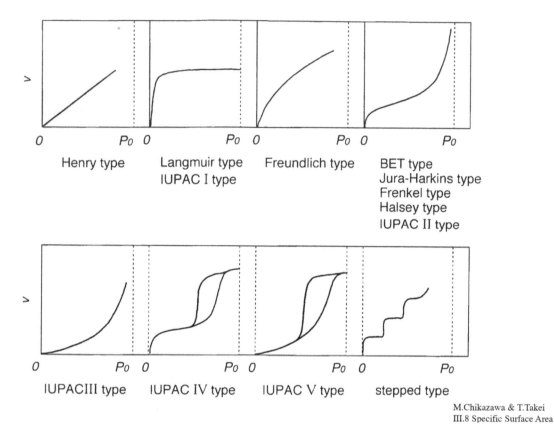

Henry type   Langmuir type   Freundlich type   BET type
              IUPAC I type   Jura-Harkins type
                                                   Frenkel type
                                                   Halsey type
                                                   IUPAC II type

IUPACIII type   IUPAC IV type   IUPAC V type   stepped type

M.Chikazawa & T.Takei
III.8 Specific Surface Area

FIGURE 1.2   Types of adsorption isotherms. This figure represents adsorption isotherms of gas–solid system. P_0, saturated vapor pressure at measurement temperature. In the case of a liquid–solid system, the abscissa indicates a concentration (C) of solute.

Equations of Adsorption Isotherms

Henry Equation

The relationship between an adsorbed amount V and an equilibrium pressure P is expressed by

$$V = k_{\mathrm{H}}P \tag{1.5}$$

where k_{H} is a constant. This equation holds for a small adsorbed amount below 1% of the mono-layer capacity. Even in the cases of the other equations, the adsorbed amounts at early stages are represented approximately by the Henry equation. The amount of O_2 or N_2 adsorbed at $0°C$ by a silica gel increases linearly with increasing pressure up to 1 atm pressure, and the isotherm is given by the Henry equation.[4]

Langmuir Equation

This equation is introduced from the most basic adsorption model, in which two assumptions are made. The first assumption is that the heat of adsorption is the same for every molecule and that there is no interaction among the adsorbed molecules. The second is that every molecule that col-lides with a molecule already adsorbed on a site reflects immediately to the vapor phase. Therefore, multilayer adsorption does not occur, and the maximum amount of adsorption is obtained at the monolayer completion, where the particle surface is just covered with closely packed molecules adsorbed on the specific adsorption sites. From such adsorption mechanisms, the Langmuir equation is derived as follows.[5] Adsorption velocity is proportional to the number of molecules that strike the bare surface of powder particles. On the other hand, desorption velocity is closely related to the adsorbed amount or the surface coverage θ. In equilibrium, the velocities of adsorption and desorption are equal to each other, and hence the following equation is derived:

$$k_{ad}(1-\theta)P = k_d\theta \tag{1.6}$$

where k_{ad} and k_d represent the rate constants of adsorption and desorption, respectively. The value of θ is given by V/V_m. Hence Equation 1.6 is rewritten as follows:

$$V = \frac{V_m(k_{ad}/k_d)P}{1+(k_{ad}/k_d)P} = \frac{AV_mP}{1+AP} \tag{1.7}$$

Especially when the equilibrium pressure is very small, Equation 1.7 will accord with the Henry equation (Equation 1.5).

Freundlich Equation

The Freundlich equation is represented by Equation 1.8 and widely used in the case of liquid–solid systems:

$$V = k_F P^{1/n} \tag{1.8}$$

where k_F is a constant and n is a value in the range 1 to 10. This equation was obtained experimentally. If the isosteric heat of adsorption decreases linearly with increase in $\ln V$, the adsorption isotherm is expressed by the Freundlich equation. It is known that the measurement temperature and pretreatment temperature of a sample affect the n value in Equation 1.8.

BET Equation

In order to understand the physical adsorption, the BET equation is the most widely used. This equation is derived by expansion of the Langmuir adsorption model to multilayer adsorption under a few assumptions. The isosteric heat of adsorption for molecules other than in the first layer is equal to the liquefaction heat of an adsorptive, interaction among the adsorbed molecules at same layers is not produced, and at saturated vapor pressure the number of adsorbed layers is infinite. From these assumptions, the BET equation is obtained[6]:

$$V = \frac{V_mCX}{(1-X)(1-X+CX)} \tag{1.9}$$

where X is the relative vapor pressure P/P_o, C is the value related to the strength of adsorption force, P_o is the saturated vapor pressure at measurement temperature, and V_m is the monolayer capacity. Detailed derivation of the BET equation is abbreviated here. If the adsorbed layer is restricted to the nth layer, the BET equation becomes

$$V = \frac{V_mCX\left[1-(n+1)X^n+nX^{n+1}\right]}{(1-X)\left[1+(C-1)X-CX^{n+1}\right]} \tag{1.10}$$

In the case of $n = 1$, Equation 1.10 accords with the Langmuir equation.

The C value, which is often used to evaluate the surface property, is expressed by

$$C = \exp\left(\frac{E_1 - E_L}{RT}\right) \tag{1.11}$$

where E_1 and E_L are the isosteric heat of the first-layer adsorption and the liquefaction heat of adsorptive, respectively. Therefore, when C is large enough, the molecules are strongly adsorbed and the adsorption isotherm at low pressure becomes steeper. When C is less than 2, the adsorption isotherm is categorized into type III, according to the classification of IUPAC. The BET equation usually holds in the relative vapor pressure range $0.05 < X < 0.35$. The adsorbed amount calculated from the BET equation at low pressure below $X = 0.05$ becomes smaller than the experimental value, but at high pressures above $X = 0.35$ it becomes larger, conversely. However, employment of an appropriate value of n in the modified BET equation (Equation 1.10) can expand its use up to $X = 0.7$.

There are many comments about the BET equation. Neglect of the interaction force produced among the adsorbed molecules, as well as the occurrence of the nth layer adsorption before completion of the $(n − 1)$th layer, are pointed out as being illogical. In the latter case, the configurational entropy increases and the adsorbed amounts calculated from the BET equation at high pressures above $P/P_0 = 0.35$ become larger than experimental values. Many elaborate amendments of the BET equation have been attempted, but a more useful and concise equation has not yet been deduced.

Jura–Harkins Equation

With respect to the adsorption layer formed on a solid surface, Equation 1.12 is proposed by Jura and Harkins[7]:

$$\pi = b - a\alpha \tag{1.12}$$

where π is the two-dimensional pressure, α the surface area occupied by one adsorbed molecule, and a and b are constants. The two-dimensional pressure π is given from the Gibbs adsorption equation.

$$\pi = RT \int_0^P \Gamma d(\ln P) \tag{1.13}$$

where Γ is the molar amount per unit surface area.

The adsorbed molar amount per unit surface area is given as

$$\Gamma = \frac{V}{V_0 S} = \frac{1}{\alpha} \tag{1.14}$$

where the values V and V_0 are the volume of adsorbed amount and one molar volume at standard pressure and temperature, respectively. From Equation 1.12 and Equation 1.13 the following equations are deduced:

$$d\pi = -a d\alpha = a V_0 S d\left(\frac{1}{V}\right) \tag{1.15}$$

$$d\pi = RT\Gamma d(\ln P) \tag{1.16}$$

Hence, using Equation 1.14 through Equation 1.16, the following equation is obtained:

$$\ln\left(\frac{P}{P_0}\right)=\left(\frac{aS^2V_0^2}{2RT}\right)\left(\frac{1}{V_0'^2}-\frac{1}{V^2}\right)$$

(1.17)

Since the value V_0' is the adsorbed amount at saturated pressure P_0, it becomes infinite, theoretically. Consequently, the following equation is obtained:

$$\ln\left(\frac{P}{P_0}\right)=-\frac{A}{V^2}$$

(1.18)

where A is constant and related to the surface area S of a powder, as follows:

$$S=k_{JH}\sqrt{A}$$

(1.19)

where k_{JH} is a constant to be corrected by using a sample of known surface area.

Because the potential energy of one molecule adsorbed on a liquid film with thickness h is proportional to h^{-3} or V^{-3}, Equation 1.20 is derived by Frenkel[8]:

$$\ln\left(\frac{P}{P_0}\right)=-\frac{C}{V^3}$$

(1.20)

On the other hand, the general formula (Equation 1.21) is proposed by Halsey,[9] assuming that the potential energy of an adsorbed molecule is proportional to $1/h^s$ (i.e., $1/V^s$):

$$\ln\left(\frac{P}{P_0}\right)=-\frac{D}{V^S}$$

(1.21)

where h is the distance between particle surface and an adsorbed molecule, and generally the value of s ranges from 2 to 3. When s is large enough, the adsorption force is specific and short ranged, whereas when the value of s is small, the force is long ranged and originates in the van der Waals force. In the case of N_2 adsorption on anatase, the s value is equal to 2.67.

Various adsorption equations are listed in Table 1.1, in which expressions convenient for plotting experimental data are also shown.

3.1.4 ADSORPTION VELOCITY

Velocity of physical adsorption is different from that of chemical adsorption, comparing the per unit surface area and at fixed temperature and pressure. Usually, the velocity of physical adsorption on a plain solid surface is faster than that of chemical adsorption. However, when a sample is porous and adsorptive pressure becomes high, capillary condensation occurs. In this case, a long time is necessary to attain an equilibrium. Therefore, as the adsorption velocity depends on such geometric properties and on the diffusion rate of adsorption heat, detailed discussion of adsorption velocity is difficult. The equation of adsorption velocity derived by Langmuir is

$$\left(\frac{dV}{dt}\right)=k_{ad}(V_m-V)-k_dV$$

(1.22)

TABLE 1.1   Equations of Adsorption Isotherms

Type of Adsorption Isotherm	Equation	Expression for Plotting Experimental Data
Henry type	$V = k_H P$	$V = k_H P$
Langmuir type	$V = \dfrac{V_m AP}{1 + AP}$	$\dfrac{P}{V} = \dfrac{1}{V_m A} + \dfrac{1}{V_m} P$
Freundlich type	$V = K_F P^{\frac{1}{n}}$	$\ln V = \ln K_F + \dfrac{1}{n} \ln P$
BET type	$V = \dfrac{V_m CX}{(1-X)(1-X+CX)}$	$\dfrac{X}{V(1-X)} = \dfrac{1}{V_m C} + \dfrac{C-1}{V_m C} X$
Jura-Harkins type Frenkel type Halsey type	$\ln X = -\dfrac{D}{V^s}$ $(2 \leq s \leq 3)$	$\ln X = -\dfrac{D}{V^s}$

As this equation is derived for a fixed pressure, the constant k_{ad} contains a pressure term, k_d represents a desorption rate constant and V_m is the monolayer capacity. Denoting the adsorbed amount in equilibrium at a fixed pressure by V_e, the adsorbed amount at time t becomes

$$V = V_e \left\{ 1 - \exp\left[-(k_{ad} + k_d)t \right] \right\} \qquad (1.23)$$

3.1.5   ADSORBED STATE OF ADSORBATE

Investigation of the state of adsorbed molecules (i.e., gaseous, liquidlike, or solidlike state) gives very important information regarding adsorption phenomena. Moreover, the determinations of adsorption structures and sites are also valuable. Chemical adsorption and physical adsorption are different from each other. This section deals with physical adsorption.

The adsorbed states are usually estimated from thermodynamic quantities such as isosteric heat of adsorption and differential molar entropy. Moreover, the adsorbed state can also be discussed based on the changes in dielectric constant and surface conductivity. On the other hand, adsorption mechanisms are studied by IR, UV, and nuclear magnetic resonance (NMR) spectroscopy methods. Especially, IR spectroscopy has become a well-established technique for the determination of adsorption sites and structures. For example, from pyridine adsorption on oxide surfaces, Brönsted and Lewis acid sites were investigated, and the adsorption mechanism was discussed.[10] Generally, the surfaces of SiO_2, TiO_2, α-$Fe2O_3$, and ZnO powders are covered with the various OH groups which are formed by chemical adsorption of H_2O under atmosphere or preparation methods (precipitation of hydroxides and sol–gel method). The effects of the quantity and quality of OH groups on adsorption phenomena are also studied by the IR spectroscopy method. Usually, samples for measuring the infrared spectrum are prepared by pressing powder samples into self-supporting disks. The sample disks thus obtained are mounted in the cell, which can be evacuated and heated in order to control the adsorption state. Various cells have been designed by many researchers.[11] When it is difficult to prepare a self-supporting disk, diffuse reflectance spectroscopy (DRS) and photoacoustic spectroscopy (PAS) are used. These methods are appropriate because of easy sample preparation without special treatments for grinding and pressing samples. Generally, the sample for DRS is diluted with KBr or KCl powder. Therefore, the measurement responsibility of DRS depends on the degree of dilution and particle sizes of diluents and samples.

In the case of PAS, a spectrum can be obtained without dilution of the sample, and this spectroscopy can be applied to any samples with different shapes (powder, pellet, and film). Moreover, PAS can obtain the depth profile of samples.

When a powder sample is very fine and well dispersed in the liquid, which is transparent for infrared rays, a spectrum of the sample can be obtained in the liquid as it is. For example, surface groups on fine silica powders were clearly observed by dispersing the powder samples in carbon tetrachloride.[12]

The interaction between the adsorbed molecules and surface functional groups such as hydroxyl groups is investigated from changes in absorption bands deduced by perturbations of adsorbed molecules. The adsorption of molecules on the surface functional groups causes the shift and decrease of the absorption peaks of surface functional groups. The correlation between differential heats (Q_a) of adsorption and peak shift ($\Delta \nu$) of the absorption band of hydroxyl groups on a silica surface is shown in Figure 1.3.[13]

The adsorption characteristics and adsorbed states largely depend on chemical composition and geometric structure of solid surfaces. The geometric structure of solid surfaces contains surface roughness, pore structure, defect sites, and difference in crystal face. Generally, the chemical composition of real solid surfaces is different from that of the bulk phase by the chemisorption of oxygen, carbon dioxide, and water molecules or by surface modification. Therefore, the real surfaces of powders will be covered with surface functional groups (e.g., hydroxyl, carbonyl, carboxyl, and various alkyl groups). These surface functional groups act as the dominant factor of the adsorption characteristics and adsorbed states.

The adsorption behavior and adsorbed states closely relate to the combination between the nature of the powder surface and that of the adsorbate molecule. Especially, in the case of polar adsorbate molecules, such effects are remarkably influenced by the surface functional groups. Figure 1.4 shows the effects of surface functional groups of silica powder treated with various silane reagents on water vapor adsorption.[14] It is important to estimate the number and types of the surface func-

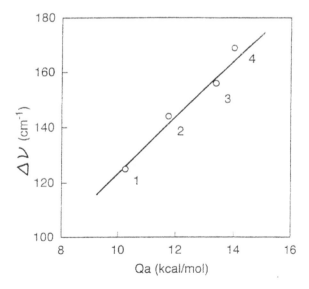

FIGURE 1.3   The correlation of stretching bands of silica surface hydroxyl groups and differential heats of adsorption. 1, benzene; 2, toluene; 3, *p*-xylene; 4, mesitylene. [From Galkin, G. A., Kiselev, A. V., and Lygin, V. I., *Trans. Faraday Soc.*, 60, 431–439, 1964. Reproduced by permission of The Royal Society of Chemistry.]

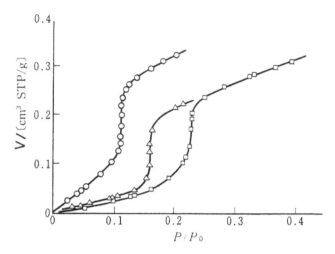

FIGURE 1.4   Water adsorption isotherms on the silane-treated silica. [From Zettlemoyer, Z. C. and Hsing, H. H., *J. Colloid Interface Sci.*, 58, 263–274, 1977. With permission.]

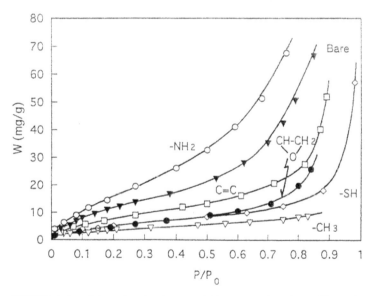

FIGURE 1.5   Adsorption isotherms of Kr on alkali halides. ○, NaCl (400°C heat treatment); △, RbCl (300°C heat treatment); □, KCl (400°C heat treatment). [Reprinted with permission from Takaishi, T. and Saito, M., *J. Phys. Chem.*, 71, 453–454, 1967, copyright (1967) American Chemical Society.]

tional groups. These groups are investigated by spectroscopic (IR, Raman, X-ray photoelectron spectroscopy [XPS], NMR) and chemical reaction methods.

If a powder sample has a pore structure, the specific surface area of this powder will be very large. Moreover, capillary condensation and micropore filling phenomena occur through gas adsorption. In these cases, the physical properties of the adsorbed layers are speculated to be different from those of adsorbed layers formed on flat surfaces. Capillary condensation is observed in mesoporous materials

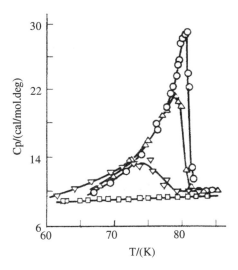

FIGURE 1.6   Change in heat capacity of Ar layer formed on rutile surface, 2.9 V_m, 4.0 V_m, 4.6V_m.

and shows the increase of adsorption capacity at a lower pressure than the saturated vapor pressure. On the other hand, the micropore-filling phenomenon is found in microporous materials (e.g., activated carbon and zeolite). The microporous materials have extremely high specific surface areas (>1000 m^2/g) and show strong adsorption force due to the effect of the pore wall. The adsorbate molecules are filled up in the micropore at very low relative pressure, and the adsorption isotherm exhibits the Langmuir-type adsorption isotherm.

Adsorption isotherms of Kr on alkali halides are depicted in Figure 1.5.[15] The sharp increases are noted in the adsorption isotherms, and they become progressively sharper with increasing pretreatment temperature. These annealing processes make the surface more homogeneous. Therefore, the sharp increase due to two-dimensional condensation is considered to be strongly affected by surface conditions.

The molar heat capacities of Ar adsorbed on rutile are illustrated in Figure 1.6.[16] The peaks in the heat capacity curves are considered to be ascribed to the phase transition. The peak becomes larger and appears at higher temperature with increased amount adsorbed. The lowering phase-transition temperature of an adsorbed layer can be explained as follows. Because the liquid structure and physical properties of the adsorbed layer are influenced by the solid surface, solidification of the adsorbed layer becomes more difficult. The same results were observed in N_2 adsorption on TiO_2[17] and H_2O adsorption on SiO_2.[18] The Clausius–Clapeyron plot and differential scanning calorimetry (DSC) measurement are other methods for determination of the phase-transition temperature.

3.1.6   ESTIMATION OF SURFACE PROPERTIES BY ADSORPTION METHOD

Physical and Chemical Properties of Powder Surfaces

Various phenomena concerning powder materials depend largely on physical and chemical properties of their surfaces. In powder processes such as classification, mixing, and comminution, adhesive force and surface energy play important roles. On the other hand, the wettability of powder surfaces is significant in filtration, sedimentation, and floatation. Moreover, reactivity, solubility, and activity of the catalyst are closely related to chemical properties of powder surfaces. Information about the chemical properties of powder surfaces is obtainable from adsorption characteristics of various adsorptives such as the isosteric heat of adsorption and adsorbed amount. Generally, the adsorption heat determined at very low coverage is large, and the adsorption is considered to occur at active sites. On the other hand, adsorption heat measured at high coverage contains lateral interaction produced among adsorbed molecules. Therefore, the adsorption heat calculated at a coverage of $\theta = 0.5$ is used to estimate the surface property. The changes in surface property of modified powder particles are shown in Figure 1.7 and Figure 1.8.[19] Their changes are expressed by the BET C value

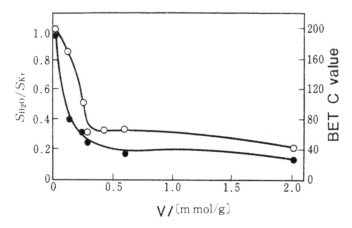

FIGURE 1.7   Change in water vapor affinity of CaF_2 coated with oleate by an amount V. S_{Kr}, S_{HO}; surface area determined by Kr and H_2O adsorption, respectively; $\bigcirc$, the surface area ratio; $\bullet$, C value for H_2O adsorption. $V_m = 0.26$ mmol/g. [From Hall, P. G., Lovell, V. M., and Frinkelstein, N. P., *Trans. Faraday Soc.*, 66, 1520–1529, 1970. With permission.]

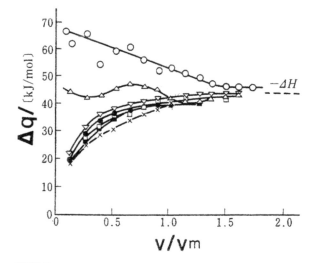

FIGURE 1.8   Isosteric heat of H_2O adsorption on CaF_2 modified with oleate. $V_m = 0.26$ mmol/g. Adsorbed amounts (mmol/g): $\bigcirc$, 0; $\triangle$, 0.125; $\triangledown$, 0.243; $\square$, 0.272; ×, 0.406; $\blacktriangle$, 0.592; $\bullet$, 2.043. [From Hall, P. G., Lovell, V. M., and Frinkelstein, N. P., *Trans. Faraday Soc.*, 66, 1520–1529, 1970. With permission.]

and isosteric heat of adsorption. The affinity of water vapor decreases markedly when the amount of potassium oleate increases and it covers the particle surfaces with a complete monolayer. The molecular diameter of a potassium oleate is calculated to be about 2.1 nm from the surface area and the adsorbed amount of oleate.

Heat treatment usually makes particle surfaces more homogeneous, and such results obtained for carbon black are shown in Figure 1.9.[20] The surface property of untreated carbon black is inhomogeneous, but it becomes more homogeneous with an increase in heat treatment temperature. These results are obtained from the comparison of adsorption heat. The adsorption heat of Ar molecules at low coverage is large and decreases gradually with the adsorbed amount. The high value of the adsorption heat at the early stage disappears with an increase in treatment temperature.

At coverage at about $\theta = 1$, lateral interactions will arise among adsorbed molecules, and an apparent maximum appears in the heat curves. The maximum due to the lateral interaction becomes increasingly prominent with progress of the graphitization by the heat treatment. From these results the lateral interaction is presumed to be effectively generated when the surface homogeneity is attained. After that, multilayer adsorption occurs, and the adsorption heat rapidly decreases approaching the liquefaction heat at adsorbed amount V above V_m.

The surface properties are also estimated from the heat of immersion. This immersion heat of particle surfaces for a certain liquid is related to the adsorption heat of the corresponding vapor molecules. The measurement of the immersion heat is relatively easy, compared with that of the adsorption heat. When water is used as the immersion liquid, the heat of immersion represents the degree of hydrophilic or hydrophobic properties of the surface. The change in the immersion heat of a silica powder modified with trimethylsilyl groups for water is shown in Figure 1.10.[21] The hydrophilic property of the surface decreases with an increase in the concentration of trimethylsilyl groups.

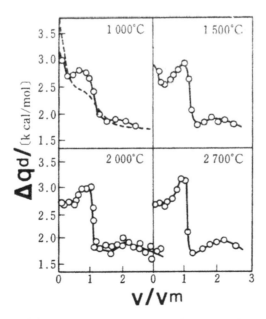

FIGURE 1.9   Differential heat of Ar adsorption on carbon black heat treated at high temperatures. [From Beebe, R. A. and Young, D. M., *J. Phys. Chem.*, 58, 93–96, 1954. With permission.]

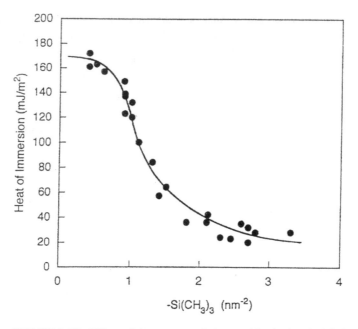

FIGURE 1.10 Effect of the amount of the combined trimethylsilyl group on the heat of adsorption. [From Tsutsumi, K. and Takahashi, H., *Colloid Polym. Sci.*, 263, 506–511, 1985. With permission.]

Surface Geometrical Properties

It is difficult to describe geometric irregularities of real particle surfaces. Avnir estimated the geometric structure of the surface using the fractal dimension D.[22,23] The fractal dimension of solid surfaces is investigated by measurements of physical adsorption of the gases, which have various molecular sizes, and is calculated as

$$\log n = (-\frac{1}{2}D)\log\sigma + C \tag{1.24}$$

where n is the monolayer capacity of adsorbate molecules on the surface, σ is the effective cross-sectional area of one adsorbate molecule, and C is a constant value. The value of D is calculated from the slope of a linear line, which is obtained by plotting $\log n$ versus $\log \sigma$. Generally, this value for the real surfaces ranges from 2 to 3. For typical solid surfaces the values are obtained as follows. In the cases of graphite and monmorillonite, which have a smooth surface, their D values are nearly 2. However, porous samples that have high specific surface areas and porous structures, such as silica gel and charcoal, are nearly 3. The study of the relationship between the fractal dimension and various adsorption phenomena will become more important in the future.

Powder with a large specific surface area often has a porous structure. The pore size, shape, and distribution are different in each sample. Generally, the diameter for a cylindrical pore and the distance between two side walls for a slit-shaped pore are used as representative pore sizes. The pores are classified into three types according to the pore size: micropores for below 2 nm, transitional or intermediate pores for 2–50 nm, and macropores for above 50 nm. In the determination of pore sizes by the adsorption method, the Kelvin equation is applied, and pore radii in the range of 0.5 to 30 nm are measured accurately. In the case of N_2 adsorption, if the pore structure is presumed to be cylindrical, the Kelvin equation holds for the radius above 2 nm but fails below 2 nm.[24] This result is considered as follows. The liquid properties of capillary condensed liquid, such as molar volume and surface tension, are different from those of bulk liquid. The Kelvin equation expresses the relationship between pore radius r and vapor pressure P at which capillary condensation occurs. Denoting the molar volume by V, the adsorbate surface tension by σ, and the contact angle of condensed liquid by θ, it becomes

$$\ln\left(\frac{P}{P_0}\right) = -\frac{2V\sigma\cos\theta}{rRT} \tag{1.25}$$

The pore size below 2 nm should be calculated by the t-method,[25] the D-R method,[26] and the H-K method.[27]

REFERENCES

1. Morimoto, T., Nagao, M., and Tokuda, F., *Bull. Chem. Soc. Jpn.,* 41, 1533–1537, 1968.
2. Morimoto, T., Kishi, J., Okada, O., and Kadota, T., *Bull. Chem. Soc. Jpn.,* 53, 1918–1921, 1980.
3. Sing, K. S. W., Everett, D. H., Haul, R. A. W., Moscou, L., Pierotti, R. A., Rouqe'rol, J., and Siemieniewska, T., *Pure Appl. Chem.,* 57, 603–619, 1985.
4. Lambert, B., and Peel, D. H. P., *Proc. Roy. Soc. London,* A144, 205, 1934.
5. Langmuir, I., *J. Am. Chem. Soc.,* 40, 1361–1403, 1918.
6. Brunauer, S., Emmett, P. H., and Teller, E., *J. Am. Chem. Soc.,* 60, 309–319, 1938.
7. Jura, G. and Harkins, W. D., *J. Chem. Phys.,* 11, 430–432, 1943.
8. Frenkel, J., *Kinetic Theory of Liquids,* Clarendon Press, Oxford, 1946, p. 332.
9. Halsey, G. D., *J. Chem. Phys.,* 16, 931–937, 1948.
10. Parry, E. P., *J. Catal.,* 2, 371–379, 1963.
11. Bell, A. T., in *Vibrational Spectroscopy of Molecules on Surfaces,* Yates, J. T. and Madey, T. E., Eds., Plenum Press, 1987, p. 116.
12. Tripp, C. P. and Hair, M. L., *Langmuir,* 8, 1961–1967, 1992.
13. Galkin, G. A., Kiselev, A. V., and Lygin, V. I., *Trans. Faraday Soc.,* 60, 431–439, 1964.
14. Zettlemoyer, Z. C. and Hsing, H. H., *J. Colloid Interface Sci.,* 58, 263–274, 1977.
15. Takaishi, T. and Saito, M., *J. Phys. Chem.,* 71, 453–454, 1967.
16. Morrison, J. A. and Drain, L. E., *J, Chem. Phys.,* 19, 1063, 1951.
17. Morrison, J. A., Drain, L. E., and Dugdale, J. S., *Can. J. Chem.,* 30, 890, 1952.
18. Plooster, M. N. and Gitlim, S. N., *J. Phys. Chem.,* 75, 3322–3326, 1971.
19. Hall, P. G., Lovell, V. M., and Frinkelstein, N. P., *Trans. Faraday Soc.,* 66, 1520–1529, 1970.
20. Beebe, R. A. and Young, D. M., *J. Phys. Chem.,* 58, 93–96, 1954.
21. Tsutsumi, K. and Takahashi, H., *Colloid Polym. Sci.,* 263, 506–511, 1985.
22. Pfeifer, P. and Avnir, D., *J. Chem. Phys.,* 79, 3558–3565, 1983.
23. Avnir, D., Farin, D., and Pfeifer, P., *J. Chem. Phys.,* 79, 3566–3571, 1983.
24. Harris, M. R., *Chem. Ind. (London),* 268–269, 1965.
25. Lippens, B. C. and Boer, J. H., *J. Catal.,* 4, 319–323, 1965.
26. Dubinin, M. M. and Stoeckli, H. F., *J. Colloid Interface Sci.,* 75, 34–42, 1980.
27. Horvath, G. and Kawazoe, K., *J. Chem. Eng. Jpn.,* 16, 470–475, 1996.

3.2 Moisture Content

Satoru Watano
Osaka Prefecture University, Sakai, Osaka, Japan

3.2.1 BOUND WATER AND FREE WATER

Water is achromatic and transparent, and it has three phases (gas, liquid, and solid), depending on the temperature and pressure. Due to its unique characteristics, various existing forms are observed.[1] In a particulate system, existing forms of water are classified as water in a single particle or water in a particulate (bulk) system, as illustrated in Figure 2.1.

Generally, water is categorized into two classes: bound water and free water. Bound water, which exists inside crystals and in aqueous solutions, gels, organisms, soil, and so forth, is stuck to materials by hydrogen bonding. As shown in Figure 2.1, bound water includes combined water, occluded water, adsorption water, and hygroscopic water. Combined water is also called crystal water, which has the following types, depending on the structure and the bonding conditions:

1. Constitution water: Water exists in compounds as a hydroxyl group.
2. Coordinated water: Water forms a chelate ion in a chelate complex.
3. Anion water: Water is stuck to anions by hydrogen bonding.
4. Lattice water: Water exists inside a crystal lattice without having direct bonding with either cations or anions.
5. Zeolite water: The vacancy of a crystal is occupied with water molecules, and the crystal structure does not change after dehydration. Occluded water is captured inside a solid (particle), and it exists as a solid solution or hydrate. Adsorption water is called either surface adsorption water or hygroscopic water; the former is bound by van der Waals force, while the latter is captured among particles.

Contrary to bound water, free water is captured by materials with relatively weak bonding strength and is called adhesive water or swelling water. In a particulate system, damping conditions by free water are categorized into funicular, capillary, and moving states. Especially in a soil particulate system, consistency of soil and the phase dramatically change depending on the water content. In general, the transition point between the different phases is called the Atterberg limit (shrinkage, plastic, and liquid limits).[2]

The condition of water distribution and bonding strength of water in a particulate system is described as a function of suction potential S(m), which is defined as

$$S = \left| -\frac{RT}{Mg} \ln\left(\frac{P}{P_0}\right) \right| \tag{2.1}$$

where P is the water vapor pressure inside the particulate system, P_0 is the vapor pressure on the free surface of pure water, R is the gas constant (= 8.315 J/mol K), M is the molecular weight of water (= 18.02 g/mol), and g is the gravitational acceleration (= 9.81×10^{-3} J/gm). Suction potential is also related to the pF value as

$$pF = \log S \tag{2.2}$$

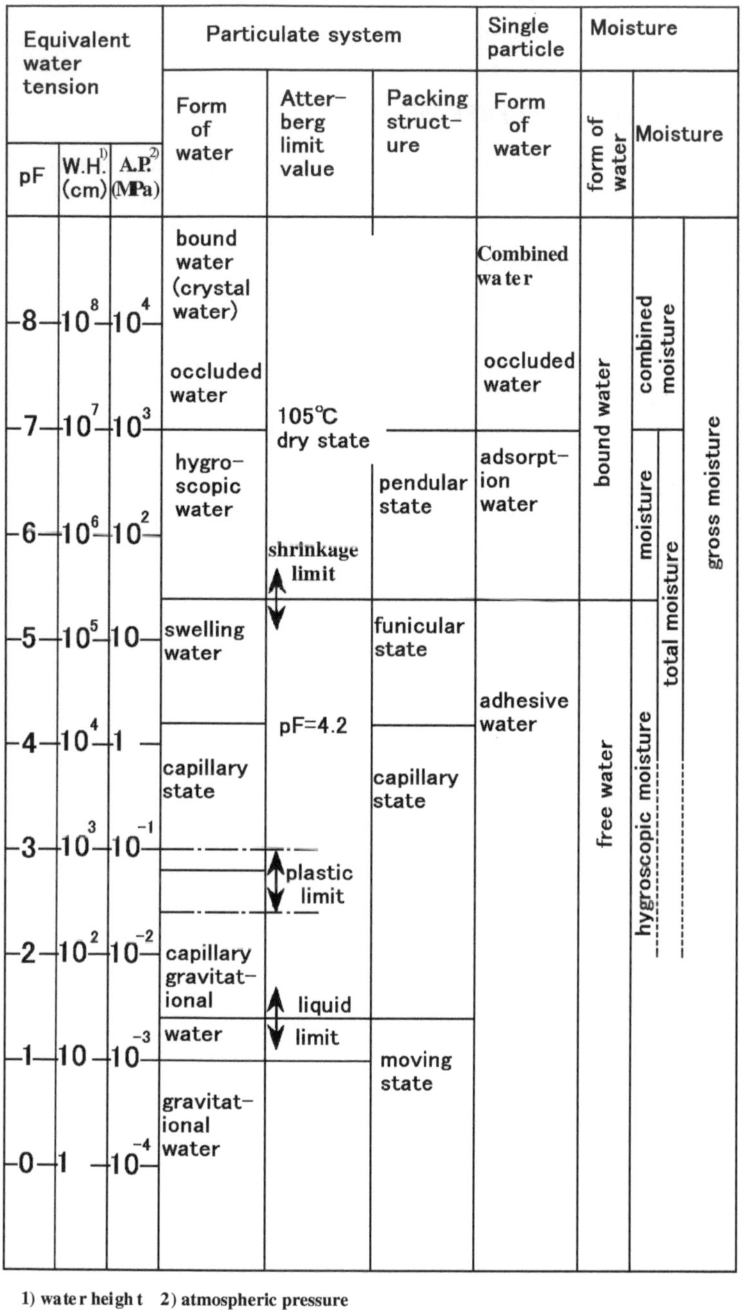

Equivalent water tension			Particulate system			Single particle	Moisture	
pF	W.H.[1] (cm)	A.P.[2] (MPa)	Form of water	Atter–berg limit value	Packing struct–ure	Form of water	form of water	Moisture
−8	10^8	10^4	bound water (crystal water)			Combined water	bound water	combined moisture / gross moisture
			occluded water			occluded water		
−7	10^7	10^3		105°C dry state				
−6	10^6	10^2	hygro–scopic water		pendular state	adsorpt–ion water		moisture
				shrinkage limit				total moisture
−5	10^5	10	swelling water		funicular state			
						adhesive water		
−4	10^4	1		pF=4.2			free water	hygroscopic moisture
			capillary state		capillary state			
−3	10^3	10^{-1}						
				plastic limit				
−2	10^2	10^{-2}	capillary gravitat–ional					
			water	liquid				
−1	10	10^{-3}		limit	moving state			
			gravitat–ional water					
−0	1	10^{-4}						

1) water height 2) atmospheric pressure

FIGURE 2.1   Existing forms of water.

TABLE 2.1   Existing Water Forms in Solid and Water Removal Temperature

Bonding Strength	Name and Bonding State	Water Removal Temperature (°C)
Weak	Water bound by adhesion or water retained through holes	Dried air at room temperature
	Occluded water Absorbed water	250~300
	Outer surface	100~130
	Submicron channels	~600
	Crystalline water	Dependent on substance
Strong	Constituent water	Dependent on substance
	Zeolitic water	600~1000

The relationship between suction potential and the existing water forms is also explained in Figure 2.1. Another way of measuring bonding strength of water in a particulate system is to use the water removal temperature in heating. Table 2.1 explains the existing water forms in a solid and their removal temperature.

3.2.2   METHODS FOR DETERMINING MOISTURE CONTENT IN A PARTICULATE SYSTEM

The appropriate method for measuring the moisture content is strongly dependent on the water form in a sample. The simplest way of measuring moisture content in a sample is to measure weight loss after drying. Other methods, such as chemical, electric, and optical methods, as well as the nuclear magnetic resonance (NMR) spectrometer method and method of applying radioactive rays, are also available.

Determination methods for water content have been established in various national and international standards such as JIS and ISO.

REFERENCES

1. Gotoh, K., Masuda, H., and Higashitani, K., Eds., *Powder Technology Handbook,* 2nd Ed. Marcel Dekker, New York, 1997, pp. 265–276.
2. Smith, K. A. and Mullins, C. E., Eds., *Soil and Environmental Analysis: Physical Methods,* Marcel Dekker, New York, 2000.
3. Kawamura, M. and Toyama, S., *J. Soc. Powder Technol. Jpn.,* 15, 292–297, 1978.

3.3 Electrical Properties

Hiroaki Masuda
Kyoto University, Katsura, Kyoto, Japan

Ken-ichiro Tanoue
Yamaguchi University, Ube, Yamaguchi, Japan

Yasufumi Otsubo
Chiba University, Image-ku, Chiba, Japan

3.3.1 IN GASEOUS STATE

The specific resistance, dielectric constant, specific charge, and contact potential difference of particles are discussed in this section. These electrical properties are necessary for studying the behavior of electrically charged particles, electrostatic precipitation, electrostatic powder coating, electrostatic powder imaging, and so on.

Specific Resistance

The electric resistance of a powder bed is one of the most important properties, especially in relation to electrostatic precipitation. If the specific resistance ρ_d of particles is lower than $10^2\ \Omega$ m, the particles will immediately be neutralized on the dust collecting electrode. Further, they are oppositely charged by the electrode and reentrained by gas flow in the presence of electrostatic repulsion between the particles and the collecting electrode. On the other hand, if the specific resistance is higher than $10^9\ \Omega$ m, back discharge from the collecting electrode will occur, resulting in a decrease in the precipitation efficiency.[1] The specific resistance of highly resistive powder is strongly dependent on the ambient temperature and moisture. It takes a maximum at some temperature (100 to 200°C) and decreases with increasing absolute humidity.[2] It also depends on the chemical composition of particles.

The specific resistance can be measured by (a) parallel-plate electrodes, (b) cylindrical electrodes, or (c) needle-plate electrodes, as shown schematically in Figure 3.1. After particles are packed or deposited between the electrodes, the applied voltage V and electric current I through the powder bed are measured. Then the specific resistance ρ_d can be calculated from the following equations[3]:

1. Parallel-plate electrodes:

$$\rho_d = \frac{A}{d}\frac{V}{I} \qquad (3.1)$$

2. Cylindrical electrodes:

$$\rho_d = \frac{2\pi l}{\ln(b/a)}\frac{V}{I} \qquad (3.2)$$

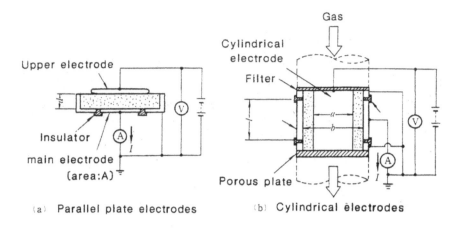

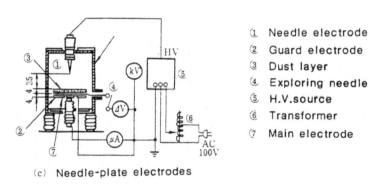

(c) Needle-plate electrodes

FIGURE 3.1   Electrodes for measuring specific resistance.

3. Needle-plate electrodes:

$$\rho_d = \frac{A}{L}\frac{\Delta V}{I} \tag{3.3}$$

where
  V = applied voltage (V)
  I = electric current (A)
  A = area of the main electrode (m^2)
  d = gap between the electrodes (m)
  l = length of the cylindrical electrode (m)
  a = outside diameter of the inner cylindrical electrode (m)
  b = inside diameter of the outer cylindrical electrode (m)
  ΔV = electrical potential of the exploring needle (V)
  L = distance between the exploring needle and the plate electrode (m)

The specific resistance depends on the packing density of the powder bed. Therefore, measurements should be carried out after precise setting of the packing density. Needle-plate electrodes are suitable for measurement under the same conditions as in an industrial electrostatic precipitator.

TABLE 3.1 $f(\bar{\varepsilon}_p) = (\bar{\varepsilon}_p - 1)/(\bar{\varepsilon}_p + 2)$

$\bar{\varepsilon}_p$	1	2	5	10	20	50	100	∞
$f(\bar{\varepsilon}_p)$	0	0.250	0.571	0.750	0.864	0.942	0.971	1

TABLE 3.2 $g(\bar{\varepsilon}_p) = 3\bar{\varepsilon}_p/(\bar{\varepsilon}_p + 2)$

$\bar{\varepsilon}_p$	1	2	5	10	20	50	100	∞
$g(\bar{\varepsilon}_p)$	1	1.50	2.14	2.50	2.73	2.88	2.94	3

The specific resistance is usually higher for a powder bed of smaller packing density. The applied voltage should be kept as low as possible to avoid a void discharge. It may take a long time to reach a constant current, especially for high-resistivity powder, because the current is partly absorbed in the powder bed as charge accumulates between particles. The ambient temperature and the humidity should be controlled. A special wind tunnel called a race track may be utilized in the needle-plate electrodes method.[3]

Dielectric Constants

The dielectric constant of particles is necessary to estimate the dielectrophoretic force or the maximum charge attainable by field charging. The dielectrophoretic force is a function of $(\varepsilon_p - \varepsilon)/(\varepsilon_p + 2\varepsilon)$, where ε_p is the dielectric constant of the particle and ε is that of the medium. If the medium is air, ε is close to ε_0 ($- 8.85 \times 10^{-12}$ F/m) of vacuum. The ratio $\bar{\varepsilon}_p = \varepsilon_p/\varepsilon_0$ is called the specific dielectric constant. Numerical values of $f(\bar{\varepsilon}_p) = (\bar{\varepsilon}_p - 1)/(\bar{\varepsilon}_p + 2)$ are listed in Table 3.1, showing its insensitivity to the specific dielectric constant. The maximum charge acquired by field charging depends on the function $g(\bar{\varepsilon}_p) = 3\bar{\varepsilon}_p/(\bar{\varepsilon}_p + 2)$, whose numerical values are listed in Table 3.2.

The dielectric constant of the powder bed is called the apparent dielectric constant ε_a, which is a function of the packing fraction f of particles as follows:

1. Maxwell model[4,5]

$$\bar{\varepsilon}_a = \frac{\varepsilon_a}{\varepsilon_0} = \frac{3 - 2(1-\phi)(\bar{\varepsilon}_p - 1)}{3 + (1-\phi)(\bar{\varepsilon}_p - 1)} \tag{3.4a}$$

2. Rayleigh model[6]

$$\frac{\varepsilon_a}{\varepsilon_0} = 1 + 3\phi\left(\frac{\bar{\varepsilon}_p + 2}{\bar{\varepsilon}_p - 1} - \phi - 1.65\frac{(\bar{\varepsilon}_p - 1)}{\bar{\varepsilon}_p + (4/3)}\phi^{10/13}\right)^{-1} \tag{3.4b}$$

3. Böettcher model[5,7,8]

$$\frac{\bar{\varepsilon}_a - 1}{3\bar{\varepsilon}_a} = \frac{\phi(\bar{\varepsilon}_p - 1)}{\bar{\varepsilon}_p + 2} \tag{3.4c}$$

4. Lichtnecker model

$$\bar{\varepsilon}_a = \bar{\varepsilon}_p{}^{\phi} \tag{3.4d}$$

5. Lichtneker–Rother model

$$\bar{\varepsilon}_a{}^k = \phi \bar{\varepsilon}_p{}^k + (1 - \phi) \tag{3.4e}$$

The apparent dielectric constant ε_a can be measured by the parallel-plate or cylindrical electrodes shown in Figure 3.1.

$$\bar{\varepsilon}_a = \varepsilon_0 \frac{C_1}{C_0} \tag{3.5}$$

where C_0 is the electric capacitance of the measuring system without powder and C_1 is the capacitance with powder between the electrodes. The capacitance C_0 is given by the following equations:

1. Parallel-plate electrodes:

$$C_0 = \frac{A}{d} \varepsilon_0 \tag{3.6}$$

2. Cylindrical electrodes:

$$C_0 = \frac{2\pi l}{\ln(b/a)} \varepsilon \tag{3.7}$$

The apparent dielectric constant depends on the packing fraction, temperature, and humidity, as in the case of the specific resistance. The humidity strongly affects the measurement because the specific dielectric constant of water is about 10 times larger (81 at 18°C) than that of particles.

The dielectric constant of a particle ε_p can be measured by use of the standard liquid listed in Table 3.3. If the capacitance of a cell filled with the reference liquid does not change by adding particles into the cell, the dielectric constant of the particles is equal to that of the liquid. The dielectric constant of the reference liquid is adjustable by blending it with different liquids.

The following relation is useful in an estimation of the dielectric constant of a multicomponent particle:

$$\frac{1}{(1/\varepsilon)_{av}} \leq \varepsilon^* < \varepsilon_{av} \tag{3.8}$$

where

$$\varepsilon_{av} = \sum v_i \varepsilon_i \text{ and } \left(\frac{1}{\varepsilon}\right)_{av} = \sum \frac{v_i}{\varepsilon_i} \tag{3.9a,b}$$

ε_i is the dielectric constant of the i component and v_i is the volume fraction.

**TABLE 3.3   Dielectric Constants
of Standard Liquids**

Liquid	Specific Dielectric Constant
Cyclohexane	2.023
Benzene	2.284
Chlorobenzene	5.708
Acetone	21.3
Nitrobenzene	35.7

Specific Charge

The particle charge q divided by the particle mass m_p is called the specific charge. The specific charge is sometimes approximated by the total charge of powder divided by the mass of the powder. Since the charge on a particle is approximately proportional to its surface area in various electrification processes, the specific charge is almost inversely proportional to the particle size. Therefore, electrostatic phenomena may become predominant for smaller particles. If a fine particle is electrically charged, its motion is usually determined only by the ratio of the electrostatic force and the fluid viscous force. The total charge of powder is measured by an electrically shielded metal vessel, known as the Faraday cage. Powder is put into the Faraday cage and the potential V is measured by a voltmeter. The charge Q is, then, given by the following equation:

$$Q = CV \tag{3.10}$$

where C is the capacitance of the Faraday cage.

The capacitance C is calibrated with a steel ball charged in an electric field supplying sufficient unipolar ions, according to the following equation for the charge of the sphere:

$$q = 3\pi\varepsilon_0 D_P^2 E \tag{3.11}$$

where E (V/m) is the electric field strength. Hence the specific charge Q/w is obtainable for known sample mass w.

The specific charge of suspended particles will be obtained through isokinetic sampling. Figure 3.2 shows the Faraday cage available for this purpose. The suspended particles are collected on a filter set in the Faraday cage. If the sampling of particles is impossible, a spherical cage made of a steel mesh may be used. The potential at the center of the spherical cage is given by the equation

$$V = \frac{Q}{8\pi\varepsilon_0 R} \tag{3.12}$$

where Q is the total charge of particles in the cage.

The specific charge of a particle is obtainable from the particle trajectory in an electric field. Figure 3.3 depicts parallel-plate electrodes where charged particles are introduced. The charge q can be obtained from the following equation:

$$q = \frac{3\pi\mu D_p ud}{C_c Ex} \tag{3.13}$$

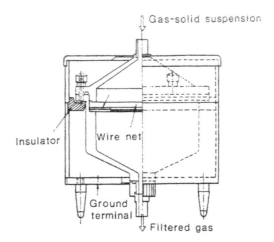

FIGURE 3.2   Faraday cage for measuring specific charge of suspended particles.

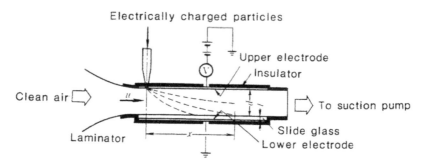

FIGURE 3.3   Parallel-plate electrodes for measuring a specific charge of individual particles.

where u is the mean air velocity, d is the distance between upper and lower electrodes, C_c is Cunningham's slip correction factor, and x is the position of deposited particle. The electric field strength E is calculated as follows:

$$E = \frac{V}{d + (t / \varepsilon_r)} \tag{3.14}$$

where V is the applied voltage, t is the thickness of the slide glass, and ε_r is the specific dielectric constant of the glass.

The particle will suspend in the space when the electric force qE balances the gravitational force $m_p g$. Hence the specific charge can be obtained from the relation $q/m_p = g/E$. If the particle sediments in a horizontal electric field, it will deposit at a point (x, y) according to the relation $x/y = qE/m_p g$. Hence the distribution of the specific charge will be obtained.[9] The particle motion in a DC electric field superimposed upon an acoustic field or AC electric field can also be utilized in the simultaneous measurement of particle size and charge distributions.[10]

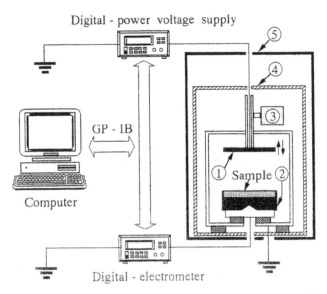

FIGURE 3.4   Measuring system for the contact potential difference.

Contact Potential Difference

The contact potential difference of particles is essential for the estimation of the charge polarities and the amount of the charge acquired in various powder handling processes. It relates to the electrostatic powder coating, electrostatic powder imaging, and so on. For pure materials, including metals, there are some reference values available. However, it should be measured for particles, because small amounts of impurities and imperfections and the surface states may change the electrostatic characteristics of the particles.

Figure 3.4 shows schematically the measurement apparatus, which is based on the Kelvin method.[11,12] It consists of an upper electrode (1), lower electrode with packed powder layer (2), motor (3), digital-power voltage supply, and a digital electrometer for electric current measurement. The upper electrode moves up and down sinusoidally with a constant frequency. The timing to supply voltage and to detect electric current is controlled by the GP-IB (General Purpose Interface Bus) of a computer. The main device was set in an autodesiccator (5) with an electric shield (4) for noise reduction. The upper electrode is gold plated while the lower electrode is nickel plated so as to have enough strength against forcible contact or friction of the packed powder.

Electric current generated by the cyclic movement of the upper electrode becomes zero, when the applied voltage V to the upper electrode is equal to the apparent contact potential difference. And the contact potential difference is obtained if the charge of the powder layer is fully diminished. The effective work function of the powder is estimated by the equation

$$\phi_P = \phi_{Au} - eV_{P/Au} \qquad (3.15)$$

where, ϕ_P is the effective work function of the powder, ϕ_{Au} the work function of Au, and $V_{P/Au}$ is the contact potential difference of powder against Au.

3.3.2   IN NONAQUEOUS SOLUTION

On the application of electric fields to a suspension of polarizable particles dispersed in an insulating liquid, each particle acquires a dipole. In high electric fields, the dipole–dipole interactions affect the particle arrangement which, in turn, leads to changes in physical properties such as rheology, electric resistance, and light transmittance. The nonlinear effects of electric fields on the physical properties of suspensions will be described in relation to the field-induced structure.

Column Structures in Electric Fields

When the dielectric constants of the particles and liquid are substantially different, an electric field generates a dipolar field around the particle. The particles attract one another if they are aligned along the field direction, whereas the particles in the plane perpendicular to the field direction repel one another. Because many polarized particles dispersed in a liquid interact in this way, chains of particles are formed to span the electrode gap. As a simplified approach, most existing models assume that all of the particles align into chains of a single particle width and equal spacing. Although the ideal single-chain model is helpful to understand the basic process of particle arrangement, the chainlike structures are unlikely to be constructed by the separated chains.

The induced structure in electric fields must have the configuration which minimizes the dipolar interaction energy. The force between two particles is attractive if the two particles are aligned within 55° of the electric field.[13] When placed near a chain, a particle will be repelled by the nearest particles in the chain. The particle will be attracted, however, by the chained particles far above and below it. When many particles are placed near the chain, they will be attracted to it and can form a second chain. Consequently, because of long-range interactions, the polarization forces between particles cause the chains to form columns. The structural changes from noninteracting particles to thick columns spanning the electrode gap are schematically shown in Figure 3.5. The column may be constructed by three types of chains: fully developed chains connecting the electrodes, chains attached at only one end to an electrode, and drifting chains.

According to the theory on the column structures by Tao and Sun,[14] the dipolar interaction energy per particle strongly depends on the lattice structure. For example, in a simple cubic lattice, a particle in one chain has four particles in other chains as its nearest neighbors, and the centers of particles lie in a plane perpendicular to the field vector. Because the forces between neighboring particles are strongly repulsive, the particles rarely construct a simple cubic lattice. In a quiescent state, the most stable structure is a body-centered tetragonal lattice. Therefore, the suspension in electric fields comprises the columns with the body-centered tetragonal lattice. The fully developed columns spanning the electrodes' gap can be achieved even at a particle concentration of 1 vol%.

Electrorheology

In response to electric fields, the suspensions often exhibit a rapid and reversible transition between viscous liquids and rigid solids. This phenomenon was first reported by Winslow[15] and is referred to as the Winslow effect or electrorheological (ER) effect. ER fluids are very attractive as effective vehicles in new devices for controlling the motion of liquids through a narrow channel. Possible devices

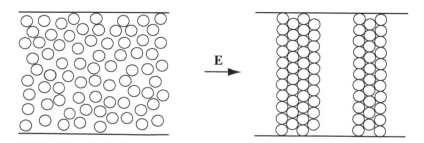

FIGURE 3.5   Column formation of particles on the application of electric fields.

include stop valves, brakes, and dampers. Figure 3.6 shows a schematic diagram of an ER clutch. The torque transmitted between two elements can be controlled by the field strength.

The columns fully developed between the electrodes can accumulate elastic energy, unless the strain exceeds some critical value. The suspensions in electric fields behave as solids under low stress or strain. When the suspensions are subjected to steady shear at low rates, the rupture and reformation of column structures are constantly repeated to produce a constant stress. In general, the constant stress in the limit of zero shear rate is considered to correspond to the yield stress. The yield stress is defined as a critical stress below which no flow can be observed under the conditions of experimentation and above which the substance is a liquid. The rheological changes of suspensions on the application or removal of electric fields can be characterized as a reversible sol–gel transition. In very high shear fields, the hydrodynamic forces dominate the electric forces. Therefore, the stress linearly increases with shear rate.

Figure 3.7 shows the flow curves in electric fields of 0 and 2.0 kV/mm for an ER fluid at a particle concentration of 30 vol%. In the absence of electric fields, the stress is proportional to shear rate, and the slope gives the Newtonian viscosity. In electric fields, significant flow occurs only after the yield stress of about 100 Pa has been exceeded. The plastic viscosity, which is determined as the slope of the flow curve for electrified suspension, is comparable to the Newtonian viscosity without an electric field. It is generally accepted[16,17] that the suspensions are converted from Newtonian liquids to Bingham bodies on the application of electric fields. The timescale of transition is on the order of 1 ms.

The deformation and rupture process of columns is very complicated. On the assumption that the particles all align into chains of single-particle width and equal spacing, Marshall et al.[18] derived a Bingham constitutive equation of ER fluids. In the analysis, all the chains spanning the electrode gap rupture in the center and immediately swing back to re-form with the nearest chain on the opposite electrode. Based on the bulk polarization theory, the yield stress σ_0 is given by

$$\sigma_0 \propto \phi \varepsilon_0 \varepsilon_c \beta^2 E^2 \qquad (3.16)$$

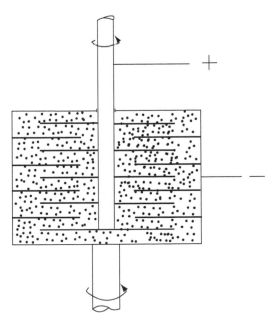

FIGURE 3.6   ER clutch.

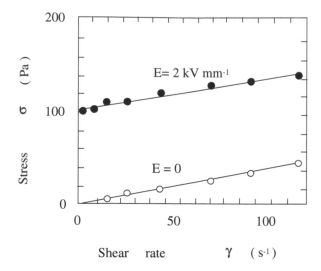

FIGURE 3.7   Transition from Newtonian flow to Bingham flow for an ER suspension.

$$\beta = \frac{\varepsilon_p - \varepsilon_c}{\varepsilon_p + 2\varepsilon_c} \qquad (3.17)$$

where ϕ is the volume concentration of particles, ε_0 is the permittivity of free space, ε_c and ε_p are the dielectric constants of medium and particles, respectively, and E is the field strength. This theory is successful in correlating the yield stress of suspensions and material functions of components. However, according to the bulk polarization theory that ignores the surface effects, the only method for formulating excellent ER fluids is a dielectric mismatch. In actual suspensions, the interparticle forces, due to surface polarization, play an essential role in the ER effects. For example, the ER behavior of silica suspensions is governed by the amount of surface silanol groups. The surface modification of particles may have great potential to improve the ER performance of suspensions.[19]

Electric Resistance

Provided that the medium is completely insulating, an electric current is attributed to the particle–medium interface or particle itself. Although the current arises only from the electrical contact of two particles, the chains spanning the electrode gap are primarily responsible for susceptible current as an overall response of static suspensions. Because the chain formation is promoted at high electric fields, the current density rapidly increases with field strength. In addition, ER fluids typically operate at electric fields of 3–5 kV/mm. The local field strength near the contact area can reach values as high as 30–100 kV/mm, where partial breakdown of oil is possible. Hence, this mechanism also leads to nonohmic behavior in which the resistance decreases with increasing field strength.

Figure 3.8 shows the current density plotted against the field strength for a 40-vol% silica suspension in a quiescent state under a steady shear rate of 200 s^{-1}.[20] The plots for each shear condition lie on a straight line. In shear fields, the rupture and reformation of columns are constantly repeated. Although the current density is decreased by the application of shear fields, the effect of shear rate is not very strong. Although the ionic and electronic migration along the columns spanning the electrodes is the main origin of current in the static suspensions, a different mechanism may be couched in the charge transfer of suspensions under shear flow.

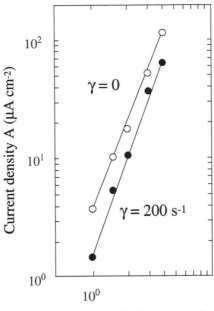

FIGURE 3.8   Current density plotted against the field strength for static and flowing suspensions.

If the suspension is completely dispersed to primary particles in electric fields, a charged particle approaches the electrode of the opposite charge. When the particle touches the electrode surface, a transfer of charge to the particle occurs until the electrode and particle reach the same potential. Then the particle is repelled from the electrode and moves toward the other electrode. Because the process is continuously repeated, the oscillatory motion of charged particles leads to the current. Presumably, the oscillatory motion of drifting chains between electrodes may contribute to conductivity. Because of combined effects of various mechanisms, the relation between the current density and field strength has not been quantitatively established. However, several authors have reported that the current density A can be expressed by

$$A \propto E^n \tag{3.18}$$

From the slopes of straight lines in Figure 3.8, the n values were determined to be 2.5 for the static suspension and 2.7 for the flowing suspension. The n value has been found to be 2.5–3.0, independent of shear rate.

Light Transmittance

In suspensions without electric fields, the particles are randomly and uniformly dispersed by Brownian motion. However, the column structures are developed and the optical anisotropy is induced in electric fields. Figure 3.9 shows the light transmittance plotted against the thickness in electric fields of 0 and 2 kV/mm for an ER suspension at a particle concentration of 7 vol%. The light emitted from a lamp is introduced parallel to the electric field. The transmittance in a zero electric field rapidly decreases with thickness. The plots are approximated by a straight line, the slope of which gives the absorption coefficient of suspension. The exponential decay of transmittance with thickness shows that in the absence of electric fields, the particles are randomly dispersed in the medium. In electric fields, the transmittance is almost constant irrespective of thickness. The most

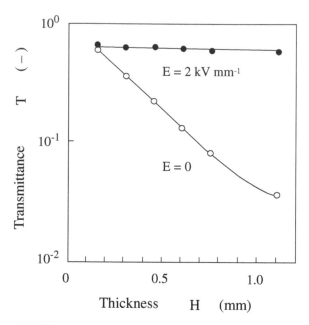

FIGURE 3.9   Thickness dependence of light transmittance for an ER suspension in electric fields of 0 and 2 kV/mm.

significant feature is that the transmittance of suspension is drastically increased in electric fields. The clearing effect is markedly enhanced as the cell thickness is increased. If the suspension does not absorb the irradiation energy, the transmittance must be equal to unity and independent of thickness. However, the results indicate that the energy is absorbed in the suspension, although the light intensity is considered to be relatively homogeneous in the direction parallel to the field vector. The light may be absorbed in a very thin layer near the surface exposed to the light source.

In electric fields, the dilute suspension consists of a collection of discrete columns; hence, the induced structures are strongly anisotropic. The plane perpendicular to the optical path is occupied by many discrete columns and a continuous medium. In ordinary suspensions, the particles are highly absorbing. Because of high optical density, the columns do not transmit light, and the vacant area without particles is regarded as transparent. Considering that the irradiation energy is absorbed mainly by the particles and the fraction absorbed by the medium is negligible, the cross section of optical path comprises dark (column) and transparent (medium) regions. Because the fraction of vacant area increases with field strength, the transmittance is markedly increased and is not strongly affected by the cell thickness. The light absorption in electric fields is a purely geometrical effect, which depends only on the cross section of columns.[21] The electro-optical effect of suspensions, due to the column formation, can provide the basis for passive display devices.

REFERENCES

1. White, H. J., in *Industrial Electrostatic Precipitation,* Addison-Wesley, Reading, Mass., 1963, p. 319.
2. Masuda, S. and Mizuno, A., *Proc. Inst. Electrostat. Jpn.,* 2, 59, 1978.
3. Masuda, S., *Denki Gakkaishi,* 80, 1790, 1960.
4. Maxwell, J. C., *Electricity and Magnetism,* Vol. 1, Clarendon, Oxford, 1892.
5. Louge, M. and Opie, M., *Powder Technol.,* 62, 85, 1990.
6. Rayleigh, W. R., *Philos. Mag. Ser. 5,* 34, 481, 1892.
7. Böettcher, C. J. F., *Rec. Trav. Chim.,* 64, 47, 1945.

8. Yadav, A. S. and Parshad, R., *J. Phys. D Appl. Phys.*, 5, 1469, 1972.
9. Masuda, H., Gotoh, K., and Orita, N., *J. Aerosol Res. Jpn.*, 8, 325–332, 1993.
10. Mazumder, M. K., *KONA*, 11, 105–118, 1993.
11. Thomson, W. (later Lord Kelvin), *Phil. Mag.*, 46, 1898.
12. Masuda, H., Itakura, T., Gotoh, K., Takahashi, T., and Teshima, T., *Adv. Powder Technol.*, 6, 295–303, 1995.
13. Halsey, T. C., *Science*, 258, 761, 1992.
14. Tao, R. and Sun, J. M., *Phys. Rev. Lett.*, 67, 398, 1991.
15. Winslow, W. M., *J. Appl. Phys.*, 20, 1137, 1949.
16. Block, H. and Kelly, J. P., *J. Phys. D Appl. Phys.*, 21, 1661, 1988.
17. Jorda, T. C. and Shaw, M. T., *IEEE Trans. Electr. Insul.*, E1–24, 849, 1989.
18. Marshal, L., Zukoski, C. F., and Goodwin, J. W., *J. Chem. Soc. Faraday Trans. 1*, 85, 2785, 1989.
19. Otsubo, Y. and Edamura, K., *J. Colloid Interface Sci.*, 168, 230, 1994.
20. Otsubo, Y., Sekine, M., and Katayama, S., *J. Rheol.*, 36, 479, 1992.
21. Otsubo, Y., Edamura, K., and Akashi, K., *J. Colloid Interface Sci.*, 177, 250, 1996.

3.4 Magnetic Properties

Toyohisa Fujita
University of Tokyo, Tokyo, Japan

3.4.1 MAGNETIC FORCE ON A PARTICLE[1]

The magnetic dipole force F_m acting on a paramagnetic particle in vacuum is given by

$$F_m = \nabla \int v\, (\boldsymbol{B} \cdot \boldsymbol{H})\, dV \qquad (4.1)$$

where $\boldsymbol{B}$ is the magnetic polarization of a particle, V is the volume of a particle, and $\boldsymbol{H}$ is the magnetic field strength. ∇ is the operator of the gradient and acting on a scalar φ that can be written as

$$\nabla \varphi = \boldsymbol{i}\delta\,\varphi\,/\delta x + \boldsymbol{j}\delta\,\varphi\,/\delta y + \boldsymbol{k}\delta\,\varphi\,/\delta z \qquad (4.2)$$

where i, j, k are the unit vectors in the directions x, y, z, respectively. If the particle is sufficiently small it can be reduced to a point dipole moment $\boldsymbol{\mu}_m = \boldsymbol{B}V$. The force on such a point dipole is

$$F^m = (\boldsymbol{\mu}_m \cdot \nabla)\boldsymbol{H} \qquad (4.3)$$

Permeability of a spherical paramagnetic or diamagnetic particle is given as

$$\mu = (1 + \kappa)\mu_0 \qquad (4.4)$$

where κ is the volume magnetic susceptibility and μ_0 is the permeability of free space numerically equal to $4\pi \cdot 10^{-7}$ H/m. The magnetic polarization of the particle is given by

$$B = \mu_0 \kappa\, H/(1 + \kappa/3) \qquad (4.5)$$

Combining Equation 4.5 and Equation 4.3, the force on a small spherical weakly magnetic particle placed in the external magnetic field can be written as

$$\boldsymbol{F}_m = [\mu_0\kappa V/(1 + \kappa/3)](\boldsymbol{H}\nabla)\boldsymbol{H} \qquad (4.6)$$

or in a simplified form (assuming that $\kappa \ll 1$),

$$\boldsymbol{F}_m = (1/2)\mu_0\kappa V\nabla(\boldsymbol{H}^2) \qquad (4.7)$$

If a paramagnetic particle (volume magnetic susceptibility: κ_p) is immersed in the fluid (volume magnetic susceptibility: κ_f), the magnetic force per unit volume acting on a particle is given by Equation 4.7 (using $\kappa = \kappa_p - \kappa_f$). For practical calculations it is sometimes advantageous to replace the magnetic field strength by the magnetic induction **B**. Then, Equation 4.7 reads as follows:

$$\boldsymbol{F}_m = (\kappa/\mu_0)V\boldsymbol{B}\nabla\boldsymbol{B} \qquad (4.8)$$

Here **B** is considered as the external magnetic induction, and $\nabla\mathbf{B}$ is the gradient of the magnetic induction. Thus, in the direction of x, the magnetic force $\mathbf{F}_{mx}$ can be written by the following equation (where $\mathbf{B} = \mu_0\mathbf{H}$ and magnetization $\kappa\mathbf{H}_x = \mathbf{I}_x$),

$$\mathbf{F}_{mx} = \mu_0 V\kappa(\mathbf{H}_x \cdot \delta\mathbf{H}_x/\delta x) = \mu_0 V\mathbf{I}_x \cdot \delta\mathbf{H}_x/\delta x \qquad (4.9)$$

Magnetic force is proportional to the product of the external magnetic field and the field gradient and has the direction of the gradient. In a homogeneous magnetic field, in which $\nabla\mathbf{B} = 0$ or $\delta\mathbf{H}_x/\delta x = 0$, the force to change the position of a particle is zero.

Along the axis of symmetry, and for values of the external magnetic field $\mathbf{H}_{x0}$ lower than the bulk saturation value $\mathbf{H}_s$ of the ferromagnetic rod filament, the magnetic field strength is given by the expression

$$\mathbf{H}_x = \mathbf{H}_{x0}(1 + a^2/r^2) \qquad (4.10)$$

where a is the rod radius and r is the radius from the center of the rod, as shown in Figure 4.1. For $\mathbf{H}_{x0} > \mathbf{H}_s$, $\mathbf{H}_x$ is given by

$$\mathbf{H}_x = \mathbf{H}_{x0} + \mathbf{H}_s(a^2/r^2) \qquad (4.11)$$

The magnetic field gradient for $\mathbf{H}_{x0} < \mathbf{H}_s$ along the x axis of the particle, when the magnetization of the particles is small, is given as

$$d\mathbf{H}_x/dx = -2\mathbf{H}_{x0}(a^2/r^3) \qquad (4.12)$$

If the Equation 4.10, Equation 4.12, and Equation 4.6 are combined, the approximate expression for the magnetic force on a pointlike spherical weakly magnetic particle is given as

$$\mathbf{F}_m = -(8/3)\pi\mu_0\kappa b^3\, \mathbf{H}_{x0}(1 + a^2/r^2)\mathbf{H}_{x0}(a^2/r^3) \qquad (4.13)$$

where b is the particle radius. Equation 4.13 can be rewritten for a matched system $a = 3b$ as

$$\mathbf{F}_m = -(75/128)\pi\mu_0\kappa b^2\mathbf{H}_{x0}^2 \qquad (4.14)$$

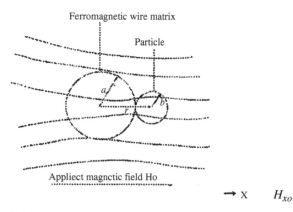

FIGURE 4.1   Cross section of spherical particle of radius b, attached to ferromagnetic wire rod of radius a (magnetized uniform magnetic field $\mathbf{H}_{x0}$).

TABLE 4.1 Characteristics of Temporary Magnetic Alloys

Alloy	Constituents[a]	Initial Permeability	Maximum Permeability	Coercive Force (Oe)	Saturated Magnetism (G)	Curie Point (°C)	Density (g/cm)
Silicon iron	4Si	500	7,000	0.5	19,700	690	7.65
45Permalloy (V)[b]	45Ni	3,000–5,000	50,000–70,000	0.06–0.12	16,000	–	8.17
78Permalloy (V)	78.5Ni	30,000–70,000	80,000–200,000	0.01–0.02	10,800	200	8.60
4–79Permalloy	4Mo, 79Ni	20,000	100,000	0.05	8,700	460	8.72
Supermalloy	5Mo, 79Ni	70,000–150,000	400,000–700,000	0.01–0.05	7,500–8,100	400	8.77
Mumetal	5Cu,2Cr,77Ni	20,000	100,000	0.05	6,500	–	8.58
Permender	50Co	800	5,000	2.0	24,500	980	8.3
Vanadium-Permender (V)	1.8V,49Co	800	4,500	0.8	24,000	980	8.2
Alperm	16Al	3,000	55,000	0.04	8,000	400	6.5
Sendust	5Al,10Si	30,000	120,000	0.05	10,000	500	7.0

[a] Residual constituent is Fe.
[b] (V) = vacuum dissolution.

Several competing forces act on the particles placed in a magnetic separator. These are the force of gravity, the inertial force, the hydrodynamic drag, and surface and interparticle forces. The force of gravity F_g can be written as

$$F_g = \rho V g \tag{4.15}$$

where ρ is the density of the particle, and g is the acceleration due to gravity. The hydrodynamic drag F_d on a particle of small radius is given by Stoke's law,

$$F_d = 6\pi \eta b v_p \tag{4.16}$$

where η is the dynamic viscosity of the fluid, b is the particle radius, and v_p is the relative velocity of the particle with respect to the fluid.

Magnetic particles will be separated from "nonmagnetic" (i.e., more magnetic particles from less magnetic particles), if the following conditions are met:

$$\mathbf{F}_m^{mag} > \sum \mathbf{F}_c^{mag} \text{ and } \mathbf{F}_m^{nonmag} < \sum \mathbf{F}_c^{nonmag} \tag{4.17}$$

where $\mathbf{F}_c$ is a competing force, while $\mathbf{F}^{mag}$ and $\mathbf{F}^{nonmag}$ are forces acting on magnetic and nonmagnetic particles respectively. Although these conditions are clearly defined, a complication arises because the relative significance of the forces is determined mainly by the particle size. It can be seen from Equation 4.8, Equation 4.15, and Equation 4.16 that, since $\mathbf{F}_m \propto b^3$ or b^2, the competing forces have the following dependence on particles size:

$$\mathbf{F}_d \propto b^1 \text{ and } \mathbf{F}_g \propto b^3 \tag{4.18}$$

In dry magnetic separation, where $\mathbf{F}_d$ is usually negligible, the particle size does not affect the efficiency of separation significantly because of the same particle size dependence of the magnetic force and of the force of gravity. On the other hand, when the hydrodynamic drag is important, selectivity of the separation will be influenced by particle size distribution. With decreasing particle size the relative importance of the hydrodynamic drag increases in comparison to the magnetic force.[1]

The nonselective nature of the magnetic force is illustrated in Table 4.1. It can be seen that the magnetic force exerted on a coarse weakly magnetic particle is similar to the one exerted on a smaller and more strongly magnetic particle. Both particles will appear in the same product of separation unless the competing forces acting on particles of different sizes is different.

3.4.2 FERROMAGNETIC PROPERTIES OF A SMALL PARTICLE[2]

Ferromagnetic material is an ensemble of small magnetic domains within which the magnetic moment is unidirectional. The boundary between these domains is called the domain wall. When a diameter d of spherical particle contains width d of several domains, the domain wall energy U_w is given by the following formula:

$$U_w = \sigma(\pi r^2)2r/d \tag{4.19}$$

where σ is a wall energy.

On the other hand, as a magnetic static energy, U_m is two times larger than the single domain's one, U_m is given by

$$U_m = (\mathbf{I}_s^2/6\mu_o)(4\pi r^3/3)(d/2r) \tag{4.20}$$

Table 4.1   The effect of particle size on separation

Particle size*	Magnetic susceptibility *	Magnetic force*
10	1	10^3
1	10^3	10^3

*Arbitrary unit

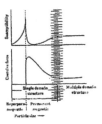

FIGURE 4.2   Variation of magnetic properties with particle size [From Mitsui, S., in *Jisei Buturi no Shinpo*, S. Chikazumi, Ed., Agne, Tokyo, 1964, p. 242. With permission.]

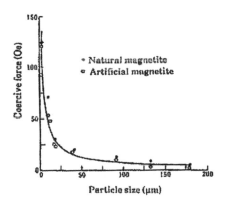

FIGURE 4.3   Dependence of coercive force of magnetite on particle size. [From Gottschalk. V H., *Physics* 6, 127, 1935. With permission.]

where I_s is saturation magnetization of ferromagnetic material. U_w in Equation 4.19 is proportional to area, while U_m in Equation 4.20 is proportional to volume. When the size of a ferromagnetic material decreases and reaches the critical small size, the wall energy becomes larger than the magnetic static energy. At this critical size the magnetic domain can not be divided. When the size of ferromagnetic particles is reduced sufficiently and reaches the critical small size, the domain wall disappears (i.e., the particles have a single domain structure), and the susceptibility and coercive force of particles vary as shown schematically in Figure 4.4. There is a particle size under which particles are super-paramagnetic, and the coercive force of particles is negligibly small by kT where k is Boltzmann constant and T is absolute temperature.

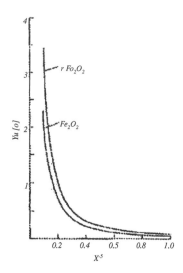

FIGURE 4.4   Critical particle size for single domain structure; *a* and *b* are the minor and major axes of the particle, respectively. [From Morrish, A. H. and Yu, S. P., *J. Appl. Phys.*, 26, 1049, 1955. With permission.]

A typical dependence of the coercive force of magnetite particles on the particle size is shown in Figure 4.3. Figure 4.4 shows the variation of critical particle size under which Fe_3O_4 and γ-Fe_2O_3 particles have a single domain structure with the aspect ratio. Magnetic fluid is one example of an apparent superparamagnetic behavior described by the same Langevin's law,[6] where the Brownian or Neel mechanisms lead.

3.4.3   MAGNETISM OF VARIOUS MATERIALS

The development of permanent magnetic materials and the improvement on their magnetic properties have been particularly studied during the last 30 years. Figure 4.5 illustrates the history of improvement in the energy product of permanent magnets.[7] Probably the most significant innovation step was made in the late 1970s when rare-earth magnets became available. Ferromagnetic materials are classified into soft and hard magnetic materials. Hard magnetic materials are used as permanent magnets, which have large coercive force. Typical properties of hard magnetic materials of artificial alloys, ferrites, and bonded in polymer are listed in Table 4.2.[2]

On the other hand, properties of soft magnetic materials are listed in Tables 4.3–4.5.[2] Soft iron as well as DC relays, plungers, pole pieces, solenoids, and brakes for intermittent use are used, since the magnetic requirements are low and the cost must also be low. The high permeability, low coercive force, and high saturation magnetization of soft magnetic materials are used in certain applications. In the past, industry employed ferrites at low power levels; however, nowadays higher power levels of ferrites have been developed.

Relative susceptibility of minerals is listed in Table 4.6.[2] Using the magnetization $\kappa H = I$, permeability μ is given in the following formula:

$$\mathbf{B} = (\kappa + \mu_0)\mathbf{H} = \mu\mathbf{H} \tag{4.21}$$

Here the relative susceptibility means κ/μ_0 (SI unit) that is $4\pi\rho$ times of susceptibility in CGS units (ρ, g/cm³).

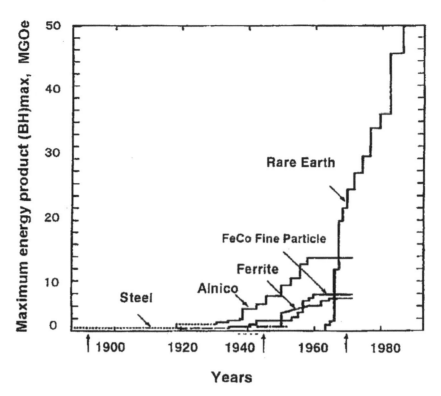

FIGURE 4.5   Progress in development of permanent magnetic materials. [Svoboda, J. and Fujita, T., in *Proceedings of the Twenty-second International Mineral Processing Congress (IMPC),* Cape Town, 2003, pp. 261–269. With permission.]

TABLE 4.2   Characteristics of Permanent Magnetic Alloys

Alloy	Constitutents[a]	Coercive Force(Oe)	Residual Magnetic Flux Density (G)	$(BH)_{max} \times 10^{-6}$ (G·Oe)	Density (g/cm³)
36Co steel	0.7C,36Co	200–250	9,000–10,000	1.0	8.2
Alnico 2	12.5Co	470–590	7,000–8,000	1.4–1.6	7.1
Alnico 5	24Co,14Ni	660–620	12,800–13,400	4.5–5.3	7.3
Alnico 5(DG)	24Co,14Ni	640	13,100	5.5	7.3
New KS steel	27Co,18Ni	950–1,050	5,500–6,300	1.6–2.1	7.4
Vicalloy 2	52Co,14V	510	10,000	3.5	—
Alnico 2(S)	2.5Co,17Ni,10Al,6Cu	540	6,900	1.4	7
Alnico 5(S)	24Co,15Ni,8Al,3Cu	575	10,000	3.5	6.6

[a] Residual constituent is Fe.

TABLE 4.3   Characteristics of Temporal Magnetic Ferrite

Ferrite		Initial Permeability	Saturated Magnetism (G)	Residual Magnetic Flux Density (G)	Coercive Force (Oe)
Mn-Zn	F	1000	3200	1200	0.3
	F_4	1300	4000	1000	0.3
	F_5	2000	3900	900	0.3
	H	2100	4000	1250	0.2
	J	800	4200	2000	0.4
	B_2	1200	4000	2500	0.3
	B_3	1400	4200	1700	0.3
	B_5	1100	4500	1500	0.3
Ni-Zn	N3b	60	4000	2300	4
	N3d	100	3000	1500	6

TABLE 4.4   Characteristics of Barium Ferrite Magnet

Barium Ferrite Magnet		Residual Magnetic Flux Density	Coercive Force (Oe)	$(BH)_{max}$ (G·Oe) × 10^{-6}
Q-1	N.O.	2300–2500	1800–2000	1.0–1.2
Q-6	O	4100–4400	1600–1900	More than 3.6
Q-7	O	3300–3700	2700–3000	More than 2.6
Q-8	O	3800–4100	2100–2500	More than 3.4
Q-10	O	3400	1900	2.4

TABLE 4.5   Characteristics of Rare Cobalt Magnet

Magnet	Magnetic Induction (kG)	Anisotropic Magnetic Field (kOe)	Curie Point (°C)	(BH_{max}) (MG·Oe)
Sm_2Co_{17}	12.5	107	920	39
$Sm_2(Co_{3.8}Fe_{0.2})_{17}$	14.0	83	890	49
$Ce(CoFe)_{17}$	11.5	15	780	—
$Pr(CoFe)_{17}$	14.0	20	880	—
$Y(CoFe)_{17}$	12.7	15–18	950	—
$SmCo_5$	11.6	290	720	31

TABLE 4.6   Susceptibility of Minerals

Minerals	Chemical Composition	Specific Susceptibility ($\times 10^{-6}$)
Franklinite	$(Fe,Zn,Mn)O(FeMn)_2O_3$	455–640
Siderite	$FeCO_2$	84.2, 142.6
Zircon	$ZrSiO_4$	−0.17, +0.732
Corundum	Al_2O_3	−0.34
Quartz	SiO_2	−0.461, −0.466
Rutile	TiO_2	1.96, 2.09
Pyrite	FeS_2	0.98
Sphalerite	ZnS	−0.264
Dolomite	$CaMgC_2O_8$	0.787, 1.20
Apatite	$Ca_5Cl(PO_4)_3$	−2.64
Chalcopyrite	$CuFeS_2$	0.85
Spinel	$MgAl_2O_4$	0.62
Galena	PbS	−0.350
Rock salt	$NaCl$	−0.50
Tourmaline	$H_9Al_2(BOH)_2Si_4O_{19}$	1.12, 0.748
Graphite	C	−2.2, −14.2
Fluorite	CaF_2	−0.285
Aragonite	$CaCO_3$	−0.392, −0.444
Calcite	$CaCO_3$	−0.363, −0.405
Ruby	Al_2O_3	0.47
Topaz	$(AlF)_2SiO_4$	−0.42
Beryl	$Be_3Al_2(SiO_3)_6$	0.826, 0.386
Epidote	$Hca_2(Al, Fe)_3Si_3O_{13}$	23.8, 23.9
Hornblende	$(Ca, Mg, Fe)_8Si_9O_{26}$	24.0, 18.0
Augite	$CaMgSiO_6$	26.6, 22.7
Diopside	$CaMg(SiO_3)_7$	8.8
Columbite	—	33.6–43.2
Monazite	—	18.1
Pitchblende	—	13.0
Gadolinite	—	5.83
Manganite	$Mn_2O_3 \cdot H_2O$	490
Cobaltite	$CoAsS$	5.8
Feldspar	—	1.6
Limestone	—	6

REFERENCES

1. Svoboda, J., *Magnetic Methods for the Treatment of Minerals,* Elsevier, New York, 1987, pp. 3–9.
2. Yashima, S. and Fujita, T., *J. Soc. Powder Technol. Jpn.,* 28, 257–266, 1991; Ushiki, K., *Powder Technology Handbook,* 2nd Ed. Macel Dekkar, New York, 1977.
3. Gottschalk. V H., *Physics* 6, 127, 1935.
4. Mitsui, S., in *Jisei Buturi no Shinpo,* S. Chikazumi, Ed., Agne, Tokyo, 1964, p. 242.
5. Morrish, A. H. and Yu, S. P., *J. Appl. Phys.,* 26, 1049, 1955.
6. Rosensweig, R. E., *Ferohydrodynamics,* Cambridge University Press, London, 1985, pp. 61–62.
7. Svoboda, J. and Fujita, T., in *Proceedings of the Twenty-second International Mineral Processing Congress (IMPC),* Cape Town, 2003, pp. 261–269.

Acknowledgment

The author gratefully appreciates Dr. Svoboda's useful discussions.

3.5 Packing Properties

Michitaka Suzuki
University of Hyogo, Himeji, Hyogo, Japan

It is well known that the packing of particulates is one of the most fundamental and important powder-handling processes, and various properties of the particulate assembly are basically determined by the geometrical arrangement of individual particles. Void fraction or voidage is a most popular and simple expression of packing characteristics of a powder bed. In an actual powder system, void fraction ε is defined by

$$\varepsilon = 1 - \frac{\rho_b}{\rho_p} = 1 - \frac{M}{\rho_p V} \tag{5.1}$$

where ρ_b, ρ_p, M, and V are apparent density, particle density, and mass and volume of powder bed, respectively. Particle volume fraction or packing fraction defined as $\phi = 1 - \varepsilon$ is frequently used also to express the packing structure in place of the void fraction.

In actual powder, the void fraction is changing with the particle size.[1] As shown in Figure 5.1, the void fraction shows the constant value ε_c over certain critical particle size D_{pc}, and below this size, the void fraction increases with decrease in the particle size. Roller[2] proposed the following equation as a relationship between the void fraction and particle size.

$$\varepsilon = \varepsilon_c \quad (D_p > D_{pc}) \tag{5.2a}$$

$$\varepsilon = \frac{1}{1 + \left(\frac{1}{\varepsilon_c} - 1\right)\left(\frac{D_p}{D_{pc}}\right)^n} \quad (D_p < D_{pc}) \tag{5.2b}$$

Index n and the critical particle size D_{pc} change with environmental conditions, packing method, and types of powder. Generally n takes the value between 0 and 1.

3.5.1 PACKING OF EQUAL SPHERES

Many researchers have investigated the assembly of equal spheres because of its simplicity and its convenience in theoretical work. Regular packing is the easiest to use to describe internal structure as a set of unit cells. There are six typical unit cells, as shown in Figure 5.2. Graton and Fraser[3] characterized the internal structure of the unit cells, consisting of a few primary spheres in which geometrical features such as the angle and distance between neighboring sphere centers, the void fraction, and the coordination number (or the number of contact points on a particle surface) are obtained. Closest or rhombohedral packing is used frequently as a reference system.

In actual packing of particles the void fraction and coordination number vary widely and continuously, but in regular packings there exist only discrete values: void fraction = 0.2594 (rhombohedral), 0.3119 (tetragonal–sphenoidal), 0.3954 (orthorhombic), and 0.4764 (cubic) and coordination number N_c = 12, 10, 8, and 6, respectively. Heesh and Laves[4] proposed a regular packing of equal spheres supported by wire frames as a model structure of particle assembly

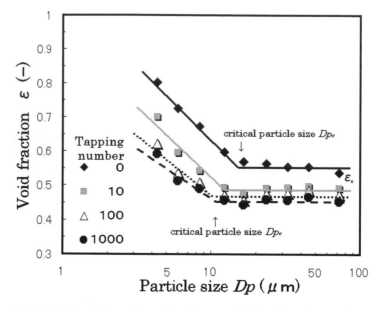

FIGURE 5.1   Effect of particle size on void fraction of monosized powder bed. [From Suzuki, M., Sato, H., Hasegawa, M., and Hirota, M., *Powder Technol.*, 118, 53–57, 2001. With permission.]

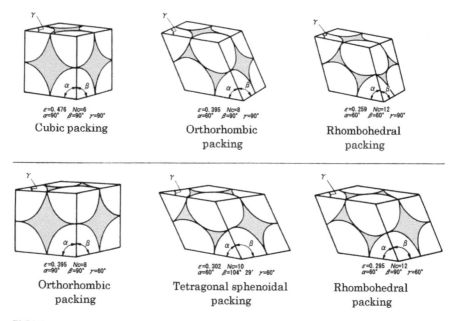

FIGURE 5.2   Regular packing structures of equal spheres.

with high void fraction. But this type of spherical structure scarcely exists in actual particulate assemblies and has fixed values of void fraction and coordination number. Smith et al.[5] proposed a hybrid model obtained by mixing the simple cubic and rhombohedral unit cells, and the model equation (7) in Table 5.1 shows the continuous relation between the void fraction ε and the coordination number N_c. Shinohara et al.[6] and Shinohara and Tanaka[7] proposed another hybrid model, which is a random mixture of the simple cubic and rhombohedral unit cells and

TABLE 5.1   Expressions of the Relationship between Void Fraction and Average Coordination Number

Expression	Equation No. in Fig. 5.3
$\overline{N}_c = 3.1/\varepsilon$	(1) Rumpf 1958
$\overline{N}_c = 2e^{2.4(1-\varepsilon)}$	(2) Meissner et al. 1964
$\varepsilon = 1.072 - 0.1193\overline{N}_c + 0.0043\overline{N}_c^{\;2}$	(3) Ridgeway et al. 1967
$\overline{N}_c = 22.47 - 33.39\varepsilon \;(\varepsilon \leq 0.5)$	(4) Haughey and Beveridge 1969
$\overline{N}_c = \left[8\pi/\left(0.727^3 \times 3\right)\right](1-\varepsilon)^2$	(5) Nagao 1978
$\overline{N}_c = 1.61\varepsilon^{-1.48} \;(\varepsilon \leq 0.82)$	(6) Nakagaki and Sunada 1968
$\overline{N}_c = 4.28 \times 10^{-3}\varepsilon^{-17.3} + 2.00 \;(0.82 \leq \varepsilon)$	
$\overline{N}_c = 26.49 - 10.73/(1-\varepsilon) \;(\varepsilon \leq 0.595)$	(7) Smith et al. 1929
$\overline{N}_c = 20.7(1-\varepsilon)/\pi - 4.35 \;(0.3 < \varepsilon \leq 0.53)$	(8) Gotoh 1978
$\overline{N}_c = 36(1-\varepsilon) \;(0.53 \leq \varepsilon)$	
$\overline{N}_c = \dfrac{2.812(1-\varepsilon)^{-1/3}}{(b/x)^2\left\{1 + (b/x)^2\right\}}$	(9) Suzuki et al. 1980
$b/x = 7.318 \times 10^{-2} + 2.193\varepsilon - 3.357\varepsilon^2 + 3.194\varepsilon^3$	
$\overline{N}_c = (32/13)(7-8\varepsilon)$	(10) Ouchiyama et al. 1980

can explain a wide and continuous change in the void fraction and the coordination number. These are useful models for close-packing structures with local order. Gotoh[8,9] and Iwata and Homma[10] made similar studies of random packing in terms of the combination of regular packings. Recently, Bargiel and Tory[11] investigated the transition from dense random packing and measured the disorder of ultradense irregular packing of equal spheres.

Many investigations have been made of the coordination number in a random assembly of equal spheres; these include studies by Smith et al.,[5] Rumpf,[12] Bernal and Mason,[13] Meissner et al.,[14] Wade,[15] Ridgway and Tarbuck,[16] Arakawa,[17] Haughey and Beveridge,[18] Bernal et al.,[19] and Nagao,[20] and a number of empirical relations have been proposed. Nakagaki and a Sunada,[21,22] Bennett,[23] and Tory et al.[24] made computer simulations of three-dimensional random packings from which they obtained the relationship between the average coordination number N_c and the void fraction. Figure 5.3 shows a comparison between the results calculated from the equations in Table 5.1 and practical experiments and computer experiments. Equation 5.1 through Equation 5.5 (the dashed curves) express the empirical relations, whereas Equation 5.6 and Equation 5.7 (the dashed curves) are obtained from the random mixture model of regular packings. Gotoh and Finney,[25] Suzuki et al.,[26] and Ouchiyama and Tanaka[27] reported the theoretical relations in Equation 5.8 through Equation 5.10 (the solid curves). Suzuki and Oshima[28] compared the calculated results obtained by Equation 5.1 through Equation 5.10 and the four kinds of computer simulated results, showing that Nakagaki et al.'s empirical formula agreed well with three simulation results over a wide range of

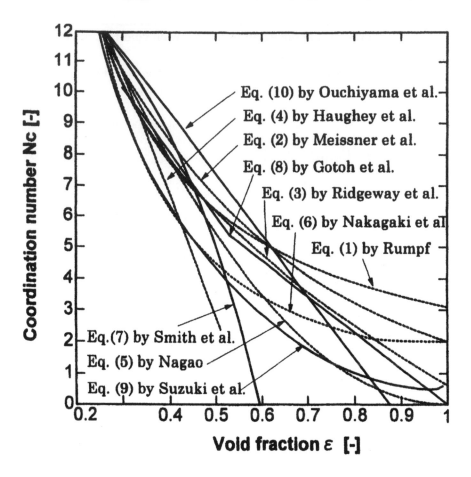

FIGURE 5.3   Relationship between void fraction and average coordination number for monosized spheres.

the void fraction. Some of the above simulation results corresponded well to Suzuki et al.'s and Gotoh's model formula and Nagao's empirical formula in a certain void fraction.

Suzuki et al.[29] investigated the effect of size distribution on the relation using the model calculation and computer simulation results. Figure 5.4 shows the relation between the void fraction and the average coordination number of the random-packed bed with log-uniform size distribution. Based on the results, the coordination number is not so affected with the size distribution of particles, but the void fraction becomes smaller for wider distribution of particle size. It means that the value of the coordination number, estimated by a conventional equation for equal spheres from void fraction data, is overestimated for randomly packed beds of multicomponent spheres with wider particle size distributions.

Knowledge of the spatial structure of particles in a random assembly such as the radial distribution function, the first-layer neighbors, and the average nearest-neighbor spacing is important in dealing with the deformation of a powder bed, the electrostatic and liquid bridge forces between particles, radiative and convective heat transfer in particle systems, and so on. The radial distribution function $g(R)$ is the local number density of particles, normalized by its bulk-mean value, at a distance R from a central reference sphere. The analytical result in the Percus–Yevick approximation is depicted in Figure 5.5, in which the damped oscillation implies the short-range order in the particle arrangement. The number density exhibits a shell-like distribution in the polar coordinate about the central sphere, and the spheres in the innermost shell are regarded as first-layer neighbors. Gotoh[30]

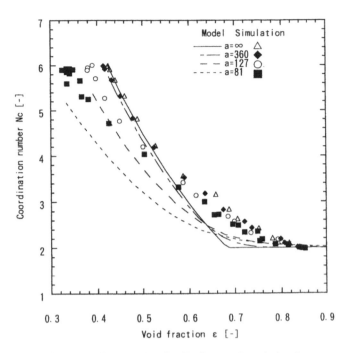

FIGURE 5.4   Effect of size distribution on the relation between void fraction and average coordination number for monosized spheres.

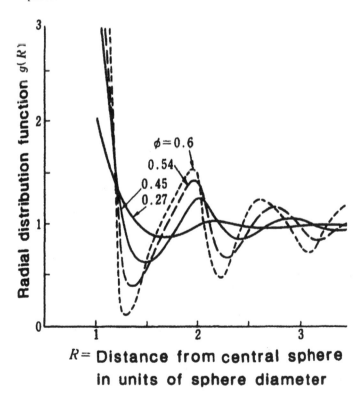

FIGURE 5.5   Radial distribution function; $\Phi(r)$, bulk-mean particle volume fraction. [From Gotoh, K., *J. Soc. Powder Technol. Jpn.*, 15, 726–733, 1980. With permission.]

and Suzuki et al.[31] discussed the structure of the first-layer neighbors. Chandrasekhar,[32] Bansal and Ardel,[33] and Gotoh et al.[34] discussed the average center-to-center distance between a central sphere and its nearest neighbor. Figure 5.6 depicts the relationship between the bulk-mean particle volume fraction $\phi = (1 - \varepsilon)$ and the distance of the first-layer neighbors from a central reference sphere in units of the sphere diameter. R_1, R_m, and R_n in Figure 5.6 express the outer edge and average positions of the first layer and the average nearest-neighbor distance, respectively. The number of first-layer neighbors is calculated from the radial distribution function as follows[30]:

$$Z = \left(\int_1^{R_1} 4\pi R^2 \, dR \right) N_v g(R) \tag{5.3}$$

in which $N_v = 6\phi/\pi = 6(1-\varepsilon)/\pi$, N_v is the number density of the particles, and Φ is the bulk-mean particle volume fraction.

The unit cell of the particle assembly can be expressed by a Voronoi polyhedron as illustrated in Figure 5.7. A plane bisects each line joining the central sphere to its neighbors; the innermost volume enclosed by these planes is the Voronoi polyhedron. Tanemura[35] discussed random tessellation using the Voronoi polyhedron. In random close packing the Voronoi polyhedron has 14.25 faces on average.

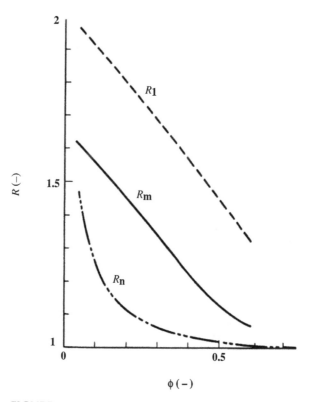

FIGURE 5.6 Relationship between packing fractions; Φ and distance of the first-layer neighbors from a central sphere in units of the sphere diameter. R_1, outer edge; R_m, average position; R_n, nearest neighbor. [From Gotoh, K., *J. Soc. Powder Technol. Jpn.*, 16, 709–713, 1979. With permission.]

Fractal dimensions of particle beds are also used to describe the particulate assemblies. Witten and Samder[36] and Tohno and Takahashi[37] expressed the particle arrangement in an aerosol agglomerate by the fractal dimension. Kamiya et al.[38] used it to characterize the packing structure of an aggregate in a particle bed. Suzuki and Oshima[39] obtained a fractal dimension of a packed particle system by the scaling method. The fractal dimension enables us quantitatively to distinguish the difference in packing structures that cannot be expressed by the relation between coordination number and void fraction.

3.5.2 PACKING OF MULTISIZED PARTICLES

Actual powders have wide distributions in size; hence, a packed bed can be regarded as a mixture of different-sized particles. Horsfield,[40] White and Walton,[41] and Hudson[42] have discussed the regular packings of multisized spheres, in which the interstices of an array of spheres are filled with smaller spheres. The void fractions of the Horsfield packings are listed in Table 5.2. Although the

TABLE 5.2   Calculated Results of the Horsfield Model for Multisized Regular Packing of Spheres

Fraction	D_{p1}/D_{p2}	Relative No. of Spheres	Void
First sphere	1	1	0.2595
Second spheres	0.414	1	0.207
Third spheres	0.225	2	0.190
Fourth spheres	0.177	8	0.158
Fifth spheres	0.116	8	0.149
Filler	Fine	Infinite number	0.039

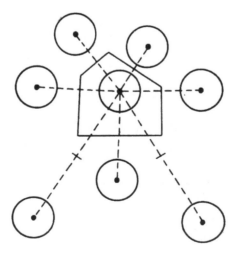

FIGURE 5.7   Two-dimensional illustration of Voronoi polyhedron.

packing densities are expected to be very high theoretically, in practice it is difficult to construct these packings.

Generally speaking, powder has a random structure, and regular packing hardly exists in practice. Experimental data of the void fraction and coordination number in a multisized particle bed have been reported by Furnas,[43] Sohn and Moreland,[44] Arakawa and Nishino,[45] and Suzuki and Oshima.[46] Powell,[47] Dinger and Funk,[48] Suzuki and Oshima,[49] and Ito et al.[50] devised a computer simulation for random packing of multisized spheres and obtained the simulation data.

Many different models have been conceived for the random packing of multisized particles, by Furnas,[51] Westman,[52] Tokumitsu,[53] Kawamura et al.,[54] Abe et al.,[55] Ouchiyama and Tanaka,[27,56,57] Suzuki and Oshima,[46] and Suzuki et al.,[58,59] and the void or packing fraction is estimated.

For example, Furnas estimated the void fraction of a binary powder mixture as follows:

$$\varepsilon = 1 - \frac{1-\varepsilon_1}{S_{v1}} \tag{5.4a}$$

$$\varepsilon = 1 - \frac{1-\varepsilon_2}{1-S_{v1}\varepsilon_2} \tag{5.4b}$$

where ε_1 is the void fraction of a coarse monosized powder bed, ε_2 is the void fraction of the fines, respectively, and S_{v1} is the fractional volume of coarse powder 1 in the binary mixture. Equation 5.3a means that the interstices between coarse particles are filled with fine particles, and Equation 5.3b means that the coarse particles are dispersed in the fine powder bed. If the two calculated voidages are not equal to each other, the higher voidage should be chosen. The typical results calculated from these equations are shown in Figure 5.8 for $\varepsilon_1 = \varepsilon_2 = 0.35, 0.4, 0.45$, and 0.5. The voidage of the binary mixture becomes much smaller than that of monosized powder bed and exhibits the minimum value, because the finer powder fills the interstices of the coarse powder.

Most investigators, except Ouchiyama and Suzuki et al., treated the packing as a continuous medium and did not consider the relation between the void or packing fraction and the coordination number. Furthermore, Ouchiyama and Tanaka's model assumes an equal void fraction for the packing of each size of sphere, but this is not the case in the actual powders. Cross et al.[60] modified the model for expressing the matrix structure of composite materials.

Here Suzuki's model for estimating the void fraction of a packed bed of m different-sized spheres is explained. There are $m \times m$ basic types of contact between a reference sphere and others (see Figure 5.9). The void fractions about the central spheres are denoted by $\varepsilon_{(1,1)}, \varepsilon_{(1,2)}, \ldots, \varepsilon_{(m,m)}$, which are defined by Equation 5.7 below. The bulk-mean void fraction ε of the bed becomes

$$\varepsilon = \sum_1^m S_{vj}\beta_j\varepsilon_j, \quad \varepsilon_j = \sum_1^m S_{ak}\varepsilon_{(j,k)} \tag{5.5}$$

where S_{vj} is the fractional volume of sphere j, ε_2 is the partial void fraction about sphere j, S_{ak} is the fractional area of sphere k about sphere j, and β_j is a constant that is calculated from the measured void fraction ε of the bed composed of sphere j by putting $S_{ak} = 1$ and $S_{vj} = 1$ in Equation 5.4:

$$\beta_j = \frac{\overline{\varepsilon}_j}{\varepsilon_{(j,j)}} \tag{5.6}$$

The next step is to derive the partial void fraction $\varepsilon_{(j,k)}$. $N_{(j,k)}$ of sphere k of diameter D_{pk} are in direct contract with a reference sphere j of diameter D_{pj}, as shown in Figure 5.10. A spherical region of radius $\overline{OD}$ (shown by the dashed curve) is considered, where O is the center of the reference sphere j and D is

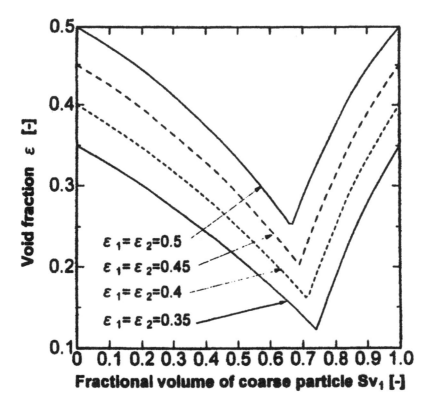

FIGURE 5.8   Calculated results from Furnas's equations.

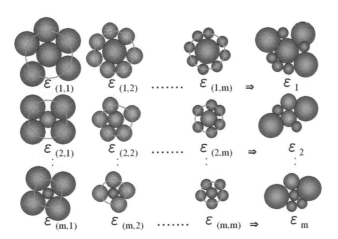

FIGURE 5.9   Fundamental combinations of contact in multi-component random mixture of spheres. [From Suzuki, M., Yagi, A., Watanabe, T., and Oshima, T., *Int. Chem. Eng.*, 26, 491–498, 1986. With permission.]

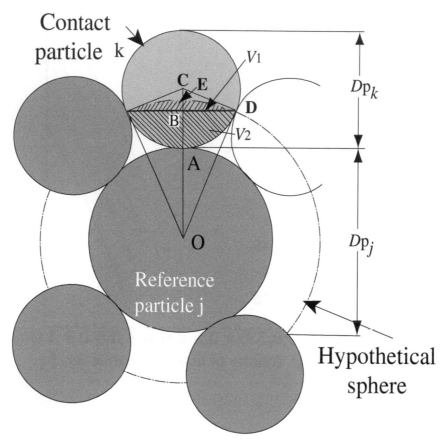

FIGURE 5.10 Model of a reference particle j in direct contact with particle k. [From Suzuki, M. and Oshima, T., *Powder Technol.*, 43, 147–153, 1985. With permission.]

the contact point of the two adjacent k spheres. The void fraction $\varepsilon_{(j,k)}$ of the spherical region can be determined as follows:

$$\varepsilon_{(j,k)} = 1 - \frac{(\pi/6)D_{pj}^3 + N_{(j,k)}(V_1 + V_2)}{V_s} \tag{5.7}$$

where V_1 is the partial volume of the spherical region, V_2 is that of the contact particle (see the cross-hatched area in Figure 5.9), and V_s the volume of the spherical region of radius $\overline{OD}$. The coordination number $N_{(j,k)}$ can be calculated from Suzuki and Oshima's model[61] for random packing of different-sized spheres.

A packed bed of two sizes of glass beads is formed for the particle size ratio D_{p1}/D_{p2} of 2, 2,83, 3.99, and 8.02. Plotted in Figure 5.11 are the experiments, and the curves depict the results calculated by Equation 5.5.

Figure 5.12 shows the relationship between the standard deviation($\ln\sigma_g$) of log-normal size distribution and the void fraction ε. The void fraction becomes smaller for wider particle size distributions. Figure 5.13 shows the relationship between the Fuller constant q of the Gaudin–Schuhmann (Andreasen) size distribution and the void fraction ε. The minimum void fraction is obtainable for

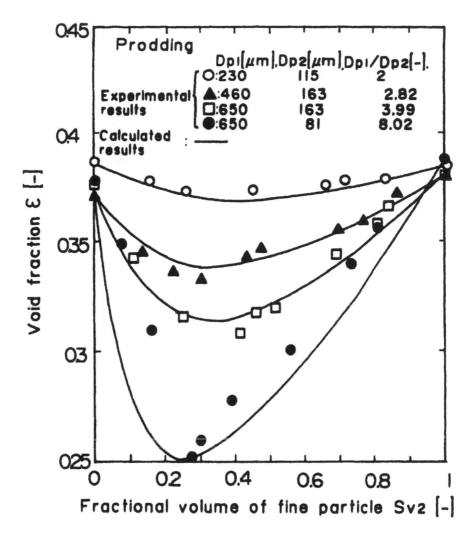

FIGURE 5.11 Relationship between fractional volume of fine spheres and void fraction of two-component mixture. [From Suzuki, M., Yagi, A., Watanabe, T., and Oshima, T., *Int. Chem. Eng.,* 26, 491–498, 1986. With permission.]

the Fuller constant of 0.5–0.8 in this case. This is because the interstices among particles decrease for $q = 0.5$ to 0.8. It was also reported by Kawamura et al.[54] experimentally that the void fraction could become minimum at $q = 0.6$, and Dinger and Funk[62] showed that optimum packing occurs for the distributions with $q = 0.37$. Suzuki and Oshima[49] reported that the q value at minimum void fraction changes with the adhesive property of fine powder and the width of size distribution. Because the calculated results are in good agreement with the experimental and simulation data, the present model may be used in the practical applications.

Notation

b	constant in Equation (h) in Table 5.1 (m)
D_{p1}, D_{p2}	diameter of coarse particle 1 or fine particle 2 in a binary mixture (m)

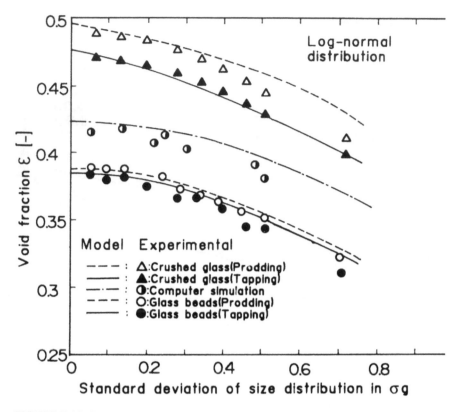

FIGURE 5.12 Relationship between standard deviation of lognormal size distribution and void. [From Suzuki, M., Oshima, T., Ichiba, H., and Hasegawa, I., *KONA Powder Sci. Technol. Jpn.*, 4, 4–12, 1986. With permission.]

D_{pc}	critical particle size of Roller's equation (m)
D_{pj}, D_{pk}	diameter of particle j or k in a multicomponent mixture (m)
$g(R)$	radial distribution function
M	mass of powder in a packed bed (kg)
m	number of components in a multicomponent mixture
n	index of Roller's equation
$\overline{N}_{(j,k)}$	coordination number for particle j in direct contact with particle k
N_c	coordination number in a regular packed bed of equal-sized spheres
$\overline{N}_c$	average coordination number in a bed of equal-sized spheres
N_v	number density of spheres
q	Fuller constant of Andreasen (Gaudin–Schumann) distribution
R	distance from a central reference spheres in units of sphere diameter
R_1	outer edge of the first layer in units of sphere diameter
R_m	average position of the first-layer neighbors in units of sphere diameter
R_n	average nearest-neighbor distance in units of sphere diameter
$S_a k$	fractional area of particle k in a multicomponent mixture
S_{v1}	fractional volume of coarse powder in a binary mixture
$S_v k$	fractional volume of particle k in a multicomponent mixture
V	volume of powder (m^3)

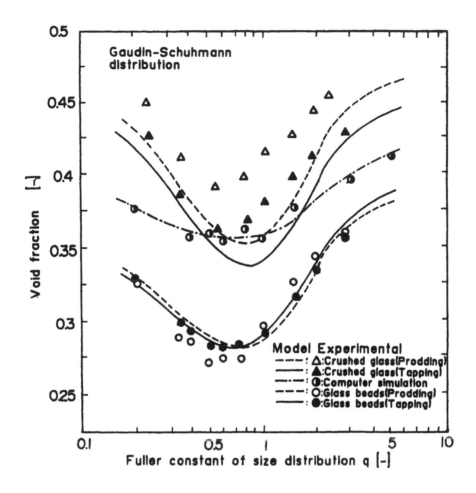

FIGURE 5.13 Relationship between Fuller constant of Gaudin–Schuhmann size distribution and void fraction. [From Suzuki, M., Oshima, T., Ichiba, H., and Hasegawa, I., *KONA Powder Sci. Technol. Jpn.*, 4, 4–12, 1986. With permission.]

V_s	volume of a hypothetical sphere in Figure 5.7 (m³)
V_1, V_2	volume of a spherical segment in Figure 5.7 (m³)
Z	number of the first-layer neighbors
β_j	proportionality constant in Equation 2.2
ε	void fraction
$\varepsilon_1, \varepsilon_2$	void fraction of coarse monosized powder bed or void fraction of fine one
$\overline{\varepsilon}_c$	void fraction of coarse monosized powder over critical particle size
$\varepsilon_{(j, k)}$	partial void fraction around particle j in direct contact with particle k
ε_j	partial void fraction around particle j
ε_j	void fraction in a bed of uniformly sized particles
ρ_b, ρ	apparent and particle density of powder (kg/m³)
σ_g	geometric standard deviation of log-normal distribution
φ	bulk-mean particle volume fraction or packing fraction ($= 1 - \varepsilon$)
$\varphi (R)$	radial distribution function

REFERENCES

1. Suzuki, M., Sato, H., Hasegawa, M., and Hirota, M., *Powder Technol.*, 118, 53–57, 2001.
2. Roller, P. S., *Ind. Eng. Chem.*, 22, 1206–1208, 1930.
3. Graton, L. C. and Fraser, H. J., *J. Geology*, 43, 785–909, 1935.
4. Heesh, H. and Laves, F., *Z. Kristallor*, 85, 433–453, 1933.
5. Smith, W. O., Foote, P. D., and Busang, P. F., *Phys. Rev.*, 34, 1271–1274, 1929.
6. Shinohara, K., Kobayashi, H., Gotoh, K., and Tanaka, T., *J. Soc. Powder Technol. Jpn.*, 2, 352–356, 1965.
7. Shinohara, K. and Tanaka, T., *Kagaku Kogaku*, 32, 88–93, 1968.
8. Gotoh, K., *Nature Phys. Sci.*, 231, 108–110, 1971.
9. Gotoh, K., *J. Soc. Powder Technol. Jpn.*, 15, 220–226, 1978.
10. Iwata, H. and Homma, T., *Powder Technol.*, 10, 79–83, 1974.
11. Bargiel, M. and Tory, E. M., *Adv. Powder Technol.*, 12, 533–557, 2001.
12. Rumpf, H., *Chem. Eng. Technol.*, 30, 144–158, 1958.
13. Bernal, D. J. and Mason, J., *Nature*, 188, 910–911, 1960.
14. Meissner, H. P., Michael, A. S., and Kaiser, R., *Ind. Eng. Chem. Process Des. Dev.*, 3, 202–205, 1964.
15. Wade, W. E., *J. Phys. Chem.*, 69, 322–326, 1965.
16. Ridgway, K. and Tarbuck, K. J., *Br. Chem. Eng.*, 12, 384–388, 1967.
17. Arakawa, M., *J. Soc. Mater. Sci. Jpn.*, 16, 319–321, 1967.
18. Haughey, D. P. and Beveridge, G. S. G., *Can. J. Chem. Eng.*, 47, 130–140, 1969.
19. Bernal, D. J., Cherry, I. A., Finney, J. L., and Knight, K. R., *J. Phys.*, E3, 388–390, 1970.
20. Nagao, T., *Trans. J.S.M.E.*, 44, 1912, 1978.
21. Nakagaki, M. and Sunada, H., *Yakugaku Zasshi*, 83, 73–78, 1963.
22. Nakagaki, M. and Sunada, H., *Yakugaku Zasshi*, 88, 705–709, 1968.
23. Bennett, H., *J. Appl. Phys.*, 43, 2727–2734, 1972.
24. Tory, R. M., Church, B. H., Tam, M. K., and Ratner, M., *Can. J. Chem. Eng.*, 51, 484–493, 1973.
25. Gotoh, K. and Finney, J. L., *Nature*, 252, 202–205, 1974.
26. Suzuki, M. Makino, K., Yamada, M., and Iinoya, K., *Int. Chem. Eng.*, 21, 482–488, 1981.
27. Ouchiyama, N. and Tanaka, T., *Ind. Eng. Chem. Fundam.*, 19, 555–560, 1980.
28. Suzuki, M. and Oshima, T., *KONA Powder Sci. Technol. Jpn.*, 7, 22–28, 1989.
29. Suzuki, M., Kada, H., and Hirota, M., *Adv. Powder Technol.*, 10, 353–365, 1999.
30. Gotoh, K., *J. Soc. Powder Technol. Jpn.*, 16, 709–713, 1979.
31. Suzuki, M., Makino, K., Tamamura, T., and Iinoya, K., *Int. Chem. Eng.*, 21, 284–293, 1981.
32. Chandrasekhar, S., *Rev. Mod. Phys.*, 15, 1–89, 1943.
33. Bansal, P. P. and Ardel, A. J., *Metallography*, 5, 97–111, 1972.
34. Gotoh, K., Jodrey, W. S., and Tory, E. M., *Powder Technol.*, 21, 285–287, 1978.
35. Tanemura, M., *J. Microscopy*, 151, 247–255, 1988.
36. Witten, T. A. and Samder, L. M., *Phys. Rev. Lett.*, 47, 1400–1403, 1981.
37. Tohno, T. and Takahashi, K., *Aerosol Res.*, 2, 117–127, 1987.
38. Kamiya, H., Yagi, E., Jimbo, G., *J. Soc. Powder Technol. Jpn.*, 30, 148–154, 1993.
39. Suzuki, M. and Oshima, T., *J. Soc. Powder Technol. Jpn.*, 26, 250–254, 1989.
40. Horsfield, H. T., *J. Soc. Chem. Ind.*, 53, 107–115, 1934.
41. White, H. E. and Walton, S. F., *J. Am. Ceram. Soc.*, 20, 155–166, 1937.
42. Hudson, D. R., *J. Appl. Phys.*, 20, 154–162, 1949.
43. Furnas, C. C., *U.S. Bur. Mines Rep. Invest.*, No. 2894, 1928.
44. Sohn, H. Y. and Moreland, C., *Can. J. Chem. Eng.*, 46, 162–167, 1968.
45. Arakawa, M. and Nishino, M., *J. Soc. Mater. Sci. Jpn.*, 22, 658–662, 1973.
46. Suzuki, M. and Oshima, T., *Powder Technol.*, 43, 147–153, 1985.
47. Powell, M. J., *Powder Technol.*, 25, 45–52, 1980.
48. Dinger, D. R. and Funk, J. E., *Interceram.*, 42, 150–152, 1993.
49. Suzuki, M. and Oshima, T., *Powder Technol.*, 44, 213–218, 1985.
50. Ito, T., Wanibe, Y., and Sakao, H., *J. Jpn. Inst. Met.*, 50, 740, 1986.
51. Furnas, C. C., *Ind. Eng. Chem.*, 23, 1052–1058, 1931.
52. Westman, A. E. R., *J. Am. Ceram. Soc.*, 19, 127–129, 1936.
53. Tokumitsu, Z., *J. Soc. Mater. Sci. Jpn.*, 13, 752–758, 1964.
54. Kawamura, J., Aoki, E., and Okuzawa, K., *Kagaku Kogaku*, 35, 777–783, 1971.

55. Abe, E., Hirosue, H., and Yokota, A., *J. Soc. Powder Technol. Jpn.,* 15, 458–462, 1978.
56. Ouchiyama, N. and Tanaka, T., *Ind. Eng. Chem. Fundam.,* 22, 66–71, 1981.
57. Ouchiyama, N. and Tanaka, T., *Ind. Eng. Chem. Fundam.,* 25, 125–129, 1986.
58. Suzuki, M., Yagi, A., Watanabe, T., and Oshima, T., *Int. Chem. Eng.,* 26, 491–498, 1986.
59. Suzuki, M., Oshima, T., Ichiba, H., and Hasegawa, I., *KONA Powder. Sci. Technol. Jpn.,* 4, 4–12, 1986.
60. Cross, M., Douglas, W. H., and Fields, R. P., *Powder Technol.,* 43, 27–36, 1985.
61. Suzuki, M. and Oshima, T., *Powder Technol.,* 35, 159–166, 1983.
62. Dinger, D. R. and Funk, J. E., *Interceram.,* 43, 87–89, 1994.

3.6 Capillarity of Porous Media

Hironobu Imakoma
Kobe University, Nada-ku, Kobe, Japan

Minoru Miyahara
Kyoto University, Katsura, Kyoto, Japan

A macroscopic body composed of fine particles naturally possesses microscopic void spaces with a size several to ten times smaller than the particle diameter. Sometimes particles themselves may have some porosity. Such pore spaces can bring functional characteristics such as adsorptive capacity, selective permeation, and dielectric properties, while in some cases they stand as a deficit. With increasing demand for functionality, recent powder-based manufacturing has tended to use finer particles of submicron size down to the nanometer range. As the result the characterization of capillarity in the nanometer range is getting more and more important.

In this section, two typical pore-size characterization methods, nitrogen adsorption and mercury porosimetry, are described with their theoretical basis and with attention paid especially to the nanometer range. Some other methods are explained briefly.

3.6.1 COMMON PHENOMENON: YOUNG–LAPLACE EFFECT

An important phenomenon that commonly applies to nitrogen adsorption and mercury porosimetry is the so-called Young–Laplace effect. As shown in Figure 6.1a, the surface of liquid that wets the wall will form a hemispherical meniscus in a sufficiently thin tube to satisfy that the contact angle θ is zero. The Young–Laplace equation describes the pressure difference ΔP brought by the pulling force of surface tension γ, which is given by Equation 6.1, including the case of partial wetting:

$$\Delta P \equiv P - P' = \frac{2\gamma\cos\theta}{r} \tag{6.1}$$

More generally, menisci with arbitrary shape can be described with two principal curvature radii, r_1 and r_2.

$$\Delta P = \gamma(\frac{1}{r_1} + \frac{1}{r_2})\cos\theta \tag{6.2}$$

For a hemi-cylindrical meniscus formed in slit space of width W, for example, Equation 6.2 will be $\Delta P = 2\gamma\cos\theta/W$ because the two principal radii are $2/W$ and 0. This equation resembles Equation 6.1, but note that r is the radius of the space while W denotes the span between walls.

The above has macroscopic and classical understandings. On the nanoscale, especially the so-called single-nano length, it fails to express the phenomena because of the hindering characteristics of a liquid surface and the effects of potential energy exerted from pore walls, for instance. With attention paid to the single-nano scale, each method is described in the following.

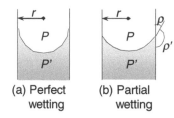

(a) Perfect
    wetting

(b) Partial
    wetting

FIGURE 6.1   Young–Laplace effect.

3.6.2   NITROGEN ADSORPTION METHOD

Pore size distributions, as well as the specific surface area, can be estimated from nitrogen adsorption isotherms at liquid-nitrogen temperature, or the amount adsorbed against relative pressure [= (equilibrium pressure p)/(saturated vapor pressure p_s)]. Measurement is done either by the volumetric method, which detects pressure variation of nitrogen gas introduced in an adsorption system with constant volume, or the gravimetric method, which measures weight variation of a sample contacting with gas of given relative pressure. In the following, description relating to measurement refers to the former method unless otherwise specified, since the great majority of commercially available automated apparatuses are based on it.

Principle: Capillary Condensation and the Kelvin Equation

Suppose that liquid nitrogen exists in a cylindrical pore as shown in Figure 6.1a. The pressure in liquid phase is lowered by $\Delta P = 2\gamma/r$ more than that in gas phase, which results in a lower free energy than a normal liquid with flat surface by $v\Delta P$, (v being the molar volume of liquid). Then the equilibrium vapor pressure of pore liquid p_g must be smaller than the saturated vapor pressure p_s. Conversely, the vapor with a pressure smaller than p_s can condensate if it goes into a pore, which is termed the capillary condensation phenomenon.

The equilibrium pressure p_g must satisfy Equation 6.3 since the difference of free energy from p_s equals $v\Delta P$.

$$\ln \frac{p_g}{p_s} = -\frac{2v\gamma}{rRT} \tag{6.3}$$

This is the so-called Kelvin equation, which was derived originally by Lord Kelvin for vapor pressure of a small droplet without the negative sign: the difference apparently comes from the concave geometry. The equation gives the pore radius r from the relative pressure p_g/p_s at which the condensation occurs. The above stands as the principle for estimating pore size distribution from a nitrogen isotherm.

Further to be considered for an adsorption system, however, is the adsorbed film with thickness t on the walls contacting with nitrogen vapor (see Figure 6.2). The condensation phase is formed within the core space excluding the films, with a contact angle of zero. The radius for a cylindrical pore is then given by

$$r_p = r + t = \frac{2v\gamma}{RT \ln(p_s / p_g)} + t \tag{6.4}$$

and the width of a slit pore will be

FIGURE 6.2 Surface adsorption film and condensation phase.

$$W = 2r + 2t = \frac{2v\gamma}{RT \ln(p_s / p_g)} + 2 \qquad (6.5)$$

Many detailed calculation schemes were proposed by, for example, Cranston and Inkley,[1] Dollimore and Heal,[2] and Barell et al.[3] Major differences between these methods are the data or the equation for adsorption thickness t, and the schemes of calculation themselves do not produce significant differences if common data for t are used. Therefore one need not worry which method is installed in a commercial apparatus; one should pay more attention to the t-data in the software and consider how precise and how abundant for various materials they are.

Measurable Range of Pore Size

Many automated adsorption apparatuses declare the upper limit of the pore size to be approximately 100 nm, which corresponds to a relative pressure of approximately 0.98. In general, however, the accuracy of the adsorption measurement decreases as the equilibrium pressure approaches the saturated vapor pressure, because the liquid nitrogen temperature will fluctuate following variation in atmospheric pressure. For safety one should understand the upper limit of reliability to be approximately 0.95 in relative pressure, corresponding to a pore size of about 40 nm. Measurement for larger pores should be done with mercury porosimetry.

As for the lower limit, pore size analysis based on the capillary condensation will lose its basis if the size goes down below about 2 nm, because the condensation phenomenon itself (or the first-order phase transition) does not occur in such small space. Much of the research conducted in the 1970s and 1980s discussed pore sizes in the range of 1–2 nm for, for example, activated carbons, but nowadays such analysis and discussion cannot be accepted. Instead, the present understanding of adsorption phenomena in such small pores is the so-called micropore filling, meaning that the gas molecules are gradually filled into the pore space by strongly attractive potential energy exerted by pore walls. Some pore analysis methods based on this mechanism are available in the literature,[4,5] among which the so-called t-plot method[6] and related ones[7,8] are suitable for simple and reliable analysis of micropore size distribution. Refer to the literature for details of these methods. Further rigorous analysis of micropore size distribution is presently under development by many researchers. Most of the approaches are based on the statistical thermodynamics method, such as molecular simulation and the density functional theory, which produce so-called local adsorption isotherms for a series of various sizes of micropore. An experimental nitrogen isotherm is expressed as a convolution integral of the local adsorption isotherms and the pore size distribution function, the latter thus can be determined through minimization of the error between measured and predicted isotherms. Note that, however, this kind of technique needs further development. Sometimes calculated pore size distribution suffers from an artifact of nonexistent pores, which may result from surface heterogeneity. One has to pay much attention to this kind of artifact, especially if the resultant distribution has bimodal or multimodal distribution.

Kelvin Model's Deficit: Underestimation in Single-Nano Range

The estimation based on the Kelvin model as given by Equation 6.4 and Equation 6.5 works well for pores larger than 10 nm. It has was pointed out in the late 1980s, however, that the model underestimates the so-called single-nano range of pores.[9,10] The research in that area was rather scientific and showed this deficit by complicated statistical thermodynamics techniques or molecular simulations with great computational costs. No method based on simplicity and convenience was available even in the 1990s, which forced people to use the Kelvin model, though knowing its inaccuracy.

A condensation model with a simple concept and easy calculation has been recently proposed[11–14] and is explained briefly below. The point is that the attractive potential energy from pore walls and the stronger surface tension of a curved interface will enhance the condensation in nanoscale pores. The basic equation is

$$RT \ln \frac{p_g}{p_s} = -\frac{2v\gamma(\rho)}{\rho(r)} + \Delta\phi(r) \tag{6.6}$$

in which the free energy for condensation (right-hand side) is compensated not only by the Young–Laplace effect with local curvature dependent surface tension $\gamma(\rho)$, but also by the attractive energy of pore walls relative to the condensing liquid $\Delta\phi(r)$. The latter effect can be determined[13] from adsorption thickness data that are usually included in the automated adsorption apparatuses. An iterative calculation is needed to determine the relation between pore size and relative pressure because the curvature of the liquid surface depends on location in the pore, but it remains an easy calculation at a pocket-computer level. For further details, see the original papers.

The degree of underestimation by the Kelvin model stays almost constant regardless of the pore size but varies depending on the pore-wall potential energy. Some examples of difference between Kelvin-based prediction and the true pore size are about 1 nm or slightly greater for carbon materials, about 1 nm or less for silica materials, and 0.5–0.7 nm for ordered mesoporous silicates (MCM-41). Thus one should understand that, if the BJH method gives peak pore size as 3 nm for a silica gel, the true size is about 4 nm or slightly less than that. The model based on Equation 6.6 can give the true pore size from the nitrogen isotherm without any additional measurement.

Hysteresis

If an adsorption isotherm goes with capillary condensation, the hysteresis between adsorption process and desorption results in most cases. The classification of the hysteresis into four types is given by IUPAC, as shown in Figure 6.3. There have been long discussions on which branch should be used for pore-size determination, but it is still quite difficult to obtain a general conclusion. Neimark and coworkers,[15,16] for example, reported that the adsorption branch for MCM-41, which possesses almost ideally cylindrical pores, is of spinodal process with higher condensation pressure than thermodynamic equilibrium because the surface adsorption layer has to grow up locally into the central portion of the pore space. The desorption process, on the other hand, does not suffer from such an energy barrier, resulting in an equilibrium desorption. This mechanism seems to be accepted by many researchers recently, but some groups have raised questions about it. By limiting the topic into the structure made up by aggregated particles or a sintered porous body, however, the following understanding can be achieved.

The characteristics of this kind of structure would be that there must exist a particles contact point at the end of the pore space. Then the contact points and their vicinity are able to provide

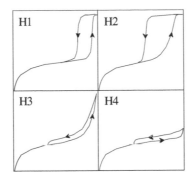

FIGURE 6.3 Classification of hysteresis defined by IUPAC.

nuclei for the condensation, and an energy barrier, as is the case for MCM-41, would not stand upon the condensation process. Therefore the condensation process for this kind of materials follows an equilibrium path. Another feature is that pores are connected through narrower spaces between particles, or the pore network is formed with connecting "necks," In this case, the desorption process itself is in equilibrium, but the so-called ink-bottle effect gives hysteresis in which the evaporation of condensate in a pore space is not possible until the pressure goes down to a value corresponding to the size of the neck. The evaporated volume at this hindered process, then, does not provide the pore volume of the pore size corresponding to this pressure. This process exhibits H2-type hysteresis, and the adsorption branch should be used to calculate the size distribution of the pore space. Examples of materials include silica gels and porous polymer gels. Many providers of such materials use desorption branches for showing porous characteristics to users because it gives sharp distribution. One should notice, however, that the peak in the distribution simply gives the neck size, and the real distribution would be broader in most cases.

As clarified by percolation theory, the extrusion process for wetting fluid is the same as the intrusion process for nonwetting fluid. Therefore the intrusion curve of mercury porosimetry, which is more often used than the extrusion, similarly gives the neck size.

Another important topic related to hysteresis is the end-closure point of the desorption branch. As seen in clay materials and activated carbons, a hysteresis loop of types H3 and H4 often closes at the relative pressure of 0.40–0.45. As described in detail in the literature,[17] this closure of hysteresis will result from the spinodal evaporation of condensed liquid when exceeding its tensile limit, which is determined not by the pore size but merely by the nature of liquid. This phenomenon should be considered when one uses the desorption branch for characterization.

3.6.3 MERCURY INTRUSION METHOD (MERCURY POROSIMETRY)

An evacuated sample is immersed in mercury, and pressurization of the system cause mercury's intrusion into the pore space. The detection and analysis of the intruded volume against applied pressure gives the pore size distribution. Many automated apparatuses are commercially available.

Principle: Nonwetting Fluid and Washburn Equation

Mercury does not wet almost all of solid surface, which corresponds to the upper portion in Figure 6.1b. Also, $P' = 0$ stands because the pore space is evacuated before measurement. Equation 6.1 then will be

$$P = \frac{2\gamma\cos\theta}{r}, \text{ or } D = 2r = -\frac{4\gamma\cos\theta'}{P} \qquad (6.7)$$

which gives the relation between applied pressure P and the pore diameter D. θ' is the contact angle for mercury, which is set to be 140° in most cases. The above equation is essentially the same as the Young–Laplace equation, but the right-hand side is called the Washburn equation. Since the intruded volume directly means the volume of pores larger than a size corresponding to the pressure, the differentiation of the volume with respect to the pore size gives the pore-size distribution.

Measurable Range of Pore Size

Many automated porosimetry apparatuses declare the lower limit of the pore size to be 3–4 nm, which corresponds to the applied pressure of about 4000 atm. At this high pressure, the porous framework might be deformed or some other influence may occur. It is also important to point out that the surface of mercury in a single-nanometer range may be different from the surface of bulk liquid, which may result in a hindered contact angle. Much attention, then, should be paid to the reliability of the data in this range, for which the gas adsorption method has far superior accuracy and reliability. The upper limit of mercury porosimetry would be around several hundred microns. The detection of cracks or supermacro pores, which is difficult to measure by gas adsorption, is precisely done by the intrusion method.

Hysteresis

The extrusion process with decreasing pressure generally gives a different path, or hysteresis. Further, it would be almost always the case that a certain amount of mercury remains in the pores even after complete release of the pressure. The extrusion process, therefore, would not be suitable for analysis, and the intrusion branch is used in general. This corresponds to the analysis of the desorption branch in gas adsorption, and one should notice that the analysis gives the neck size for aggregated or sintered bodies.

3.6.4   OTHER TECHNIQUES OF INTEREST

Bubble-Point Method

This technique detects perforating pores while the gas adsorption method and mercury porosimetry cannot distinguish those from dead-end pores. Because of this feature the method is often applied to filters, membranes, cloths, or those porous materials whose permeation properties are of importance. Depending on the wettability, the porous material is immersed in freon or water. Pressurized air or nitrogen is then introduced to one side of the material. At a pressure corresponding to maximum size of the perforating pore, the gas starts to permeate. Other than the detection of the maximum pore size, the size distribution can be estimated by applying higher pressures, because smaller pores start to open with increased pressure. Detectable pore size is usually above several tens of nanometers with freon, or a few hundred nanometers with water.

Thermoporometry: Detection of Freezing Point Depression

Based upon the Gibbs–Thomson equation, which assumes that freezing point depression in a pore from bulk temperature is inversely proportional to the pore radius, the size distribution is estimated from calorimetric measurement. It may sometimes be the case that the pore structure when wetted varies after drying because of capillary suction pressure or de-swelling of the base material. The gas adsorption or mercury porosimetry cannot characterize such porous materials because both methods need evacuation before measurement. This method may be used to overcome the above problem: the measurement goes as wetted. However, not much has been clarified for the freezing behavior in confined space. Recent study, for example, has clarified that the freezing point may even be higher than the bulk freezing point, depending on the physicochemical nature of the pore walls.[18] The method is thus especially controversial if the single-nano range is concerned, and it is better not to rely on it for the smaller range of pore sizes.

Small Angle X-ray Scattering

X-ray scattering with angles smaller than 10° can probe porous characteristics in the single-nanometer range. Since X-rays can detect not only open pores but also closed (or isolated) pores, measurements for low-permittivity materials are often seen as an application. One has to be careful, however, because the resulting space distribution or correlation length does not necessarily relate to the scale of the pores but has resulted from the electron density distribution. Further, an ordered material such as MCM-41 would give clear signals showing its lattice size or the periodicity of the regular pores, but not much sensitivity can be expected for materials with a disordered or random nature.

REFERENCES

1. Cranston, R. W. and Inkley, F. A., *Adv. Catal.,* 9, 143, 1957.
2. Dollimore, D. and Heal, G. R., *J. Appl. Chem.,* 14, 109, 1964.
3. Barell, E. P., Joyner, L. G., and Halenda, P. P., *J. Am. Chem. Soc.,* 73, 373, 1951.
4. Horvath, G. and Kawazoe, K., *J. Chem. Eng. Jpn.,* 16, 470, 1983.
5. Saito, A. and Foley, H. C., *AIChE J.,* 37, 429, 1991.
6. Lippens, B. C. and de Boer, J. H., *J. Catal.,* 4, 319, 1965.
7. Sing, K. S. W., in *Surface Area Determination,* Everett, D. H. and Ottewill, R. H., Eds., Butterworths, London, 1970, p. 25.
8. Mikhail, R. S., Brunauer, S., and Bodor, E. E., *J. Colloid Interface Sci.,* 26, 45, 1968.
9. Evans, R., Marconi, U. M. B., and Tarazona, P., *J. Chem. Phys.,* 84, 2376, 1986.
10. Miyahara, M., Yoshioka, T., and Okazaki, M., *J. Chem. Phys.,* 106, 8124, 1997.
11. Miyahara, M., Yoshioka, T., and Okazaki, M., *J. Chem. Eng. Jpn.,* 30, 274, 1997.
12. Miyahara, M., Kanda, H., Yoshioka, T., and Okazaki, M., *Langmuir,* 16, 4293, 2000.
13. Miyahara, M., Yoshioka, T., Nakamura, J., and Okazaki, M., *J. Chem. Eng. Jpn.,* 33m 103, 2000.
14. Kanda, H., Miyahara, M., Yoshioka, T., and Okazaki, M., *Langmuir,* 16, 6622, 2000.
15. Vishnyakov, A. and Neimark, A. V., *J. Phys. Chem. B,* 105, 7009, 2001.
16. Neimark, A. V., Ravikovitch, P., and Vishnyakov, A., *Phys. Rev. E,* 65, 2002.
17. Ravikovitch, P. and Neimark, A. V., *Langmuir,* 18, 1550, 2002.
18. Miyahara, M. and Gubbins, K. E., *J. Chem. Phys.,* 106, 2865, 1997.

3.7 Permeation (Flow through Porous Medium)

Chikao Kanaoka
Ishikawa National College of Technology
Ishikawa, Japan

Permeation is fluid flow through interstices among many discrete particles. Flows of groundwater, crude oil, and natural gas are typical permeations taking place in nature. This type of flow is widely utilized in industrial processes in which both particles and fluid flow play important roles simultaneously, such as a catalytic reaction, filtration, and so on. Thanks to the extensive number of studies on porous media since D'arcy, macroscopic flow characteristics in a porous medium can be evaluated fairly well, although the microscopic flow pattern inside the medium is complex and not amenable to rigorous solution by the Navier–Stokes equation. In this section, general theories developed to explain resistance to flow through a porous medium are described, and the resistances of granular and fibrous packed beds are outlined as typical examples of high and low packing densities, respectively.

3.7.1 RESISTANCE TO FLOW THROUGH A POROUS MEDIUM

A porous medium is composed of many discrete particles with different shapes and sizes, as shown in Figure 7.1a. The packing density of the medium is high and its structure is sophisticated. Porous media can be simplified as an assembly of many tiny channels or uniformly spaced spherical particles of equal size, as shown in Figure 7.1b and 7.1c.

For the case of the former model, flow resistance can be predicted by analogy with that of a straight circular tube, and for the case of the latter, it can be calculated from the fluid drag acting on individual particles. These two models are usually called the "channel" and "drag" theories, respectively. Of course, pressure drop can be predicted by either model, but, in general, the former is believed to be suited for a flow-through porous medium with high packing density and the latter for one with low packing density.

Channel Theory

Because the porous system is assumed to be a bundle of tiny channels with noncircular cross sections, the pressure drop of the system is identical with that for one of those channels.

When fluid with density ρ and viscosity μ flows through a straight tube with diameter D and length L, the pressure drop ΔP is expressible by the so-called Hagen–Poiseuille equation for laminar flow and by Fanning's equation in general:

Hagen–Poiseuille equation:
$$\Delta P = \frac{32L\mu u}{D^2} \tag{7.1}$$

Fanning's equation:
$$\Delta P = 4f\frac{L}{D}\frac{\rho u^2}{2} \tag{7.2}$$

317

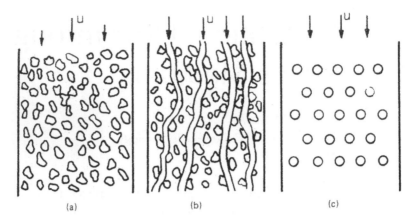

FIGURE 7.1   Models of a porous medium.

Here, Fanning's friction factor f is expressed by Equation 7.3 in a laminar flow region, which becomes dependent not only on the Reynolds number Re but also on the surface roughness in the turbulent flow region:

$$f = 16\frac{\mu}{Du\rho} = \frac{16}{Re} \tag{7.3}$$

Equation 7.1 through Equation 7.3 are applicable to noncircular channels by introducing the hydraulic radius m defined as

$$m = \frac{\text{Cross-sectional area of a channel}}{\text{Wetted periphery}} \tag{7.4}$$

The value of m calculated from Equation 7.4 equals one quarter of the tube diameter for circular tubes and about one quarter of the equivalent diameter for other cross-sectional shapes. The definition above is not applicable, as it is to the porous medium; hence, it is modified as follows.

$$
\begin{aligned}
m &= \frac{\text{Cross-sectional area of a channel} \times \text{tube length}}{\text{Wetted periphery} \times \text{tube length}} \\[6pt]
&= \frac{\text{Tube volume}}{\text{Wetted area of a tube}} \\[6pt]
&= \frac{\text{Total pore volume}}{\text{Total surface area of particles in a porous medium}} \\[6pt]
&= \frac{\text{Porosity}}{\text{Specific surface area based on the volume of porous medium}} \\[6pt]
&= \frac{\varepsilon}{S_{\mathrm{B}}} = \frac{\varepsilon}{S_v (1-\varepsilon)}
\end{aligned} \tag{7.5}
$$

where S_{B} and S_v are the specific surface areas based on packed layer volume and particle volume, respectively.

According to Carman's tortuosity model,[1] the time required for fluid to travel an equivalent channel is equal to the actual penetration time to pass through the bed. Hence, the following relation is obtained:

$$\frac{L_e}{u_e} = \frac{L}{u/\varepsilon} \tag{7.6}$$

Substituting Equation 7.5 and Equation 7.6 into Equation 7.1 gives

$$\Delta P = \frac{32 L_e \mu u_e}{(4m)^2} = \frac{2 L_e \mu S_B^2 L_e / L(u/\varepsilon)}{\varepsilon^2} = 2\left(\frac{L_e}{L}\right)^2 \frac{S_B^2 L \mu u}{\varepsilon^3}$$

$$= 2\left(\frac{L_e}{L}\right)^2 \frac{S_v^2 (1-\varepsilon)^2 L \mu u}{\varepsilon^3} \tag{7.7}$$

As mentioned earlier, the hydraulic radius calculated from Equation 7.5 is one fourth of the tube diameter for circular tubes. Hence, four times the hydraulic radius is used in Equation 7.7. L_e/L is the ratio of the equivalent channel length to the bed thickness, and it is called the tortuosity. For noncircular channels, the numerical constant in the last two expressions in Equation 7.7 differs from 2. Therefore, we write the following expression for the general case of a channel model by replacing the numerical constant 2 by κ_o:

$$\Delta P = \kappa_0 \left(\frac{L_e}{L}\right)^2 \frac{S_B^2 L \mu u}{\varepsilon^3} = \kappa_0 \left(\frac{L_e}{L}\right)^2 \frac{S_v^2 (1-\varepsilon)^2 L \mu u}{\varepsilon^3} \tag{7.8}$$

Equation 7.8 is the expression based on the channel model, and the pressure drop can be evaluated by knowing the values κ_o and L_e/L; $\kappa = \kappa_o (L_e/L)^2$ is usually called the Kozeny constant and is dependent on particle shape, packing structure, and so on. κ is about 5 for most cases. Carman obtained the following well-known expression called the Kozeny–Carman equation:

$$\Delta P = 5 \frac{S_B^2 L \mu u}{\varepsilon^3} = 5 \frac{S_v^2 (1-\varepsilon)^2 L \mu u}{\varepsilon^3} \tag{7.9}$$

Because Equation 7.9 is derived from the Hagen–Poiseuille equation, Equation 7.1, its applicable range is limited to the laminar flow region. In the turbulent flow region, Fanning's equation, Equation 7.2, has to be used by modifying the Reynolds number and friction factor as follows[2]:

$$Re = \frac{m(u/\varepsilon)\rho}{\mu} = \frac{\varepsilon}{S_B} \frac{u}{\varepsilon} \frac{\rho}{\mu} = \frac{u\rho}{\mu S_B} = \frac{u\rho}{\mu S_v (1-\varepsilon)} \tag{7.10}$$

$$2f = \frac{Pm}{L\rho(u/\varepsilon)^2} = \frac{P\varepsilon/S_B}{L\rho(u/\varepsilon)^2} = \frac{P\varepsilon^3}{L\rho u^2 S_B} = \frac{P\varepsilon^3}{L\rho u^2 S_v (1-\varepsilon)} \tag{7.11}$$

Equation 7.11 can also be applicable to the laminar flow region. Substituting Equation 7.10 and Equation 7.11 into Equation 7.9, one obtains

$$2f = \frac{P\varepsilon^3}{L\rho u^2 S_v (1-\varepsilon)} = 5\frac{\mu S_v (1-\varepsilon)}{u\rho} = \frac{5}{Re} \tag{7.12}$$

In Equation 7.12 the friction factor is inversely proportional to Reynolds number. This agrees well with the experiments for $Re < 2$, as shown in Figure 7.2. The friction factor for $Re > 2$ has to be determined experimentally.[3,4]

Drag Theory

The pressure drop of a porous medium has to be equal to the fluid drag experienced by particles inside the medium:

$$\Delta PA = \sum_{i=1}^{N} R_{mi} \tag{7.13}$$

where A is the cross-sectional area of a porous medium, R_{mi} is the fluid drag experienced by the ith particle, and N is the total number of particles in the medium.

Let us consider a uniformly packed bed of spherical particles with diameter D_p, depth L, and porosity ε, as shown in Figure 7.1c. It is considered that fluid flow around each particle in the bed is identical, and hence the fluid drag acting on it is also the same, yielding $R_{mi} = R_m$. R_m is expressed by the following equation:

$$R_m = Rf(x) \tag{7.14}$$

R expresses the fluid drag acting on a single spherical particle in infinite field and $f(\varepsilon)$ is the correction term due to the change in packing density and usually referred to as the porosity function. R and N are given by the following equations:

$$R = C_D \frac{\pi D_p^2}{4} \frac{\rho u^2}{2} \tag{7.15}$$

$$N = \frac{6AL(1-\varepsilon)}{\pi D_p^3} \tag{7.16}$$

where C_D is the drag coefficient of a single sphere.

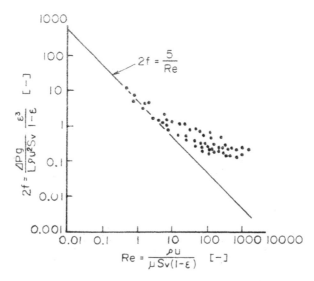

FIGURE 7.2 Correlation of experimental friction factor and Reynolds number.

Substituting Equation 7.14 through Equation 7.16 into Equation 7.13, it becomes

$$\Delta P = \frac{3}{2} C_D (1-\varepsilon) f(\varepsilon) \frac{L}{D_p} \frac{\rho u^2}{2} \tag{7.17}$$

By adopting Steinour's[6] expression for $f(\varepsilon)$, which was obtained for hindered settling, Equation 7.17 becomes

$$f(\varepsilon) = \frac{25(1-\varepsilon)}{3\varepsilon^3} \tag{7.18}$$

$$\Delta P = \frac{25}{2} C_D \frac{(1-\varepsilon)^2}{\varepsilon^3} \frac{L}{D_p} \frac{\rho u^2}{2} \tag{7.19}$$

For turbulent flow, the following Burke and Plummer expression[7] obtained from the dimensional analysis is popular:

$$\Delta P = \frac{3.5}{4} \frac{1-\varepsilon}{\varepsilon^3} \frac{L}{D_p} \frac{\rho u^2}{2} \tag{7.20}$$

One of the most effective expressions for both laminar and turbulent flow regions is the following Ergun's equation 7, which is a sum of fluid drags proportional to the first and second powers of the fluid velocity:

$$\frac{\Delta P}{L} = 150 \frac{(1-\varepsilon)^2}{\varepsilon^3} \frac{\mu u}{D_p^2} + 1.75 \frac{1-\varepsilon}{\varepsilon^3} \frac{\rho u^2}{D_p} \tag{7.21}$$

In the form of a modified friction factor, it becomes

$$2f = \frac{150(1-\varepsilon)}{Re_p} + 1.75 \tag{7.22}$$

where

$$Re_p = \frac{D_p u \rho}{\mu} \tag{7.23}$$

Figure 7.3 is the correlation between $2f$, and calculated values from the Kozeny–Carman and the Burke–Plummer equations are also shown. As one can see from the figure, Ergun's expression agrees well with the Kozeny–Carman equation in the laminar flow region[1,8] and with the Burke–Plummer equation in the turbulent flow region.

3.7.2 PRESSURE DROP ACROSS A FIBROUS MAT

A fibrous mat can be considered as a special kind of porous medium composed of particles with an extremely large aspect ratio, to which both channel and drag theories are applicable. The porosity of a fibrous mat is usually higher than 85%; thus, the average interfiber distance becomes larger than several times the fiber diameter. This means that interference effects among neighboring fibers are

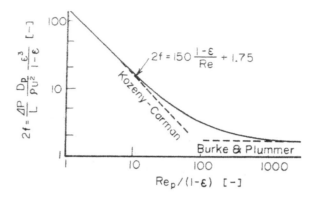

FIGURE 7.3 Comparison of Ergun's correlation with others.

substantially smaller compared with the ordinary granular packed bed. Hence, most of the equations proposed have been derived based on the drag theory.

According to the drag theory, the pressure drop ΔP across a fibrous mat with cross-sectional area A and thickness L is related to the fluid drag acting on fibers in the mat as follows:

$$\Delta PA = Fl_f AL \tag{7.24}$$

where F and l_f denote the fluid drag acting on a fiber with unit length and the total fiber length in a unit mat volume, respectively, and they are defined by the following equations:

$$F = C_D D_f \frac{\rho u^2}{2} \tag{7.25}$$

$$l_f = \frac{4\alpha}{\pi D_f^2} \tag{7.26}$$

Substitution of the preceding two equations into Equation 7.24 yields

$$\Delta P = 4C_D \frac{\alpha L}{D_f} \frac{\rho u^2}{2} \tag{7.27}$$

where u denotes the average fluid velocity in the mat and C_D is usually given as a function of the Reynolds number based on the approaching velocity u_e. Furthermore, there exist some fibers parallel to the fluid flow. Hence, it would be better in practice to define ΔP by

$$\Delta P = 4C_{De} \frac{\alpha L}{D_f} \frac{\rho u^2}{2} \tag{7.28}$$

Table 7.1 summarizes the proposed effective drag coefficients C_{De}. In the table, the expression of Langmuir is derived from the channel theory.

TABLE 7.1 Drag Coefficient of Fiber

Researchers	Drag Coefficient	Remarks
Kozeny and Carman	$\dfrac{(8_{11}K_f/Re_f)0}{(1-\alpha)^2}$	Channel theory
Langmuir (1942)	$\dfrac{8\pi B}{Re_f}\dfrac{1}{\ln\alpha + 2\alpha - \alpha^2/2 - 3/2}$	Channel theory, B = 1.4
Davies (1952)	$\dfrac{32\pi}{Re_f}\alpha^{0.5}(1+56\alpha^2)$	Dimensional analysis
Lamb (1932)	$\dfrac{8\pi}{Re_f}\dfrac{1}{2-\ln Re_f}$	Perpendicular to the flow, isolated fiber
Iberall (1950)	$\dfrac{8\pi}{Re_f}$	Parallel to the flow, isolated fiber
Iberall (1950)	$\dfrac{4.8\pi}{Re_1}\left(\dfrac{2.4 - \ln Re_f}{2 - \ln Re_f}\right)$	Semi-empirical equation
Chen (1955)	$\dfrac{2}{Re_f}\dfrac{k_2}{\left|\ln k_3\alpha^{-0.5}\right|}$	$k_2 = 6.1, k_3 = 0.64$
Happel (1959)	$\dfrac{8\pi}{Re_f}\dfrac{1}{-\ln\alpha + 2\alpha - \alpha^2/2 - 3/2}$	Tube bank parallel to the flow
Happel (1959)	$\dfrac{16\pi}{Re_f}\dfrac{1}{-\ln\alpha + (1-\alpha^2)/(1+\alpha^2)}$	Tube bank perpendicular to the flow
Kuwabara (1959)	$\dfrac{16\pi}{Re_f}\dfrac{1}{-\ln\alpha + 2\alpha - \alpha^2/2 - 3/2}$	Tube bank perpendicular to the flow
Kimura and Iinoya	$\dfrac{0.6 + 4.7/\sqrt{Re_f} + 11/Re_f}{1-\alpha}$	Empirical $10^{-3} < Re_f < 10^{-2}$, $3 < D_f < 270\ \mu m$

Notation

A	Cross-sectional area of porous medium (m²)
C_D	Drag coefficient
D	Diameter (m)
D_f	Fiber diameter (m)
D_p	Particle diameter (m)
F	Fluid drag acting on a fiber with unit length (N/m)
f	Fanning's friction factor
$f(\varepsilon)$	Porosity function
k	Kozeny constant
k_o	Numerical constant used in Equation 7.8
L	Thickness (m)
L_e	Equivalent channel length (m)

l_f	Total fiber length per unit filter volume (m/m^3)
m	Hydraulic radius (m)
N	Total number of spheres in a porous medium (m^{-3})
ΔP	Pressure drop (Pa)
R	Fluid resistance acting on an isolated particle (N)
R_m	Fluid resistance acting on a particle in a porous medium
Re	Reynolds number ($Du\rho/\mu$)
Re_f	Reynolds number based on fiber diameter ($D_f u\rho/\mu$)
Re_p	Reynolds number based on particle diameter ($D_p u\rho/\mu$)
S_B	Specific surface of particle based on the bed volume (m^2/m^3)
S_v	Specific surface of particle based on the particle volume (m^2/m^3)
u	Fluid velocity (m/s)
u_e	Effective fluid velocity in a porous medium (m/s)
α	Fiber packing density
ε	Porosity
μ	Viscosity (Pa·S)

REFERENCES

1. Carman, P. C., *Trans. Inst. Chem. Eng.*, 15, 150–156, 1937.
2. Blake, F. E., *Trans. Am. Inst. Chem. Eng.*, 14, 415–421, 1922.
3. Chilton, T. H. and Colburn, A. D., *Ind. Chem. Eng.*, 23, 913–918, 1931.
4. Brownell, L. E. and Katz, D. L., *Chem. Eng. Prog.*, 43, 537–548, 1947.
5. Leva, M., *Fluidization*, McGraw-Hill, New York, 1959.
6. Steinour, H., *Ind. Eng. Chem.*, 36, 618–624, 1944.
7. Burke, S. P. and Plummer, W. B., *Ind. Eng. Chem.*, 20, 1196, 1928.
8. Ergun, S., *Chem. Eng. Prog.*, 48, 89–94, 1952.
9. Kozeny, J., *Sitzungsber. Akad. Wissensch. Wien*, 136, 271–306, 1927.
10. Langmuir, I., *OSRD Rep. No. 865*, 1942.
11. Davies, C. N., *Proc. Inst. Mech. Eng. (London)*, 131, 185–213, 1952.
12. Iberall, A. S., *J, Res. Natl. Bur. Stand.*, 45, 85–108, 1950.
13. Chen, C. Y., *Chem. Rev.*, 55; 595, 1955.
14. Happel, J., *AIChE J.*, 5, 174–177, 1959.
15. Kuwabara, S., *J. Phys. Soc. Jpn.*, 14, 527–532, 1959.
16. Kimura, N. and Iinoya, K., *Kagaku Kogaku*, 33, 1008–1013, 1969.

3.8 Specific Surface Area

Masatoshi Chikazawa and Takashi Takei
Tokyo Metropolitan University, Hachioji, Tokyo, Japan

3.8.1 DEFINITION OF SPECIFIC SURFACE AREA

Specific surface area S_w (m²/kg) of a powder is one of the basic properties of the powder and is generally represented by the surface area of total particles contained in a unit mass of powder. This value implies the internal and external surfaces which can be measured using various probes, such as gases and liquids. In the case of adsorption methods, adsorptive gas as a probe molecule must be accessible to all of the surfaces in cavities, cracks, and micropores. Nowadays, powders used as raw materials and intermediate manufactured goods have become more important, and industrial needs concerning particle size, particle shape, purity and uniformity, of powders have become more stringent. For example, for fine powders, purity and uniformity are necessary for manufacturing of precision materials such as electronics and fine ceramics. When the particle size decreases markedly, powdery phenomena depend largely on surface properties. Therefore, characterization of the powder surface becomes increasingly significant with a decrease in the diameter of powder particles.

The surface area of spherical particle type having a diameter of D_i is obtained from

$$S = \pi D_i^2 \tag{8.1}$$

The mass of the particle is

$$W = \frac{\pi D_i^3 \rho}{6} \tag{8.2}$$

If distributions of particle size and shape in a powder sample are known, the surface area of the powder can be calculated. For example, when the particle is spherical and the number of particles having diameter of D_i is n_i, the specific surface area of the powder is given by

$$S_w = \frac{\sum n_i \pi D_i^2}{\sum n_i \pi D_i^3 \rho / 6} \tag{8.3}$$

If the particle size is assumed to be uniform as for the spherical or cubic type, Equation 8.3 is reduced to

$$S_w = \frac{6}{D_m \rho} \tag{8.4}$$

where D_m represents the specific surface area diameter. Generally, the specific surface area of a powder is determined by gas adsorption, permeability, or heat-of-immersion methods.

3.8.2   ADSORPTION METHOD

The surface area of a powder is generally determined by the adsorption method. In this case, it is necessary to obtain the monolayer capacity and cross-sectional area of an adsorbate molecule. To estimate the monolayer capacity V_m, the Langmuir or BET (Brunauer-Emmett-Teller) equation has been applied to experimental adsorption data. The Langmuir equation can be rewritten as

$$\frac{P}{V} = \frac{1}{bV_m} + \frac{P}{V_m}P \tag{8.5}$$

where V and V_m are the adsorbed amount at vapor pressure P and the monolayer completion, respectively. The b value is a constant. Therefore, the plot of P/V against P should be a straight line with slope $1/V_m$. The V_m value is obtained from the reciprocal of the slope.

On the other hand, the BET equation can be represented by

$$\frac{P}{V(P_0 - P)} = \frac{1}{V_m C} + \frac{C-1}{V_m C}\frac{P}{P_0} \tag{8.6}$$

The plots of $P/V(P_0 - P)$ against P/P_0 should therefore exhibit a straight line with slope $(C - 1)/(V_m C)$ and intercept $1/(V_m C)$. These plots are called BET plots. The BET plots usually hold within the relative pressure range 0.05–0.35. The slope $(C - 1)/(V_m C)$ and intercept values $1/V_m C$, are determined graphically or by linear regression. Therefore, the monolayer capacity V_m can be derived from the reciprocal of the summation between the slope and intercept values, as shown in

$$V_m = \frac{1}{\left(\dfrac{C-1}{V_m C} + \dfrac{1}{V_m C}\right)^{-1}} \tag{8.7}$$

To calculate the surface areas of powders, another important value is the cross-sectional area A of the adsorbate molecules. The surface area S (m²) of the powder can be obtained from the product of the monolayer capacity V_m and cross-sectional area A of an adsorbate molecule, using

$$S = \frac{V_m A N_A}{22.4 \times 10^{-3}} \ (\text{m}^2) \tag{8.8}$$

where N_A is Avogadro's number. A reasonable estimation of the cross-sectional area of an adsorbate molecule is generally obtained from the liquid density of the adsorptive at measurement temperature using Equation 8.9 with several assumptions: (a) the shape of an adsorptive molecule is spherical, (b) the liquid structure of the adsorptive is the closest packing structure with 12 nearest neighbors, and (c) the adsorptive molecules are adsorbed on the particle surface with the 6 nearest neighbors in the close-packed hexagonal arrangement:

$$A = 1.091 \left(\frac{M}{\rho N_A}\right)^{2/3} \tag{8.9}$$

The factor 1.091 in Equation 8.9 is a packing factor deduced from the assumptions described above.

The recommended molecular cross-sectional areas of adsorbate gases are summarized in Table 8.1. The nitrogen molecule is generally used as an adsorptive molecule. If the surface area is relatively small, the argon or krypton molecule is used in order to precisely measure adsorbed amounts.

Single-Point Method

The BET plots usually show a linear relationship in the pressure range of $0.05 < P/P_0 < 0.35$. C in the BET equation is a constant that expresses a magnitude of the interaction between solid surface and adsorptive molecules. When the value of C is reasonably high, the intercept value $1/(V_m C)$ becomes zero. Then the BET plots are recognized to be a linear line passing close to the origin. In this case, Equation 8.6 is rewritten as

$$\frac{P}{V(P_0 - P)} = \frac{1}{V_m} \frac{P}{P_0} \tag{8.10}$$

Equation 8.10 is reduced to

$$V_m = V\left(1 - \frac{P}{P_0}\right) \tag{8.11}$$

Consequently, the surface area can be determined by measuring one adsorbed amount at vapor pressure P. This method is called the "single-point method." Usually, inorganic materials have large C values for N_2 adsorption. However, because the single-point method is deduced by simplifying the BET equation, there is some intrinsic error. The error in the single-point method can be evaluated using the BET equation. Table 8.2 summarizes the difference in the specific surface areas determined by the single-point method and the multipoint method. These results suggest that the adsorption measurement should be done at high relative pressure in which the BET plots exhibit a straight line.

Volumetric Method

This method is the most standard method. A simplified volumetric apparatus is shown in Figure 8.1. A dosing volume in an adsorption apparatus must be determined so that adsorbed amounts can be calculated. The volume is obtained by expanding noble gases such as He and N_2 from known volume

TABLE 8.1   Cross-Sectional Areas of Adsorbate Gases

Gas	Measurement Temperature (°C)	Bulk State*	σ_2 (nm²)	Gas	Measurement Temperature (°C)	Bulk State*	σ_2 (nm²)
N_2	−196	L	0.162	CO_2	−56.5	G	0.170
	−183	G	0.170		−78	S	0.208
O_2	−183	L	0.141	NH_3	−32.5	L	0.154
Ar	−183	L	0.143		−78	S	0.141
	−196	S	0.138	Ethane	−183	S	0.220
Kr	−196	S	0.202	n−Butane	0	L	0.444
Xe	−183	S	0.232		20	G	0.479
CO	−183	L	0.166		25	G	0.510

* Phase of bulk state at measurement temperature: G, gas; L, liquid; S, solid.

TABLE 8.2   Difference, Represented by S_s/S_m Values, in Specific Surface Areas Determined by Single-Point and Multipoint Methods Using the BET Equation

			S_s/S_m		
C Value	$P/P_0 = 0.35$	$= 0.3$	$= 0.25$	$= 0.2$	$= 0.1$
10	0.843	0.811	0.769	0.714	0.526
20	0.915	0.896	0.870	0.833	0.690
30	0.942	0.928	0.909	0.882	0.769
50	0.964	0.955	0.943	0.926	0.847
70	0.974	0.968	0.959	0.946	0.886
100	0.982	0.977	0.971	0.962	0.917
200	0.991	0.988	0.985	0.980	0.957
500	0.996	0.995	0.994	0.992	0.982

Note: The S_s and S_m values are surface areas determined by the single-point method and multipoint method, respectively. The values P/P_0 are the relative pressure at which single-point measurings are performed.

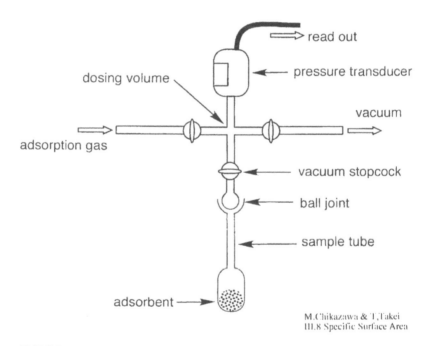

M.Chikazawa & T.Takei
III.8 Specific Surface Area

FIGURE 8.1   Diagram of a volumetric apparatus.

to a volume of an objective part. It is necessary to measure the pressures before and after expanding the noble gas. An equilibrium pressure is determined using a pressure transducer, for example, a capacitance manometer (baratron).

To obtain an adsorption isotherm, a known amount of adsorptive gas is introduced into a sample tube. The sample adsorbs the adsorptive molecules, and the vapor pressure decreases gradually until equilibrium is attained. The introducing pressure and equilibrium pressure must be measured precisely. The adsorbed amount is determined from the difference between the introduced amount and the residual amount. The residual amount is evaluated from the same procedure using He gas. In this case, adsorption of He molecules does not occur. Therefore, the adsorbed amount is determined from the difference between the amount of introduced adsorptive molecules and the amount of He gas introduced into the sample tube, comparing at the same equilibrium pressure. By repeating this procedure, adsorbed amounts at various pressures can be obtained. Generally, estimation of residual amounts of adsorptive molecules is performed using He gas, then a dead volume is determined. The residual amounts of adsorptive molecules at the other vapor pressures are calculated using this dead volume.

The dead volume must be measured immediately before or after the adsorption measurement. During the measurement, the surface level of liquid nitrogen in a cooling bath should be maintained constant within 1 mm at least 5 cm above the powder sample.

Carrier Gas Method

A mixture gas of known concentration of adsorptive in a nonadsorbable gas such as He is passed through a sample cell which is cooled in a liquid nitrogen bath. As a result of adsorption, the concentration of adsorptive decreases. This concentration change was first detected by Nelsen and Eggertsen,[1] using a thermal conductivity detector. The nitrogen molecule is generally used as an adsorptive. When the sample cell is immersed in the liquid nitrogen, only nitrogen gas is adsorbed and its concentration changes. This concentration change generates an adsorption peak as a function of time. After removing the coolant, the adsorbed molecules are degassed, and a desorption peak is recorded in the opposite direction. The adsorption and desorption processes are recorded as shown in Figure 8.2.

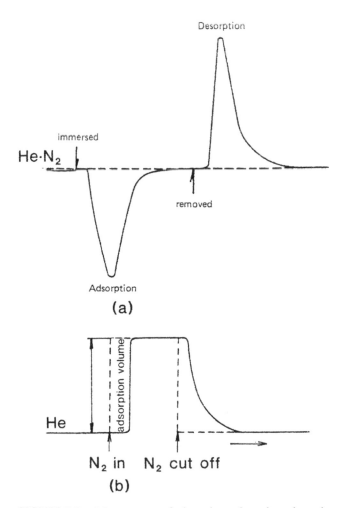

FIGURE 8.2   Measurement of adsorption volume by a dynamic method. (a) Nelsen method, (b) elution method.

To calculate the surface area from the adsorption or desorption peak, it is necessary to calibrate the gas detector in an adsorption apparatus. The calibration is performed by injecting a known amount of adsorptive gas and by comparing this amount with a peak area generated. Adsorbed amounts are determined at various partial pressures of mixed gas, and using the BET equation, the monolayer capacity V_m is calculated. However, the carrier gas method is generally used for the single-point method. The apparatus developed by Suito et al.[2] is shown in Figure 8.3.

Gravimetric Method

In this method, the adsorption amount is determined by measuring the increase in a sample weight. For example, when a monolayer adsorption is accomplished on a powder surface by nitrogen molecules, the increasing weight is calculated to be 0.286 mg/m². If the specific surface area of a powder is 100 m²/g, the 1-g powder sample adsorbs 28.6 mg nitrogen molecules. This weight change can be measured sufficiently well by an ordinary balance. However, when the surface area is small, a correction due to buoyancy produced by volumes of sample powder and sample supporter is necessary. In the case of nitrogen adsorption, the buoyancy of a 1-ml volume at a relative pressure of 0.3 and at temperature 77 K is calculated to be 0.0594 mg.

3.8.3   HEAT OF IMMERSION

When a solid surface is immersed in a liquid that does not dissolve the solid, heat is generated as a result of the interaction between the solid surface and the liquid molecules. This heat is called the heat of immersion. A diagram of the enthalpy changes deduced by immersion and heat of adsorption is shown in Figure 8.4.

The value h_i represents the heat of immersion of the solid and is equal to the enthalpy change between the solid surface and solid–liquid interface. The Q and Q' values are heat of adsorption and real heat of adsorption, respectively. The symbol H_L is the liquefaction heat of adsorptive gas. When powder particles are placed in an adsorptive vapor, adsorption of the vapor molecules occurs. If the resulting particles are immersed in the liquid which consists of the adsorptive molecules, the adsorption

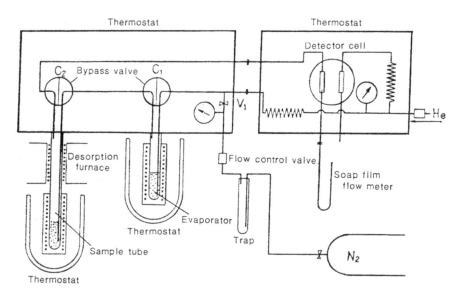

FIGURE 8.3   Schematic diagram of a dynamic method. (An evaporator is necessary in order to use an organic adsorptive with a high boiling point.)

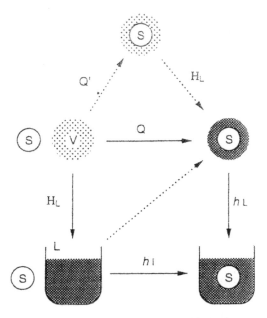

FIGURE 8.4 Schematic diagram of heat change.
S, solid; L, liquid; V, vapor; h_i, immersion heat; H_L,
liquefaction heat of vapor; h_L, enthalpy of liquid surface; Q, adsorption heat; Q', real adsorption heat.

layer disappears, and only a solid–liquid interface appears instead. In such a case, the heat of immersion per unit surface area is shown by applying the Gibbs–Helmholtz equation as follows:

$$h_i = \gamma_s - \gamma_{sl} - T\left(\frac{\partial(\gamma_s - \gamma_{sl})}{\partial T}\right)_p \qquad (8.12)$$

where γ_s and γ_{sl} are the free energies of solid surface having a adsorbed layer and solid–liquid interface, respectively. On the other hand, Equation 8.13 is well known as the Young equation:

$$\gamma_s = \gamma_{sl} + \gamma_l \cos\theta \qquad (8.13)$$

where θ is the contact angle of a liquid droplet formed on a solid surface, and γ_l represents the
free energy of liquid surface. From Equation 8.12 and Equation 8.13, the following equation is
obtained:

$$h_i = \gamma_l \cos\theta - T\left(\frac{\partial \gamma_l \cos\theta}{\partial T}\right)_p \qquad (8.14)$$

when θ is zero, Equation 8.14 is rewritten

$$h_i = \gamma_l - T\left(\frac{\partial \gamma_l}{\partial T}\right)_p \qquad (8.15)$$

The right-hand side expresses the surface enthalpy of a liquid. The condition $\theta = 0$ means that the
solid surface is covered with the liquid layers whose properties are the same as those of the immer-

sion liquid. If the entire surface of the powder particles is covered with adsorbate molecules thick enough, the external surface of the liquid layer newly formed by adsorption would be identical in nature with the surface of bulk liquid adsorptive. The thickness of the liquid layer necessary for the application of this method is five to seven layers. However, Partyka et al.[3] reported that two molecular layers was adequate in the case of water layers. When the particle surfaces thus treated are immersed in the liquid adsorptive, the enthalpy change per unit surface area is considered to be equal to the enthalpy h_L per unit surface area of the liquid adsorptive.

Using the principle of this immersion heat, Harkins and Jura[4] developed an "absolute method" to determine the surface areas of powders. The surface area of the powder is regarded to be in accord with the external surface of the liquid adsorbed layer, if the particle diameters are large enough comparing the thickness of the liquid layer. This method is independent of adsorption isotherms and is only based on calorimetry. The magnitude of the immersion heat depends only on the surface area and is independent of the surface properties of powder particles. Therefore the surface area is obtained by

$$S = \frac{h_i}{h_L} \tag{8.16}$$

A practical measuring procedure is as follows: Adsorptive molecules are adsorbed on powder particles under appropriate vapor pressure. After adsorption equilibrium, a sample tube is sealed and then the sample tube is placed in a calorimeter. After attainment of temperature equilibrium, the sample tube is broken by pressing a breaking rod. The heat evolved by immersing the powder particles into the adsorptive liquid is measured.

3.8.4 PERMEAMETRY

The surface area of powder can be determined by measuring the pressure drop in the course of fluid flow through a packed powder bed. This method has been widely used on account of the simple apparatus, facility of operation, and rapidness of measuring, compared with other methods.

The measuring method is based on the Kozeny–Carman equation,[5]

$$S_w = \frac{1}{\rho}\left[\left(\frac{\Delta P t A}{k_k \eta L V}\right)\left(\frac{\varepsilon^3}{(1-\varepsilon)^2}\right)\right]^{1/2} \tag{8.17}$$

where S_w is the specific surface area of the sample powder, ΔP is the pressure drop, t is time, A is cross-sectional area of the powder bed, L is length of the powder bed, ε is the porosity of the powder bed, ρ is the true density of the sample powder, k_k is the Kozeny constant which is usually taken as 5, η is the viscosity coefficient of the fluid, and V is the volume of fluid passed through the powder bed during time t.

Therefore, the specific surface area S_w of the powder bed whose porosity is known can be determined by measuring the flow rate V/t and pressure drop ΔP of fluid. However, in order to use this equation, it is necessary that the stream of permeating fluid is a viscous flow and the Hagen–Poseuilles law can be applied. The surface area thus obtained expresses the external surface area and does not contain internal surfaces in the micro pores and cracks.

When air is used as a fluid material, the mean free path of gas molecules is calculated to be about $\lambda = 1.1 \times 10^{-5}$cm at 15°C under 1 atm. In this system, it is necessary for the viscous flow that the diameter D of the flow path is about 10 times larger than λ, that is, larger than about 1 μ. The gas flow mechanism changes with decrease in the particle size of powders from viscous flow to

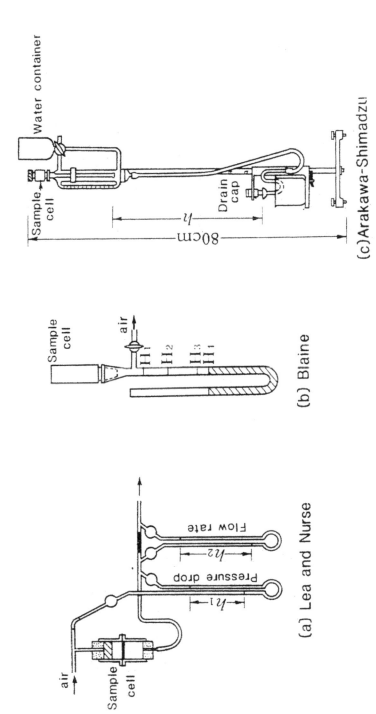

FIGURE 8.5   Schematic apparatus for permeability measurement.

Knudsen flow which is governed by the rate of molecular diffusion. In the range of $1 < D/\lambda < 10$, both mechanisms coexist, and the following equation holds:

$$\frac{V'L}{\Delta PA} = \frac{P'}{5\eta S_v^2 RT}\frac{\varepsilon^3}{(1-\varepsilon)^2} + \frac{4Z}{3}\frac{v}{S_v}\frac{\varepsilon^2}{(1-\varepsilon)} \tag{8.18}$$

$$\rho S_w = S_v \tag{8.19}$$

where V' is the velocity (mol/s) of fluid gas and P' is the average pressure in the powder bed. Z is a constant and is related to the shape of the path. v is the rate of molecular diffusion. The first term of the right-hand side is the contribution of viscous flow, and the second term is that of Knudsen flow. The plots of $V'L/\Delta PA$ versus P' should show a linear line. Then, the surface area estimated from a slope of the linear line originates in the viscous flow, and it is assumed to be related with an outer surface area of agglomerate particles. This value depends on the porosity of a powder bed. On the other hand, the surface area calculated from the intercept value is attributed to the Knudsen flow and is considered to represent the total surface area of primary particles. The specific surface area determined by this method agrees well with that calculated from the particle size distribution obtained by electron microscopy.[6]

The determination of Z is difficult, so this value is obtained experimentally. Derjaguin[7] determined the Z to be 0.69, and Arakawa and Suito[6] obtained 0.47.

Apparatuses and a practical measuring procedure are as follows. Gas and liquid are used as fluid materials. Apparatuses commercially available are based on air permeametry, and many of them are applicable for viscous flow using the Koseny–Carman equation. Typical apparatuses are shown in Figure 8.5. The surface areas measured by various apparatuses do not agree each other, and the differences attain about 20–30% in some cases. However, reproducibility in each apparatus is very high.

In this method, the specific surface area S_w varies with the porosity of the sample bed. In general, S_w increases with a decrease in the porosity and reaches a constant value below a certain porosity. The Kozeny constant k_k in Equation 8.14 is taken as 5. However, this value is affected by tortuosity of the flow path, which depends largely on the packing degree of the sample bed.

REFERENCES

1. Nelsen, G. and Eggertsen, F. T., *Anal. Chem.,* 30, 1387–1390, 1958.
2. Suito, E., Arakawa, M., Sakata, M., and Natsuhara, Y., *Zairyo,* 15, 178–183, 1966.
3. Partyka, S., Rouquerol, F., and Rouquerol, J., *J. Colloid Interface Sci.,* 68, 21–31, 1979.
4. Harkins, W. D. and Jura, G., *J. Am. Chem. Soc.,* 66, 1362–1366, 1944.
5. Carmen, P. C., *Flow of Gases through Porous Media,* Butterworth, London, 1956.
6. Arakawa, M. and Suito, E., *Kogyo Kagaku Zasshi,* 63, 556, 1960.
7. B Derjaguin, *C. R. Acad. Sci. URSS,* 53, 623, 1946.

3.9 Mechanical Properties of a Powder Bed

Michitaka Suzuki
University of Hyogo, Himeji, Hyogo, Japan

The mechanical properties of powder such as adhesion, cohesion, and flowability may affect powder handling. These properties of powder can be measured by a tensile, shearing, and compression test of the powder bed. In this section, we discuss how to measure the mechanical properties of a powder bed and how to describe the mechanical strength or flowability of a powder bed based on the powder mechanics.

Various types of shear testers have been devised to measure the powder yield locus (PYL), consolidation yield locus (CYL), and critical state line (CSL), which are projections of the flow surface, consolidation surface, and CSL, respectively, onto the σ-τ plane in the Roscoe diagram (see Figure 9.9). Several kinds of tensile testers have been also developed to measure the tensile property, which is represented by a curve $t_1 t_2$ in the Roscoe diagram.

3.9.1 SHEARING STRENGTH OF A POWDER BED

Triaxial Compression Test

The schematic diagram of a triaxial compression tester is shown in Figure 9.1. A cylindrical specimen is placed into the space surrounded by two end plates and a rubber side membrane, to which vertical and lateral pressures are applied independently. From the pressures at the break point, the Mohr circle with the maximum principal stress σ_1 and the minimum principal stress σ_3 can be obtained, as shown later in Figure 9.13. The yield locus is obtainable as an envelope of a family of the Mohr circles at the critical stress condition. Although the triaxial compression tester is used widely in soil mechanics,[1,2] it is complex in structure and has the disadvantage that the failure surface of test samples becomes irregular. Hence, it is not widely used in the field of powder technology except for a few cases.[3,4] Aoki et al.[5] modified the tensile tester of the split-cell type, called the Cohe tester, and made a simplified triaxial tester to measure the PYL of loosely packed powders in a low-stress region. Arthur et al.[6] designed a new biaxial tester with flexible membranes stretched to the retaining guides. The tester imposes known uniform principal stresses on a precisely compacted cubical sample powder and measures the resulting strains in a simple and immediate manner.

Direct Shear Test

A direct shear tester has a simple structure and the testing procedure is also simple. Hence, it is widely used in powder technology, especially in the field of bulk materials handling. The direct shear tester usually consists of two cells: one cell is placed on the other cell, which is fixed. They are filled with the particulate material under examination, the test sample is consolidated to attain a prescribed packing density, and then under a compression load, a horizontal shear force is applied to the upper cell so as to measure the yield point. The relation between the compression load and shear force at the yield point gives the yield locus for the prescribed packing density. Various types of testers have been proposed for this purpose.

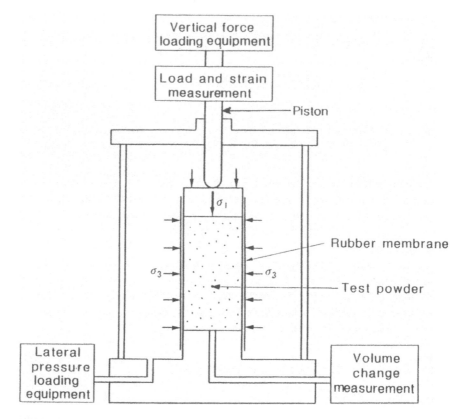

FIGURE 9.1 Triaxial compression tester.

Jenike's Shear Tester

The shear tester developed by Jenike et al.[7] is the most popular and widely used of those designed with a hopper, bin, or silo.[8–13] Recently, Jenike's shear tester and test procedure became an international standard.[14] A cross-sectional view of Jenike's shear cell is shown in Figure 9.2. A test sample is filled in the cell and consolidated by twisting under normal stress σ_p. Then the ring is sheared under the same normal stress σ_p until 95% of the yield force measured beforehand is attained, and then the normal stress is replaced by a smaller stress, σ ($<\sigma_p$). The cell is sheared again until the specimen reaches yield. In this way, a relationship between σ and τ is obtained. Hirota et al.[15] reported coincidence of the results obtained by the Jenike shear tester and his own tester, which is a parallel-plate tester, using the same procedure. On the Jenike shear tester, Haaker[16] investigated the effect of the ratio of externally applied load during consolidation and preshearing.

Richmond and Gardner[11] devised shear blades in the cell to avoid the slip of powder on the wall while in shearing. Tsunakawa and Aoki[17,18] modified the Jenike cell to keep a constant volume and obtained the yield locus from a single shear test by recording the normal and shear stresses simultaneously on an X-Y recorder. Kirby[19] subsequently presented a query for their measuring procedure. Terashita et al.[20] measured the normal stress distribution in the cell and noted that the normal stress does not apply accurately on the shear surface because of the wall friction. After correcting the normal stress on the shear surface, they obtained the CSL. Matsumoto et al.[21] modified the direct shear tester to measure the actual normal stress on the shear surface. These types of shear tester have limitations in effective shear surface, and Hirota et al.[22] pointed out that the size of the shear cell affects the measuring results substantially. Hidaka et al.[23] investigated the particle assemblies in the shearing

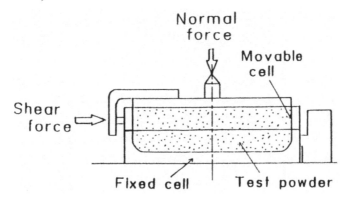

FIGURE 9.2   Jenike's shear tester. [From Jenike, A. W., Elsey, P. J., and Wolley, R. H., *Am. Soc. Test. Mater. Proc.*, 60, 1168–1190, 1960. With permission.]

process by computer simulation based on the distinct element method, and discussed the acoustic emission from the shear flow of granular materials.

Annular Ring Shear Tester

Annular ring or torsion shear testers are used for cylindrical powder beds. The major advantages of this type are that the area of the shear surface does not change during the test, there is no limitation in the amount of angular displacement, and the shear characteristics after the yield can also be measured. Walker[24] made the annular shear tester shown in Figure 9.3. Suzuki et al.,[25] Gotoh et al.,[26] and Yamada et al.[27] used similar shear testers. Scarlett and Todd[28] designed a split annular ring shear cell to achieve a simple flow profile of powder, but its structure was complicated.

Simple Shear Tester

Roscoe et al.[29] made a simple shear tester in which the stress can be determined accurately. As shown in Figure 9.4, the guided plunger stresses the test powder vertically, and the shear force is applied to the bottom unit of the tester. This shear cell is the only one of the translational type that can determine the void fraction of the powder bed in the shear zone. Enstad[30] developed an annular tester to achieve an unlimited amount of displacement and to make the shear zone wider, as with the simple shear tester, but its structure and measuring procedure are complicated.

Parallel-Plate Shear Tester

Hiestand and Wilcox[31] and Budny[32] devised a parallel-plate tester, and Hirota et al.[33] installed electric rod heaters in the plates, as shown in Figure 9.5, in order to measure the shear properties at high temperatures up to 800 K. The test powder is placed between two solid plates, the bottom plate is fixed, and the normal and shear forces are applied to the upper movable plate. No sidewall exists, so that it is unnecessary to consider the wall friction. Small expansion and contraction on the order of 1 μm are detected during the shear process.[34] Hirota et al.[35,36] also devised tensile and compression testers similar to the parallel plates shear tester.

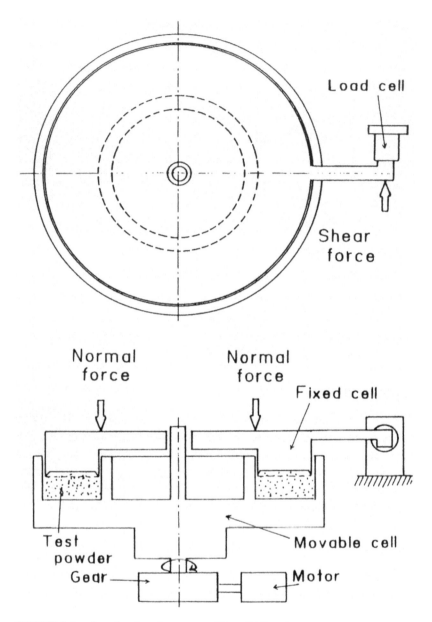

FIGURE 9.3   Annular ring shear tester. [From Walker, D. M., *Powder Technol.*, 1, 228–236, 1967. With permission.]

3.9.2   ADHESION OF A POWDER BED

The tensile strength of powder bed σ_t can be measured by several kinds of tensile testers, which are classified to two types: one is based on a vertical tensile method and the other a horizontal tensile method.

Horizontal Tensile Method

The tensile force in this method is applied perpendicular to the compression load for consolidation of powder. Ashton et al.[37] developed a split-cell-type tensile tester, which is shown in Figure 9.6.

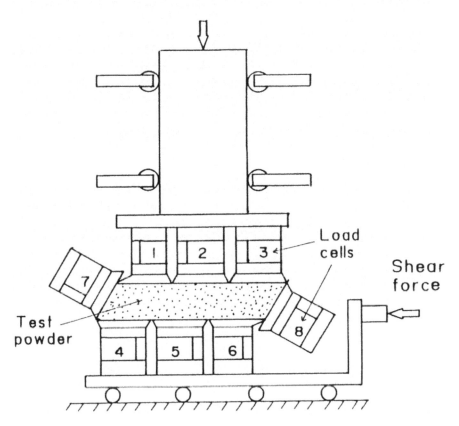

FIGURE 9.4 Simple shear tester. [From Schwedes, J., *Powder Technol.*, 11, 59–67, 1975. With permission.]

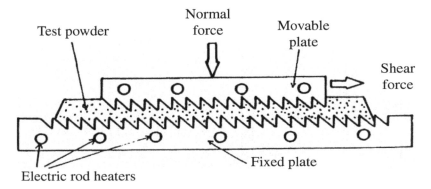

FIGURE 9.5 Parallel-plate shear tester for high temperature shear tests. [From Hirota, M., Oshima, T., and Naitoh, M., *J. Soc. Powder Technol. Jpn.*, 19, 337–342, 1982. With permission.]

Test powder is compacted into the split cell, and half of the cell is fixed on a plate. The other half of the cell is put on ball bearings for movement. The tensile strength of the test powder is measured by a load cell connected to the movable cell. The tensile strength of a loosely packed powder bed can also be measured using this type of tester because the friction of the movable cell is reduced by the ball bearing, but the cleaning of the balls after each test is time-consuming. Schotton and Harb[38] measured the tensile strength of powder using the same kinds of tester on a slanted plate by changing the angle of the plate. Jimbo and Yamazaki [39] developed a modified split-cell bearing-type tester, which consists of two movable cells. The friction of the tester can be reduced, but the structure of the tester is complicated. Yokoyama et al.[40] devised a frictionless split cell by hanging one half of the cell, like a swing, as shown in Figure 9.7, the Cohe tester. This is a tester to measure the tensile strength of a loosely packed powder bed in a high void fraction region.

Vertical Tensile Method

The tensile load can be applied vertically to the compacted powder; it means that the direction of the tensile force is the same as that of the compression force. Shinohara and Tanaka,[41] Arakawa,[42] and Suzuki et al.[25] developed the tensile testers based on this method. Arakawa[42] developed the vertical-type tensile tester from an electric balance, as shown in Figure 9.8. He also tried to explain the effect of a particle packing structure on tensile strength of a powder bed by a two-dimensional particle model experiment. Hirota et al.[36] developed a parallel-plate-type shear and tensile tester to obtain the yield loci under the normal stress ranging from positive to negative (including the adhesion force) and measured the PYL in the tensile region, but the operation of the tester was not so easy.

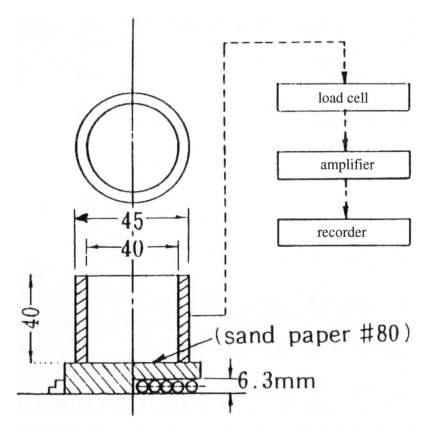

FIGURE 9.6   Horizontal tensile tester. [From Ashton, M. D., Farley, R., and Valentin, F. H. H., *J. Sci. Instrum.*, 41, 763–765, 1964. With permission.]

The tensile strength measured by the vertical tensile method is usually larger than that measured by the horizontal tensile method, because the direction of compression force at preconsolidation is the same as the direction of tensile force in the vertical method and is perpendicular to the tensile force in the horizontal method. The relation between the tensile strengths by these methods is not clear yet.

3.9.3 YIELDING CHARACTERISTICS OF A POWDER BED

Figure 9.9 shows the concept of the Roscoe diagram or Cambridge model proposed by Roscoe et al.,[43,44] which expresses the yielding characteristics of a powder bed in relation to shear, tensile, and compression strengths (see Section 3.9.1). The coordinates of the diagram are the shear stress τ, the normal stress σ, and the void fraction ε of the powder bed. The curve $t_1 t_2$ on the plane $\tau = 0$, called the tensile property, depicts the relationship between tensile strength and void fraction. The curve $f_1 f_2$ on the plane $\tau = 0$, called the compressive property or consolidation line, is the relation between compression strength and void fraction. Only the region between these two curves is possible.

In the three-dimensional diagram, two failure surfaces, called the flow surface and the consolidation surface, exist between these two curves. The intersection of them is the curve $e_1 e_2$ in Figure 9.9, which is called the CSL. The flow surface between curves $t_1 t_2$ and $e_1 e_2$ corresponds to the plastic flow with expansion, and the projection of the flow surface onto the σ-τ plane is called the powder yield locus, or simply the yield locus. The consolidation surface between curves $f_1 f_2$ and $e_1 e_2$ corresponds to plastic deformation with contraction, whose projection onto a σ-τ plane is called

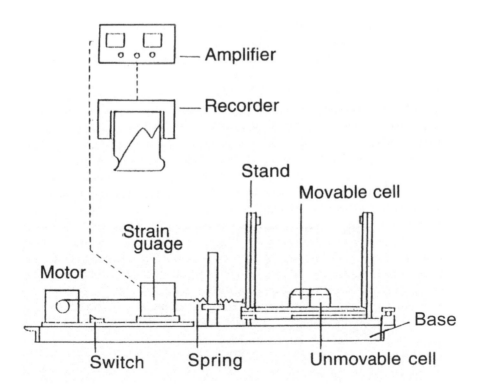

FIGURE 9.7   Horizontal tensile tester called the Cohe tester. [From Yokoyama, T., Fujii, K., and Yokoyama, T., *Powder Technol.*, 32, 55–62, 1982. With permission.]

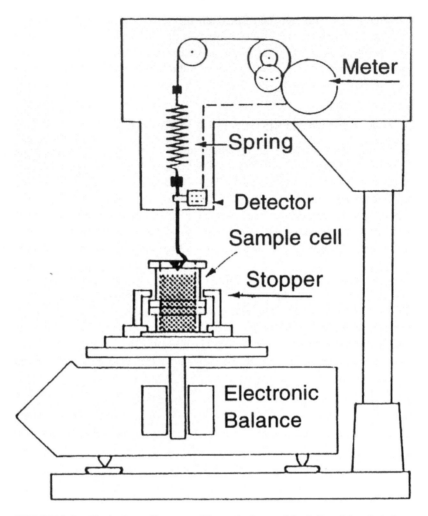

FIGURE 9.8 Vertical tensile tester. [From Arakawa, M., *J. Soc. Mat. Sci. Jpn.*, 29, 881–886, 1980. With permission.]

the CYL. The elastic condition corresponds to the region below the surface in Figure 9.9, while the region above the surfaces is actually impossible.

Consider the shear process of a powder bed with the void fraction ε_b in the three-dimensional diagram. At the normal stress σ_b, the shear stress increases from point C to point E in Figure 9.9, where the steady flow occurs. In this case the stress versus strain curve passes over the line OPES in Figure 9.10. A set of τ and σ on the CSL can be obtained from τ_b and σ_b of point E in Figure 9.10.

When $\sigma = \sigma_a$ ($\sigma_a < \sigma_b$), τ increases from point H to point P in Figure 9.9, where yielding starts with expansion, and then it decreases gradually along PZ with increasing void fraction from ε_b to ε_a until the powder bed begins to flow steadily at point Z. A set of τ and σ on the PYL can be obtained from τ_d and σ_a of point P in Figure 9.10.

When $\sigma = \sigma_e$ ($\sigma_e > \sigma_b$), τ increases from point F to Q in Figure 9.9, where the yielding starts with contraction, and then it increases gradually along QK with decreasing void fraction from ε_b to ε_e until the powder bed starts to flow steadily at point K. A set of τ and σ on the CYL can be obtained from τ_e and σ_e of point Q in Figure 9.10. As a result, the yield locus, which is the relation between τ and σ at the yield point of the powder bed, can be depicted by the curve I'A'P'SQ', which is the

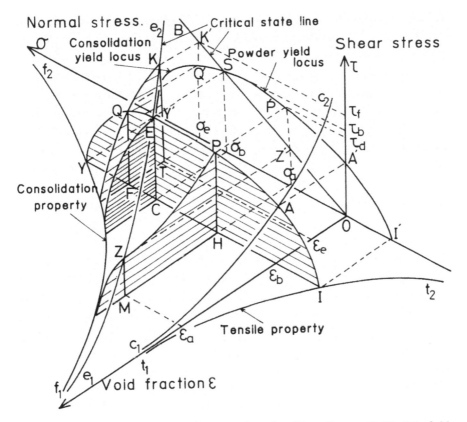

FIGURE 9.9 Roscoe's condition diagram of powder. [From Roscoe, K. H., Schofield, A. N., and Wroth, C. P., *Geotechnique*, 8, 22–53, 1958. With permission.]

projection of the curve IAPEQ in Figure 9.9 onto the σ-τ plane. Point Y expresses the preconsolidation state at the starting point of yielding.

The yield locus at the state of steady flow leads the curve OZ'SK'B, which is the projection of the curve e_1ZEKe$_2$ in Figure 9.9 onto the σ-τ plane and equals the CSL proposed by Schwedes.[45] The void fraction ε decreases with increasing σ, and τ and σ are specified for a given value of ε.

If the Roscoe diagram is obtained for a powder bed, the static mechanics of the bed can be discussed quantitatively.

The examples of the experimental data are shown in Figure 9.11 and Figure 9.12. These diagrams were obtained from the compression, shear, and tensile testers of parallel-plate type by use of the three-dimensional computer graphic technique.[46] Figure 9.11 is for flyash and Figure 9.12 for precipitated calcium carbonate powder. Because the yield locus becomes a straight line on the σ-τ plane, the flyash might be a Coulombic powder, while the upper surface of the yield locus of the calcium carbonate powder looks like a mountain, and it is found to be a cohesive bulk material.

If a powder is a rigid plastic or Coulombic solid, its yield locus can be expressed by

$$\tau = \mu_i \delta + \tau_c = \sigma \tan \phi_i + \tau_c \tag{9.1}$$

where τ and σ are the shear and normal stresses, μ_i the internal friction coefficient, ϕ_i the internal friction angle, and τ_c the cohesion (kPa). Using Equation 9.1, the mechanical properties of powder, including the CSL, can be expressed by two parameters, μ_i or ϕ_i and τ_c. The Coulomb equation is applicable in general to cohesionless coarse powders. Because the yield locus is the envelope of a

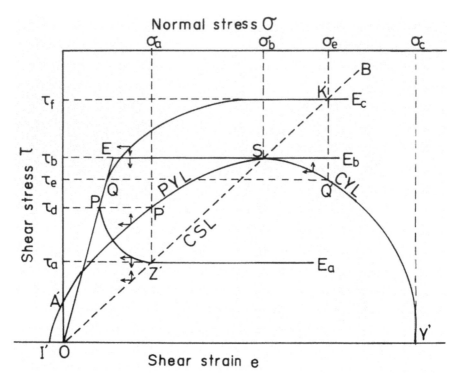

FIGURE 9.10 Conceptual diagram of shearing process. [From Hirota, M., Oshima, T., and Naitoh, M., *J. Soc. Powder Technol. Jpn.*, 19, 337–342, 1982. With permission.]

family of yield stress circles (Mohr circles), it should cross the σ axis perpendicularly at point I', which corresponds to the tensile strength of the powder bed. As one can see from Figure 9.13, however, the Coulomb equation does not cross the σ axis at a right angle, a contradiction explained by Jenike et al.,[7] Umeya et al.,[47] and Makino et al.[48]

On the assumption that the particle arrangement is a face-centered cubic and the attractive force between particles is a function of the separation of nearest neighbors with a maximum limit of separation, Ashton et al.[49] derived the following relation:

$$\left(\frac{\tau}{\tau_c}\right)^n = \frac{\sigma}{\sigma_t} + 1 \tag{9.2}$$

where σ_t is the tensile strength, τ_c is the cohesion of the powder bed, and n is the shear index. In the case of $n = 1$, Equation 9.2 reduces to Equation 9.1; n is found to be independent of the void fraction and it ranges from about 1 for free-flowing powders up to 2 for very cohesive ones. Umeya et al.[47] derived the same result experimentally.

The values of σ_t and τ_c in Equation 9.1 and Equation 9.2 vary with the void fraction. Williams and Birks[50] suggested that if a family of yield loci are plotted on the τ/σ_L versus σ/σ_L diagram, where σ_L is the normal stress at the end point of a yield locus, they come to lie on a single PYL curve, which is a straight line on logarithmic scales. Hirota et al.[36] pointed out that it is difficult to measure σ_L; also, Williams's linearization method cannot apply to the CYL. Therefore, they recommended to use a $\tau/(\sigma_c - \sigma_t)$ versus $(\sigma - \sigma_t)/(\sigma_c - \sigma_t)$ diagram in order to express various PYL and CYL curves with a single curve.

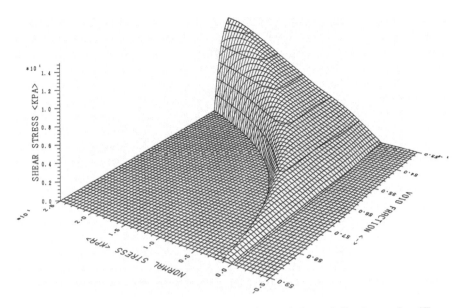

FIGURE 9.11 Roscoe diagram based on experimental data of flyash powder. [From Suzuki, M., Hirota, M., and Oshima, T., *J. Soc. Powder Technol. Jpn*, 24, 311–314, 1987. With permission.]

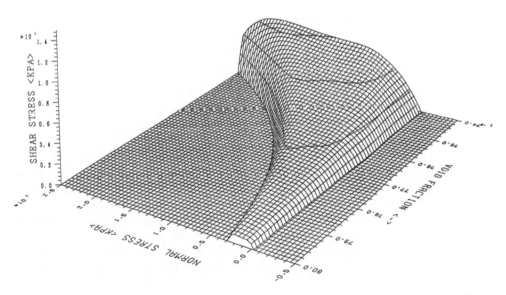

FIGURE 9.12 Roscoe diagram based on experimental data of calcium carbonate powder. [From Suzuki, M., Hirota, M., and Oshima, T., *J. Soc. Powder Technol. Jpn*, 24, 311–314, 1987. With permission.]

Based on a theoretical model, Rumpf[51] derived the following expression for the tensile strength of randomly packed bed of equal spheres:

$$\sigma_t = \frac{1-\varepsilon}{\pi} N_c \frac{H}{D\pi^2} \tag{9.3}$$

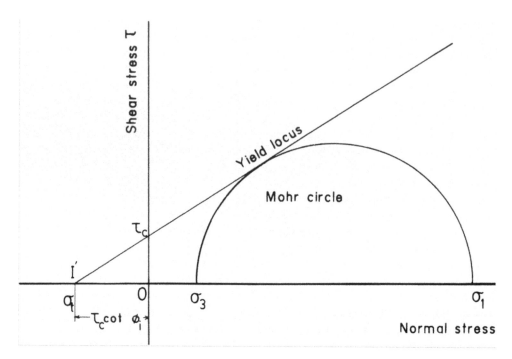

FIGURE 9.13 Yield locus of a Coulomb powder.

where ε is the void fraction, N_c is the average coordination number. H is the cohesion force at a contact point, and D_p the particle diameter. Molerus[52] supposed that Equation 9.3 holds for the relation between the stress and the applied force under the isotropic or hydrostatic pressure, and he analyzed the shear mechanism of a powder bed.

Makino et al.[48] proposed a model for the yield of a powder bed, taking into account a particle arrangement and interparticle force of a Lennard–Jones type. They discussed a yield condition under constant strain, and the PYL and CYL were made obtainable for any void fraction from only four parameters determined by tensile and compression tests. Yamada et al.[27] subsequently reduced the number of parameters required to estimate the PYL and CYL of a powder bed. But, it is difficult to express the interparticle friction by the Lennard–Jones type of potential.

Nagao[53] proposed a more precise theory of powder mechanics, including equation of stress and moment equilibrium, equation of continuity, a geometric relation, stress–strain relations, and boundary conditions. Based on the fundamental equations, he calculated the stress distribution in compressed powder beds by use of the finite-element method.[54] A principle of similarity in the mechanics of granular materials was also investigated.[55]

Kanatani[56] established a theory of continuum mechanics for the flow of granular materials, which is an application of the theory of polar fluids. He derived theoretically various characteristics of granular materials, such as the critical stress condition, angle of repose, and the effects of coupling stress in the Couette flow between two parallel plates.

Tsubaki and Jimbo[57] reported that Rumpf's equation for tensile strength of a powder bed coincides with Nagao's theory for the stress–strain relation, Molerus's research on the yield of cohesive powders[52] and Kanatani's theory for flow of granular materials. Hence it can be said that Rumpf's equation is applicable not only to the tensile strength but also to the shear and compression strengths of the powder bed.

REFERENCES

1. Newland, P. L. and Allely, B. H., *Geotechnique,* 7, 17–34, 1957.
2. Rowe, P. W., *Proc. R. Soc. London A,* 269, 500–527, 1962.
3. Masuda, Y. and Sakai, N., *J. Jpn. Soc. Powder and Powder Metall.,* 19, 1–16, 1972.
4. Takagi, F. and Sugita, M., *J. Soc. Powder Technol. Jpn.,* 16, 277–282, 1979.
5. Aoki, R., Suganuma, A., and Motone, M., *J. Soc. Powder Technol. Jpn.,* 9, 538–542, 1982.
6. Arthur, J. R. F., Dunstan, T., and Enstad, G. G., *Int. J. Bulk Solids Storage Silos,* 1, 7–10, 1985.
7. Jenike, A. W., Elsey, P. J., and Wolley, R. H., *Am. Soc. Test. Mater. Proc.,* 60, 1168–1190, 1960.
8. Jenike, A. W., *Trans. Inst. Chem. Eng.,* 40, 264–271, 1962.
9. Jenike, A. W., *Trans. Soc. Min. Eng.,* 235, 267–275, 1966.
10. Jenike, A. W., *Powder Technol.,* 11, 89–91, 1975.
11. Richmond, O. and Gardner, G. C., *Chem. Eng. Sci.,* 17, 1071–1078, 1962.
12. Williams, J. C., *Chem. Process Eng.,* 46, 173–179, 1965.
13. Eisenhart-Rothe, M. and Peschl, I. A. S. Z., Powder testing techniques for bulk materials equipment, in *Solid Handling,* McGraw-Hill, New York, 1981.
14. ISO 11697, 1994.
15. Hirota, M., Kobayashi, T., Tajiri, H., Murata, H., Wakabayashi, M., and Oshima, T., *J. Soc. Powder Technol. Jpn.,* 22, 144–149, 1985.
16. Haaker, G., *Powder Technol.,* 51, 231–236, 1987.
17. Tsunakawa, H. and Aoki, R., *J. Soc. Powder Technol. Jpn.,* 11, 263–268, 1974.
18. Tsunakawa, H. and Aoki, R., *Powder Technol.,* 33, 249–256, 1982.
19. Kirby, J. M., *Powder Technol.,* 39, 291–292, 1984.
20. Terashita, K., Miyanami, K., Yamamoto, T., and Yano, T., *J. Soc. Powder Technol. Jpn.,* 15, 526–534, 1978.
21. Matsumoto, K., Yoshida, M., Suganuma, A., Aoki, R., and Murata, H., *J. Soc. Powder Technol.,* 19, 653–660, 1982.
22. Hirota, M., Oshima, T., and Hashimoto, S. *J. Soc. Powder Technol. Jpn.,* 16, 198–206, 1979.
23. Hidaka, J., Kinboshi, T., and Miwa, S., *J. Soc. Powder Technol. Jpn.,* 26, 77–84, 1989.
24. Walker, D. M., *Powder Technol.,* 1, 228–236, 1967.
25. Suzuki, M., Makino, K., Iinoya, K., and Watanabe, K., *J. Soc. Powder Technol. Jpn.,* 17, 559–564, 1980.
26. Gotoh, A., Kawamura, M., Matsushima, H., and Tsunakawa, H., *J. Soc. Powder Technol. Jpn.,* 21, 131–136, 1984.
27. Yamada, M., Kuramitsu, K., and Makino, K., *Kagaku Kogaku Ronbunshu,* 12, 408–413, 1986.
28. Scarlett, B. and Todd, A. C., *Trans. ASME Ser. B,* 91, 478–488, 1969.
29. Roscoe, K. H., Arthur, J. R. F., and James, R. G., *Civil Eng. Public Works Rev.,* 58, 873–876, 1963.
30. Enstad, G. G., *Proc. Eur. Symp. Particle. Technol.,* B, 997, 1980.
31. Hiestand, E. N. and .Wilcox, C. J., *J. Pharm. Sci.,* 58, 1403–1410, 1969.
32. Budny, T. J., *Powder Technol.,* 23, 197–201, 1979.
33. Hirota, M., Oshima, T., and Naitoh, M., *J. Soc. Powder Technol. Jpn.,* 19, 337–342, 1982.
34. Hirota, M., Kobayashi, T., Sano, O., and Oshima, T., *J. Soc. Powder Technol. Jpn.,* 21, 137–142, 1984.
35. Hirota, M., Oshima, T., Ishihara, T., and Kanazawa, A., *J. Soc. Mater. Sci. Jpn.,* 33, 1125–1129, 1984.
36. Hirota, M., Kobayashi, T., and Oshima, T., *J. Soc. Powder Technol. Jpn.,* 22, 271–277, 1985.
37. Ashton, M. D., Farley, R., and Valentin, F. H. H., *J. Sci. Instrum.,* 41, 763–765, 1964.
38. Schotton, E. and Harb, N., *J. Pharm. Pharmacol.,* 18, 175–178, 1966.
39. Jimbo, G. and Yamazaki, R., *European Symposium Particle Technology Preprints,* Vol. B, 1064–1074, 1980.
40. Yokoyama, T., Fujii, K., and Yokoyama, T., *Powder Technol.,* 32, 55–62, 1982.
41. Shinohara, K. and Tanaka, T., *J. Chem. Eng. Jpn.,* 8, 46–50, 1975.
42. Arakawa, M., *J. Soc. Mat. Sci. Jpn.,* 29, 881–886, 1980.
43. Roscoe, K. H., Schofield, A. N., and Wroth, C. P., *Geotechnique,* 8, 22–53, 1958.
44. Roscoe, K. H., Schofield, A. N., and Thurairajah, A., *Geotechnique,* 13, 211–240, 1963.
45. Schwedes, J., *Powder Technol.,* 11, 59–67, 1975.

46. Suzuki, M., Hirota, M., and Oshima, T., *J. Soc. Powder Technol. Jpn*, 24, 311–314, 1987.
47. Umeya, K., Kitamori, N., Araki, Y., and Mima, H., *J. Soc. Mater. Sci. Jpn.*, 15, 166–171, 1966.
48. Makino, K., Saiwai, K., Suzuki, M., Tamamura, T., and Iinoya, K., *Int. Chem. Eng.*, 21, 229–235, 1981.
49. Ashton, M. D., Cheng, D. C. H., Farley, R., and Valentine, F. H. H., *Rheol. Acta*, 4, 206–218, 1965.
50. Williams, J. C. and Birks, A. H., *Powder Technol.*, 1, 199–206, 1967.
51. Rumpf, H., *Chem. Eng. Technol.*, 42, 538–540, 1970.
52. Molerus, O., *Powder Technol.*, 12, 259–275, 1975.
53. Nagao, T., *Trans. Jpn. Soc. Mech. Eng.*, 33, 229–241, 1967.
54. Nagao, T. and Katayama, S., *Jpn. Soc. Mech. Eng.*, 46, 355–366, 1980.
55. Nagao, T., *J. Soc. Powder Technol. Jpn.*, 21, 398–405, 1984.
56. Kanatani, K., *J. Soc. Powder Technol. Jpn.*, 16, 445–452, 1979.
57. Tsubaki, J. and Jimbo, G., *Powder Technol.*, 37, 219–227, 1984.

3.10 Fluidity of Powder

Toyokazu Yokoyama

Hosokawa Powder Technology Research Institute,
Hirakata, Osaka, Japan

3.10.1 DEFINITION OF FLUIDITY

The fluidity of powder is defined intuitively as the ease of flow and relates to the change of mutual position of individual particles forming the powder bed. The fluidity of powder is strongly related to physical properties such as frictional force and cohesive force of the particles. The dynamic behavior of powder seems to be determined basically by interparticle forces and packing structure.

Powder flow in various industrial processes takes place in different ways that can hardly be described in a universal form. Table 10.1 classifies the type of powder flow from the practical viewpoint.[1] Based on the source of energy exerted on the particles, powder flow is classified as (1) gravitational flow, (2) mechanically forced flow, (3) vibration flow, (4) compression flow, and (5) fluidized flow, which appear simultaneously in actual processes in most cases.

3.10.2 MEASUREMENT OF FLUIDITY

As the phenomena of powder flow are complicated and involve the combined flow patterns listed in Table 10.1, a suitable method of measurement is required for individual cases. Although there are a variety of methods to determine fluidity and related properties, it is difficult to find a general result among them. Therefore, it is essential for measurement to clarify the measuring conditions and powder properties. Furthermore, attention should be paid to the range applicable to the powder-handling processes.

Gravitational Flow

The flow from hoppers or through an orifice is often considered to express the fluidity of powder. In this case, the magnitude and uniformity of flow rate, as well as the tendency of choking, are regarded as the criteria of the fluidity. The flow pattern and segregation are also related to the fluidity of powder.

In the field of metallurgy, the time required for a powder sample of 50 g to be discharged by gravitational force from the funnel shown in Figure 10.1 is often used as a measure to evaluate fluidity. For cohesive powders, Irani et al.[2] proposed a double-funnel method with larger openings.

Even if the flow rate is high, a flow with a large fluctuation cannot be regarded as a good measure of fluidity. Several reports have been presented on the uniformity of the discharge rate, especially in the pharmaceutical field. Gold et al.,[3] for example, studied the discharge rate and its fluctuation experimentally.

Cohesive powders tend to form a bridge near the outlet of the container. The minimum opening size free of choking corresponds to the critical condition for powder flow and can be a measure of the fluidity. For the Coulomb powders, the critical choking aperture can be obtained from Mohr's stress circle.

The mass-discharge rate is proportional to the product of the height of the surface level and the square root of the orifice diameter in the case of liquid; it is independent of the height

TABLE 10.1 Type of Flow and Fluidity

Type of Flow	Process/Phenomena	Expression of Fluidity
Gravitational flow	Discharge from bin and hopper, chute, sand-glass, tumbling mixer, moving bed, packing	Discharge rate, angle of wall friction, angle of repose, critical discharge opening
Mechanically forced flow	Powder mixing, chain conveyor, screw conveyor, table feeder, ribbon mixer, rotary feeder, extruder	Angle of internal friction, angle of wall friction, mixing resistance
Compression flow	Briquetting, tableting	Compressibility, angle of wall friction, angle of internal friction
Vibration flow	Vibrating feeder, vibrating conveyor, vibrating screen, packing, discharge	Angle of repose, discharge rate, compressibility, bulk density
Fluidized flow	Fluidized bed, pneumatic conveyor, air slide, aerated vibration dryer	Angle of repose, minimum fluidization velocity, aeration resistance, apparent viscosity

Source: Hayakawa, S., *Funtai Bussei Sokuteiho*, Asakura, Tokyo. 1973, p. 80.

of the powder bed and proportional to the orifice diameter to the power of 2.5 to 3 in the case of powders. Numerous experimental equations[4] have been proposed for the flow rate of powder discharged by gravitational force.

Matsusaka et al. investigated the flow of fine powder using a microfeeder with a vibrating capillary tube having a diameter of less than 1.6 mm.[5] Figure 10.2 shows the effect of the inner diameter of a capillary tube d on maximum powder flow rate W_{max}. It was confirmed that fine particles of about 10 µm diameter are successfully discharged using a vibrating capillary tube, and that the maximum flow rate is proportional to the 2–$2.5th$ power of the capillary diameter, which is a little low compared with that for the conventional orifice. This flow rate is also regarded as a measure of fluidity of fine powder.

The angle of repose is defined as the angle from the horizontal plane to the free surface of a pile of powder under the critical stress in the gravitational field. There are three principal ways to determine the angle of repose, as shown in Figure 10.3. The angle of repose tends to increase in the following order: injection method, discharge method, and tilting method. The major factors influencing the angle of repose are size distribution,[6] surface roughness of the particles, void fraction of the powder bed,[7] and moisture content. In the case of the injection method, the injection rate, falling distance,[8] and size of the conical pile of powder[9] influence the angle of repose.

Forced Flow

Mechanically Forced Flow

The powder flow actuated by mechanical force has been studied in relation to torque using various types of rotary viscometers: those of Stomer, Couette, Green, and MacMichael. Iiyama and Aoki[10] obtained an experimental equation to find the torque for the vortex flow of standard silica sands, in which the apparatus shown in Figure 10.4 was used with various-shaped agitating blades.

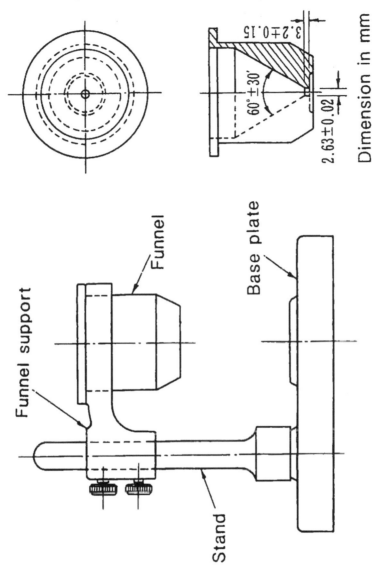

Dimension in mm

FIGURE 10.1 Device for the measurement of fluidity. [From Japanese Industrial Standard Z 2502 (1979).]

Compression Flow

In a large vessel such as a huge silo, the value of k in Janssen's equation is used as a measure of the fluidity.[11] A compression test also gives a measure for the fluidity of powder. A number of equations have been proposed for the relationship between the external pressure acting on the powder bed and the change of volume. One typical experimental equation[12] related the apparent volume reduction ratio C to the applied pressure P (Pa):

$$\frac{P}{C} = \frac{1}{ab} + \frac{P}{a} \tag{10.1}$$

where a and b are constants, and a is considered to increase with decreasing fluidity of the powder.

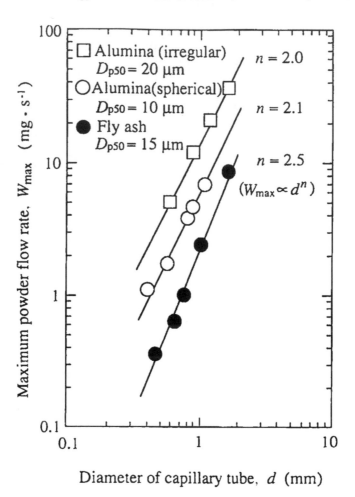

FIGURE 10.2 Effect of the inner diameter of a capillary tube d on maximum powder flow rate W_{max}.

Vibration Flow

Vibration changes the packing structure of powder bed and, hence, leads to the change of the fluidity. Miwa[13] proposed the following experimental relation between the mass flow rate through sieves and sieve openings:

$$\frac{W_s}{\rho_a} = kD_s^n D_p^{-2} \tag{10.2}$$

where W_s (tons/h) is the mass flow rate, ρ_a (tons/m³) is the particle specific gravity, D_s (mm) is the sieve opening, D_p (mm) is the particle diameter, and n is a constant (about 2.7). The factor k, called the sieve fluidity index, expresses the ease with which particles pass through sieves and is considered to be a measure of fluidity under dynamic conditions.

Arakawa[14] studied vibration fluidity in relation to the packing structure, paying attention to the vibration in the electric resistance of the electroconductive particles in vibrating motion. Under the

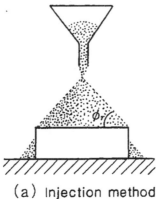

(a) Injection method

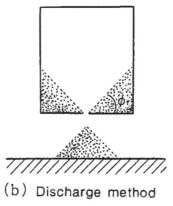

(b) Discharge method

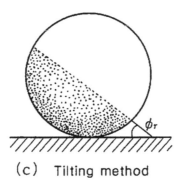

(c) Tilting method

FIGURE 10.3 Methods to measure the angle of repose.

condition of gravitational acceleration, the relation between the electric resistance ratio and the fluidity ϕ is shown in Figure 10.5 and described by the equation

$$\varphi = a \exp\left(-b\frac{R_0}{R_x}\right) \tag{10.3}$$

where a and b are constants, R_x and R_0 are electric resistances of the powder beds in flow and in the most densely packed state, respectively.

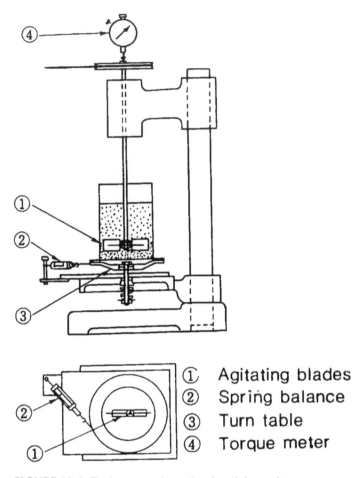

①	Agitating blades
②	Spring balance
③	Turn table
④	Torque meter

FIGURE 10.4 Equipment to determine the mixing resistance.

Sato et al.[15] evaluated the flow characteristics of powders by a vibrating powder tester. A detecting sphere attached to the force transducer is displaced horizontally within the powder bed. The fluidity was discussed in terms of an apparent viscosity obtained from the relationship between the force and displacement velocity.

Shear Properties and Fluidity

In powder mechanics, the ultimate balanced condition is estimated from the yield locus obtained by shearing tests. The fluidity defined on the basis of the yield locus indicates the condition where the force balance is about to be lost and is related strongly to the chocking in the pneumatic conveyor or the container outlet.

Jenike[16] defined the yield stress f (kPa) and the maximum compaction stress σ_1 from two Mohr's stress circles coming in contact with the yield locus, and he called the ratio σ_1/f the flow function (FF), which is regarded as a measure of the fluidity. The fluidity increases with the flow function.

The following Warren–Spring equation, proposed by Ashton et al., is widely applied to evaluate the shear stress:

$$\left(\frac{\tau}{C}\right)^n = \frac{\sigma + T}{T} \tag{10.4}$$

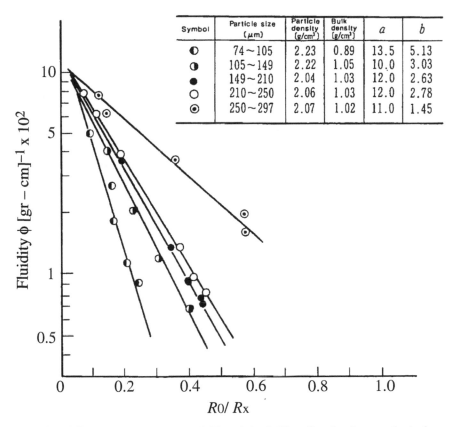

Symbol	Particle size (μm)	Particle density (g/cm³)	Bulk density (g/cm³)	a	b
◑	74~105	2.23	0.89	13.5	5.13
◐	105~149	2.22	1.05	10.0	3.03
●	149~210	2.04	1.03	12.0	2.63
○	210~250	2.06	1.03	12.0	2.78
◉	250~297	2.07	1.02	11.0	1.45

FIGURE 10.5 Relationship between R_o/R_x and the fluidity of a vibrating powder bed.

where C (kPa) is the cohesive stress, T (kPa) is the tensile strength, σ (kPa) is the normal stress, and τ (kPa) is the shear stress. Farley and Valentin[17] took the shear index n as a measure of fluidity and proposed an experimental equation. Stainforth and Berry[18] found a linear relationship between σ_1/f and σ_e/T and defined the gradient as the fluidity, where σ_e (kPa) is the normal stress at the terminal of the yield locus.

One-dimensional yield stress f (kPa) and the bulk density ρ_a (kg/m³) increase with increasing maximum compaction stress σ_1 (kPa). The ratio f/ρ_a is considered to be another measure of the fluidity:

$$\text{FI} = \frac{f}{\rho_a g} \tag{10.5}$$

where g is the acceleration due to gravity.

The flow index FI becomes larger as the fluidity decreases. Tsunakawa[19] studied the relationship between FI and the maximum compaction stress σ_1 and obtained the following experimental equation from Figure 10.6:

$$\text{FI} = a\sigma_1^b \tag{10.6}$$

where the coefficient a and index b, which are specific to each powder, increase as the fluidity decreases.

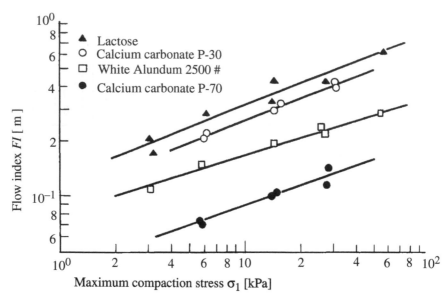

FIGURE 10.6 Relationship between the flow index and the maximum compaction stress.

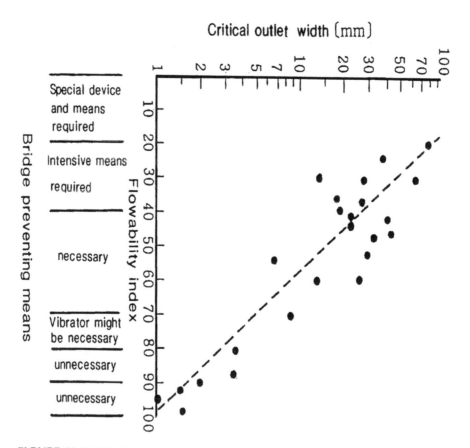

FIGURE 10.7 Relationship between the critical discharge opening and the flowability index.

Fluidized Flow

When a powder bed is aerated and agitated, the particles have less opportunity to come in contact with each other, and the powder bed exhibits fluidlike properties. The apparent viscosity of the aerated powder bed measured with a Stomer viscometer[20] decreases with increasing aeration rate. Diekman and Forsythe[21] regarded the time for the apparent viscosity to reach a certain value as a measure of the fluidity.

Total Evaluation

In practical powder-handling processes, the various types of powder flow shown in Table 10.1 occur simultaneously in most cases. Therefore, it is necessary to evaluate the fluidity from the practical viewpoint for the design of suitable processes.

Carr[22] proposed a method for the evaluation of fluidity by the addition of several indexes from his abundant experience with a variety of powders. This method was not developed on a theoretical basis but is a numerical synthetic evaluation of fluidity from several different factors. A powder tester has been developed to conduct these measurements within one unit.[23]

The flowability is obtained with the powder tester as a sum of the indexes converted from the four properties according to a chart: (1) angle of repose, (2) compressibility, (3) angle of spatula, and (4) cohesiveness (or uniformity). It gives an indication of how to select a means of preventing the formation of powder bridge.

The floodability index is related to the flushing tendency of powder. It is given as a sum of four indexes converted from the flowability index and the following three properties, using another chart: (1) angle of collapse, (2) the difference between angles of repose and collapse, and (3) dispersibility. In general, the compressibility is more reproducible than the angles and correlates well with the flowability index.

Suzuki and Maruko[24] found the relationship between the critical outlet width for the powder discharge and Carr's flowability index experimentally, as shown in Figure 10.7. The flowability index has also been investigated in relation to the sieve fluidity index[25] and the coefficient of mixing rate of the horizontal cylindrical mixer.[26]

3.10.3 FACTORS AFFECTING FLUIDITY

The fluidity of powder is influenced by various properties of the particles forming the powder, such as particle size and its distribution, shape and surface roughness of the particles, and other interparticle forces. As a general tendency, finer powders show less flowability. Tsunakawa[27,28] obtained an experimental equation to estimate the flow index of fine powders with the particle true density ρ_s (kg/m^3) and voidage ε from the specific surface diameter D_{sp} (μm) and major consolidation stress σ_1 (kPa) accounting for the mean contact number κ:

$$\text{FI} = \kappa \left(g\rho_s \right)^{-1} D_{sp}^{-2b} k \left(\frac{\tau \sigma_1^b}{k(1-\varepsilon)} \right)$$ 10.7

where κ and β are constants. Additionally, the influence of voidage has been further discussed.

3.10.4 IMPROVEMENT OF FLUIDITY

The fluidity of a powder can be improved by changing its physical properties, such as moisture content and particle size and shape, by means of drying, grinding,[29] classification, and granulation. The fluidity is also improved by altering the dynamic contact of particles, making use of pulsed air pressure. Experimental work on a vibrating two-dimensional hopper[30] and the discharge rate through an orifice by vibration[31] and compressed air have been reported.

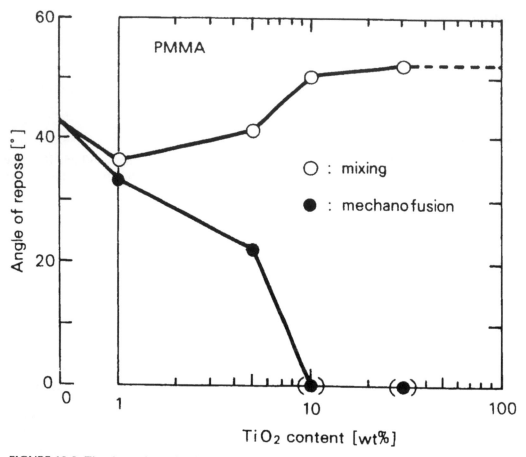

FIGURE 10.8 The change in angle of repose of composite powder produced by a mechanical treatment method compared with that produced by a simple mixing.

A small amount of additive can improve the fluidity of a powder. The mechanism is not yet elucidated clearly but is suggested to be attributable to the additives' properties, alteration of the particle arrangement, and the discharge of electrostatic charge, as confirmed experimentally by Nash et al.[32] and Jimbo.[33] Although the optimum additive fraction varies with the sort of additives, it is effective generally at a low mixing fraction in the range of 0.1–2%. It is well known that an additive has a different magnitude of influence on the fluidity, depending on the type of powder.[34]

Furthermore, it has been pointed out that the mechanical treatment for making composite particles changes the fluidity to a great extent. Figure 10.8[35] shows the changes in the angle of repose of polymethylmethacrylate particles (average particle size 5 μm) treated with TiO_2 particles (0.015 μm). It is hardly changed by the simple mixing but decreases drastically by intensive mechanical treatment forming composite particles. It has been also made clear that fluidity shows a maximum value at a certain condition, and excessive treatment deteriorates the fluidity when the surface particles are buried into the core particles.[36] It should also be noted that moisture affects the fluidity of powder considerably.

REFERENCES

1. Hayakawa, S., in *Funtai Bussei Sokuteiho,* Asakura, Tokyo. 1973, p. 80.
2. Irani, R. R., Callis, C. F., and Liu, J., *Ind. Eng. Chem.,* 51, 1285–1288, 1959.

3. Gold, G., Duvall, R. N., and Palermo, B. T., *J. Pharm. Sci.,* 55, 1133–1136, 1966.
4. Miwa, S., in *Funryutai Kogaku,* Asakura, Tokyo, 1972, p. 211.
5. Matsusaka, S., Yamamoto, K., and Masuda, H., *Adv. Powder Technol.,* 7, 141–151, 1996.
6. Pilpel, N., *Br. Chem. Eng.,* 11, 699–702, 1966.
7. Otsubo, K., *J. Res. Assoc. Powder Technol. Jpn.,* 2, 179–188, 1965.
8. Aoki, R., *J. Res. Assoc. Powder Technol. Jpn.,* 6, 3–8, 1969.
9. Train, D., *J. Pharm. Pharmacol.,* 10 (Suppl.), 127T–135T, 1958.
10. Iiyama, E. and Aoki, R., *J. Chem. Eng. Jpn.,* 24, 205–212, 1960.
11. Stepanoff, A. J., in *Gravity Flow of Bulk Solids and Transport of Solids in Suspension,* Wiley-Interscience, New York, 1969, p. 26.
12. Kawakita, K., *Funtai to Kogyo,* 9(2), 59–64, 1977.
13. Miwa, S., *Funsai (The Micromeritics),* 15, 8–11, 1970.
14. Arakawa, M., *J. Soc. Mater. Sci. Jpn.,* 20, 776–780, 1971.
15. Sato, M., Shigemura, T., Hamano, F., Fujimoto, T., and Miyanami, K., *J. Soc. Powder Technol. Jpn.,* 27, 308–314, 1990.
16. Jenike, A. W., *Utah Unlv. Eng. Exp. Station,* 108, 1961.
17. Farley, R. and Valentin, F. H. H., *Powder Technol.,* 1, 344–354, 1967.
18. Stainforth, P. T. and Berry, R. E. E., *Powder Technol.,* 8, 243–251, 1973.
19. Tsunakawa, H., *J. Soc. Powder Technol. Jpn.,* 19, 516–521, 1982.
20. Matheson, G. L., Herbst, W. A., and Holt, P. H., *Ind. Eng. Chem.,* 41, 1099–1104, 1949.
21. Diekman, R. and Forsythe, W. L., Jr., *Ind. Eng. Chem.,* 45, 1174–1177, 1953.
22. Carr, R. L., *Chem. Eng.,* 72, 163–168, 1965.
23. Yokoyama, T. and Urayama, K., *J. Res. Assoc. Powder Technol. Jpn.,* 6, 264–272, 1969.
24. Suzuki, A. and Maruko, O., *Funsai (The Micromeritics),* 18, 80–88, 1973.
25. Miwa, S. and Shimizu, S., *Funsai (The Micromerilics),* 16, 103–116, 1971.
26. Yano, T., Terashita, K., and Yamazaki, T., *Funsai (The Micromeritics),* 16, 96–102, 1971.
27. Tsunakawa, H., *J. Soc. Powder Technol. Jpn.,* 27, 4–10, 1990.
28. Tsunakawa, H., *J. Soc. Powder Technol. Jpn.,* 30, 318–323, 1993.
29. Oshima, R., Zhang, Y., Hirota, M., Suzuki, M., and Nakagawa, T., *J. Soc. Powder Technol. Jpn.,* 30, 496–501, 1993.
30. Suzuki, A. and Tanaka, T., *Powder Technol.,* 6, 301–308, 1972.
31. Yoshida, T. and Kousaka, Y., *J. Res. Assoc. Powder Technol. Jpn.,* 6, 194–201, 1969.
32. Nash, J. H., Leiter, G. G., and Johnson, A. P., *Ind. Eng. Chem. Prod. Res. Dev.,* 4(2), 140–145, 1965.
33. Jimbo, G., in *Jitsuyo Funryutai Purosesu to Gijutsu,* Kagaku Kogyo, Tokyo, 1977, p. 186.
34. Hayashi, S., *J. Res. Assoc. Powder Technol. Jpn.,* 6, 286–291, 1969.
35. Yokoyama, T., Urayama, K., Naito, M., Kato, M., and Yokoyama, T., *KONA,* 5, 59–68, 1987.
36. Terashita, K., Umeda, K., and Miyanami, K., *J. Soc. Powder Technol.,* 27, 457–462, 1990.

3.11 Blockage of Storage Vessels

Kunio Shinohara
Hokkaido University, Sapporo, Japan

Hiroshi Takahashi
Muroran Institute of Technology, Muroran, Japan

Among various kinds of bridging phenomena, blockage of storage vessels is dealt with here. It causes stoppage of solids flow and thus leads to unstable operation of powder handling processes, shortages of production capacity, and degradation of product quality. It is one of the serious problems in the operation of storage and supply, including breakage of vessels, unpredictable discharge rate of solids, nonuniform residence time and distribution of particles, and segregation of solids mixtures.

3.11.1 PHENOMENA AND FACTORS

Based on particle properties, blockage phenomena are classified into three kinds. One is a geometrical interlocking of coarse particles at an outlet of the container. Statistically, it occurs when the opening size is smaller than several times the particle diameter. It is based on friction between particles and/or a vessel wall. The particle shape also affects the flow criterion. The second kind of blockage is a mechanical stable arch of cohesive fine powders formed inside the bed. It is directly governed by solids pressure and strength of powder mass and, thus, is closely related to the dimension and shape of the container, storage weights, filling methods, and storage periods. The third blockage is caused by caking with solid bonds due to chemical reaction or physical change, or by an adhered layer to the wall due to electrostatic or physicochemical attractive force. It stems mainly from powder properties, irrespective of the vessel shape. Thus, in addition to the chemical and physical constitution, it is affected by the water content of the powder, humidity and temperature of the operating atmosphere, period of resting time, and solids pressure exerted.

From the operational viewpoint, the blockage is sometimes caused by tapping, vibration, and aeration, in addition to gravity. Hammering and vibration are often useful to promote solids flow due to impact force but are dangerous in that they can compact and strengthen the powder bed at the same time. Proper conditions of intensity and frequency of impact are necessary. Aeration from below the bed can support the arch against discharge, though it is effective to reduce wall friction and caking owing to bed expansion. The air pressure and the air flow rate or the aeration period are relevant to this type of blockage.

3.11.2 MECHANISMS AND FLOW CRITERIA

Interlocking of Coarse Particles

Even cohesionless particles are able to form a stable arch at a converging section in the container. It happens when the particles are so arranged as to support the weight of solids above the arch by the frictional force on the wall surface and still not collapse the arch by the reaction force. Thus, the critical opening size ratio to the particle diameter depends hardly at all on particle density but considerably on particle shape. The following are known experimental equations.

For general granules,[1]

$$\frac{D_s}{d_s} = 1.8 + 0.038\varphi_s^{1.8}$$ (11.1)

$$\frac{D_c}{d_s} = 2.3 + 0.071\varphi_s^{1.8}$$ (11.2)

where D_s and D_c are the minimum dimensions of a slit and a circular orifice, respectively, d_s is a specific surface diameter of the particle, and φ_s is the shape coefficient based on the specific surface of the particle of circle equivalent diameter, d_c; that is, as compared with Equation 11.1, experimental results with a two-dimensional hopper indicated more variation in $\varphi_s > 10$, and that the critical dimension was smaller for spheres and cylinders due to easy orientation for discharge around the outlet, but larger for prisms due to plane contact with each other.

For coarse sands and pebbles,[2]

$$D_o = \frac{(1+c)}{2c} j (d_p + 0.081) \tan \varphi_i$$ (11.3)

where D_o is the diameter of a circular orifice or the shortest side length of a rectangular outlet, and c is the length ratio of the long side to the short side. j is a constant of 2.6 for uniform size or 2.4 for general granules, d_p is the arithmetic mean diameter of the largest and smallest particles, and φ_i is the angle of internal friction or the repose angle.

In any case, this type of blockage is a rather particular case in storage vessels and seldom occurs in practice except on a screen of aperture size near the particle's.

Arching of Cohesive Powders

Fine powders tend to behave as a continuum due to particle cohesion, and the flow criterion is estimated on the basis of powder mechanics and flow models.

In case of gravity flow of cohesive powders, there are two types of flow obstruction, stable arching and piping. Let us consider the criterion to the formation of a cohesive arch in mass-flow bins at first. Assume the following: The solids material consists of a stack of self-supporting free arches, one upon another, and the element arch has uniform vertical thickness T, as shown in Figure 11.1. The arch is formed along the direction of maximum principal stress in the passive state of stress. Perpendicular to the direction of the arch, the minimum principal stress equals zero, because a stress-free surface is assumed. In practice, the material would collapse in the arch due to stress from above, and thus the self-supporting arch would define an upper bound of the critical span for the blockage.

Consider a conical hopper with a smooth wall. For the material of the element arch about to fail within itself at the wall, Mohr's circle M_R is shown in Figure 11.2, passing through the origin of T-σ coordinates and being tangential to the powder yield locus (PYL). The wall yield locus (WYL) is assumed to be linear. Both can be determined by shear cell test.[3] φ_w is the angle of material–wall friction. f_c is the unconfined yield strength of the material as a function of the consolidating stress, which can be determined by shear cell test. The shear stress acting on the wall surface is $\overline{ST}$. The vertical plane makes α as a half of the hopper apex angle anticlockwise from the wall plane, equivalent

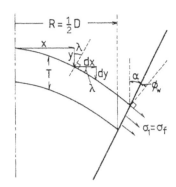

FIGURE 11.1 Arch of uniform thickness.

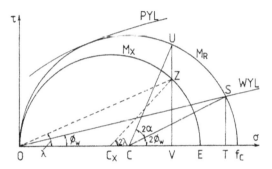

FIGURE 11.2 Mohr's circles along free arch.

to 2α from point S to point U along the circle M_R. Therefore, the shear stress on the vertical plane at the foot of the arch is $\overline{UV}$, which is expressed as $\overline{UV} = (f_o/2)\sin 2(\alpha + \varphi_w)$. Since sum of the vertical forces is given as $\pi DT(\overline{UV}) = (\pi D^2/4)T\rho_b g$, the limiting diameter D_c for a free arch is obtained as[4]

$$\frac{f_c}{\rho_b g D_c} = \frac{1}{2\sin 2(\alpha + \varphi_w)} \tag{11.4}$$

where ρ_b and g are the bulk density of the material and acceleration due to gravity, respectively. Thus, the minimum diameter for discharge D_c depends upon the values of powder properties like f_c and φ_w and hopper geometry like α. In the case of a plane-walled wedge-shaped hopper, the limiting span for the slit opening S can be derived as a half of D_c, given by Equation 11.4. In order to evaluate the critical diameter free from blockage, it is necessary that the strength–stress characteristics of powder, f_c vs. σ_s, and the major consolidating stress, σ_f, set up in the arched field are known.

In the case where both PYL and WYL are given by Coulomb-type equations, the limiting diameter is derived by the same consideration as mentioned above.[5]

$$D_c = \frac{4C_s(1+\sin\varphi_i)\sin 2(\alpha+\beta)}{\rho_b g \cos_i} \tag{11.5}$$

$$\beta = \frac{\varphi_w + \sin^{-1}(\sin\varphi_w / \sin\varphi_i)}{2} \tag{11.6}$$

where C_s is the cohesive shear strength of powder and φ_i is the angle of internal friction, defined by PYL: $\tau = \sigma \tan \varphi_i + C_s$. As for WYL, it is assumed that the cohesive tensile strength of powder and that between powder and wall plate are equal (WYL: $\tau = \sigma \tan \varphi_w + C_w$, $C_w = C_s \cot \varphi_i \tan \varphi_w$). The properties C_s, φ_i, and ρ_b are a function of the void fraction, and these increase with consolidating stress. Therefore, the limiting diameter usually increases with the hopper apex angle as well as the bed height.[5]

The horizontal normal stress σ_x is assumed to be constant across the arch. Describing Mohr's stress circle at the horizontal distance x from the apex of the arch by M_x in Figure 11.2, the vertical plane at x is characterized by the intersection Z of the line $\overline{UV}$ and M_x due to constant $\sigma_x = \overline{OV}$ across the arch. The angle λ between the major principal plane and the vertical plane in Figure 11.1 corresponds to the angle $\angle ZC_xE = 2\lambda$ on Mohr's circle M_x. Accordingly, the following relation holds.

$$\frac{dy}{dx} = \tan \lambda = \frac{\overline{ZV}}{\overline{OV}} = \frac{T_{xy}}{\overline{OV}} \tag{11.7}$$

The shear stress T_{xy} over the vertical plane at x is given by taking the vertical force balance as $T_{xy} = \rho_b g x / 2$, and the relation $\overline{OV} = (f_c/2)(1 + \cos 2(\alpha + \varphi_w))$ is derived from Mohr's circle. Hence, the profile of the free arch is parabolic, using Equation 11.4.

$$y = \frac{x^2}{D_c} \frac{\sin 2(\alpha + \varphi_w)}{1 + \cos 2(\alpha + \varphi_w)} \tag{11.8}$$

In the case of a hopper with a rough wall, the value of 45° for $\alpha + \varphi_w$ in Equation 11.4 and Equation 11.8 is chosen so as to maximize the vertical component of the supporting forces at the foot of the arch, that is, $\overline{UV}$ in Figure 11.2.

The following is Jenike's method[3,6,7] to determine the critical span of the outlet for flow–no flow criterion. Figure 11.3 illustrates the profiles of the stresses along the wall that are required for

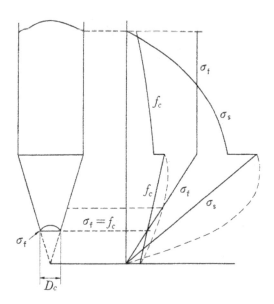

FIGURE 11.3 Determination of critical size of outlet. [From Jenike, A. W., *Powder Technol.*, 1, 237, 1967/1968. With permission.]

the design calculation, where σ_f is the major principal stress set up in a passive stress field during flow, and f_c is the unconfined yield strength representing the strength of the self-supporting arch. As the powder fills the bin, it becomes consolidated under the major consolidating stress (i.e., the major principal stress in the active stress fields σ_s). σ_s after filling is also shown in Figure 11.3. For each value of the major consolidating stress σ_s, there is a corresponding unconfined yield stress f_c. Wherever σ_f exceeds f_c, the powder does not have sufficient strength to support an arch. Therefore, the minimum size of the outlet for mass flow to continue free from arching is determined by the condition

$$\sigma_f = f_c \tag{11.9}$$

The condition that ensures steady gravity flow is $\sigma_f > f_c$, and blockage with a stable arch is $\sigma_f \leq f_c$. Jenike defined flow function (FF) and flow factor (ff) as

$$FF = \frac{\sigma_s}{f_c} \tag{11.10}$$

$$ff = \frac{\sigma_s}{\sigma_f} \tag{11.11}$$

FF depends on the powder property and is determined by shear cell tests as a function of major consolidating stress σ_s. ff represents the stress condition set up in a hopper indicating the flowability from the hopper. The critical diameter D_c is evaluated by

$$\frac{\sigma_f}{\rho_b g D_c} = \frac{1}{H(\alpha)} \tag{11.12}$$

$H(\alpha)$ is given in graphical form in Figure 11.4, which is derived from the refined arch analysis by considering the variation in thickness of the arch.[3,6,8] In the case of a slit aperture of length L longer than three times the width B, D_c represents B. ff is also given in graphical form as a function

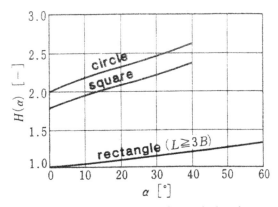

FIGURE 11.4 $H(\alpha)$ versus α for conical and pyramidal hoppers. [From Jenike, A. W., *Bull. Utah Eng. Exp. Station*, 108, 1961; Jenike, A. W., *Bull. Utah Eng. Exp. Station*, 123, 1964. With permission.]

of effective angle of internal friction δ, φ_w, and α, as shown in Figure 11.5. Reading the value of ff from Figure 11.5 with a given set of δ, φ_w, and α, we can draw a straight line, $\sigma_f = (1/ff)\sigma_s$, on $\sigma_f - \sigma_s$ rectangular coordinates. The critical value of σ_f defined by Equation 11.9 is obtained from the intersection of the above-mentioned ff straight line and the FF curve by plotting f_c against σ_s, which is obtained by shear tests. The critical span D_c is then calculated from Equation 11.12. The assumption of constant ff in the hopper means that the profiles of σ_f and σ_s are similar, as shown in Figure 11.3. However, if the profile of σ_s is given as a broken line, the value of ff would not be constant. In such a situation the critical span evaluated would considerably differ from that obtained by the above method.[9]

For a steep hopper with a smooth wall, Walker's method is also useful to predict the critical span. The major principal stress exerted during flow, σ_f, is generally proportional to the span D.[4]

$$\sigma_f = K\left(\frac{D}{2}\right)\rho_b g \tag{11.13}$$

The value of K is determined theoretically. Equation 11.4 means that the material can bridge if the unconfined yield strength develops to the critical value f_c required for arching due to consolidation during flow under the major principal stress σ_f. Based on Equation 11.4 and Equation 11.13, Walker defined the critical flow factor of the hopper FF_c for the arch to be formed by

$$FF_c = \frac{\sigma_f}{f_c} = K\sin 2(\alpha + \varphi_w) \tag{11.14}$$

The critical unconfined strength f_c is obtained from the intersection of the FF_c line and the FF curve. Substituting A into f_c in Equation 11.4, the critical diameter D_c can be predicted by

$$\frac{A}{\rho_b g D_c} = \frac{1}{2\sin 2(\alpha + \varphi_w)}; \quad \alpha + \varphi_w \le 45° \tag{11.15}$$

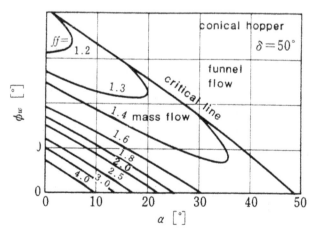

FIGURE 11.5 ff as φ_w versus α for conical hoppers. [From Jenike, A. W., *Bull. Utah Eng. Exp. Station*, 108, 1961; Jenike, A. W., *Bull. Utah Eng. Exp. Station*, 123, 1964. With permission.]

For the steep and smooth-walled hopper, such a procedure as is described above is easier than Jenike's method, because the unknown factor for the calculation is only A.

In funnel flow the solids flow down through a channel formed within the material. According to Jenike's method,[6,7] a lower bound of the critical diameter, D_f, for no-piping is estimated by

$$D_f = \frac{G(\varphi_t)}{\rho_b g}\sigma_f \tag{11.16}$$

where φ_t is a static angle of internal friction. The function $G(\varphi_t)$ and a flow factor $ff_p(=\sigma_s/\sigma_f)$ for no-piping are reproduced in Figure 11.6 and Figure 11.7. The value of σ_f in Equation 11.16 is obtained by drawing the ff_p line on the FF graph and finding the intersection, which enables D_f to be predicted. Nevertheless, piping flow would appear in practice, when the powder head descends lower than a critical level.

Based on a block-flow model,[10,11] the flow criteria are generally derived under gravity, tapping, and aeration. The model assumes that the portion of the powder directly above the outlet initiates discharge as blocks of aperture size. The minimum opening size of a conical hopper, D_o, is obtained under gravity and tapping as

$$\begin{aligned}
D_o = {} & \frac{4}{\rho_b g(1+I)}\left[C_s+\frac{k\mu_i b}{a}\left(1-\frac{1-Y_o^{a+1}}{(a+1)(1-Y_o)}\right)\right] \\
& +\frac{4k\mu_i H}{a-1}\left(\frac{1+Y_o}{2}-\frac{1-Y_o^{a+1}}{(a+1)(1-Y_o)}\right)
\end{aligned} \tag{11.17}$$

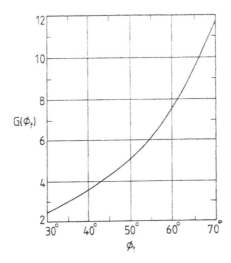

FIGURE 11.6 Function $G(\varphi_t)$. [From Arnold, P. C., McLean, A. G., and Roberts, A. W., in *Bulk Solids,* TUNRA Bulk Solids Handling Research Associates, Australia, 1980, p. 3E11, p. 3E30. With permission.]

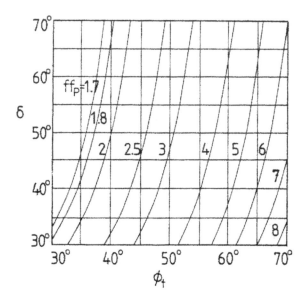

FIGURE 11.7 Flow factor for no piping, ff_p. [From Arnold, P. C., McLean, A. G., and Roberts, A. W., in *Bulk Solids,* TUNRA Bulk Solids Handling Research Associates, Australia, 1980, p. 3E11, p. 3E30. With permission.]

where

$$a = 2\mu_w \cot\alpha\left(\sin^2\alpha + k\cos^2\alpha\right) \qquad (11.18)$$

$$b = -2C_w \cot\alpha \qquad (11.19)$$

$$k = \frac{1 - \sin\varphi_i}{1 + \sin(\varphi)} \qquad (11.20)$$

and Y_o = (height of outlet from hopper apex)/(height of solid surface from hopper apex, H), I is intensity of tapping impact, $\mu_i = \tan\varphi_i$, $\mu_w = \tan\varphi_w$, and C_w is the cohesive shear strength between the powder and the wall. The equation indicates that a higher bed requires a larger outlet and higher intensity of impact.

3.11.3   METHODS OF PREVENTING BLOCKAGE

The methods of preventing blockage can be devised on the basis of the above-mentioned mechanisms, and some of them have been adopted industrially. But, sometimes, in reality they may require inverse treatments. Careful and perspective attention to driving discharge against resistance should, therefore, be paid in practice, so that the most appropriate way can be chosen for individual purposes.

In designing storage vessels, a bigger outlet is the most effective, and other contrivances are also useful, such as asymmetric shapes with a vertical wall, a steeper wall for the hopper portion, smooth surfaces with smaller angles of wall friction, avoidance of acute corners, and proper inserts.

Adjustments of particle properties are preferable, such as uniform and large size, less moisture content, and increased flowability with additives and particle surface treatments, besides smaller amounts and shorter times of storage, if possible.

Mechanical operations and devices are also helpful, such as arch breakers with fluidization and air jets, eccentric vibrating mortors, hammering, agitating breakers with screws, chain conveyors and so forth.

REFERENCES

1. Langmaid, R. N. and Rose, H. E., *J. Inst. Fuel*, 30, 166, 1957.
2. Pikon, J., Sasiadek, B., and Drozdz, M., *Process Technol. Int.*, 17, 888, 1972.
3. Jenike, A. W., *Bull. Utah Eng. Exp. Station*, No. 108, 1961.
4. Walker, D. M., *Chem. Eng. Sci.*, 20, 975, 1966.
5. Aoki, R., in *Kagaku Kogaku Binran*, Maruzen, Ed., Society of Chemical Engineers, Tokyo, p. 1002.
6. Jenike, A. W., *Bull. Utah Eng. Exp. Station*, No. 123, 1964.
7. Jenike, A. W., *Powder Technol.*, 1, 237, 1967/1968.
8. Arnold, P. C., McLean, A. G., and Roberts, A. W., in *Bulk Solids*, TUNRA Bulk Solids Handling Research Associates, Australia, 1980, p. 3E11, p. 3E30.
9. Tsunakawa, H., *J. Soc. Powder Technol. Jpn.*, 18, 405, 1981.
10. Shinohara, K. and Tanaka, T., *Chem. Eng. Sci.*, 30, 369, 1975.
11. Shinohara, K. and Tanaka, T., *Ind. Eng. Chem. Process Des. Dev.*, 14, 1, 1975.

3.12 Segregation of Particles

Kunio Shinohara

Hokkaido University, Sapporo, Japan

3.12.1 DEFINITION AND IMPORTANCE

In almost all industrial processes in which particulate materials are handled, mixtures that differ in particle properties are often subjected to relative movements. When each component (which may have the same physical properties but different chemical properties) is randomly dispersed throughout the particle bed in a microscopic sense, the result is referred to as "well mixed." But if one of the key components congregates or is condensed locally in the macroscopic sense, it is called "segregation," as a sort of separation phenomenon in the dense phase. The mechanisms by which mixing and segregation occur are identical in principle. In mixing, the scale of scrutiny and the partial composition are important as an index of the quality of mixing, but in segregation the location of segregated species is of main concern.

Segregation is usually undesirable when product homogeneity is required, but it may be helpful for separation. Especially in the chemical, pharmaceutical, agricultural, and smelting industries, segregation causes serious problems such as uneven quality of fertilizers and tablets, fluctuating packet weight, low mechanical strength of compacts and abrasives, poor refractory materials, and low rates of contact and reaction. In sieving or classification, segregation during stratification or dispersion of particles facilitates subsequent separation.

3.12.2 RELATED OPERATIONS

Segregation is encountered in various operations accompanied by deformation or flow of solids mixtures.

Gravitational Operations

In piling up solids on a plane for storage, the component of lower flowability, being smaller, denser, more angular, more frictional, more cohesive, or less resilient particles of relatively smaller fraction (here called a segregating component), collects around the pouring point and forms an inverted conical portion inside the conical heap.[1] While, in the initial feeding of solids mixture at a greater rate from a higher position into the narrower cylinder, the larger component of continuous size distribution becomes rich at the center of the bottom plate due to stronger impact.[2] In moving beds and storage vessels such as hoppers and bins, the segregating component is deposited in a core region of shape similar to the container after central filling. The mixture flows down with a nearly flat top surface in a mass-flow hopper of apex angle smaller than about 30° and exhibits little segregation during discharge. But in a funnel-flow hopper of larger apex angle, the central region above the outlet first discharges with an excess of the segregating component collected due to segregation in filling, and then the other component in the peripheral portions follows. This causes serious separation with a lapse of discharge time. In the operation of a blast furnace the combined segregation takes place in part of a moving bed and a kind of segregation in filling occurs in the bottom dead zone. The size segregation is usually remarkable.[3]

Flowing out from a chute linked perpendicularly to a lower conveyor belt, the segregating component tends to settle down near the bottom and the remainder to the outer side of the belt.[4]

Within a shear testing cell, finer particles move through a deforming bed under large shear strain and congregate near the bottom. Particle size ratio and shape affect the percolation velocity markedly, but normal stress does not.[5]

Rotational Operations

In a rotating horizontal cylinder at a low rate of revolution, the segregating component congregates in the core region of its cross section, and segregation bands extend in the axial direction with stripes of the excess component and the mixture with critical composition. In the case of a rotating horizontal cone, the fine component collects near the wide end and the coarse one near the opposite end, but at a high speed of revolution the order becomes reversed, as is often the case with a conical ball mill.[6] Inside a V-shaped mixer the denser component is apt to settle down in any stagnant region.[7]

Vibratory Operations

In a container under tapping[8] or an inclined trough under vibration[9] smaller particles tend to sink to the bottom, while larger ones float up through the bed. An intermediate layer is present in the mixture. Inverse density segregation of a mixture as lower glass and upper lead particles of the same size will take place in a vertically vibrated cylinder at a higher critical velocity amplitude of 1.0 m/s below the frequency of 90 Hz.[10] Differential decrease in mass median diameter of a particle mixture is proportional to the root of the falling height due to tapping impact. Particles circulate upward at the center of the deeper bed. In contrast to larger particles moving to the center, smaller ones descend faster in the vicinity of the cylindrical wall.[11]

Aeration Operations

In a fluidized bed at low air velocity the larger or denser particles become the segregating component as in a liquid. The segregation is caused by a difference in density rather than in size of particles, and it is considerably reduced at larger superficial velocities.[12]

In a spouted bed, larger or denser particles congregate near the center of a circulating annular region above an inlet nozzle. At higher gas velocity the situation in the radial direction becomes inversed, but the segregating component is still abundant around the bottom.[13] In pneumatic conveying pipes similar segregation occurs according to particle density and shape.

3.12.3   FUNDAMENTAL MECHANISMS

Since segregation is the inverse of mixing, the mechanism seems to involve the three basic processes of mixing: diffusion, shear, and convection. It is governed by the physical properties of particles, the dimensions of the vessels, and the operational conditions; but these have not yet been elucidated. As a basis for the general case, some simple segregation mechanisms under gravity alone are therefore considered here.

Trajectory Effect

When a particle of size d and density ρ is projected into a fluid of viscosity μ with initial horizontal velocity v_h, larger or denser ones can fly farther horizontally by a distance L[14]:

$$L = \frac{v_h \rho d^2}{18\mu} \tag{12.1}$$

Thus it is possible to cause size or density segregation due to air drag during flying or feeding. But, the denser particle will collect near the feed point in practice.

Rolling Effect

After a single particle of radius r moves down a plate of height H and inclination I, it travels a distance L on a horizontal plane. L is written for the rolling and sliding motions, respectively, as follows:

$$L = rH(\cos^2 I)\frac{1 - f_r/(r\tan I)}{f_r} \tag{12.2}$$

$$L = H(\cos^2 I)\frac{1 - f_s\tan I}{f_s} \tag{12.3}$$

where f_r and f_s are friction coefficients of rolling and sliding motions, respectively. These suggest that the larger or less frictional particles roll down farther and that only friction segregation takes place during sliding motion, irrespective of particle density.[15] According to the motion analysis of an ellipsoidal particle over an inclined plate, it is possible for the smaller or more spherical particle to fly farther from the bottom end of the plate.[16] But smaller particles will collect near the feed point in practice.

Stumbling Effect

It is easier for larger particles to roll over obstacles of height k and to travel farther due to the larger angular velocity w:

$$w = \left(\frac{2g\left[r - (\sin I)\sqrt{r^2 - (r-k)^2} - (\cos I)(r-k)\right](r^2 + j^2)}{r(r-k) + j^2}\right)^{1/2} \tag{12.4}$$

where j is the radius of gyration. Hence size segregation may occur on the heap surface.[17]

Push-Away Effect

A sphere of diameter d_t and density ρ_t placed over two particles on a two-dimensional plane can push away the bottom ones of d_b and ρ_b in inverse directions against the sliding friction. The criterion is represented by[18]

$$u = \frac{(d_t/d_b)^3(\rho_t/\rho_b)\tan\theta_a}{(d_t/d_b)^3(\rho_t/\rho_b) + 2} \tag{12.5}$$

where θ_a is a half of the top angle of the isosceles triangle formed by the centers of three spheres. This suggests the possibilities of density and size segregation in terms of size and density ratios. But, larger particles usually tend to go farther.

Percolation

Whenever voidage and flowability increase with deformation of the particle bed, the smaller component may easily percolate through the interstices between larger particles. Such size segregation happens under shear strain or while in flow, even for a size ratio as small as 1.53. For the radial dispersion of small particles through the packed bed, the number of small particles, N, having centers within radius r_d at time t is expressed by[19]

$$\frac{r_d}{4E_r t} = \ln \frac{N_o}{N_o - N}$$

(12.6)

where E_r is the radial dispersion coefficient and N_0 is the total number of percolated particles. A discharge equation for hypothetical hoppers formed within a flowing bed gives the relative falling velocity of small particles.[20]

Combined Effects

With respect to surface segregation of a binary solids mixture in a horizontally rotating cylinder, the combined effect of pushing away and percolation is analyzed to give the relationship among size ratio, density ratio, and volume fraction of coarse particles.[21] The ability for a coarse particle to sink in a solids bed based on the potential energy to the work to open a void must be combined with the probability of not finding a void of size large enough for percolation of fines to give a segregation parameter, S,

$$S = \frac{\rho_{rc}}{d_{rc}} \left(\frac{1 + V_c (d_{rc} - 1)}{1 + V_c (\rho_{rc} - 1)} \right) \times \left(1 - \varepsilon \exp \left\{ -\frac{1 - \varepsilon}{\varepsilon} \right. \right.$$

$$\left[\left(1 + \frac{1}{1 + V_c (d_{rc} - 1)} \right)^2 - 1 \right] \right\} \times \left[1 - \varepsilon \exp \left(-3 \frac{1 - \varepsilon}{\varepsilon} \right) \right]^{-1} \right)$$

(12.7)

where ρ_{rc} and d_{rc} are the density and size ratios of a coarse particle to a fine one, respectively, V_c is the volume fraction of coarses, and ε is the voidage of coarses. S will correlate with the mixing index, M.

$$M = 1 + \frac{V_{ci} - V_{co}}{\Delta V_c}$$

(12.8)

where subscripts i and o denote inner and outer regions of the cascading layer in the cylinder, and ΔV_c is the maximum difference of V_c in both regions. Thus, $M = 1$ means perfect mixing, $0 \leqq M \geqq 1$ segregation with excess coarse particles or percolation of fines, and $1 \leqq M \geqq 2$ segregation due to sinking of coarse particles.

Screening Model

When multicomponent mixtures of different sizes flow down an inclined heap surface, the smaller component is screened with the larger one to settle down on the stationary underlayer. Then the oversize fraction f_{ox} of the surface layer is represented by applying Equation 12.9 for continuous sieving as

$$\log \frac{f_{ox,n}}{f_{ox,n-1}} = \log \frac{f_{i,n}}{f_{i,n-1}} - (c_n - c_{n-1}) x$$

(12.9)

where f_i is the oversize fraction in the feed, c is a constant, and the subscripts x and n denote, respectively, the distance from the feed point along the surface and the screen number from the top of the standard screen piled up. The finer component becomes dilute close to the periphery, and size segregation becomes conspicuous with a large c value that increases with smaller feed rate, larger flowability, and wider size range.[1]

Screening Layer Model

When a binary mixture of solids differing in particle size, density, or shape flows down a heap surface of inclination I, the segregating component of initial mixing ratio M_i is depleted in the mixture at the penetration rate Q by a screening process and descends over the stationary underlayer until it reaches the vessel wall. Then the smaller particles are packed into the interstices in the underlayer at the packing rate P. The segregation process is illustrated in Figure 12.1 and described by material balance in each sublayer of thickness h, velocity v, and voidage ε[22] as

$$\frac{\partial h_r}{\partial t} = -\frac{\partial (h_r v_r)}{\partial x} + \frac{Q \cos I}{1 - \varepsilon_r} \frac{1 - M_i}{M_i}$$

$$\frac{\partial h_{rs}}{\partial t} = -\frac{\partial (h_{rs} v_r)}{\partial x} - \frac{Q \cos I}{M_i (1 - \varepsilon_{rs})}$$

$$\frac{\delta h_s}{\delta t} = -\frac{\delta (h_s v_s)}{\delta x} + \frac{Q - P}{1 - \varepsilon_s} \cos I$$

$$\frac{\partial h_{sp}}{\partial t} = -\frac{P \cos I}{1 - \varepsilon_s} \qquad (P = 0 \text{ except size segregation}) \qquad (12.10)$$

As a result, general segregation patterns during the filling of vessels are obtained as the mixing ratio versus the distance from the feed point, x.

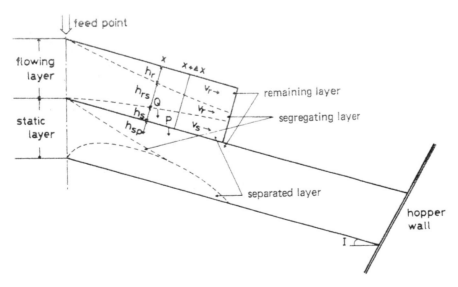

FIGURE 12.1 Segregation process during flow on a solids heap. [From Shinohara, K. and Miyata, S. *Ind. Eng. Chem. Process Des. Dev.*, 23, 423–428, 1984. With permission.]

3.12.4 PATTERNS AND DEGREES

The mechanisms leading to the prediction of patterns or degrees of segregation are useful in engineering applications. For storage and supply, there are three cases or typical patterns of segregation: over heap surface, inside layer, and their combination.

Surface segregation occurs when feeding the mixture onto the heap. The segregation patterns in the heap are described by variations in the mixing fraction of the segregating component with the traveling distance from the feed point along the heap surface. Figure 12.2 depicts a pattern of density segregation by the screening layer model. Then the degree of segregation is represented approximately by the distance ratio of the segregating core region to the heap periphery. The region increases with higher mixing fraction and feed rate. A bigger difference in particle properties yields the smaller region. In the case of a dead man at the bottom of a blast furnace the multipoint feeding generates the peak of the mixing fraction at the middle of the inclined heap surface, as shown in Figure 12.3.[23]

A typical segregation pattern of a five-component mixture is illustrated in Figure 12.4 for the case where the feed composition consisted of equal proportions of all five components. The smallest particles (GB1) were found to collect close to the central feed line of a vessel, and the largest particles (GB5) to collect close to the vessel wall.[24] Thus, the mixture could be considered to consist of a key component of smallest particles (GB1) and another "pseudo" component containing the remaining larger ones. In the case of multicomponent density segregation, the segregation index I_{s} will experimentally be correlated with operational conditions and particle properties.[25]

$$I_{\mathrm{s}} = 1.25 M^{-0.653} F^{-0.150} C^{-0.219} D^{0.631} S^{-0.125} \tag{12.11}$$

where F is the normalized feed rate, C is the number of components, D is the density ratio of densest to pseudolight components, and S is the size ratio of smallest to pseudolarge components.

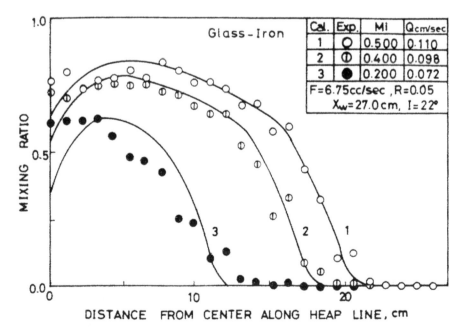

FIGURE 12.2 Density segregation pattern in a two-dimensional bin by central feed. [From Shinohara, K. and Miyata, S. *Ind. Eng. Chem. Process Des. Dev.*, 23, 423–428, 1984. With permission.]

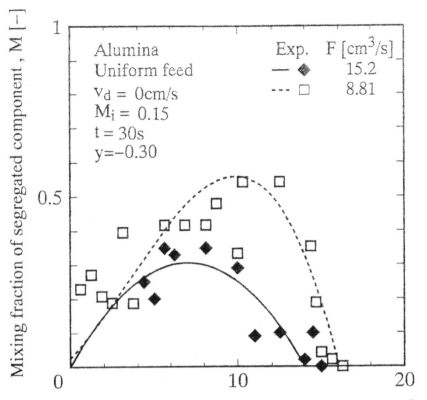

FIGURE 12.3 Size segregation pattern over a static dead man in a blast furnace by uniform feed. [From Shinohara, K. and Saitoh, J. *ISIJ Int.*, 33, 672–680, 1993. With permission.]

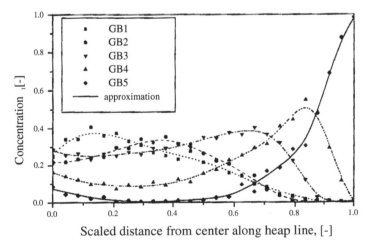

FIGURE 12.4 Size segregation pattern for five-component mixture of equal proportions at $F_t = 12$ cm³/s. [From Shinohara, K., Golman, B., and Nakata, T., *Adv. Powder Technol.*, 12, 33–43, 2001. With permission.]

A kind of segregation index of a multisized particle mixture during the filling of a two-dimensional hopper could, for example, be defined by the concept of partial separation efficiency.[26] The sharpness of separation, which is defined as a slope at the complete or initial mixing fraction of each component, will change along the heap surface. Then the slope of the sharpness of separation at the intersect x_{is} on the scaled distance line will give the intensity of segregation as well as S_{sb}, as illustrated in Figure 12.5. x_{is} corresponds to the segregation zone.

Inside layer segregation refers to separation of the mixture during emptying vessels. The segregation patterns are expressed as the change in mixing fraction of the discharge with time. Figure 12.6 shows one of the patterns where small particles are initially placed on large ones separately. Size segregation of a binary mixture during discharge is affected by the size ratio of particles and the initial and dynamic mixing ratio as well.[27] This could be explained through the microstructure by defining the limiting fines fractions on a number basis, where large particles are completely covered by the fines. The maximum packing density during flow is attained in the range 40–60 wt% of fines, and the peak becomes low and sifts to the larger fraction with a smaller size ratio of coarses to fines. As a result, the discharge rate of the binary mixture, W_{mix}, could be expressed by modification of monosized granules with respect to the so-called empty annulus.[28]

$$W_{mix} = 0.58\rho_{mix}\left(V_f, d_r\right)g^{1/2}(D_o - k_{mix}d_{mix})^{5/2} \qquad (12.12)$$

where k_{mix} varies between 1 and 2 depending on the kind of continuous phase and on the weight mean diameter of the mixture, d_{mix}.

In the case of a ternary mixture of different-sized particles, the continuous phase boundaries can be drawn on a triangle diagram as a function of different size ratios.[29] The discharge rate is estimated by substituting the mean diameter of the continuous phase into d_{mix} in Equation 12.12. Thus, the discharge rate under size segregation will change with the particle size ratios and the mixture

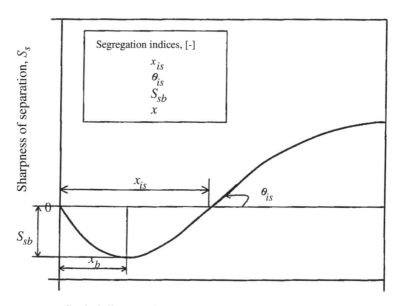

FIGURE 12.5 Definition of segregation indices by sharpness of separation with distance along heap line. [From Shinohara, K. and Golman, B. *Adv. Powder Technol.*, 13, 93–107, 2002. With permission.]

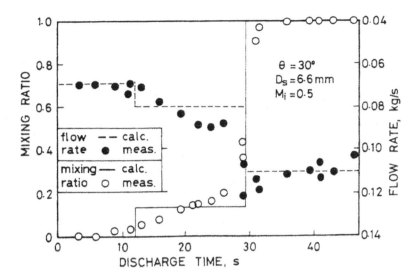

FIGURE 12.6 Size segregation pattern during discharge from a two-dimensional hopper. [From Shinohara, K., Shoji, K., and Tanaka, T. *Ind. Eng. Chem. Process Des. Dev.*, 9, 174–180, 1970. With permission.]

composition through microstructural phase transformation in the flow field. But, the rate or extent of segregation during discharge is not predicted here.

A combination of these two mechanisms gives the general patterns of segregation. They depend largely on the flow pattern inside the bin and include mixing as well as segregation. In these cases the standard deviation of the segregating component or various other definitions of the degree of mixing are usually adopted as a measure. In the case of a moving bed, where the continuous heaping from above and withdrawal from below the bed occur simultaneously, the segregation pattern generated by the filling of the hopper will give rise to the initial condition of a segregation pattern during withdrawal on the basis of the velocity profile assumed by Equation 12.13 and the penetration effect described by Equation 12.14 and Equation 12.15.

$$V_r = -\frac{A}{r}\left[\cos\left[\left\{\varphi_w + \arcsin\left(\frac{\sin \varphi_w}{\sin \varphi_i}\right)\right\}\left(\frac{\theta}{\theta_w}\right)\right]\right]^{\theta_w/c_w} \tag{12.13}$$

$$\frac{\partial}{\partial r}\left(r\left(M\rho_{ti}\right)V_r\right) + \frac{\partial}{\partial r}\left(rQ_m \cos \theta\right) + \frac{\partial}{\partial \theta}\left(Q_m \sin \theta\right) = 0 \tag{12.14}$$

$$Q_m = q_m(M\rho_{ti})^n \tag{12.15}$$

where V_r is the radial velocity, Q_m is the penetration flux, ρ_{ti} true density of the ith component, and q_m and n are constants. As a result, the segregation profiles in the circumferential direction θ and the radial one r are drawn in Figure 12.7.[30]

In the case of a batch operation, where the particles are withdrawn after the filling of the hopper without feeding, for example, the segregation pattern of a binary mixture of different densities during emptying of a two-dimensional hopper, the segregation time zone expands and the peak decreases with increasing initial mixing ratio of the denser component and the feed rate and with decreasing density ratio, as shown in Figure 12.8.[31]

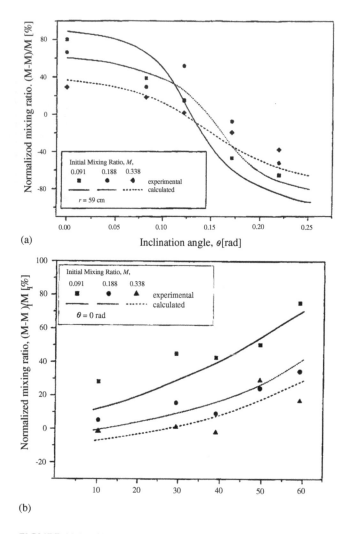

(a)

(b)

FIGURE 12.7 Size segregation profile inside moving bed (a) in the θ direction and (b) in the radial direction. [From Shinohara, K. and Golman, B. *Chem. Eng. Sci.*, 57, 277–285, 2002. With permission.]

3.12.5 MINIMIZING METHODS

As long as solids mixtures are handled, segregation is inevitable. Hence some ways to minimize segregation are proposed on the basis of the segregation mechanisms. The particle properties are usually adjusted beforehand, and if possible, it is effective to make them uniform by agglomeration with mixing, surface treatment, humidification, and so on. It may also be useful to increase the initial mixing fraction of the segregating component.

Regarding design, tall bins or hoppers of small apex angle are preferable to shorten the flow length during filling and to promote mass flow during discharge. But for already installed bins, vertical partitions or proper insertions will be of some help.[32] Multiple outlets, slit openings, and multiple dicharge pipes will improve the flow pattern. Multiple fixed feeding spouts, fixed deflectors, or horizontally moving feeders also prevent heap formation. The following will be useful to reduce segregation such as higher feed rate: less handling, avoidance of vibration, and shorter falling distance. Multiple passes of the mixture through funnel-flow hoppers also results in mixing or reducing segregation to some extent.[33]

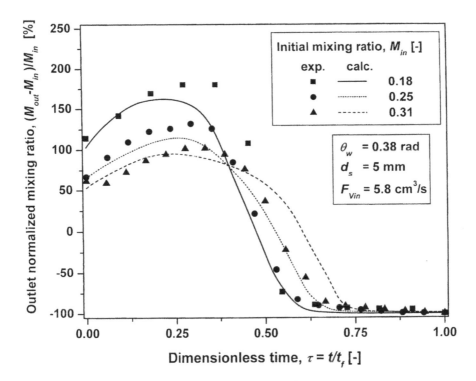

FIGURE 12.8   Effect of initial mixing ratio on outlet time segregation during discharge. [From Shinohara, K. and Golman, B. *Adv. Powder Technol.*, 14, 333–347, 2003. With permission.]

REFERENCES

1. Miwa, S., *Res. Assoc. Powder Technol. Jpn.*, 26, 1–14, 1960.
2. Gotoh, K., Maki, T., and Masuda, H., *J. Soc. Powder Technol. Jpn.*, 31, 842–849, 1994.
3. Van Denburg, J. F. and Bauer, W. C., *Chem. Eng.*, (Sept. 28), 135–142, 1964.
4. Johanson, J. R., *Chem. Eng.*, (May 8), 183–188, 1978.
5. Bridgwater, J., Cook, H. H., and Drahun, J. A., *Ind. Chem. Eng. Symp. Ser.*, 69, 171–191, 1983
6. Sugimoto, M. and Yamamoto, K., *Kagaku Kogaku Ronbunshu*, 5, 335–340, 1979.
7. Yamaguchi, K., *J. Res. Assoc. Powder Technol. Jpn.*, 14, 520–529, 1977.
8. Hayashi, T., Sasano, M., Tsutsumi, Y., Kawakita, K., and Ikeda, C., *J. Soc. Mater. Sci.*, 19, 84–92, 1970.
9. Williams, J. C. and Shields, G., *Powder Technol.*, 1, 134–142, 1967.
10. Ohyama, Y. and Uchidate, I., *J. Soc. Powder Technol. Jpn.*, 35, 218–221, 1998.
11. Yubuta, K., Gotoh, K., and Masuda, H., *J. Soc. Powder Technol. Jpn.*, 32, 89–96, 1995.
12. Rowe, P. N. and Nienow, A. W., *Powder Technol.*, 15, 141–147, 1976.
13. Uemaki, O., Yamada, R., and Kugo, M., *Can. J. Chem. Eng.*, 61, 303–307, 1983.
14. Williams, J. C., *Powder Technol.*, 15, 245–251, 1976.
15. Matthee, H., *Powder Technol.*, 1, 265–271, 1967/1968.
16. Shinohara, K., *Powder Technol.*, 48, 151–159, 1986.
17. Miwa, S., in *Funryutai Kogaku*, Asakura, Tokyo, 1972, p. 222–230.
18. Tanaka, T., *Ind. Eng. Chem. Process Des. Dev.*, 10, 332–340, 1971.
19. Bridgwater, J., Sharpe, N. W., and Stocker, D. C., *Trans. Inst. Chem. Eng.*, 47, T114–119, 1969.
20. Shinohara, K., Shoji, K., and Tanaka, T., *Ind. Eng. Chem. Process Des. Dev.*, 9, 174–180, 1970.
21. Alonso, M., Satoh, M., and Miyanami, K., *Powder Technol.*, 68, 145–152, 1991.
22. Shinohara, K. and Miyata, S., *Ind. Eng. Chem. Process Des. Dev.*, 23, 423–428, 1984.

23. Shinohara, K. and Saitoh, J. *ISIJ Int.,* 33, 672–680, 1993.
24. Shinohara, K., Golman, B., and Nakata, T., *Adv. Powder Technol.,* 12, 33–43, 2001.
25. Shinohara, K., Golman, B., and Mitsui, T., *Powder Hand. Proc.,* 14, 91–95, 2002.
26. Shinohara, K. and Golman, B., *Adv. Powder Technol.,* 13, 93–107, 2002.
27. Arteaga, P. and Tuzun, U., *Chem. Eng. Sci.,* 45, 205–223, 1990.
28. Beverloo, W. A., Leniger, H. A., and Van de Velde, J., *Chem. Eng. Sci.,* 15, 260–269, 1961.
29. Tuzun, U. and Arteaga, P., *Chem. Eng. Sci.,* 47, 1619–1634, 1992.
30. Shinohara, K. and Golman, B., *Chem. Eng. Sci.,* 57, 277–285, 2002.
31. Shinohara, K. and Golman, B. *Adv. Powder Technol.,* 14, 333–347, 2003.
32. Peacock, H. M., *J. Inst. Fuel,* 11, 230–239, 1938.
33. Kawai, S., *Bull. Fac. Eng. Kanazawa Univ.,* 2, 187–194, 1959.

3.13 Vibrational and Acoustic Characteristics

Jusuke Hidaka
Doshisha University, Kyotanabe, Kyoto, Japan

In many treatments of powder in industrial processes, mechanical vibration and sonic energy have been used to increase the efficiency of the operation, and to enhance the transport rates in various industrial operations such as emulsification, drying, and agglomeration.

Furthermore, acoustic noise from powder industrial processes has been used for instrumentation of the state variables in the powder industrial processes. Sound, generally, is radiated by the vibration of a surface, the rapid dilation of fluid medium, and the formation of fluctuating eddies in a fluid. In the instrumentation of a particulate system, impact and frictional sound between particles are important to measure the state variables of the processes.

3.13.1 BEHAVIOR OF A PARTICLE ON A VIBRATING PLATE

A plate with an incline of β to the horizontal plane is vibrating with amplitude A, and angular frequency ω in direction AB as shown in Figure 13.1.

Then, the x and y components of displacement of the plate are given, respectively, by

$$x = A \sin \omega t \cos(\alpha + \beta), \quad y = A \sin \omega t \sin (\alpha + \beta) \qquad (13.1)$$

and the velocity and acceleration of the plate are described as follows:

$$
\begin{aligned}
x &= A\omega \cos \omega t \cos(\alpha + \beta) \\
y &= A\omega \cos \omega t \sin(\alpha + \beta) \\
x &= -A\omega^2 \sin \omega t \cos(\alpha + \beta) \\
y &= -A\omega^2 \sin \omega t \sin(\alpha + \beta)
\end{aligned}
\qquad (13.2)
$$

The acceleration b of a plate along the direction AB and the component c of the acceleration in gravitational direction are described as follows:

$$x^2 + y^2 = b = \pm a\omega^2 \sin \omega t$$
$$c = b\frac{\sin(\alpha + \beta)}{\cos \beta} \qquad (13.3)$$

The maximum value of acceleration of the plate is $r\omega^2$ when $\omega t = \pi/2$, and then

$$c = \pm r\omega^2 \frac{\sin(\alpha + \beta)}{\cos \beta} \qquad (13.4)$$

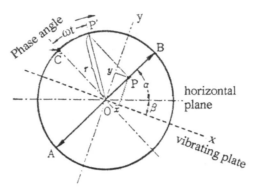

FIGURE 13.1 Vibrating plate.

The ratio of the component c in the gravitational direction of acceleration of the plate to gravitation acceleration g is given by

$$K_v = K \frac{\sin(\alpha+\beta)}{\cos\beta} \tag{13.5}$$

where k is centrifugal effect$(= K = r\omega^2/g)$. The relation between displacement and acceleration of the vibrating plate is illustrated in Figure 13.2.

For $K_v > 1$, a particle on the vibrating plate leaves the plate with the velocity of the plate at that time. When K_v equals one, the phase angle ϕ_L is given by

$$r\omega^2 \sin\phi_L \frac{\sin(\alpha+\beta)}{\cos\beta} = g$$

$$\sin_L = \frac{1}{\left(r\omega^2/g\right)\langle\sin(\alpha+\beta)/\cos\beta\rangle} = \frac{1}{K_v} \tag{13.6}$$

The particle contacts with the vibrating plate at $0 < \varphi < \varphi_L$ and jumps out of the plate at $\varphi_L < \varphi$. The coordinate (x_L, y_L) of the colliding point between the flying particle and the plate is given by

$$x_L = r \sin \varphi_L \cos(\alpha+\beta)$$
$$y_L = r \sin \varphi_L \sin(\alpha+\beta) \tag{13.7}$$

The coordinate (x', y') of the flying particle that jumped out of the vibrating plate at velocity v_L $(= r \omega\cos\varphi_L)$ is described as follows

$$x' = x_L + v_L (\cos\alpha/\cos\beta)$$
$$y' = y_L + v_L \sin(\alpha+\beta)\theta' - (1/2)g\,\theta'^2 \cos\beta \tag{13.8}$$

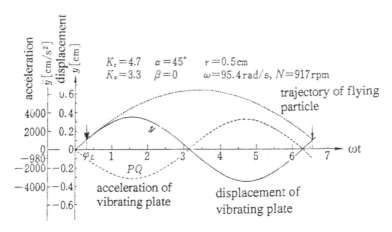

FIGURE 13.2 Acceleration, displacement of vibrating plate, and motion of particle on the plate.

where θ' is the time when the particle jumps out of the plate.

We obtained the time and phase angle at the maximum height of the flying particle as follows:

$$\theta'_m = \frac{v_L \sin(\alpha + \beta)}{g \cos \beta}$$

$$\varphi_m = \varphi_L + \frac{v_L \sin(\alpha + \beta)}{g \cos \beta}\omega = \varphi_L + \frac{\cos \varphi_L}{\sin \varphi_L} \qquad (13.9)$$

Then, the flying duration of particle is given by $(\varphi_0 - \varphi_L)/\omega$, and during the flying, the particle takes the distance along the vibrating plate as follows:

$$v_L \left(\frac{\cos \alpha}{\cos \beta}\right)\frac{(\varphi_0 - \varphi_L)}{\omega} \qquad (13.10)$$

where φ_0 is the phase angle at colliding point of the flying particle with vibrating plate.

If the phase angle at the colliding time equals $\varphi_L\ (=\varphi_0)$, the particle jumps again out of the vibrating plate as soon as the flying particle collides with the plate. The displacement of vibrating particle is shown in Figure 13.2.

3.13.2 BEHAVIOR OF A VIBRATING PARTICLE BED

The behavior of a vibrating bed depends on the particle size distribution, particle density, frequency, and amplitude of the applied vibration.[1] The flow behavior of a vibrating bed can be classified into four patterns, as shown in Figure 13.3. The type in Figure 13.2a denotes uniform consolidation, the type in Figure 13.3b denotes the surface flow near the side wall, the type in Figure 13.3c shows inward circulation, and the type in Figure 13.3d shows outward circulation.

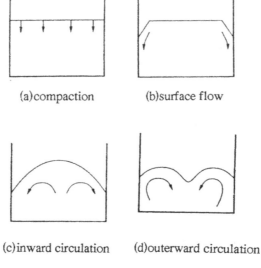

(a)compaction   (b)surface flow

(c)inward circulation   (d)outerward circulation

FIGURE 13.3 Flow pattern of a vibrating powder bed.

3.13.3   GENERATING MECHANISM OF IMPACT SOUND BETWEEN TWO PARTICLES

Impact sound from spherical particles stems from the particle surfaces suddenly accelerated by the colliding particles.[2-4] An acceleration waveform acting on two particles colliding elastically can be obtained using Hertz's elastic contact theory described below. When spherical particles 1 and 2, having diameters of a_1 and a_2, respectively, as shown in Figure 13.4, collide with each other at relative speed v_0, the following equation holds for the elastic deformation $\xi(t)$ and deformation acceleration $\ddot{\xi}(t)$:

$$\ddot{\xi}_{(t)} = - q_1 q_2 \xi_{(t)}^{3/2} \tag{13.11}$$

where

$$q_1 = \frac{m_1 + m_2}{m_1 m_2}, q_2 = \frac{4}{3\pi}\left(\frac{1}{\delta_1 + \delta_2}\right)\sqrt{\frac{a_1 a_2}{a_1 + a_2}}$$

$$\delta_{1,2} = \frac{1 - v_{1,2}^2}{\pi E_{1,2}} \tag{13.12}$$

and m_1 and m_2 are the masses of particle 1 and 2, v_1 and v_2 are their Poisson's ratios, and E_1 and E_2 are Young's moduli. Elastic deformation $\xi(t)$ can be expressed approximately as

$$\xi_{(t)} = \xi_m \sin \frac{1.068 v_0 t}{\xi_m} \tag{13.13}$$

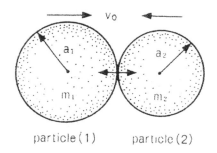

FIGURE 13.4 Collision between two elastic particles.

where ξ_m is the maximum deformation of the colliding particles and can be expressed as

$$\xi_m = \left[\frac{15\pi v_0^2 (\delta_1 + \delta_2) m_1 m_2}{16(m_1 + m_2)}\right]^{\frac{2}{5}} \left(\frac{a_1 + a_2}{a_1 a_2}\right)^{\frac{1}{5}}$$

(13.14)

Assuming that particle deformation acceleration equals gravity center acceleration, the acceleration waveform $[\alpha_{(t)} = 1/2 \cdot \xi_{(t)}]$ of the particle surfaces due to collision can be obtained from Equation 13.11 and Equation 13.13.

Let us determine the sound pressure waveform (impulse response) radiated by a particle when unit impulse acceleration acts on it. The velocity potential $\phi(r,\theta,t)$ of sound generating when a spherical particle Figure 13.5b oscillates reciprocatingly along the z axis at a speed of $U_0 \exp(j\omega t)$ satisfies the following equation:

$$\frac{\partial^2 \varphi}{\partial r^2} + \frac{2\partial\varphi}{r\partial r} + \frac{1}{r^2 \sin\theta} \frac{\partial}{\partial\theta}\left(\sin u \frac{\partial\varphi}{\partial\theta}\right) + k^2 \varphi = 0$$

(13.15)

where k is the wave number. When one has solved Equation 13.15 using the boundary condition that indicates that the speed of the particle surface equals that of the medium, the following velocity potential is obtained:

$$\varphi' = \frac{a^3 U_0}{r^2} \frac{(1+jkr)\cos\theta \cdot e^{j[\omega t - k(r-a)]}}{\left[2(1+jka) - k^2 a^2\right]}$$

(13.16)

Based on Equation 13.16, we can obtain the sound pressure radiated by the particle subjected to unit impulse acceleration as follows, using the method employed by Koss et al.:

$$P_{imp} = \frac{\rho_0 Ca \cos\theta}{r}\left[\cos\left(\frac{C}{a}t'\right) - \left(1 - \frac{a}{r}\right)\sin\left(\frac{C}{a}t'\right)\right]e - \left(\frac{C}{a}t'\right)$$

(13.17)

where $t' = t - (r - a)/C$, ρ_0 is medium density, while C is sound speed.

The pressure waveform of sound radiated by a particle when subjected to a given acceleration $\xi(t)$ can generally be expressed as follows:

$$P(r, \theta, t) = \int_0^1 P_{imp}(t-\tau)^* \xi(\tau) d\tau \tag{13.18}$$

If we approximate the acceleration resulting from Equation 13.11 and Equation 13.13 by $\xi(t) = 1/2 \cdot \xi_m \sin bt$ for easier calculation, we obtain the following equation from Equation 13.17 and Equation 13.18 where $b = \pi/T$, and T is the contact time between two colliding particles:

When $0 < t' < T$,

$$P(r_i, \theta_i, t_i) = \frac{\rho_0 \alpha_m a_i^3 \cos \theta_i}{8(b^4 + 4l_i^4)} \frac{1}{r_i^2} \left\{ \left(\frac{2r_i}{a_i} - 1 \right) \left[(8l_i^3 b - 4l_i b^3) \cos bt_i' + 8l_i^2 b^2 \sin bt_i' \right] \right.$$

$$-4b^4 \sin bt_i' - (8l_i^3 b + 4l_i b^3) \cos bt_i'$$

$$+\left(\frac{2l_i}{a_i} - 1 \right) \left[(4b^3 l_i - 8bl_i^3) \cos l_i t_i' - (8bl_i^3 + 4b^3 l_i) \sin l_i t_i' \right] \exp(-l_i t_i')$$

$$+\left[(4b^3 l_i - 8bl_i^3) \cos l_i \left(t_i' - \frac{\pi}{2l_i} \right) - (8bl_i^3 + 4b^3 l_i) \sin l_i \left(t_i' - \frac{\pi}{2l_i} \right) \right] \exp(-l_i t_i')$$

$$+\frac{\rho_0 \alpha_m a_i^3 \cos \theta_i}{2r_i^2} \sin bt_i' \tag{13.19}$$

When $T < t'$,

$$P(r_i, \theta_i, t_i) = \frac{\rho_0 \alpha_m a_i^3 \cos \theta_i}{8(b^4 + 4l_i^4)} \frac{1}{r_i^2} \left\{ \left(\frac{2r_i}{a_i} - 1 \right) \left\{ \left[(4b^3 l_i - 8bl_i^3) \cos l_i (t_i' - T) \right. \right. \right.$$

$$-(8bl_i^3 + 4b^3 l_i) \sin l_i (t_i' - T) \right] \exp\left[-l_i (t_i' T) \right]$$

$$+\left[(4b^3 l_i - 8bl_i^3) \cos l_i t_i' - (4b^3 l_i + 8bl_i^3) \sin l_i t_i' \right] \exp(-l_i t_i') \}$$

$$-\left[(8bl_i^3 - 4b^3 l_i) \cos l_i \left(t_i' - T - \frac{p}{2l_i} \right) \right.$$

$$+(8bl_i^3 + 4b^3 l_i) \sin l_i \left(t_i' - T - \frac{p}{2l_i} \right) \right] \exp\left[-l_i (t_i' - T) \right]$$

$$-\left[(8bl_i^3 - 4b^3 l_i) \cos l_i \left(t_i' - \frac{\pi}{2l_i} \right) + (8bl_i^3 + 4b^3 l_i) \right.$$

$$\left. \sin l_i \left(t_i' - \frac{\pi}{2l_i} \right) \right] \exp(-l_i t_i') \right\} \tag{13.20}$$

where the subscript i is an index of the spherical particle, $l_i = C/a_i$. Sound pressure $P(r, \theta, t)$ from two colliding particles at the observation point $M(r, \theta)$, shown in Figure 13.5, must allow for the sum of sound pressure radiated from particles 1 and 2.

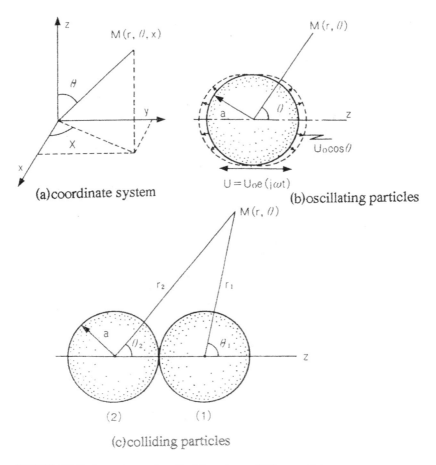

FIGURE 13.5 Impact sound emitted from two colliding particles.

$$P\left(r,\theta,\text{t}\right) = P\left(r_1,\theta_1,t_1\right) + P\left(r_2,\theta_2,t_2\right) \tag{13.21}$$

Figure 13.6 shows the comparison of a measured pressure waveform and a calculated one. The relationship between the peak pressure of sound P_m and the impact velocity v_0 is shown by the following equation:

$$P_m \propto v_0^{6.5} \tag{13.22}$$

The frequency of maximum peak f_m in the frequency spectrum of the impact sound is shown experimentally by the following equation as shown in Figure 13.7.

$$f_m \propto D_p^{-1} \tag{13.23}$$

where D_p is particle diameter ($a = D_p/2$).

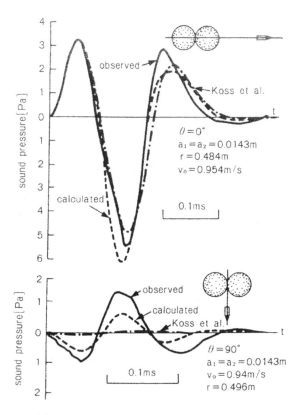

(a)calculated pressure waveforms of impact sound

FIGURE 13.6 Pressure waveforms and frequency spectra of impact sound.

Based on these relations, we can measure the state variables of powder industrial processes with the acoustic noise emitted from the processes.

3.13.4 FRICTIONAL SOUND FROM A GRANULAR BED

The sound pressure radiated from the friction between two particles is ordinarily too small to be detected with a microphone, but a granular bed of particles in contact with each other can radiate the frictional sound of which the parameters relate to the frictional properties of the bed.[5,6]

In the flow and deformation of a granular bed, successive intermittent slip lines are formed in the bed, and the periodicity of the intermittent shear yields relates to the frictional properties of the bed. The frictional sound results from the dilation of particles in the vicinity of the slip line. As results of the dilation, the surface of a granular bed vibrates as a piston, and the sound is radiated into space.

Consider, for the sake of simplicity, a plate penetrating at constant velocity into a dense, infinitely wide bed. In the bed, the successive slip lines as shown in an X-ray radiograph (Figure 13.8) are formed, and the intervals, Δh, between the slip lines are shown by the following equation:

$$\Delta h = \frac{B}{2}\left\{ \tan\left(\frac{\pi}{4} + \frac{\varphi_s}{2}\right) - \tan\left(\frac{\pi}{4} + \frac{\varphi_e}{2}\right)\right\} \qquad (13.24)$$

where B is the diameter of the penetrating plate, ϕ_s is the static angle of internal friction, and ϕ_e is the dynamic angle of internal friction. The sound pressure from the surface of the bed can be

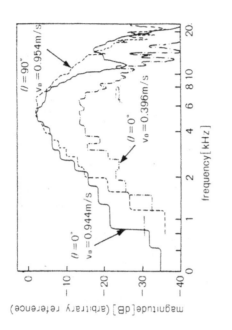

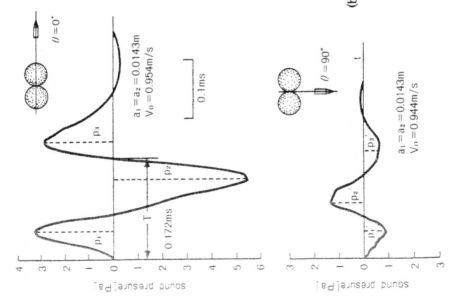

(b) measured pressure wave forms and frequency spectra of impact sound

FIGURE 13.6 (Continued)

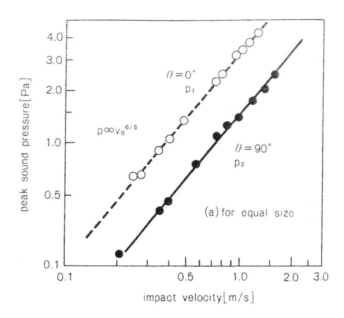

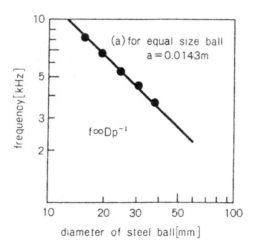

FIGURE 13.7 Relationship among peak pressure and frequency of impact sound, impact velocity, and particle diameter.

estimated by the model of sound emission from a piston in an infinite wall. Let the circular piston of radius, a, as shown in Figure 13.9, oscillate with a small amplitude and with a velocity given by $U = U_0 \exp(-j\omega t)$. The sound pressure p along the center axis is given by the following equation:

$$\widetilde{P} = \mathrm{j}^2\, \mathrm{r}\, CU\, \sin\frac{kr}{2}\left(\sqrt{1+\alpha^2-1}\right)\exp\left\{\omega t - \mathrm{j}\frac{kr}{2}\left(\sqrt{1+\alpha^2-1}\right)\right\} \qquad (13.25)$$

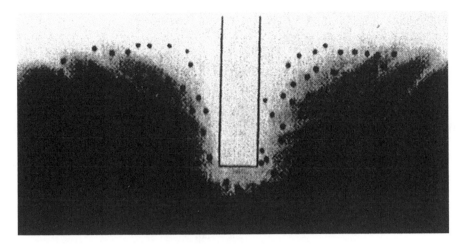

FIGURE 13.8 X-ray radiograph of slip lines in a powder bed.

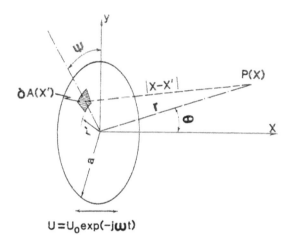

FIGURE 13.9 Sound emission by a flat, circular piston in a finite wall.

where k is the wave number, $\alpha = a/r$. Thus, the relationship between the peak sound pressure P_m and the radius a is shown as follows:

$$P_{\mathrm{m}} = 2\rho\, CU_0 \left| \sin\frac{kr}{2}\left(\sqrt{1+\alpha^2}-1\right)\right| \qquad (13.26)$$

The increase Δa in the radius of the oscillating surface of the bed at each shear yield is given as follows:

$$\Delta a = \frac{\Delta h}{\sin\left(\dfrac{\pi}{4}-\dfrac{\varphi_{\mathrm{s}}}{2}\right)} \qquad (13.27)$$

The amplitude of a vibrating surface is shown by the equation

$$A = \frac{\Delta h}{b}\left(\frac{1}{\lambda_2 - \lambda_1}\right) \tag{13.28}$$

where b is the coefficient for describing the average distance between particles in the bed, λ_1 is the linear concentration of particles before rupture, and λ_2 is the linear concentration of particles after rupture.

Figure 13.10 is a comparison the measured sound pressure waveform with the calculated one. The frequency of the frictional sound can be estimated by the following equation:

$$f_a = \frac{\omega}{2\pi} = \frac{V_p}{\Delta h} \tag{13.29}$$

where V_p is the penetrating speed of the plate into the powder bed. The parameters of this sound closely relate to the frictional properties and the formation of rupture layers in the granular bed.

3.13.5   VIBRATION OF A SMALL PARTICLE IN A SOUND WAVE

The vibration of a small particle in a sound wave is of some importance in a variety of areas, such as in acoustic coagulation of aerosols.[7,8] This motion has been studied by Brandt and Hiedemann.[9]

The amplitude X_m of a fluid medium can be written in terms of angular frequency ω, wavelength λ, maximum amplitude A_m, and the distance from the node of the wave to the specified particle l.

$$X_m = A_m \, \sin\, \omega t \, \sin\left(\frac{2\pi l}{\lambda}\right) \tag{13.30}$$

The force acting on the oscillating sphere can be presented by the well-known Stokes law for drag as follows:

$$R = 3\pi\mu\, D_p\Delta = 3\,\pi\mu\, D_p\left(\frac{dx_m}{dt} - \frac{dx_p}{dt}\right) \tag{13.31}$$

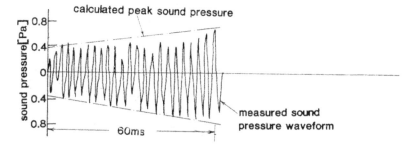

FIGURE 13.10   Comparison of the measured sound pressure waveform with the calculated one.

where μ is the coefficient of viscosity of a medium, Δv is the relative velocity between the particle and the fluid, X_p is the amplitude of the particle.

The motion equation of the particle in a sound wave is given by

$$\rho_p \frac{\pi D_p^3}{6} \frac{d^2 X_p}{dt^2} = 3\pi\mu D_p \left[\omega A \cos \omega t \sin\left(\frac{2\pi l}{\lambda}\right) - \frac{dX_p}{dt} \right] \tag{13.32}$$

The solution yields

$$X_p = \frac{A_m \sin(2\pi l / \lambda) \sin(\omega t - \beta)}{\left[1 + \left(\frac{\pi\rho D_p^2 f}{9\mu}\right)^2 \right]^{0.5}} \tag{13.33}$$

where f is the frequency of the sound wave, and β is the phase difference of the particle and medium in oscillation. The ratio X_p/X_m yields

$$\frac{X_p}{X_m} = \frac{1}{\left[1 + \left(\frac{\pi\rho D_p^2 f}{9\mu}\right)^2 \right]^{0.5}} \tag{13.34}$$

where X_m is the amplitude of the medium. The phase difference β is given by

$$\beta = \tan^{-1}\left(\frac{\pi\rho_p D_p f}{9\mu}\right) \tag{13.35}$$

Figure 13.11 shows the calculated values of X_p/X_m.

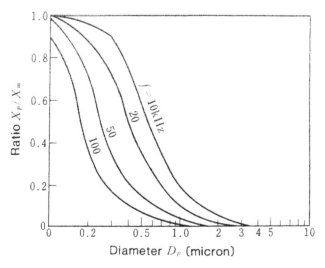

FIGURE 13.11 Ratio of the amplitude of a vibrating particle and medium.

3.13.6 ATTENUATION OF SOUND IN A SUSPENSION OF PARTICLES

When a sound wave propagates in a dilute suspension, it is attenuated by the particle-wave interaction, such as the oscillation of the particle and the scattering of the incident wave.[10,11] If a plane sound wave propagates in the suspension, then the decay of sound pressure is given by

$$P = P_0 \exp(-\gamma x) \tag{13.36}$$

where P_0 is the sound pressure of incident wave at $x = 0$, g is the damping coefficient, and x is the distance along the path. The decay of energy can also be expected to be exponential:

$$E(t) = E_0 \exp(-2\gamma_e t) \tag{13.37}$$

where E_0 is the value of E at $t = 0$.

If κ is the energy removal rate due to one sphere, the energy dissipation rate per unit volume E is given by

$$E = n \cdot k \tag{13.38}$$

where n is the number of particles per unit volume. For very small particles, the energy loss rate due to scattering is very small. Therefore the power that must be spent instantaneously to maintain the motion is represented by $\eta p(\nu p - \nu)$ in terms of the velocity of particle v_p and that of fluid v. Thus, the average energy dissipation rate per unit volume is

$$E = n \cdot \eta_p (v_p - v) \tag{13.39}$$

For a plane sound wave, the energy per unit volume E_0 equals $(1/2)\rho v_0^2$ where v_0 is the velocity amplitude of the medium. Thus the coefficient of amplitude attenuation is obtained by Equation 13.37

$$\gamma = \frac{R(v_p - v)}{\rho C v_0^2} = n \cdot \frac{Re\left[\tilde{R}(\tilde{v}_p - \tilde{v}) \right]}{2\rho C v_0^2}$$

$$\sqrt{\omega a^2 / 2\lambda} \le 1, \; \rho_p \, \rho \le 1 \tag{13.40}$$

For

$$\tilde{R} = 6\pi\mu (\tilde{v}_p - \tilde{v}), \; \tilde{v}_p - \tilde{v} = v_0 \frac{j\omega\tau}{1 - j\omega\tau} \tag{13.41}$$

From Equation 13.40 and Equation 13.41, we obtain

$$\gamma = \frac{3\eta\mu_k na}{C} \frac{\omega^2 \tau^2}{1 + \omega^2 \tau^2} \tag{13.42}$$

where μ_k is the coefficient of kinetic viscosity of the medium. Rewriting Equation 13.42 in terms of and $\tilde{\alpha} = \gamma C / \omega$ and $\psi = (4/3)a^3 n(\rho_p/\rho)$, Equation 13.42 gives

$$\tilde{\alpha} = \frac{1}{2} \psi \frac{\omega^2 \tau^2}{1 + \omega^2 \tau^2} \qquad (13.43)$$

The above equation shows that the attenuation per wavelength has a maximum at $\omega\tau = 1$. Thus, the sound waves having the frequency of the order of τ^{-1} are damped the most by the attenuation. The frequency for maximum attenuation per wavelength is given by the following equation:

$$f = \frac{1}{4\pi} \frac{9\mu_k}{a^2} \frac{\rho_p}{\rho} \qquad (13.44)$$

REFERENCES

1. Tamura, S. and Aizawa, T., *Int. J. Mod. Phys.*, 7, 1829, 1993.
2. Hidaka, J., Shimosaka, A., and Miwa, S., *KONA*, 7, 4, 1989.
3. Hidaka, J., Shimosaka, A., Ito, H., and Miwa, S., *KONA*, 10, 175, 1992.
4. Hidaka, J. and Shimosaka, A., *Int. J. Mod. Phys.* 7, 1965, 1993.
5. Hidaka, J., Kirimoto, Y., Miwa, S., and Makino, K., *Int. Chem. Eng.*, 27, 514, 1987.
6. Hidaka, J., Miwa, S., and Makino, K., *Int. Chem. Eng.*, 28, 99, 1988.
7. Hoffman, T. L. and Koopman, G. H., *J. Acoust. Soc. Am.*, 101, 3421, 1997.
8. Song L., Koopman, G. H., and Hoffman, T. L., 116, 208, 1994.
9. Brandt, O. and Hiedemann, *Trans. Faraday Soc.*, 32, 1101, 1936.
10. Temkin, S., *Elements of Acoustics*, John Wiley & Sons, New York, 1981, p. 455.
11. Richard, S. D., Leighton, T. G., and Brown, N. R., *J. Acoust. Soc. Am.*, 114, 1841, 2003.

Part IV

Particle Generation and Fundamentals

4.1 Aerosol Particle Generation

Richard C. Flagan
California Institute of Technology, Pasadena, California, USA

An aerosol is a suspension of solid or liquid particles in a gas. Aerosols are generated by a wide range of methods involving either mechanical dispersion of condensed phase materials or condensation from the vapor phase. In this section, we focus on the formation of particles by physical processes, spray technologies, and condensation. The formation of aerosol particles by chemical reactions is discussed in the next section. Table 1.1 classifies the different physical methods for generation of aerosol particles. Representative methods are described below.

4.1.1 CONDENSATION METHODS

Generation of Mists

Mists are suspensions of liquid particles produced by vapor condensation onto foreign nuclei. Mists may be generated inadvertently, for example, by releasing hot, vapor-laden gases into a cooler atmosphere, leading to visible plumes. Mists are also generated in the laboratory as a source of relatively uniform organic particles. Table 1.2 lists some of the organic liquids that are commonly used for laboratory mists, along with their physical properties. The classical apparatus for production of laboratory mists is the Sinclair–LaMer aerosol generator[1] that is illustrated in Figure 1.1. The apparatus consists of a vapor source, a source of foreign nuclei, a reheater, and a condenser. The vaporizer is maintained at a constant temperature, typically 100°C to 200°C, to provide a uniform concentration of vapor in the carrier gas. Solid nuclei are generated by a spark source, as illustrated, or by vaporization of mineral salts such as NaCl or AgCl, or metals, commonly silver. The seeded vapor then passes through the reheater to homogenize the mixture. Subsequent cooling in the condenser causes the vapor to condense on the seed particles. The size distribution of the resulting aerosol droplets depends on the temperatures of the vaporizer, reheater, and cooling tube, the number concentration of the seed nuclei, and the gas flow rate. The breadth of the size distribution is influenced primarily by the temperature of the reheater. This generator can produce relatively monodisperse aerosols with geometric standard deviations, σ_G, on particle diameter, D_p, less than 1.2 can be produced with this generator. Typical particle sizes range from 0.03 μm to 2 μm. Number concentrations may be as high as 10^7 to 10^8 particles/cm³.

One important variant of the Sinclair–LaMer generator is the Rapaport–Weinstock generator,[2] which uses atomized droplets as the liquid to be evaporated. Impurities or seed materials in the atomized solution cause each droplet to leave a nonvolatile residue after evaporation. Each such residue particle then acts as a condensation nucleus for subsequent vapor condensation.

Fume Generation

Fumes are aerosols of ultrafine solid particles produced by evaporation of metals or salts and subsequent condensation of their vapors, usually in high-temperature systems. Electrical resistance heaters,

TABLE 1.1   Aerosol Generation Methods

Methods	Particles	Material
Condensation aerosols		
Seeded liquid Evaporation/condensation	Droplets	DOP, DOS, DBP, glycerol
Solid evaporation Condensation	Solid particles	Metals, mineral salts, stearic acid
Gas evaporation method (low pressure)	Nanoparticles	Metals
Liquid atomization		
Pressure atomizers Nebulizers Ultrasonic nebulization Electrospray Spray pyrolysis	Droplets	DOP, mineral oil
Mechanical power resuspension	Solid particles	Powders, beads, fibers
Chemical reaction		
Gas-phase chemical reaction	Droplets	H_2SO_4, photochemical smog
	Solid particles	NH_4Cl, GaAs, Si, SiC, TiO_2
Combustion	Solid particles	Carbon black, TiO_2, SiO_2, smoke soot

TABLE 1.2   Organic Liquids Used in the Laboratory Generation of Mists

Compound	Formula	Molecular Weight	Specific gravity	Melting point (°C)	Boiling point (°C)
Dibutylphthalate (DBP)	$C_{16}H_{22}O_4$	278.4	1.05	-35	340
Dioctylphthalate (DOP)	$C_{24}H_{38}O_4$	390.6	0.99	-55	386
Dioctylsebacate (DOS)	$C_{26}H_{50}O_4$	426.7	0.92	-60	377
Linoleic acid (LA)	$C_{18}H_{30}O_2$	278.5	0.91	—	194
Oleic acid (OA)	$C_{18}H_{34}O_2$	282.5	0.89	14	360
Stearic acid (SA)	$C_{18}H_{36}O_2$	284.5	0.85	71	370

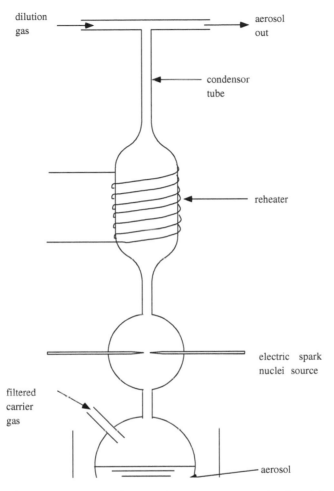

FIGURE 1.1 Evaporation condensation generator developed by Sinclair and LaMer.

high-frequency induction furnaces, and infrared ovens are used to heat materials to the temperature at which they will evaporate. Thermal plasmas and pulsed laser ablation are also used, particularly for compound materials for which composition control becomes a problem when less vigorously heated sources are employed. Homogeneous nucleation and condensation of the vapors occur when the hot, vapor-laden gases are cooled. Particle sizes are generally below 1 μm. Rapid cooling, often by dilution with a cold gas, favors the formation of ultrafine particles ($D_p < 0.1$ μm). A major mechanism of particle growth is Brownian coagulation. In addition to promoting condensation, dilution serves to slow particle coagulation. Nonetheless, agglomerate particles are commonly produced by this method.

Fumes produced by direct evaporation of metals have recently received considerable attention in the production of so-called nanostructured materials. Metals or ceramics produced by consolidation of particles finer than 20 nm diameter have a large fraction of their atoms in grain boundary regions, leading to physical properties that differ dramatically from those of materials with coarser grain structures. These properties have been demonstrated using particulate materials generated by the so-called gas evaporation method in which a metal is evaporated directly into a low-density gas,

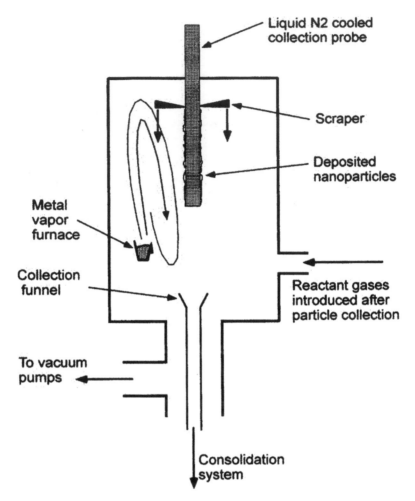

FIGURE 1.2   Gas evaporation method apparatus for synthesizing nanoparticles by evaporation of a metal into a low-density, inert gas. Particles are deposited by thermophoretic diffusion on a cold substrate. Following collection of the nanoparticles, reactant gases such as oxygen may be introduced to produce oxides or other materials. Particles are scraped from the deposition substrate for collection and consolidation.

as illustrated in Figure 1.2. Vapors diffuse rapidly from the hot source into the cold surrounding gas where they homogeneously nucleate to form nanoparticles. The particles grow by coagulation, forming agglomerates that are collected by thermophoretic deposition on a cold substrate. Although the gas evaporation method is a valuable laboratory tool, particle production rates are small. Numerous efforts are focusing on continuous flow systems that operate at higher pressures with short residence times to increase rates of production of nanoparticles.

4.1.2   LIQUID ATOMIZATION

Liquid atomization is the production of droplets in the act of dispersing a bulk liquid into a gas. The production of droplets from a bulk liquid requires work to produce the additional surface area. Fluid energy is supplied either by the fluid being sprayed or by a second fluid. The ratio of the kinetic energy of the surrounding fluid to the energy required to form a droplet is the Weber number

$$We = \frac{\rho V^2 D_p}{\sigma}$$

(1.1)

where V is the relative velocity between the liquid and the surrounding gas, and ρ is the gas density. Droplets form when the Weber number exceeds a critical value. The droplet diameter scales inversely with the square of the velocity and is proportional to the surface tension.

Atomization of a liquid can produce liquid particles directly, or solid materials by evaporation of a liquid solvent to leave a residue of a nonvolatile, solid solute or by chemical reactions. Evaporation may also play a critical role in the production of liquid droplets since solvents may be added to reduce the viscosity or to lower the concentration of the solute residue. The final size of the dried particles is directly related to that of the solution being sprayed and the densities of the two phases, that is,

$$d_{pf} = d_d \, (C_{vs} + I_v)^{1/3}$$

(1.2)

where C_{vs} and I_v are the volume fractions of solute and impurities in the solution, respectively. Impurities in the solution limit the size reduction that can be achieved by spraying dilute solutions.

Pressure Atomizers

Release of a pressurized liquid through an appropriately designed nozzle disperses the liquid as droplets. The range of sizes produced depends on the nozzle design, the fluid properties, and the flow rate of the liquid. Such atomizers are routinely used to deliver liquid fuels to combustion systems ranging from gas turbines to diesel engines to large boilers and furnaces. Pressure atomizers generally produce relatively large droplets, $d_p > 10$ µm, and often much larger, due to the difficulty in producing large velocities in spraying a relatively viscous liquid. Pressure atomizers can, however, efficiently process large volumes of liquid.

Two-Fluid Atomizers and Nebulizers

One way to increase the energy imparted to the liquid and, thereby, to reduce the droplet size is to use a high-velocity gas flow. Air assist and two-fluid atomizers use this technology for large-scale dispersal of a liquid phase into fine droplets. Nebulizers are the laboratory implementation of this technology. The operating principles of the two classes of devices are the same: a high velocity, often sonic or near-sonic, gas flow is blown past the liquid to produce the capillary instabilities that break the liquid into fine droplets. A common laboratory nebulizer is illustrated in Figure 1.3. In this device the Bernouli effect of a high-velocity gas jet blowing past a small tube that supplies the liquid reduces the pressure at the capillary outlet, allowing the liquid to be siphoned from a reservoir. Alternatively, the liquid may be fed by a pump to ensure a constant flow rate. The gas jet breaks the liquid into fine droplets and then impinges on a target. Large particles impact on the target for collection or recycle, so only droplets below a critical size are conveyed from the nebulizer in the carrier gas flow. While properly designed air assist atomizers can disperse the entire liquid feed into a gas flow, nebulizers disperse only a fraction of the liquid. Because of their simplicity of design and operation, however, nebulizers find numerous applications in the laboratory

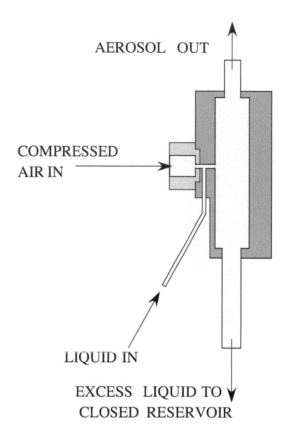

AEROSOL OUT

COMPRESSED
AIR IN

LIQUID IN

EXCESS LIQUID TO
CLOSED RESERVOIR

FIGURE 1.3   Constant rate nebulizer of Liu and Lee[4]
for dispersal of liquid as fine droplets.

and in medicine. Nebulizers typically produce particles in the 1 to 10 μm diameter size range at number concentrations between 10^6 and 10^7 cm^{-3}.

Ultrasonic Nebulizers

Ultrasonic nebulizers such as the one illustrated in Figure 1.4 use high-frequency acoustic energy to disperse a liquid as fine droplets. The size of droplets depends on the applied frequency f (Hz). Mercer[3] suggests an empirical relation for the droplet diameter,

$$d_{\mathrm{d}} = 0.34 \left(\frac{8\pi\sigma}{\rho f^2} \right)^{1/3}$$

(1.3)

where σ and ρ_{l} are the surface tension and the density of the liquid, respectively. For frequencies of 0.1 to 10 MHz, droplet diameters are typically 5 to 10 μm. Ultrasonic nebulizers and atomizers have been applied to aqueous solutions, organic liquids, and even molten metals.

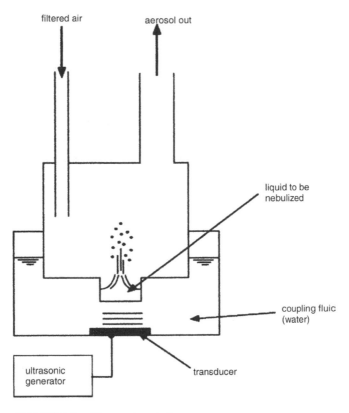

filtered air   aerosol out

liquid to be
nebulized

coupling fluic
(water)

ultrasonic   transducer
generator

FIGURE 1.4   Ultrasonic nebulizer.

Electrospray

The energy needed to produce the droplet surface area can also be supplied electrostatically. When a liquid is sprayed into an electric field, charge separation occurs. The extreme limit of this charge separation is the electrospray, which has found extensive use as a method for the introduction of fragile, high molecular weight species into mass spectrometers, but which has also been proposed for the production of particulate materials,[5] spraying of liquid fuels,[6] and the production of calibration particles. When an electric field is applied to a droplet at the end of a capillary, the droplet will distort into a so-called Taylor cone. This sharp cone is established in the hydrostatic limit when there is no flow through the capillary and it has a half-angle of 49.3. If the electric field is increased above a critical value, flow is induced. Ultimately, this leads to the ejection of liquid from the apex of the cone. That liquid breaks into small droplets that are highly charged. Evaporation of the liquid increases the surface charge density on the droplets. The repulsive forces on a particle carrying q charges balance the surface tension forces at the so-called Rayleigh limit, that is,

$$q = 8\pi \left(\epsilon_0 \sigma R_p^3 \right)^{\frac{1}{2}} \tag{1.4}$$

where ε_0 is the permittivity of free space. As the droplet shrinks below this critical size, it becomes unstable, forms a cusp, and then emits a significant fraction of its charge and a small fraction of its mass as even smaller droplets.[7] A cascade of these Coulombic explosion events may produce finer and finer droplets. The electrostatically driven flow through the electrospray is small. This has limited the use of the electrospray to applications in instruments of analytic chemistry, although arrays of electrosprays can be used to enhance the throughput.[5]

4.1.3   POWDER DISPERSION

Powders can be dispersed into an air flow by a variety of methods. Most systems that are employed for such dispersion consist of a constant rate feeder and an entrainment apparatus. One such system is the Wright dust feeder that is illustrated in Figure 1.5. In this device, powder that has been compacted into a rotating cup is scraped from the compact and entrained into a carrier gas flow. Large agglomerates are removed from the gas flow by an impactor. Another approach to powder dispersion is a fluidized bed in which glass or metal beads of about 100 μm in diameter are fluidized to facilitate the dispersal of a finer, dry powder that is continuously fed into the bed.[8] Fine particles are entrained into the carrier gas and carried out as an aerosol, while coarser particles fall back into the bed. The aerosol generation rate can be controlled through the feed rate of the powder and the carrier gas flow rate.

4.1.4   GENERATION OF MONODISPERSE PARTICLES

Uniformly sized particles are needed to calibrate aerosol measurement instruments, to evaluate filter efficiencies, and for inhalation studies and a wide range of research applications. A variety of sources are used to produce so-called monodisperse particles. Each produces a range of particle sizes. For many applications such as filter testing, a polydisperse aerosol with a geometric standard deviation of 1.2 or 1.3 is sufficiently monodisperse. Such aerosols in the 0.1 to 1 μm size range can readily be generated by vapor condensation (Section 4.1.1). Some nebulizers can produce particles with this degree

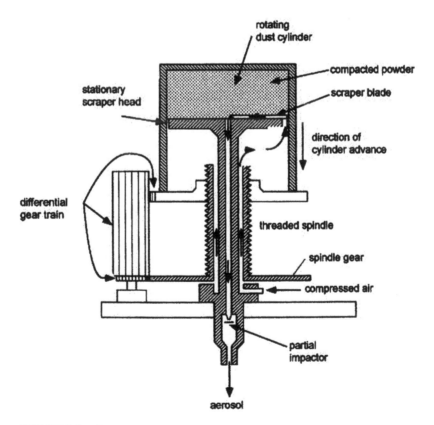

FIGURE 1.5   Schematic of the Wright dust feeder.

of uniformity for sizes ranging from 0.5 μm to several microns in diameter. Production of aerosols of greater uniformity requires special methods.

Polystyrene Lattices

Instrument calibration frequently requires highly monodisperse aerosols. Light-scattering instruments are frequently calibrated in the 0.08 to 10 μm size range using aerosols produced by spraying dilute suspensions of commercially available uniformly sized polymeric particles, most commonly polystyrene latex (PSL). To prevent the formation of agglomerate particles, the PSL suspension must be sufficiently dilute to ensure that the probability is low that a droplet produced during atomization will contain multiple PSL spheres. Furthermore, very high purity solvents (typically water) should be used to dilute the concentrated suspension since nonvolatile solutes may alter the sizes of the dried particles. Even when high purity solvents are used, surfactants that are added to the suspension to prevent agglomeration during storage and other solutes produce small residue particles, so the resulting aerosol generally has a peak at the size of the PSL particles and a broad residue distribution.

Spinning Disk Generators

Spinning disk aerosol generators disperse a liquid that is supplied to the top surface of a rapidly rotating disk.[9] The spinning disk may be driven by a motor or pneumatically. The size of the droplets, D_d, produced is related to the angular frequency, ω and diameter, d, of the disk, that is.,

$$D_d \omega \sqrt{\frac{\rho d}{\sigma}} = K$$

(1.5)

where K is an empirically determined constant, ranging from 3.8 to 4.12. Disks ranging from 2 to 8 cm in diameter are typically operated at angular velocities of 10^2 to 10^5 rad/s. Droplets are typically produce at a rate of 10^6 to 10^8 particles/min and dispersed into a gas flow smaller than 1 m^3 min^{-1}. The standard deviation of the droplet size is often quite small, although much smaller "satellite" particles are sometimes generated as the droplets break from the spinning disk. Some versions of the device eliminate satellite droplets from the product aerosol by surrounding the disk with a small sheath flow. Only those particle that are large enough to penetrate through the sheath flow are carried from the generator in the main flow.

Vibrating Orifice Aerosol Generator

A capillary jet is unstable and will disintegrate into droplets given sufficient time. Although a range of particle sizes is generally produced, a very uniform aerosol can be generated as the capillary jet is excited at an appropriate frequency. The resulting droplet size is determined by the volumetric flow rate of the liquid, Q (cm^3 s^{-1}), and the excitation frequency, f(Hz), namely,

$$D_d = \left(\frac{6Q}{\pi f}\right)^{1/3}$$

(1.6)

Not all frequencies will produce monodisperse particles. Some frequencies will produce one or more satellite particles. A small jet of air impinging on the aerosol jet from the side is commonly used to determine whether a single droplet size is produced (indicated by a single stream in the deflected jet) or multiple sizes (indicated by multiple streams in the deflected jet).

A variety of systems has been constructed to produce monodisperse particles by this method. The most common version, which is shown in Figure 1.6, is the vibrating orifice aerosol generator (VOAG) introduced by Berglund and Liu.[10] In the VOAG, a piezoelectric crystal directly vibrates the orifice, thereby exciting the capillary jet. Numerous other modes of excitation have been used in

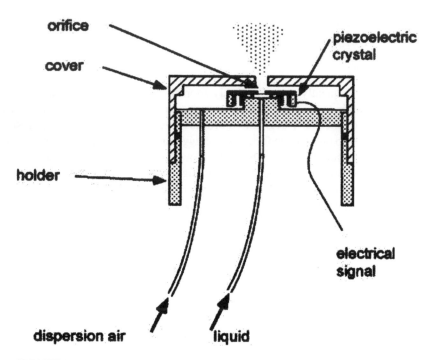

FIGURE 1.6   Vibrating orifice aerosol generator developed by Berglund and Liu. [From Berglund, R. N. and Liu, B. Y. H., *Environ. Sci. Technol.*, 7, 147–153, 1973. With permission.]

other implementations of this technology. Once a suitable frequency is found, an air flow is introduced along with the aerosol through a small orifice, dispersing the aerosol throughout the gas and minimizing coagulation of the uniform particles produced by the VOAG.

Polydisperse Aerosol Classification

Another way to produce an aerosol comprised of uniformly sized particles is to classify a polydisperse aerosol. Below 0.08 µm, the primary method for generating monodisperse aerosols is differential mobility classification.[11,12] In this method, charged particles are caused to migrate across a particle-free sheath flow by an imposed electric field. Particles are classified in terms of their electrical mobilities,

$$Z = \frac{qD}{k_B T}$$

(1.7)

where q is the charge on the particle, D is the particle diffusivity, k_B is the Boltzmann constant, and T is the temperature. A small flow that is extracted through an opening in the electrode opposite from the aerosol entrance downstream of the aerosol entrance carries particles that migrate across the gap between electrodes during the flow time. Ideally, the particles carry only one elementary charge ($q = \pm e$), so the particle mobility provides a direct measure of the diffusivity and, indirectly, of the particle size through the Stokes–Einstein relation

$$D = \frac{kTC_c}{3\pi\mu\delta_p}$$

$$\tag{1.8}$$

where μ is the viscosity, and

$$C_c = 1 + Kn\left(\alpha + \beta\exp\left[-\frac{\gamma}{Kn}\right]\right)$$

$$\tag{1.9}$$

is a slip correction factor that accounts for noncontinuum effects for particles with sizes comparable to or smaller than the mean-free-path of the gas molecules, λ. The empirically determined coefficients are $\alpha = 1.207$, $\beta = 0.596$, and $\gamma = 0.999$.[13] The differential mobility analyzer can produce aerosols ranging from 1 μm to as small as 0.001 μm with geometric standard deviations of 1.1 or smaller, although the distributions broaden for very small particle sizes.[14-16] For particle sizes above about 0.1 μm diameter, multiple charging leads to multiple peaks in the size distribution of the transmitted aerosol, although methods have been developed to minimize this effect.

Classification of particles at larger sizes can be accomplished by inertial separation using a combination of inertial impaction ontos a substrate and a virtual impactor. The size distribution produced by this method is generally broader than that produced by the differential mobility analyzer, although there is no fundamental reason why sharper inertial classification cannot be achieved.

REFERENCES

1. Sinclair, D. and LaMer, V. K., *Chem. Rev.,* 44, 245–267, 1949.
2. Rapaport, E. and Weinstock, S. E., *Experimentia,* 11, 364–363, 1955.
3. Mercer, T. T., *Arch. Intern. Med.,* 13, 39–50, 1973.
4. Liu, B. Y. H. and Lee, K. W., *Am. Ind. Hygiene Assoc.,* 36, 861–865, 1975.
5. Rulison, A. J. and Flagan, R. C., *Rev. Sci. Instrum.,* 64, 683–686, 1993.
6. Gomez, A. and Chen, G., *Combust. Sci. Technol.,* 96, 47–59, 1994.
7. Smith, J. N., Flagan, R. C., and Beauchamp, J. L. J., *Chem. Phys. A,* 106, 9957–9967, 2002.
8. Guichard, J. C., in *Fine Particles,* Liu, B. Y. H., Ed., Academic Press, New York, 1976, p. 173.
9. Mitchell, J. P. J., *Aerosol Sci.,* 15, 35–45, 1984.
10. Berglund, R. N. and Liu, B. Y. H., *Environ. Sci. Technol.,* 7, 147–153, 1973.
11. Knutson, E. O. and Whitby, K. T. J., *Aerosol Sci.,* 6, 443–451, 1975.
12. Knutson, E.O. and Whitby, K. T. J., *Aerosol Sci.,* 6, 453–460, 1975.
13. Allen, M. D. and Raabe, O. G., *Aerosol Sci. Technol.,* 4, 269–286, 1985.
14. Kousaka, Y., Okuyama, K., and Adachi, M., *Aerosol Sci. Technol.,* 4, 209–225, 1985.
15. Kousaka, Y., Okuyama, K., Adachi, M., and Mimura, T. J., *Chem. Eng. Jpn.,* 19, 401–407, 1986.
16. Stolzenburg, M. R., Ph.D. Thesis, University of Minnesota, 1986.

4.2 Generation of Particles by Reaction

Richard C. Flagan
California Institute of Technology, Pasadena, California, USA

Yasushige Mori
Doshisha University, Kyotanabe, Kyoto, Japan

4.2.1 GAS-PHASE TECHNIQUES

Particle synthesis by reactions in the gas phase begins with seed generation, generally by a burst of homogeneous nucleation when reactions produce condensible products or when volatile products are quenched. This may be physical nucleation of a single condensible species, but often involves complex chemical reactions. Once seed particles are present, they may grow by condensation or physical vapor deposition, chemical vapor deposition, or coagulation. The latter mechanism is the collisional aggregation of small particles to form larger ones. The processes involved in particle synthesis from the vapor phase are summarized in Figure 2.1. Because it is a second-order process, coagulation dominates when the number concentrations of particles in the nucleation burst is large, which is normally the case in industrial powder synthesis reactors. Since liquid particles coalesce upon coagulation, spherical particles result unless the viscosity is extremely high. Coagulation of solid particles or of very high viscosity liquids, notably silica, often results in the formation of agglomerate particles comprised of a large number of smaller primary particles. Such agglomerate formation frequently begins after a period of coalescent coagulation in high-temperature regions of the reactor, leading to a relatively uniform primary particle size that is sometimes misinterpreted as evidence that coagulation has ceased. Instead, coagulation accelerates once agglomerates begin to form due to the increased collision cross-section that result from agglomerate formation. Strong bonds may form between the primary particles by solid-phase sintering, leading to hard agglomerates that are often difficult to disperse. Such agglomerates are undesirable for many applications but impart desirable properties in others, notably when used to reinforce soft polymers.[1]

Coagulation and Aggregation Kinetics

Particle growth by coagulation is described by the coagulation equation, also known as the Smoluchowski equation,

$$\frac{dn}{dt} = \frac{1}{2}\int_0^v \beta(\tilde{v}, v-\tilde{v})n(\tilde{v})n(v-\tilde{v})d\tilde{v} - \int_0^\infty \beta(\tilde{v}, v)\, n(\tilde{v})n(v)d\tilde{v} \qquad (2.1)$$

where $n(v)$ is the number concentration of particles with volumes between v and $v + dv$, $\beta(v, \tilde{v})$ is the collision frequency function for particles of volume v with those of volume $\tilde{v}$. The structure of the particles, and their size relative to the mean-free-path of the gas molecules, λ, determines the form for β. Most aerosol synthesis reactors produce small particles at elevated temperatures, so that $R_p < \lambda$ throughout the growth process. Such particles are in the free-molecular regime, where descriptions derived from the kinetic theory of gases apply.

413

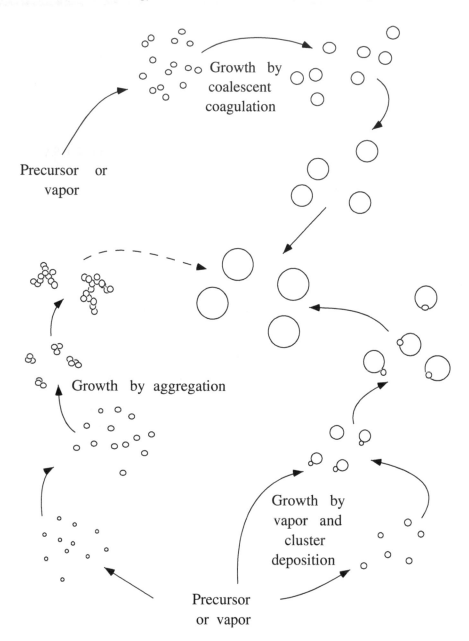

FIGURE 2.1   Mechanisms of particle formation and growth from vapor-phase precursors.

Coagulating liquid particles are generally spherical, but solid aggregates form more complex structures. In many cases the structure of particulate aggregates approaches a self-similar form wherein the particle mass scales with the radius of gyration of the agglomerate, R_g, as

$$m_p \approx m_0 \left(\frac{R_g}{R_0} \right)^{D_f} \tag{2.2}$$

where D_f is the mass fractal dimension, which is generally smaller than the Euclidean dimension of 3, often about 1.8 for highly agglomerated particles, and $m_0 = \rho_p v_0$ is the mass of the primary particle.

The particle size distribution that results when particles grow by coagulation approaches an asymptotic shape known as the self-preserving particle size distribution. For particles that grow as dense spheres, the size distribution is frequently approximated as log normal (even though it is notably asymmetric), with a geometric standard deviation of approximately 1.45 if the particles are much smaller than the mean free path of the gas molecules (free-molecular regime), and 1.35 if they are larger than the mean free path (continuum regime). Agglomerate particles also approach a self-preserving particle size distribution, albeit with a different geometric standard deviation that depends on the agglomerate structure.

The self-preserving particle size distribution model allows one to predict the decay in the total particle number concentration as[2]

$$\frac{dN}{dt} = -2^{1-v} \alpha \kappa \rho v_0^{\frac{2}{3} - \frac{v}{D_f}} \phi^{\frac{v}{D_f} - \frac{1}{2}} N^{\frac{5}{2} - \frac{v}{D_f}} \tag{2.3}$$

where α is a dimensionless constant whose value depends on the fractal dimension and the shape of the self-preserving size distribution ($\alpha = 6.67$ for dense spheres, $D_f = 3$),

$$\kappa = \left(\frac{6 k_b T}{\rho_p} \right)^{\frac{1}{2}} \left(\frac{3}{4\pi} \right)^{\frac{v}{D_f} - \frac{1}{2}} R_g^{2 - \frac{6}{D_f}} \tag{2.4}$$

$v = Min(2, D_f)$, k_b is the Boltzmann constant, and T is the temperature, and ϕ is the volume fraction of particulate material. The mean particle volume is $\bar{v} = \phi/N$.

Aerosols with size distributions that are narrower than the self-preserving distribution can be generated if growth by vapor deposition or condensation dominates over coagulation. This generally requires relatively low particle concentrations, which leads to low production rates, or very short growth times. Charging the particles can inhibit coagulation to facilitate growth by vapor deposition to produce a narrow size distribution.

Agglomeration Control

Highly desirable for applications such as polymer reinforcement, rigid agglomerate particles are deleterious for many others. Because agglomerate particles have a much larger collision cross section than do dense spherical particles of the same volume, once agglomerates begin to form, they quickly scavenge the remaining small dense particles and become the dominant form of particle. In many particle synthesis technologies, particles are initially formed at high temperature where coagulating particles quickly sinter into dense spheres. If the timescale for that coalescence is small compared to the time between collisions, dense particles will be grown. If the coalescence time is very long compared to the time between collisions, aggregates will form, but the necks between the primary particles will remain small, and the agglomerates can be broken apart in later processing. If, however, the two times are comparable, significant necking will occur, and hard agglomerates will be formed. Following high temperatures during the growth phase by rapid quenching to a temperature so low that sintering is insignificant enables formation of powders that can readily be dispersed as primary particles.

Types of Particles Synthesized

Table 2.1 lists some of the many approaches to powder synthesis and the powders produced by those methods. Major commercial aerosol–synthsized particulate materials include carbon black, titanium dioxide, and fume silica, although a much wider range of compositions can be generated. Carbon black is used as structural reinforcement in polymer composites employed in tires and a wide range of other

TABLE 2.1   Methods for Production of Fine and Ultrafine Powders

Method	Description	Powder Charactereristics	Ref.
Fine Powder			
Evaporation and condensation	Vapor of a low-boiling-point metal is cooled.	Zn	16
Melt atomization	Molten metal is sprayed.	Metals: $D_p > 10\ \mu m$	
Spray pyrolysis	Solution containing salt or other precursor compound is sprayed. Solvent is evaporated and precursor is reacted at elevated temperature.	Metal oxide	
Ultrafine Powder			
Gas evaporation method	Metal is vaporized into cool, low-pressure gas.	Metals: $D_p > 20\mu m$	7, 10
Thermal decomposition aerosol reactor	Thermal decomposition of precursor vapor in heated reactor.	Fe, Ni: from carbonyls; Si from SiH_4; Al_2O_3, SiO_2, TiO_2 from alkoxides 5 nm $< D_p < 1\mu m$	19, 20
Aerosol flow reactor	Reaction of vapor-phase precursors in heated flow reactor.	S, SiC, Si_3N_4 AIN, TiN, BC, etc. 5nm $< D_p < 1\mu m$	2, 3
Flame synthesis	Combustion of precursor vapor in hydrocarbon flame. Combustion of precursor vapor in Na flame	Carbon black, TiO_2, SiO_2 10 nm $< D_p < 1\mu m$	25
Thermal plasma reactor	Decomposition of vapor or liquid precursor or direct evaporation of material in thermal plasma leading to particle formation in postplasma region.	Metals, carbides, nitrides, borides, oxides	13, 14, 27, 28
Laser synthesis	Laser-induced reactions of vapor-phase precursors.	Si, SiC, Si_3N_4, WC	4, 5

applications. Titanium dioxide particles of the rutile phase in the near-submicron size range (typically 0.2 to 0.3 μm) have a high refractive index that scatters visible light efficiently. Such particles are used to increase the opacity of paints and pigments. The structures of the particles can vary widely: structural carbon blacks are generally ramified agglomerate particles, while the dense nonagglomerated particles are generally preferred for pigment applications. With the exception of TiO_2 pigment particles, most particles produced from the vapor phase are in the ultrafine size range ($D_p < 0.1\ \mu m$). With potential applications in magnetic storage media, catalysis, ceramics, and electronic devices among many others, ultrafine particles are the subject of considerable interest.

Even smaller particles in the nanometer size range ($D_p < 0.05\ \mu m$) have attracted considerable interest because they can impart unique physical, chemical, electronic, and optical properties to structures fabricated from them. Ceramics and metals produced from such nanoparticles acquire unusual physical properties as a result of the large numbers of atoms that reside in grain boundary regions when the particles are consolidated to form a macroscopic structure. The high surface energies of such materials lower melting points and allow consolidation at much lower temperatures than would be required for composites of larger particles. Quantum confinement of charge carriers in nanoparticles increase the band gaps of semiconductor materials, altering optical, electronic, and chemical properties. These special properties have greatly increased interest in nanoparticle synthesis.

Flame Synthesis

The predominant technology used to produce particles from the vapor phase is flame synthesis. Carbon black, titanium dioxide, and fume silica particles are produced by this method in which a precursor vapor-phase compound is burned to produce condensible liquid or solid-phase products. For carbon black synthesis, combustion of a hydrocarbon fuel (either fuel oil or natural gas) heats the reactants to the high temperature required to drive the chemical reactions and provides the material from which the particles are formed. Soot particles are rapidly produced in the fuel-rich flame and then coagulate to generate agglomerate particles that are quenched and then separated from the gas flow. Titanium dioxide particles are produced by combustion of titanium tetrachloride ($TiCl_4$). The oxidation of $TiCl_4$ is not sufficiently exothermic to raise the reactants to a high enough temperature to achieve rapid reactions, so another energy source is generally required. Most commonly, that energy is supplied by combustion of a hydrocarbon fuel, although thermal plasmas have been used to supply the energy while avoiding the introduction of hydrogen that leads to production of large quantities of HCl vapor that severely aggravate corrosion problems in TiO_2 synthesis reactors. The slow reaction kinetics of the $TiCl_4$ precursor allow chemical vapor deposition to play a more important role in the growth of TiO_2 particles than it does in carbon black synthesis. Combined with the low melting point of TiO_2, that aids in the production of dense particles. Flame synthesis methods can be extended to a wide range of oxide materials. Metallic, carbide, boride, and nitride particles can be produced using more exotic flames. For example, combustion of $TiCl_4$ by sodium vapor has been used to produce $TiBr_2$ particles.[3]

Thermal Reactions of Volatile Precursors

A wide variety of particulate materials can be synthesized by gas-phase reactions of volatile precursors. Combustion reactions were discussed above. Thermal decomposition of metal carbonyls has been used to generate ultrafine Ni and Fe powders. Silicon particles ranging in size from 1 nm to 5 μm have been generated by thermal decomposition of silane gas (SiH_4). Silicon carbide and silicon carbide, which have promising applications for ceramics, have been synthesized by vapor-phase reactions. SiC has been produced by gas-phase reactions of CH_3SiCl_3 or of SiH_4 with C_2H_4. Si_3N_4 is produced by reaction of SiH_4 with NH_3. Both amorphous and highly crystalline particles with sizes ranging from 1 nm to 0.2 μm have been produced by these routes. Moreover, a wide range of compound particles have been synthesized, including particles with a core of one material and a shell of another. As one example, Si nanoparticles encapsulated within SiO_2 have been produced by first growing silicon nanoparticles by silane pyrolysis and then forming an oxide coating on that core. Two coating methods were demonstrated: (1) chemical vapor deposition of tetraethyl orthosilicate and (2) partial oxidation of the outer portion of the silicon particle.

Laser Synthesis of Ultrafine Particles

Laser-induced reactions of precursor gases have been used to produce a variety of particulate materials including Si, SiC, and Si_3N_4.[4,5] Laser irradiation is used to provide the energy needed to drive the chemical reactions. Ultrafine particles have been generated by this route. Nonagglomerated SiC and Si_3N_4 powders were produced when the SiH_4 precursor was first decomposed by laser irradiation, producing molten silicon particles that were later reacted with C_2H_4 or NH_3 that was injected downstream of the primary reaction zone. Agglomerate particles were produced when the reactants were injected simultaneously.

Gas Evaporation Method

Evaporation of a metal into a low-density gas has long been used to produce ultrafine metal particles.[6–10] The material is evaporated into an inert gas where ultrafine particles are produced by vapor condensation as the vapors diffuse into the cool gas surrounding the vapor source. The method has been applied to a wide range of materials. Pure metals have been produced by direct evaporation, as has SiC,[11] producing particles ranging in size from 2 to 200 nm.[7] The size of particles produced depends on the gas and the pressure, with finer particles being produced at lower pressures, consistent with particle growth primarily by coagulation.

Plasma Synthesis of Ultrafine Particles

Thermal plasmas produce temperatures in the range of 5000 to 10,000 K. At such high temperatures, precursor materials can be vaporized, or precursor compounds can be decomposed.[12–14] The resulting vapors condense when the vapors cool, generating large numbers of ultrafine particles by homogeneous nucleation. Gas-phase reactions can be carried out to produce compound materials such as SiC, Si_3N_4, or TiO_2. Particles can be produced in large quantities by this method, with particle sizes ranging from 10 nm to nearly 1 μm.

Lower-intensity plasmas provide a highly reactive environment that has been employed in synthesis of particles with a wide range of compositions from gas-phase precursors. Inductively coupled plasmas have also been used to synthesize highly uniform silicon nanoparticles at slightly reduced pressures.[15] Microwave plasmas have been applied to a wide range of metal and ceramic nanoparticle syntheses.

4.2.2   LIQUID-PHASE TECHNIQUES

When a gradual increase in the concentration of the required materials for particles in solution is produced by the chemical reaction, or the physical techniques, precipitation usually does not occur as soon as the concentration reaches the saturation value.[29] As the concentration reaches a certain degree of supersaturation value, nucleation is induced and then the nuclei ultimately grow into the particles. Methods of preparing powders by the supersaturation of the concentration of the required materials could be classified into the techniques listed in Table 2.2 where inorganic or organic materials are separated.

The chemical technique is the precipitation from homogeneous solution by chemical reaction or chemical equilibrium. On the other hand, the physical technique means that the solution is first divided into small isolated droplets, and then the powder is made by the evaporation of the solvent or by the chemical reaction in the individual droplets. The solvent evaporation methods in inorganic powder and the mechanical dispersion methods in organic powder in Table 2.2 may be of the physical technique.

Inorganic Powder

If the generation rate of inorganic powder is fast, most of the particles will be in the amorphous state or the aggregate state consisting of many small crystals, so that the shape of the powder is nearly spherical. The particles will grow to nearly a single crystal by a slow generation speed.[30] The generation method of the oxide powder generally consists of two steps: formation of precursor salt or hydrated oxide, and thermal decomposition.

Solvent Evaporation Methods

These methods are of physical techniques; that is, the source solution is divided into small droplets of liquid, and the powder is formed by the evaporation of the solvent in the liquid droplets. These methods are applicable for any multicomponent solution and are superior to the precipitation methods, because

TABLE 2.2   Methods of Preparing Powders from Solution

Inorganic Powder

Solvent evaporation methods	Freeze drying
	Spray drying
	Spray pyrolysis
Precipitation methods	Addition of precipitating agent
	Hydrolysis
	Redox reaction
	Decomposition of compound
Special reaction field	Hydrothermal
	Supercritical fluid
	Microemulsion

Organic Powder

Mechanical dispersion methods Mixing methods interfacial reaction	Suspension polymerization
	Emulsion polymerization
	Mini-emulsion polymerization
Capillary methods	Vibration nozzle
	Porous membrane
Phase-separation methods	Soap-free polymerization
	Micro-emulsion polymerization
	Inverse micelle polymerization
	(Nonaqueous) dispersion polymerization
	Coacervation
Other method	Seeding polymerization

they keep homogeneity in the source solution if the solute cannot evaporate.[31] The powders produced are usually spherical and porous, and their size can be controlled by the solute concentration in the feed solution. Spray methods are the usual way to disperse to small droplets. These are called freeze drying, spray drying, and spray pyrolysis, depending on the operating temperature. In the freeze-drying technique, the solutions are atomized into the refrigerant, and then the sublimation of the solvent produces the dry powders. The specific surface area of the powder decreases as the temperature increases in the sublimation process. The spray-drying technique uses the chemical reaction between the droplets and the environmental hot gas. In spray pyrolysis, a solution is atomized in an atmosphere of temperature high enough to decompose the metal salt into the final oxide. Roy[32] reported a laboratory-scale process for the production of powders using this method. Sometimes the high temperature is provided by burning combustible solvents such as alcohol. The metal or alloy powder can be produced directly from a solution by the reduced reaction in hydrogen or nitrogen gases.

Precipitation Methods

In the precipitation methods, homogeneous nucleation and growth can occur in the solution, and then the LaMer model[33] is taken as the principle to produce monodispersed particles, in which the nucleation stage should be separated from the growth stage. Sugimoto[34] reviewed the particle formation process to control size and shape based on this model, with many experimental examples, including the process for organic powders. Recently, some studies for particle formation used new analytical techniques, and the results that were found could be explained by the aggregation model rather than the LaMer model.[35]

Precipitation methods are classified into the addition of precipitating agents, hydrolysis, redox reaction, and decomposition techniques by the reaction type.

Techniques of the addition of precipitating agents are further classified as to the formation of metal hydroxides, coprecipitation, and homogeneous precipitation. The metal hydroxide precipitates from an aqueous solution of metal salt by adjusting certain pH values. When the simultaneous precipitation occurs at a certain pH value from the solution mixed with various metal salts, all metal hydroxides are precipitated, and the polycomponent oxide powder is prepared from the metal hydroxides mixture by thermal decomposition. This technique is called coprecipitation. However, homogeneity of the metal hydroxides mixture is difficult to achieve because of the different conditions for precipitation of each hydroxide. It is necessary to make a multicomponent powder where the precipitated conditions are nearly the same among the metal components, and each precipitation rate is fast. Moreover, it may be necessary to have operating conditions such as the addition of a large amount of precipitating agent and vigorous agitation in the solution. In the special case of compound precipitation, there is a useful precursor compound with a desired composition such as $BaTiO(C_2O_4)_2$ or $CaZrO(C_2O_4)_2$. The homogeneous precipitation technique is a different operation from the above techniques of the addition of precipitating agents, where precipitating agents are produced by a chemical reaction instead of addition in the solution. This technique is superior to precipitated metal hydroxide, because the concentration of the precipitating agent becomes more homogeneous in the solution. For example, urea, thioacetamide, or dimethyl oxalate can be used in hydrolysis as a precipitating agent, instead of ammonia, hydrogen sulfide, or oxalic acid. Sacks et al.[36] reported the formation of aluminum oxide by homogeneous precipitation and thermal decomposition techniques from the mixture of aluminum sulfate hydrate and urea solutions.

In the category of hydrolysis, there is formation from the corresponding metal alkoxides in alcohol solution as well as from metal salt solutions. The technique to produce particles by hydrolysis of metal alkoxides is called the alkoxide method. The particles prepared by this technique are normally amorphous metal oxide and hydrated. This technique has several advantages, such as simple and rapid reaction around room temperature, the easy achievement of high purity of the product because it is free from inorganic ions of all kinds, and the strong possibility of the formation of a complex component product from the mixture of metal alkoxides. Uniform and spherical silica particles are often prepared by this technique, as reported by Stöber et al.[37] The metal hydroxide solution is easily generated by hydrolysis from the metal salt solution, especially with a small ionic radius metal such as Si^{4+} or Al^{3+}, or with a high ionic charged metal such as Fe^{3-} or Ti^{4+}. Metal oxide particles are produced by thermal decomposition. When metal salts are highly purified, the particles with high purity will be produced very easily, and the average size and size distribution of particles can be controlled by an aging process. Matijevic developed many procedures for uniform metal (hydrous) oxide particles, and some are reviewed, for example, the formation of iron oxide.[38]

In the redox reaction technique,[39] fine metal particles, such as noble metals and sulfur, are produced by reducing or oxidizing the metal salt or the metal chelate complex solutions. Many kinds of complex or protective colloid agents are frequently employed to moderate the reaction speed and to stabilize the generated particles, so that highly monodispersed particles are obtained. Matijevic also reported that preparation of fine powders of some noble metals and copper, using this technique, was combined with the following technique, called decomposition of the compound.[40] Some organic compounds such as EDTA, triethanol amine, thioacetamide, and urea can be used in the technique of decomposition of compound. Haruta et al. obtained molybdenum and cobalt sulfide particles by hydrolysis of thioacetamide promoted with hydrazine in ammonium orthomolybdate and cobalt acetate, respectively.[41]

Special Reaction Field

The hydrothermal method is one method for producing metal (hydrous) oxide crystalline particles in aqueous solution under high temperature and high pressure. This method can be classified as hydrothermal oxidation, hydrothermal precipitation, hydrothermal synthesis, hydrothermal decomposition, hydrothermal crystallization, or hydrothermal reduction, according to the reaction mechanism. When an autoclave is used as a reaction vessel, it takes several hours or days to produce fine metal oxide particles, because of the slow dehydration rate. Recently, Adschiri et al. proposed a new process using supercritical water and showed the synthesis time of AlOOH particles was a few minutes.[42]

The methods of particle formation using supercritical fluids are the rapid expansion of the super-critical solution method[43] and the gas antisolvent recrystallization method reported by Gallagher et al.[44] As the operating temperature is relatively low when carbon dioxide supercritical fluid is used, these methods are now being applied to make the products for the medical and food industries.

Ultrafine particles, which sometimes have quantum size effects, can be produced by the reaction in pools of water in (water-in-oil) microemulsions. The redox reactions can be used by mixing more than two kinds of microemulsion with the reactant in the water pool, or by adding gas or an aqueous solution into the microemulsion. Nagy et al. prepared uniform and very fine nickel boride and cobalt boride particles for catalysis, and also reviewed the formation of many metal particles.[45] Fendler's review is focused on novel particle formation by the use of a surfactant assembly, such as reversed micelles, microemulsions, vesicles, and bilayer liquid membranes.[46]

Organic Powder

Organic polymer particles are produced by chemical reactions among the monomers, which are dissolved or dispersed in a solution. Dispersion methods to make monomer droplets are roughly classified into mechanical dispersion methods and phase-separation methods. The former methods need mechanical energy from agitation or pressure difference, and a dispersing agent, such as a surfactant or polymer, is often used for easy dispersion. On the other hand, the droplets of the monomer can be made spontaneously and quite stable in the latter methods.

Seeding polymerization cannot be classified within the two above methods, because the polymer particles already exist in the system. These polymer particles are used as seeds and grow by polymerization between the surface of the particles and the monomer supplied in the system.[47]

Mechanical Dispersion Methods

Suspension polymerization, emulsion polymerization, and mini-emulsion polymerization are included in the mechanical dispersion methods, depending on the droplet size of the monomer. The interfacial reaction technique, which acts at the surface of the dispersed droplets, is also in this category. In this technique, microcapsules can be easily produced, because the chemical reaction rate to make solids decreases rapidly for the high diffusion resistance of the solid film already produced at the interface, and the inner materials can not be completely converted to solid. Sometimes this technique is applied to inorganic particles, as Nakahara et al. reported of the microcapsule formation of spherical calcium carbonate particles.[48]

Suspension polymerization is one of the methods of polymer formation that uses vigorous agitation of the aqueous solution contained in the dispersing agent and hydrophobic monomer contained in the oil-soluble initiator for polymerization.[49,50] The polymerization occurs in the dispersed small droplets of the monomer. This technique operates easily, but the products have average diameters of 2–10 μm and wide size distributions. Sometimes fine particles or water-soluble polymers can be used as the dispersing agent. In the case of emulsion polymerization, a water-soluble initiator is used with a surfactant, of which concentration is more than the critical micelle concentration. The polymerization is initiated at the surface of the monomer solubilized in micelle, and a fresh monomer comes from the large-sized organic droplets stabilized by the surfactant.[50] The products are typically 0.5-μm-diameter particles with narrow distribution.

In the mini-emulsion polymerization technique, fine oil droplets are made with a small amount of the surfactant, adding to the slightly water-soluble materials like hexadecane or long-chain alcohol.[51] The initiator of both water-soluble types and oil-soluble types can be used. The polymer produced by this technique is 0.1 1μm in diameter and relatively polydispersed.

Mixing methods essentially lead to particle size distribution. On the other hand, capillary methods are one idea to control particle size. The droplets of uniform size are produced by the vibration from the breakup of the liquid jet through a nozzle or a capillary. This vibration is usually created by mechanical systems such as piezoelectric, ultrasonic, or pressure difference units. Sometimes an electrostatic spray can be used. Most droplets, however, are dispersed into the gas to make a uniform-size polymer.[52] The porous membrane or tube can be also used to produce the uniform droplets. Omi et al. reported how to make polymer particles of about 8 μm with uniform size by using a porous tube called Shirasu porous glass.[53]

Phase-Separation Methods

The dispersion phase for the reaction can be created spontaneously or by adding a small amount of mechanical energy. Soap-free polymerization is one of the phase-separation methods. In this technique, a very small amount of monomer dissolved in the aqueous solvent becomes the nuclei for the polymer particles. The residue of the initiator for polymerization, which has a charge in the aqueous solution, remains on the surface of the products, and so the stability of the dispersion state of the products is kept due to the repulsion for the surface charge. The particle size of the products is around 1 μm and monodispersed. The greatest advantage for this technique is that the surface of the particles is free of the surfactant.[54]

On the other hand, the micro-emulsion polymerization technique uses a large amount of surfactant with the monomer and the solvent (usually water). The polymerization occurs in a micelle system of high concentration. The particle size of this product is in the range from 5 to 50 nm. Inverse micelle polymerization is the same technique as micro-emulsion polymerization, but a water-soluble monomer is used and produces polymer particles 5–100 nm in diameter in the reversed micelle.[55]

In the dispersion polymerization technique including nonaqueous types, the amphiphilic polymer is used as the dispersing agent with the organic solvent, for example, an alcohol, which is a good solvent for the monomer but is a poor solvent for the polymer. The particle size of the products is from a few micrometers to a few tens of micrometers. They are relatively uniform and can be controlled by the degree of the compatibility between the polymer and the solvent.[56]

The coacervation technique also uses a poor solvent for the deposition of the polymer phase, and it can be controlled by the operated temperature or by the progress of the polymerization. Monodispersed but porous particles, of from a few micrometers to a few tens of micrometers, are produced.[57] The polymer produced by the coacervation has a high surface activity, so that when the other materials are dispersed in the system, the coacervation occurs on the surface of the other materials and polymer membrane covering it. This technique can be applied to make microcapsules.[58]

REFERENCES

1. Friedlander, S. K., Ogawa, K., and Ullmann, M. J., *Polym. Sci. Polym. Phys.,* 38, 2658–2665, 2000.
2. Rogak, S. N. and Flagan, R. C. J., *Colloid Interface Sci.,* 151, 203–224, 1992.
3. Dufaux, D. P. and Axelbaum, R. L., *Combust. Flame,* 100, 350–358, 1995.
4. Cannon, W. R., Danforth, S. C., Flint, J. H., Haggerty, J. S., and Marra, R. A., *J. Am. Ceram. Soc.,* 65, 324–330, 1982.
5. Fantoni, R., Borsella, E., Piccirillo, S., Ceccato, R., and Enzo, S., *J. Mater. Res.,* 5, 143–150, 1990.
6. Pfund, A. H., *Rev. Sci. Instrum.,* 1, 398, 1930.
7. Kimoto, K., Kamiya, Y., Nonoyama, M., and Uyeda, R., *J. Appl. Phys.,* 2, 702, 1963.
8. Yatsuya, S., Kasukabe, S., and Uyeda, R., *Jpn. J. Appl. Phys.,* 12, 1675–1684, 1973.
9. Granqvist, C. G. and Buhrman, R. A., *J. Appl. Phys.,* 47, 2200–2219, 1976.
10. Birringer, R., *Mater. Sci. Eng.,* A117, 33–43, 1989.
11. Ando, Y. and Uyeda, R., *Commun. Am. Ceram. Soc.,* 1, C12–C13, 1981.
12. Uda, M., *Bull. Jpn. Inst. Met.,* 27, 412, 1983.
13. Chang, Y. I. and Pfender, E., *Plasma Chem. Plasma Process,* 7, 299–316, 1987.
14. Girshick, S. L., Chiu, C. P., and McMurry, P. H., *Plasma Chem. Plasma Process.,* 8, 145–157, 1989.
15. Shen, Z., Kim, T., Kortshagen, U., McMurry, P. H., and Campbell, S. A., *J. Appl. Phys.,* 94, 2277–2283, 2003.
16. Messing, G. L., Zhang, S. C., and Jayanthi, G. V., *J. Am. Ceram. Soc.,* 76, 2707–2726, 1993.
17. Xia, B., Lenggoro, I. W., and Okuyama, K., *J. Mater. Sci.,* 36, 1701–1705, 2001.

18. Zachariah, M. R. and Huzarewicz, S., *J. Mater. Res.,* 6, 264–269, 1991.
19. Alam, M. K. and Flagan, R. C., *Aerosol Sci. Technol.,* 5, 237–248, 1986.
20. Okuyama, K., Kousaka, Y., Tonge, N., Yamamoto, S., Wu, J. J., Flagan, R. C., and Seinfeld, J. H., *AIChEJ,* 32, 2010–2019, 1992.
21. Flagan, R. C. and Lunden, M. M., *Mater. Sci. Eng. A Struct.,* 204, 113–124, 1995.
22. Wu, J. J., Flagan, R. C., and Gregory, O. J., *Appl. Phys. Lett.,* 49, 82–84, 1986.
23. Griffin, G. L., *J. Am. Ceram. Soc.,* 75, 3209–3214, 1992.
24. Akhtar, M. K., Pratsinis, S. E., and Mastrangelo, S. V. R., *J. Am. Ceram. Soc.,* 75, 3408–3416, 1992.
25. Ulrich, G. D., *Combust. Sci. Technol.,* 4, 47, 1971.
26. Pratsinis, S. E., *Prog. Energy Combust. Sci.,* 24, 197–219, 1998.
27. Holmgren, J. D., Gibson, J. O., and Sheer, C., *J. Electrochem. Soc.,* 111, 362–369, 1964.
28. Yoshida, T., Kawasaki, A., Nakagawa, K. J., and Akashi, K., *J. Mater. Sci.,* 14, 1624–1630, 1979.
29. Sugimoto, T., Ed., *Fine Particles: Synthesis, Characterization, and Mechanisums of Growth,* Marcel Dekker, New York, 2000.
30. Sugimoto, T., *Monodispersed Particles,* Elsevier Science, Amsterdam, 2001.
31. Neilson, M. L., Hamilton, R. J., and Walsh, R. J., in Kuhn, B., Lamprey, W. H., and Sheer, C., Eds., *Ultrafine Particles,* John Wiley & Sons, New York, 1963, p. 181.
32. Roy, D. M., Neurgaonkar, R. R., O'Holleran, T. P., and Roy, R., *Am. Ceram. Soc. Bull.,* 56, 1023–1024, 1977.
33. LaMer, V. K. and Dinegar, R. H., *J. Am. Chem. Soc.,* 72, 4847, 1950.
34. Sugimoto, T., *Adv. Colloid Interface Sci.,* 28, 65, 1987.
35. Calvert, P., *Nature,* 367, 119, 1994.
36. Sacks, M. D., Tseng, T. Y., and Lee, S. Y., *Am. Ceram. Soc. Bull.,* 63, 301–310, 1984.
37. Stöber, W., Fink, A., and Bohn, B., *J. Colloid Interface Sci.,* 26, 62, 1968.
38. Blesa, M. A. and Matijevic, B., *Adv. Colloid Interface Sci.,* 29, 173–221, 1989.
39. House, H. O., *Modern Synthetic Reactions,* 2nd Ed., W.A. Benjamin, Redwood, CA, 1972.
40. Matijevic, B., *Faraday Discuss.,* 92, 229–239, 1991.
41. Haruta, M., Lamaitre, J., Delannay, F., and Delmon, B., *J. Colloid Interface Sci.,* 101, 59, 1984.
42. Adschiri, T. K., Kanazawa, K., and Arai, K., *J. Am. Ceram. Soc.,* 75, 1019–1022, 1992.
43. Mohamed, R. S., Halverson, D. S., Debenedetti, D. G., and Prud'homme, R. K., in Johnston, K. P. and Penninger, J. M. L., Eds., *Supercritical Fluids: Science and Technology,* ACS Symposium Series, No. 406, American Chemical Society, Washington, DC, 1989, p. 355.
44. Gallagher, P. M., Coffey, M. P., Krukonis, V. I., and Klasutis, N., in Johnston, K. P. and Penninger, J. M. L., Eds., *Supercritical Fluids: Science and Technology,* ACS Symposium Series, No. 406, American Chemical Society, Washington, DC, 1989, p. 334.
45. Nagy, J. B., Derouane, B. G., Gourgue, A., Lufimpadio, N., Ravct, I., and Verfaillie, J. P., In Mittal, K. L., Ed., *Surfactants in Solution,* Vol. 10, Plenum Press, New York, 1989, pp. 1–43.
46. Fendler, J. H., *Chem. Rev.,* 87, 877, 1987.
47. Ugeistad, J., Mork, P. C., Mufutakhamba, H. R., Soleimy, B., Nordhuns, I., Schmid, R., Berge, A., Ellingson, T., and Aune, O., *Science and Technology of Polymer Colloids,* Vol. 1, NATO ASI Series B, No. 67, Kluwer Academic, Dordrecht, 1983.
48. Nakahara, Y., Mizuguchi, M., and Miyata, K., *J. Colloid Interface Sci.,* 68, 401–407, 1979.
49. Bishop. R. B., *Practical Polymerization of Polystyrene,* Cahners Publ., Des Plaines, IA, 1971.
50. Blackley, D. C., *Emulsion Polymerization: Theory and Practice,* Applied Science Publ., London, 1970.
51. Barnette, D. T. and Schork, F. J., *Chem. Eng. Prog.,* 83(6), 25–30, 1987.
52. Panagiotou, T. and Levendis, Y. A., *J. Appl. Polymer Sci.,* 43, 1549–1558, 1991.
53. Omi, S., Katami, K., and Iso, M., *J. Appl. Polymer Sci.,* 51, 1, 1994.
54. Ceska, G. W., *J. Appl. Polymer Sci.,* 18, 2493–2499, 1974.
55. Candau, F. and Orrewill, R. H., Eds., *Scientific Methods for the Study of Polymeric Colloids and Their Applications,* NATO ASI Series C, No. 303, Kluwer Academic Publ., Dordrecht, 1990.
56. Barrett, K. B. J., Ed., *Dispersion Polymerization in Organic Media,* John Wiley & Sons, New York, 1970.
57. Hou, W. H. and Lloyd, T. B., *J. Appl. Polymer Sci.,* 45, 1783–1788, 1992.
58. Nixon, J. R., Ed., *Microencapsulation,* Marcel Dekker, New York, 1976.

4.3 Crystallization

Matsuoka Masakuni

Tokyo University of Agriculture and Technology, Koganei, Tokyo, Japan

Industrial crystallization to produce particulate products with desired mean size and size distribution from a liquid phase is reviewed briefly here. By controlling operating conditions and using properly designed apparatuses, desired products are, in principle, obtainable. In this section, fundamentals of crystallization kinetics and phenomena are described, together with their applications to practical operations.

4.3.1 CRYSTALLIZATION PHENOMENA AND KINETICS

In order for crystallization to proceed, the establishment of supersaturation is essential. This is done either by the cooling or evaporating of solvents, or a combination of these two operations. Chemical reactions to form insoluble materials are also applied for this purpose. Crystallization phenomena usually include nucleation and growth of crystals.

Nucleation

The formation of crystal nuclei can be classified into primary and the secondary nucleation categories, as shown in Table 3.1.[1] Primary nucleation refers to the spontaneous formation of crystals from clear supersaturated solutions and is further classified into homogeneous and the heterogeneous nucleation. The latter occurs when foreign solid surfaces, such as vessel walls, stirrer blades, and/ or dust particles, exist in the solution and act as nucleation sites. Theoretical investigations have mainly been developed for homogeneous primary nucleation; however, quantitative correlations without empirical constants are not yet possible. The classical theory predicts the dependence of the nucleation rate B [#/(m^3 s)] on the solution saturation ratio S ($= x/x^*$), in which x is the solution concentration in terms of mole or mass fraction, and x^* is the saturation concentration at the operating temperature as

$$B = A_1 \exp[-A_2 T^{-3} (\ln S)^{-2}] \approx A_1 \exp(-A_2 T^{-3} \sigma^{-2}) \qquad (3.1)$$

where A_1 and A_2 are constants, T [K] is the solution temperature, and σ ($= S - 1 = (x - x^*)/x^*$) is the supersaturation ratio or simply the supersaturation of the solution. The last equality is held only if the supersaturation is so small that $\ln S = \ln(1 + \sigma) = \sigma$.

The secondary nucleation predominates in practical crystallizers where a large number of crystal particles are present in the solution. Although a number of mechanisms of the secondary nucleation have been proposed, only empirical correlations are available to correlate the rate of nucleation and operating conditions such as solution supersaturation σ, flow conditions in terms of stirring rates N [rps], and slurry density M_T [kg/m^3]. A typical correlation is as follows:

$$B = k_b N^a M_T^i \sigma^b = k_b' N^a M_T^j G^i \qquad (3.2)$$

where k_b and k_b' are the coefficients for secondary nucleation and the exponents a, j, b, and i are experimentally determined for each system and crystallizer. G [m/s] stands for the growth rate of

TABLE 3.1   Classification of Nucleation and Growth Phenomena

	System	
Phenomena	Clear	Suspension
Nucleation	Primary nucleation	Secondary nucleation
	Homogeneous nucleation	Contact nucleation
	Heterogeneous nucleation	Fluid shear nucleation
		Breaking/attrition
		Agglomeration (negative)
Growth	Primary growth	Secondary growth
		Agglomeration
		Growth enhancement by
		microcrystals
		Breakage/attrition (negative)

Source: Matsuoka, M., *J. Chem. Eng. Jpn.,* 35, 1025–1037, 2002.

crystalline particles as described in the following section. Typical values of these exponents are summarized by van Rosmalen and van der Heijden[2] as $a = 1.5 - 4$, $j = 1$ or 2 (depending on the collision mechanisms involved), $b = 1 - 3$, and $i (= b/g) = 0.5 - 3$. The second equality is held because the growth rate is a function of the solution supersaturation and will be mentioned in the next section.

Crystal Growth

Crystal particles grow with the driving force of solution supersaturation. The rate processes involved in crystal growth kinetics are the mass transfer of crystallizing component(s) from the bulk to the surface in the solution, the surface integration in which the crystallizing components are incorporated into the crystal lattice, and, finally, the heat transfer of the latent heat of crystallization. The first two processes occur in series, whereas the last parallels them.

For mass transfer rates, the conventional treatment is applicable, and the following expression can be used:

$$\frac{dW}{dt} = k_d A C \ln \frac{1 - w_i}{1 - w} \tag{3.3}$$

where W [kg] and A [m²] are the mass and the surface area of a crystal, C is molar or mass density of the solution, and x_i denotes the solution concentration at the surface and can be equal to the bulk concentration (x) when the solution is completely agitated.

Theoretical treatments of the surface integration rate have been developed for single crystals. These treatments include the Burton–Cabrera–Frank (BCF) theory and the birth and spread (B&S) model, which correlate the linear growth rate R_{hkl} [m/s] of a specific crystallographic surface of (hkl) with the surface supersaturation $\sigma_s (= (x_i - x^*)/x^*)$. The dependence of R_{hkl} on σ_s is different for each model, as shown in Figure 3.1, so that from an engineering viewpoint, it is enough to express the relation by the power law with $g = 1$–2:

$$R_{hkl} = k_g \sigma_s^g \tag{3.4}$$

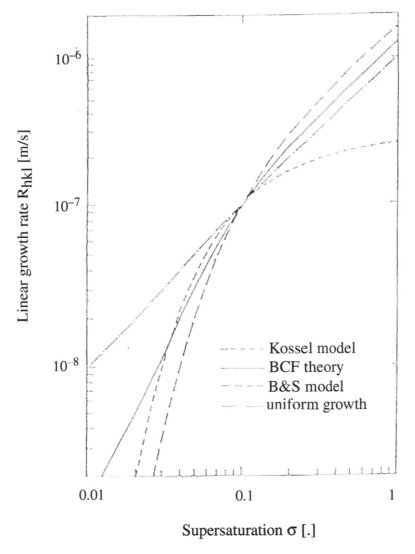

FIGURE 3.1 Linear grown rate versus supersaturation for several models of crystal growth. Coefficients and constants are taken to satisfy $G = 10^{-7}$ m/s when $\sigma = 0.1$.

As the mass and the surface area of a crystal particle can be expressed by the use of its characteristic dimension (diameter) L [m], and the shape factors ϕ_v and ϕ_s as $W = \rho_s \phi_v L^3$ and $A = \phi_s L^2$, the mass rate of crystal growth can be converted into the linear growth rate G ($= dL/dt$):

$$\frac{1}{A}\frac{dW}{dt} = 3\frac{\rho_s \phi_v}{\phi_s}\frac{dL}{dt} = 3\frac{\rho_s \phi_v}{\phi_s}G \tag{3.5}$$

The two linear growth rates are different, as R_{hkl} is the advancement of the crystallographic surface (hkl), corresponding to the increasing rate of a radius, whereas G is considered to be that of a diameter. Hence, $G \approx 2R_{hkl}$.

In addition, heat transfer plays an important role in particular systems with high solution concentration, large heat of crystallization, and large temperature-dependent solubility.[3] Therefore, the

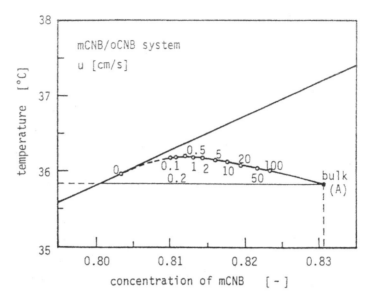

FIGURE 3.2   Solution conditions at the growing crystal surface at different flow rates. [From Matsuoka, M. and Garside, J., *J. Cryst. Growth*, 129, 385–393, 1993. With permission.]

solution conditions at the crystal surface can be a measure of the relative importance of the three rate processes. For example, they approach the bulk conditions as the relative velocities between the crystal and the solution increase as shown in Figure 3.2.[4]

4.3.2   OPERATION AND DESIGN OF CRYSTALLIZERS

Analysis with Population Balance Equation

Besides respective measurements of the nucleation and growth rates, the rates can be determined experimentally if the crystal size distribution (CSD) of the product is known. For a simple well-stirred tank crystallizer, operated continuously at steady state, the number balance of particles of size L leads to the following mathematical relation between the growth and the nucleation rates[5]:

$$n(L) = n(0)\exp\left(-\frac{L}{G\tau}\right) = \frac{B}{G}\exp\left(-\frac{L}{G\tau}\right) \tag{3.6}$$

where $n(L)$ [#/(m³m)] denotes the population density of particles of size L, and τ [s] is the mean resident time (i.e., the crystallizer volume divided by the volume flow rate). This type of crystallizer is known as a mixed-suspension mixed-product removal (MSMPR) crystallizer, in which complete mixing is assumed for both the slurry in the vessel and the product.

Equation 3.6 was derived from the population balance with the assumptions of steady state, size-independent growth rate, zero size nuclei, no particles in the feed stream, and no agglomeration or breakage of crystal particles in the crystallizer. By plotting the population density of products against size on a semilogarithmic paper, and knowing the residence time, one can determine the growth and the nucleation rates simultaneously from the slope and the intersect of the straight line, respectively, as shown as a bold line in Figure 3.3.

The growth rate thus determined is often different from the one discussed in the preceding section, in that effects of neglected phenomena of agglomeration or breakage are all included. Product

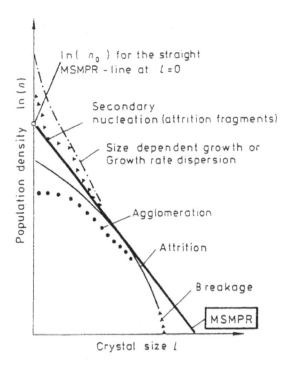

FIGURE 3.3 Ideal semilogarithmic population density plot (**bold line**) with some deviations from ideality. [Modified from Garside, J., Mersmann, A., and Nyvlt, J., Eds., *Measurement of Crystal Growth Rates,* FECE Working Party on Industrial Crystallization, München, 1990, p. 160. With permission.]

crystals from industrial crystallizers are usually agglomerated, and their corners are rounded due to collisions between the particles or impeller blades. However, the growth and the nucleation rates thus determined are the effective parameters that control the product particles having the CSD measured.

From Equation 3.6, physically meaningful particle characteristics can be obtained: The mean size L_M is given as $G\tau$, that is, (growth rate) $\times$ (residence time); and the total number of particles per unit volume N_T is equal to $B\tau$, that is, (nucleation rate) $\times$ (residence time). For mass-based mean size, $3G\tau$ can be derived as the modal or dominant size where the distribution is maximum.

Crystallization Phenomena in Crystallizers

In actual crystallizers, crystallization phenomena can be very different from those occurring during the growth of single crystalline particles. Crystals of the same size can grow at different rates, or the rates can be different among the particles of different sizes under the same supersaturation. These are known as growth rate dispersion (GRD) or size-dependent growth rates (SDG), respectively.[2]

Recently, experimental evidence has been reported for the effects of microcrystals on larger crystal particles. During the growth of a crystal particle from a clear supersaturated solution, induced nucleation causes a sudden increase in the growth rate. This phenomenon is called "growth rate enhancement (GRE) by microcrystals"[6] and is considered to be common in actual crystallizers where nucleation is continuously occurring and large numbers of particles coexist in the mother liquor. The agglomeration of particles is a similar but different phenomenon and results in polycrystals, whereas GRE can produce perfect, single crystal-like particles.

These phenomena result in curved population density plots. The curves in small size regions are often concave upward, suggesting either slower growth rate, from Equation 3.6, or secondary

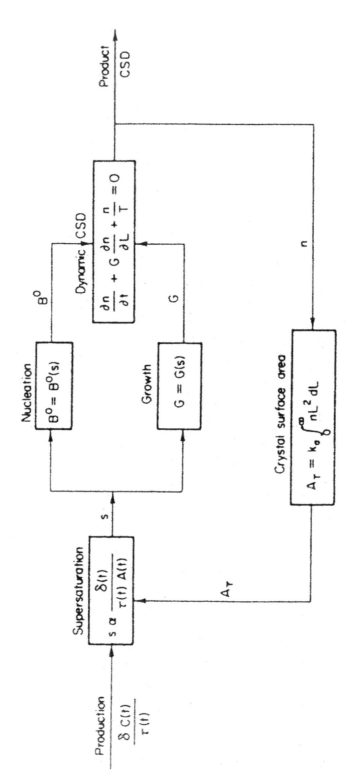

FIGURE 3.4   Factors affecting the CSD and their interrelations. [From Randolph, A. D. and Larson, M. A., in *Theory of Particulate Processes*, 2nd Ed., Academic Press, San Diego, 1988, p. 201. With permission.]

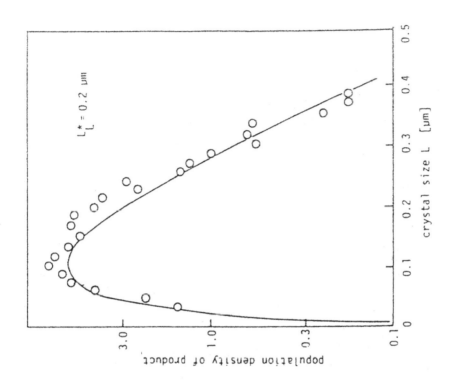

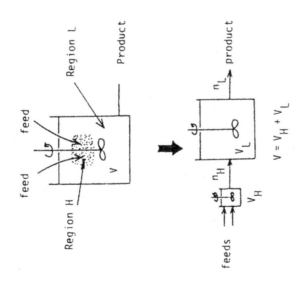

FIGURE 3.5   Control of CSD utilizing ripening phenomena. [From Garside, J., personal communication (special lecture at TUAT), 1989. With permission.]

nucleation by attrition, although the same effects can be obtained by poor achievement of MSMPR conditions. Some typical effects of factors causing deviations from the ideal density plot are shown in Figure 3.3. The diagram is modified from that of Garside et al.[7] Accordingly the growth of crystals in suspension systems may be different from the one in clear solution, and therefore regarded as secondary growth in contrast to the primary growth where single crystalline particles are growing from clear solutions. This classification is already given in Table 3.1 (see work by Matsuoka[8]).

Design of Crystallizers

With an MSMPR crystallizer, the interrelations among the factors influencing the product CSD are illustrated in Figure 3.4.[9] Many factors such as feed concentration, residence time, and operating temperature can change the product CSD through the solution supersaturation with feedback loops from the CSD, which defines the suspension density and total surface areas. Still, it is not easy to tell what happens on the CSD when one of these factors is changed. For example, the answer to a question such as, "Does the mean size increase when the residence time increases?" depends strictly on the dependencies of both the nucleation and growth rates on the supersaturation. In most cases, where $b > g$ (i.e., $i > 1$), the answer is "Yes," provided the solution supersaturation is kept constant.

As long as MSMPR crystallizers are used, the product possesses a broad CSD, and in order to obtain uniformly sized particles, classification, in principle, is essential. Garside,[10] however, illustrated a process to produce very fine particles with a narrow CSD by the use of a series of MSMPR crystallizers. This can be achieved by utilizing size-dependent solubilities of crystallizing component(s) and controlled supersaturations in the crystallizers (Figure 3.5). In the first small tank, with higher supersaturation, rapid nucleation occurs with very small nuclei, whereas in the second tank, operated at lower supersaturation with longer residence time, smaller nuclei tend to dissolve, and relatively large particles can survive, so that sharply distributed particles are produced. This occurs because of the Ostwald ripening phenomena; the size-dependent solubilities given by the Gibbs–Thomson equation can cause small particles to dissolve even in the supersaturated solutions. The Gibbs–Thomson equation is written as

$$\ln\frac{x(L)}{x(\infty)} = \frac{4\Delta G_a M}{RT\rho_s L} \tag{3.7}$$

where $x(L)$ is the solubility in terms of mole fraction of particles of diameter L, ΔG_a [J/m^2] is the surface tension, M is the molecular weight, and ρ_s is the density of crystals. For moderately soluble materials, particles of which size is a few microns are large enough to assume that their solubility is size independent. For sparingly soluble materials of that size, however, they usually show size-dependent solubilities, hence the ripening can effectively proceed.

REFERENCES

1. Mullin, J. W., in *Crystallization*, 3rd Ed., Butterworth-Heinemann, Oxford, 1993, p. 172.
2. Van Rosmalen, G. M. and van der Heijden, A. E., in *Science and Technology of Crystal Growth*, van der Eerden, J. P. and Bruinsma, O. S. L., Eds., Kluwer Academic Publishers, Dordrecht, 1995, p. 259.
3. Matsuoka, M. and Garside, J., *Chem. Eng. Sci.*, 46, 183–192, 1991.
4. Matsuoka, M. and Garside, J., *J. Cryst. Growth*, 129, 385–393, 1993.
5. Randolph, A. D. and Larson. M. A., in *Theory of Particulate Processes*, 2nd Ed., Academic Press, San Diego, 1988, p. 80.

6. Matsuoka, M. and Eguchi, N., *J. Phys. D Appl. Phys.,* 26, B162–B167, 1993.
7. Garside, J., Mersmann, A., and Nyvlt, J., Eds., *Measurement of Crystal Growth Rates,* FECE Working Party on Industrial Crystallization, München, 1990.
8. Matsuoka, M., *J. Chem. Eng. Jpn.,* 35, 1025–1037, 2002.
9. Randolph, A. D. and Larson, M. A., in *Theory of Particulate Processes,* 2nd Ed., Academic Press, San Diego, 1988, p. 201.
10. Garside, J., personal communication (special lecture at TUAT), 1989.

4.4 Design and Formation of Composite Particles

Hideki Ichikawa and Yoshinobu Fukumori
Kobe Gakuin University, Kobe, Japan

Particles have been processed by various methods for composite particle formation depending on required size and function (Figure 4.1). Particles larger than 200 μm can be successfully processed by a fluid bed. Many kinds of composite particles prepared by a fluid bed are on the market; for example, most recent pharmaceuticals are composites for controlled drug release, including sustained release, prolonged release, delayed release, and taste masking. For smaller particles, agglomeration has been successfully carried out to get free-flowing coarse particles, but their coating for forming multilayered structures without core-particle agglomeration is difficult, because of the adhesive and cohesive properties of such fine particles.[1] So, liquid-phase processes, including emulsifying processes and phase separation, have been applied to make fine composite particles. In this section, the fluid bed process[2] and the emulsifying process are described, mostly with examples limited to the pharmaceutical field, since the methods and applications are diverse among industrial fields.

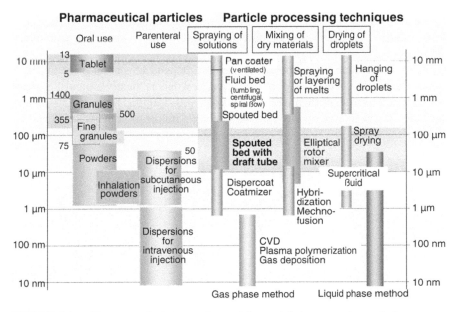

FIGURE 4.1 Pharmaceutical composite particles and their preparation techniques.

4.4.1 COMPOSITE PARTICLE FORMATION IN THE FLUID BED PROCESS

Apparatus in the Fluid Bed Process

Fluid bed processors have been used for drying, agglomeration, and coating. In addition to a simple fluid bed, the tumbling, agitating, centrifugal, and spiral flow fluid bed and the spouted bed with or without draft tube have been developed for improving process performance. Among many types of surface modification or composite formation processes, fluid bed processes are characterized by their simple, easy formation of multilayers on particles, leading to commercial production of many types of functional particulate systems.

Typical fluid bed processors are illustrated in Figure 4.2. Figure 4.2A shows a simple fluid bed, usually with a conical shape at the bottom of the chamber, inducing spouted particle flow that more or less depends on the angle of the conical chamber. The spray is supplied from the top toward the fluid bed in agglomeration, but into the bed in a tangential direction in coating. Figure 4.2B shows an example of a tumbling fluid bed processor. The fluidization air is supplied through the slit between the turntable and the chamber. The tumbling forces particles to be centrifuged toward the chamber wall; then, the particles are blown up by the slit air. These make a circulating fluid bed, in which the spray is supplied from the top or in tangential mode. The rolling of particles on the turntable makes the wet particles roundish and compact. Figure 4.2C shows a typical assembly of the Wurster process, a kind of spouted bed process assisted with a draft tube. The particles fluidized in the annular part between the draft tube and the chamber are introduced into the draft tube due to accelerated air flow from the bottom and, then, spouted from the draft tube. The particle velocity is reduced in the upper expansion chamber, leading to return of particles to the fluid bed in the annular part. During this circulating flow, the particles are sprayed in the draft tube; at the same time, they have a chance to be exposed to the spray-air jet, which can exert a strong separation force on the particles. Usually, the particles are easily agglomerated or coagulated during spraying, but the above strong air-jet can disintegrate the agglomerates. These characteristics of the Wurster process make it possible to coat finer particles[3] or to produce finer agglomerates,[4] compared with the other types of fluid bed processes.

The types of fluidization, as proposed by Geldart, depend on particle size and density.[5] For example, since the particle density of pharmaceutical powders is mostly around 1.5 g/cm³, the particles are categorized into larger than 900 μm (D type), 900–150 μm (B type), 150–20 μm (A type), and smaller than 20 μm (C type) particles from Geldart's fluidization map. The A and C types are

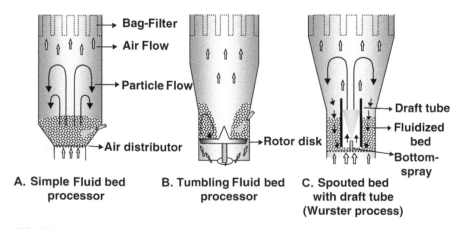

FIGURE 4.2   Typical fluid bed equipment for film-coating of pharmaceutical particulates.

cohesive and adhesive; therefore, they are agglomerated into D and B types in order to achieve free-flowing properties. This is the main purpose of agglomeration. The D type cannot be fluidized due to too large particle size; therefore, the coating is simply processed using a rotating pan. The B type always exhibits bubbling fluidization, which induces inhomogeneous particle flow; in practical terms, this is disadvantageous because such a flow, especially in the spray zone, leads to poor coating performance in yield and homogeneity of the product. In order to achieve homogeneous particle flow of B-type particles, the particle flow patterns that are different from simple fluidization, such as those in the tumbling, centrifugal, and spiral fluid bed and the spouted bed, have been required. This is the reason that many different types of fluid bed processor have been developed so far. The A-type particles can be fluidized homogeneously without air bubbles, but the separation force from fluidization is not sufficient to avoid agglomeration during spraying the liquid solution. This prevents the fluid bed coating process from being industrially applied to these small particles in spite of much research attempting fine particle coating technologies. The C-type particles cannot be fluidized due to their small size.

When particles are fluidized or spouted under spraying, separation force is exerted to the particles and they are more or less agglomerated by the spray solution. Balance between the separation force and the binding strength of the binder or coating material determines the degree of agglomeration, that is, agglomerate size.[6,7] Each fluid bed processor can generate its inherent separation force, surely depending on the operating conditions such as air flow pattern and inlet air flow rate. The excessive separation force can even disintegrate core particles. Thus, depending on the requisite of final agglomerate size or the core particle size to be discretely coated, the optimal processor should be selected (Figure 4.1).

Material in the Fluid Bed Process

Table 4.1 lists typical binders and coating materials for agglomeration and coating. The water-soluble polymers are used as binders in agglomeration, and some of them are also used as coating materials for water-proofing or taste masking. As each type of water-soluble polymer becomes higher in molecular weight, it contributes more to the increase in the viscosity of its solution and the interparticle binding strength, leading to more enhanced particle growth in agglomeration and coating.[6,7]

The commercially available polymeric dispersions, which have been most widely used as coating material, are classified into three types based on the preparation methods[8]: (1) latexes synthesized by emulsion polymerization; (2) pseudolatexes prepared by emulsifying processes such as emulsion-solvent evaporation, phase inversion, and solvent change; and (3) dispersions of micronized polymeric powders.

Eudragit L30D and NE30D are acrylic copolymer latexes synthesized by emulsion polymerization.[9] The particle sizes of these latexes are in a submicron order. L30D is a copolymer of ethyl acrylate (EA) as an ester component with methacrylic acid (MA) (MA/EA 1:1). It is used for enteric coating because of the presence of carboxyl groups in the copolymer. NE30D is a copolymer of ester components only, EA and MMA (2:1). The films formed from NE30D have a very low softening temperature and hence are flexible and expandable even under indoor conditions.

Cellulose derivatives cannot be synthesized directly in latexes; therefore, they are prepared as pseudolatexes (Aquacoat, Aquateric,[10] Surelease[11]), or micronized powders (Aqoat [HPMCAS],[12] EC N-10F). While the pseudolatexes can be prepared as submicron particles mostly by emulsifying processes, the micronized powders have mean particle sizes of a few micrometers. Poly(VAP) is also supplied as a micronized powder (Coateric).[11] Eudragit RS and RL are terpolymers of EA and MMA as ester components with trimethylammonioethyl methacrylate chloride (TAMCl) as hydrophilic quaternary ammonium groups; RS and RL are 1:2:0.1 and 1:2:0.2 terpoly (EA/MMA/TAMCl), respectively. Because Eudragit RS and RL contain MMA-rich ester components (EA/MMA 1:2), they have softening temperatures higher than those of NE30D (EA/MMA 2:1) and form hard films

TABLE 4.1 Examples of Commercially Available Coating Materials

Type	Trade name	Supplier	Solubilization	Application form	Components
Solutions	Kollidon VA64	BASF	Water-soluble	Aqueous solution	6:4 Poly(Vinylpyrrolidone/Vinylacetate)
	Kollidon 20, 30	BASF	Water-soluble	Aqueous solution	Polyvinylpyrrolidone (PVP)
	Opadry	Coloracon	Water-soluble	Aqueous solution	Mix of water- soluble polymer, plasticizer and pigment
	TC-5	Shin-Etsu	Water-soluble	Aqueous solution	Hydroxypropylmethylcellulose (HPMC)
	HPC-SL, SSL, SSM	Nippon Soda	Water-soluble	Aqueous-solution	Hydroxypropylcellulose (HPC)
	CMEC	Freund	Enteric	Aqueous ethanolic soln	Carboxymethylethylcellulose (CMEC)
	HPMCP	Shin-Etsu	Enteric	Aqueous ethanolic soln	Hydroxypropylmethylcellulose phthalate (HPMC)
	Eudragit E 100	Rhom Pharm	Acid-soluble	Organic solvent soln	Poly(BMA/MMA/DAEMA)
	Eudragit L/S	Rhom Pharm	Enteric	Organic solvent soln	Poly(MMA/MAA)
	Eudragit RS 100/RL100	Rhom Pharm	Water-soluble	Organic solvent soln	1:2:0.1/1:2:0.2 poly(EA/MMA/TAMCl)
Dispersions	EC N-10F	Shin-Etsu	Water-insoluble	Aqueous dispersion	Ethylcellulose (EC)
	Aquacoat	Asahi-Kasei	Water-insoluble	Pseudo-latex	Ethylcellulose (EC)
	Surerease	Coloracon	Water-insoluble	Pseudo-latex	Ethylcellulose (EC)

	Product	Manufacturer	Solubility	Form	Composition
	Eudragit RS30D	Rhom Pharm	Water-insoluble	Pseudo-latex	1:2:01 poly(EA/MMA/TAMACl)
	Eudragit RL30D	Rhom Pharm	Water-insoluble	Pseudo-latex	1:2:02 poly(EA/MMA/TAMACl)
	Eudragit NE30D	Rhom Pharm	Water-insoluble	Latex	2:1 Poly(EA/MMA)
	Aqoat	Shin-Etsu	Enteric	Aqueous dispersion	Hydroxypropylmethycellulose acetate succinate
	Aquateric	FMC	Enteric	Aqueous dispersion	Cellulose acetate phthalate (CAP)
	Kollicoat MAE30D/DP	BASF	Enteric	Aqueous dispersion	Poly(EA/MMA)
	Eudragit L30D	Rhom Pharm	Enteric	Latex	1:1 Poly(MMA/EA)
Wax coat	Lubri Wax 101/103	Freund	Water-insoluble	Powder	Hydrogenated caster oil/hydrogenated rapessed oil
	Polishing wax	Freund	Water-insoluble	Powder	Carnauba wax
	PEP 101	Freund	Water-insoluble	Powder	Poly (Ethylene oxide/propyloene oxide)
Cores	Nonpareil101/103/105	Freund		Granule	Granules of Sucrose-Starch/Sucrose/Lactose Microcrystalline cellulose
	Celphere 102/203/305/507	Asahi-Kasei		Granule	Granules of microcrystalline cellulose

MAA, methacrylic acid; MMA, methyl methacrylate; EA, ethyl acrylate; TAMCl, trimethylammonioethylmethascrylate chloride; DAEMA, dimethylaminoethylmethacrylate; BMA, butyl methacrylate.

under indoor conditions. Eudragit RS and RL powders are easily transformed into pseudolatexes by emulsifying their powders in hot water without additives.[9] It is costly to ship aqueous dispersions around the world; therefore, the Aquateric pseudolatex is supplied as a spray-dried powder; it is redispersed just before use.[10]

The film formation process from aqueous dispersion is shown schematically in Figure 4.3. The mechanisms of film formation from aqueous polymeric dispersions have been discussed for a long time, and many theories have been proposed.[8–10,13] Fusion and film formation of polymeric particles during the coating process can be explained by the wet sintering theory for particles suspended in water, the capillary pressure theory for particle layers containing water in various degrees of saturation, and the dry sintering theory for dry particle layers.

Particle Design in the Fluid Bed Process

The typical particle structures produced from the agglomeration process are shown in Figure 4.4. Simple agglomerates (Figure 4.4A) are prepared by a fine powder mixed with fine additive powders in a fluid bed, and then, binder solution is sprayed from the top of the simple fluid bed (Figure 4.2A), followed by drying. In this case, since the particles are agglomerated in the fluidization air flow, they are usually very porous. The tumbling fluid bed or the spouted bed process assisted with a draft tube (Figure 4.2B and 4.2C) can be applied if more compact agglomerates are required. When a fine powder is agglomerated with coarse carrier particles (cores in Table 4.1), fine-powder-layered agglomerates shown in Figure 4.4B can be prepared.[14] In the latter process, the tumbling fluid bed is often used, because tumbling on the turntable contributes to efficient layering of the fine powder. The fine-powder-layered agglomerates are often used as core particles in the coating process.

Among many kinds of surface modification process, the fluid bed processes can most easily produce multilayered particle structures with each layer being monolithic, a random multiphase structure, an ordered multiphase structure, and so on (Figure 4.5). Combination of different components

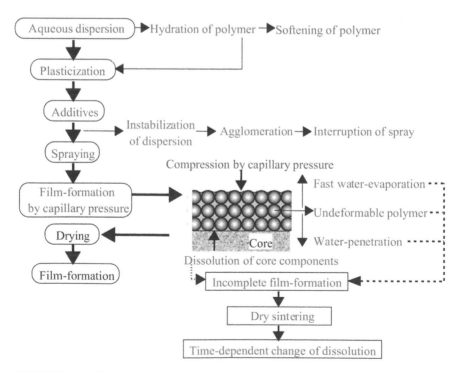

FIGURE 4.3 Film formation from latex.

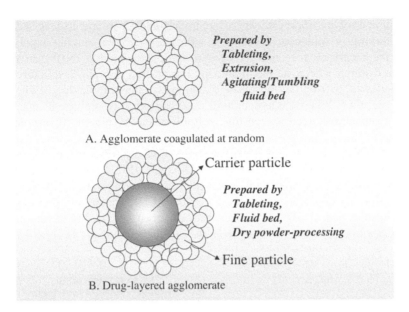

FIGURE 4.4   Structure of agglomerates.

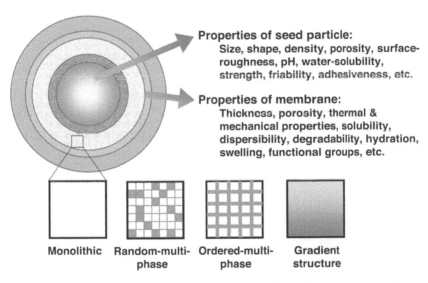

FIGURE 4.5   Possible structure and characteristics of microencapsulated particles.

and layers can produce almost infinite types of functional particles. As an example, designs and preparations of thermosensitive controlled-release particles are described below.

Controlled-release technology based on externally temperature-activated release can find applications in diverse industrial fields.[1,15-17] In the pharmaceutical area, for example, deviation of the body temperature from the normal temperature (37°C) in the physiological presence of pathogens or pyrogens can be utilized as a useful stimulus that induces the release of therapeutic agents from a thermosensitive controlled-release system. Physically controlled temperature using a heat source such as microwaves from outside the body can also be used for temperature-activated antitumor drug release, combined with local hyperthermic treatment of cancer.

The membranes of the thermosensitive controlled-release microcapsules were constructed by a random mixing of Aquacoat (Table 4.1) with latex particles having a poly(EA/MMA/2-hydroxyethyl methacrylate) core and a poly(N-isopropylacrylamide (NIPAAm)) shell (Figure 4.6). This is an example where the membrane has the random two-phase structure in Figure 4.5. The microcapsules exhibited a thermosensitive release of water-soluble drug.[18] The mechanism is explained in Figure 4.6. When the temperature was changed in a stepwise manner from 30°C to 50°C, the microcapsules showed an "on-off" pulsatile release. This on-off response was reversible.

Alternatively, this type of thermosensitive microcapsule can be prepared even with already established pharmaceutical ingredients. As is well known, HPC, the commonly used binder and coating substrate (Table 4.1), has a lower critical solution temperature (LCST), around 44°C, and its water solubility drastically changes across the LCST. Negatively thermosensitive drug-release microcapsules with an HPC layer were thus designed by utilizing the thermally reversible dissolution/precipitation process resulting from the LCST phenomena[19] (Figure 4.7). In this case,

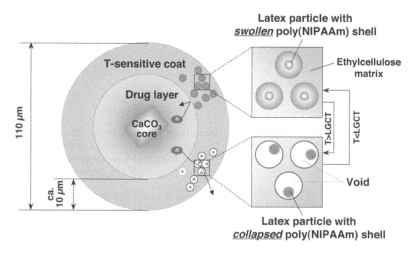

FIGURE 4.6   Composite particle exhibiting positively thermosensitive drug release.

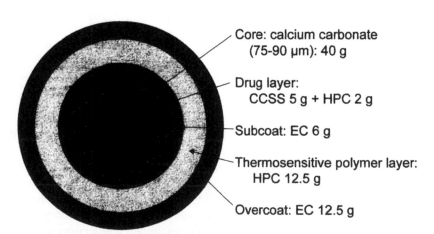

FIGURE 4.7   Composite particle exhibiting negatively thermosensitive drug release.

particles are constructed with multilayers of monolithic structure (Figure 4.5). The release rate from the microcapsules with a sandwiched HPC layer was found to be drastically decreased when the temperature came near the LCST: the release rate at 50°C was approximately 10 times lower than that at 30°C.

Unlike many types of millisized devices that have been developed so far, the microparticles mentioned here were around 100 μm in diameter. Such small dimensions provided a sharp release rate change in response to temperature change. This may be one of advantages of fine particle coating technology. The above demonstrated flexibility in designing membrane structures also offers considerable advantages over conventional microencapsulation methods.

4.4.2 PARTICLE DESIGN IN THE EMULSIFYING PROCESS

Particulate systems, including microcapsule, liposome, microemulsion, polymeric nanoparticles, and micelles, have been developed for pharmaceutical use. Their application as drug carriers for cancer therapies has attracted much attention recently. The particle structures to be constructed are essentially the same as those of microparticles produced by the fluid bed process. The present authors have been developing particulate drug delivery systems for gadolinium neutron-capture therapy (Gd-NCT) for cancer. Gd-NCT is a cancer therapy that utilizes γ-rays, X-rays, and electrons emitted *in vivo* as a result of the nuclear neutron-capture reaction with administered gadolinium-157.[20–24]

A key factor for success in Gd-NCT at the present stage is a device whereby Gd is delivered efficiently and is retained in tumor for a sufficient period during thermal-neutron irradiation. Here, as examples of composite formation in liquid phase, we describe preparation of gadolinium-containing lipid nanoparticles and chitosan nanoparticles, which have been designed and prepared in order to accumulate gadolinium in a tumor.

Lipid Nanoparticles

Gadolinium-diethylenetriaminepentaacetic acid (Gd-DTPA) is traditionally used clinically as an MRI contrast agent. It is reported that Gd-DTPA is rapidly eliminated from systemic circulation and hardly accumulates in tumors because of its high hydrophilicity. Meanwhile, the Gd concentration in tumors required to obtain the minimum tumor inactivation effect was estimated to be as high as about 100 μg Gd-157/ml.[22]

One of the major obstacles to using lipid particles as drug carriers is their rapid removal from systemic circulation by the reticuloendothelial system (RES). To avoid RES uptake, attempts have been made to coat lipid particles with a hydrophilic polymer such as polyethylene glycol (PEG). In addition, it is reported that drug carriers smaller than 100 nm are expected to more easily avoid uptake by RES. From these perspectives, a hydrophobic stearylamide (SA) derivative of Gd-DTPA was synthesized. Soybean oil and lecithin were employed to be the standard components of the parenterally administerable nanoparticulate systems because of their biocompatible properties. Gd-containing nanoparticles were surface modified with a cosurfactant having polyoxyethylene (POE) units. They were designed and prepared to fulfill the following requirements: particle size smaller than 100 nm; high Gd content so as to achieve a Gd concentration of 100 μg/g in the tumor *in vivo;* and surface properties giving rise to prolonged circulation in the blood (Figure 4.8).[25,26]

Lipid nanoparticles containing Gd-DTPA-SA (Gd-nano-LPs) were prepared by a thin-film hydration method combined with a sonication (bath-type sonication) method. Formulations and the particle size of lipid emulsions are listed in Table 4.2.

As shown in Table 4.2, use of a cosurfactant led to a marked reduction in the size of the resultant nanoparticles: the particle size was less than 100 nm. Additionally, the use of HCO-60 was found to be effective to reduce the particle size even in the high-Gd formulation. At 48 h after intraperitoneal injection of the lipid nanoparticles containing HCO-60 in tumor-bearing hamsters, the Gd

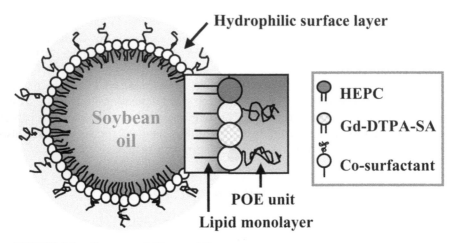

FIGURE 4.8   Structure of Gd-nano-LP.

TABLE 4.2   Formulation of Gd-DTPA Derivative-Containing Nano-LPs

| | Standard-Gd Formulation | | High-Gd Formulation |
	Plain Gd-nano-LP	Cosurfactant Gd-nano-LP	
HEPC[a]	500 mg	500 mg	250mg
Gd-DTPA-SA	250 mg	250 mg	500 mg
Soybean oil	2 ml	2 ml	2 ml
Cosurfactant	—	750 mg[b]	750 mg[c]
Water	23 ml	23 ml	23 ml
Particle size	250 nm	92,76,78 nm	84 nm

[a] $L_{-\alpha}$ Phosphatidylcholine hydrogenated (from egg yolk).

[b] Myrj 53, Brij 700, or HCO-60.

[c] HCO-60.

concentration in the tumors reached 107 µg/g tumor (wet basis). This result suggested that the potent antitumor effect might be obtained using the HCO-60 Gd-nano-LPs in the high-Gd formulation as a Gd carrier in Gd-NCT.

Chitosan Nanoparticles

As an alternative approach, biodegradable and highly gadopentetate-loaded chitosan nanoparticles (Gd-nano-CP) were prepared by a novel emulsion-droplet coalescence technique.[27–29] Chitosan, a polysaccharide, has been widely studied as a material for drug delivery systems due to its bioerodible, biocompatible, bioadhesive, and bioactive characteristics. These distinctive features led us to use chitosan as a promising material for design and preparation of Gd-loaded nanoparticulate devices. Further, its amino pendants dangling from sugar backbones may be useful for enrichment of Gd-DTPA into the nanoparticles by a possible ionic interaction between two free-carboxylic groups of Gd-DTPA and the amino groups of chitosan.

The preparation process of Gd-nano-CPs is shown in Figure 4.9. Chitosan was dissolved in an aqueous solution of Gd-DTPA. This solution was added to the paraffin liquid containing Arlacel C

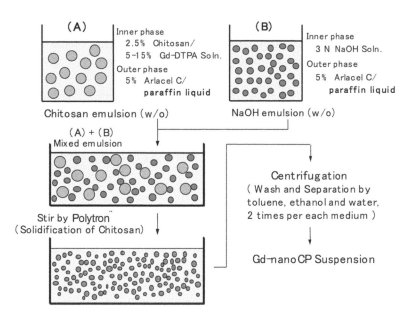

FIGURE 4.9   Preparation of Gd-nano-CP by emulsion-droplet coalescence technique.

and was stirred using a high-speed homogenizer to form a water-in-oil (w/o) emulsion A. Similarly, a w/o emulsion B consisting of sodium hydroxide solution and paraffin liquid containing Arlacel C was prepared. The emulsions A and B were mixed and stirred to solidify chitosan as a result of collision and coalescence between droplets of each emulsion.

With 100% deacetylated chitosan, mean particle size and Gd content were 426 nm and 9.3% (corresponding to 32.4% Gd-DTPA), respectively. The mechanism of Gd-nano-CP generation is unclear in detail, but it is presumed that the incorporation of Gd resulted from ionic interaction between two free carboxylic groups in Gd-DTPA molecules and polyamino groups in the chitosan molecules.

Release of Gd-DTPA from Gd-nano-CPs in an isotonic phosphate buffer saline solution (PBS) at 37°C was hardly recognized during 7 days (1.8%) in spite of high water solubility of Gd-DTPA. In contrast, in human plasma, Gd-nano-CPs released 55% and 91% of Gd-DTPA during 3 and 24 h at 37°C, respectively.

Neutron irradiation testing was carried out *in vivo* to confirm its therapeutic potential in Gd-NCT, using B16F10 melanoma-bearing male C57BL/6 mice. Gd-nano-CP suspension or Gd-DTPA solution as a control was intratumorally injected twice at a dose of Gd 1200 μg 24 and 8 hours before thermal-neutron irradiation. The thermal-neutron irradiation was performed only once for each tumor site with a flux of 2×10^9 n/cm^2/s for 60 min at Kyoto University Research Reactor (Japan).

When the change of tumor volume in mice after the thermal-neutron irradiation was examined, the tumor growth was significantly suppressed in the Gd-nano-CP-administered group despite the radio-resistive melanoma model, compared with the cases of no Gd–administered or Gd-DTPA solution–administered group. As is generally recognized, Gd-DTPA in solution used in ordinary Gd-NCT trials can be eliminated very rapidly from tumor tissue. The enhanced Gd-NCT effects by Gd-nano-CP might result from the high gadolinium quantity retained in tumor tissue.

4.4.3   SUMMARY

Using fluid bed processes and processors along with the appropriate materials and their well-designed formulation and particulate structure make it possible to prepare highly functional particles, as demonstrated here. However, this method has an unavoidable limit in the size of particles that it can efficiently process because of easy agglomeration of particles smaller than 200 μm during spraying. In order to expand applications of this process, some new or improved fluidization technologies will be required.

Very fine, especially nanosized, composite particles can be produced through liquid-phase processes. When applied to cancer therapy, there would be many factors affecting their biodistribution. Their size and surface properties have to be regulated depending on treatment strategy, but this cannot yet be done well. Since these particles have many attractive applications, as demonstrated here, development of novel technologies of composite formation is expected.

REFERENCES

1. Jono, K., Ichikawa, H., Miyamoto, M., and Fukumori, Y., *Powder Technol.*, 113, 269–277, 2000.
2. Fukumori, Y. and Ichikawa, H., in *Encyclopedia of Pharmaceutical Technology,* DOI:10.1081/E-EPT 120018218, Marcel Dekker, New York, 2003, pp. 1–7.
3. Ichikawa, H., Tokumitsu, H., Jono, K., Osako, Y., and Fukumori, Y., *Chem. Pharm. Bull.*, 42 (6), 1308–1314, 1994.
4. Ichikawa, H. and Fukumori, Y., *Int. J. Pharm.*, 180, 195–210, 1999.
5. Geldart, D., *Powder Technol.*, 7, 285–292, 1973.
6. Fukumori, Y., Ichikawa, H., Jono, K., Takeuchi, Y., and Fukuda, Y., *Chem. Pharm. Bull.*, 40 (8), 2159–2163, 1992.
7. Fukumori, Y., Ichikawa, H., Jono, K., Fukuda, T., and Osako, Y., *Chem. Pharm. Bull.*, 41 (4), 725–730, 1993.
8. Fukumori, Y., in *Multiparticulate Oral Drug Delivery,* Marcel Dekker, New York, 1994, pp. 79–111.
9. Lehmann, K. O. R., in *Aqueous Polymeric Coatings for Pharmaceutical Dosage Forms,* Marcel Dekker, New York, 1989, pp. 153–246.
10. Steuernagel, C. R., in *Aqueous Polymeric Coatings for Pharmaceutical Dosage Forms,* Marcel Dekker, New York, 1989, pp. 1–62.
11. Moore. K. L., in *Aqueous Polymeric Coatings for Pharmaceutical Dosage Forms,* Marcel Dekker, New York, 1989, pp. 303–316.
12. Nagai, T., Sekikawa, F., and Hoshi, N., in *Aqueous Polymeric Coatings for Pharmaceutical Dosage Forms,* Marcel Dekker, New York, 1989, pp. 81–152.
13. Muroi, S., *High Polymer Latex Adhesives,* Kobunnshi-Kankokai, Kyoto, Japan, 1984.
14. Fukumori, Y., Yamaoka, Y., Ichikawa, H., Fukuda, T., Takeuchi, Y., and Osako., Y., *Chem. Pharm. Bull.*, 36 (4), 1491–1501, 1988.
15. Kaneko, Y., Sakai, K., and Okano, T., *Biorelated Polymers and Gels,* Academic Press, Boston, 1998, pp. 29–70.
16. Ichikawa, H. and Fukumori, Y., *STP Pharm.*, 7 (6), 529–545, 1997.
17. Peppas, N. A., Bures, P., Leobandung, W., and Ichikawa, H., *Eur. J. Pharm. Biopharm.*, 50 (1), 27–46, 2000.
18. Ichikawa, H. and Fukumori, Y., *J. Controlled Release,* 63, 107–119, 2000.
19. Ichikawa, H. and Fukumori, Y., *Chem. Pharm. Bull.*, 47 (8), 1102–1107, 1999.
20. Allen, B. J., Mcgregor, B. J., and Martin, R. F., *Strahlenther. Onkol.*, 165, 156–158, 1989.
21. Akine, Y., Tokita, N., Matsumoto, T., Oyama, H., Egawa, S., and Aizawa, O., *Strahlenther. Onkol.*, 166, 831–833, 1990.
22. Akine, Y., Tokita, N., Tokuye, K., Satoh, M., Fukumori, Y., Tokumitsu, H., Kanamori, R., Kobayashi, T., and Kanda, K., *J. Cancer Res. Clin. Oncol.*, 119, 71–73, 1992.
23. Fukumori, Y., Ichikawa, H., Tokumitsu, H., Miyamoto, M., Jono, K., Kanamori, R., Akine, Y., and Tokita, N., *Chem. Pharm. Bull.*, 41 (6), 1144–1148, 1993.
24. Miyamoto, M., Ichikawa, H., Fukumori, Y., Akine, Y., and Tokuye, K., *Chem. Pharm. Bull.*, 45 (12), 2043–2050, 1997.
25. Miyamoto, M., Hirano, K., Ichikawa, H., Fukumori, Y., Akine, Y., and Tokuye, K., *Chem. Pharm. Bull.*, 47 (2), 203–208, 1999.

26. Miyamoto, M., Hirano, K., Ichikawa, H., Fukumori, Y., Akine, Y., and Tokuuye, K., *Biol. Pharm. Bull.,* 22 (12), 1331–1340, 1999.
27. Tokumitsu, H., Ichikawa, H., and Fukumori, Y., *Pharm. Res.,* 16 (12), 1830–1835, 1999.
28. Tokumitsu, H., Hiratsuka, J., Sakurai, J., Kobayashi, T., Ichikawa, H., and Fukumori, Y., *Cancer Lett.,* 150, 177–182, 2000.
29. Tokumitsu, H. et al., *STP Pharm. Sci.,* 10 (1), 39–49, 2000.

4.5 Dispersion of Particles

Kuniaki Gotoh
Okayama University, Okayama, Japan

Hiroaki Masuda and Ko Higashitani
Kyoto University, Katsura, Kyoto, Japan

4.5.1 PARTICLE DISPERSION IN GASEOUS STATE

In order to disperse the agglomeration of particles in a gaseous state, external force larger than the adhesive force between primary particles in the agglomeration should be applied to the agglomeration. The dispersio methods can be classified by the methods of applying dispersion forces. Here, we outline methods for dispersion of agglomerated particles in a gaseous state, with particular attention to the external forces that can be utilized for the dispersion of agglomerates.

Dispersion Force

Dispersion Force Induced by Acceleration or Deceleration of Airflow

One of the external forces that can be utilized for the dispersion of agglomerated particles is a fluid resistance induced by the acceleration or deceleration of a fluid. Here, it is assumed that agglomerated particle can be modeled by a doublet consisting of two primary particles having different sizes, as shown in Figure 5.1. When the agglomerated particle is in a uniform flow field, the motion equation of each primary particle can be written as follows[1,2]:

$$m_A \frac{du_p}{dt} = R_{fA} - F_d$$
$$m_B \frac{du_p}{dt} = R_{fB} + F_d$$

(5.1)

where R_{fA} and R_{fB} are the fluid resistance forces acting on each particle, u_p is particle velocity, and F_d is the force acting on each particle. For the agglomeration, the force F_d acts as a dispersion force.

m_A and m_B are the mass of particle A and particle B respectively. These masses can be calculated by the particle diameters D_{pA}, D_{pB} and the particle mass density ρ_{pA}, ρ_{pB}.

$$m_A = \frac{\pi}{6} \rho_{pA} D_{pA}^3$$
$$m_B = \frac{\pi}{6} \rho_{pB} D_{pB}^3$$

(5.2)

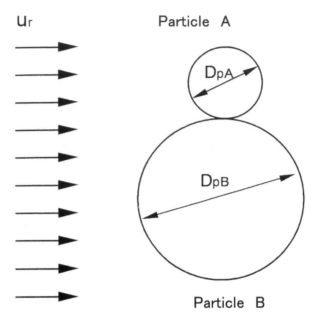

FIGURE 5.1   Agglomerated particle in uniform flow.

If the particle Reynolds number is in the range of the Stokes law, the resistance force R_f can be expressed by

$$R_{fA} = 3\pi\mu u_r D_{pA}^3$$
$$R_{fB} = 3\pi\mu u_r D_{pB}^3 \tag{5.3}$$

where u_r is the relative velocity between particle and fluid. Substituting Equation 5.2 and Equation 5.3 into Equation 5.1 and rearranging under the assumption that the mass density of the agglomerated particle is the same as that of the primary particles ($\rho_p = \rho_{pA} = \rho_{pB}$) leads to the following equation expressing the dispersion force F_d.

$$F_d = 3\pi\mu u_r D_{pA} D_{pB} \frac{\left(D_{pB} - D_{pA}\right)}{\left(D_{pA}^2 - D_{pA}D_{pB} + D_{pB}^2\right)} \tag{5.4}$$

When we consider the resistance force beyond the Stokes regime, the resistance force R_f is expressed by the following equation[2]:

$$R_f = \frac{p}{8} D_v^2 \rho_f u_r^2 \left(0.55 + \frac{4.8}{\sqrt{Re}}\right) \tag{5.5}$$

$$D_v \equiv D_{pA}/D_{pB}$$
$$Re = D_v \rho_f u_r / m$$

$$\rho_f : \text{fluid density}$$

In this case, the dispersion force F_d can be expressed by

$$
\begin{aligned}
F_d = \{ &0.119\rho_f u_r^2 D_{pB}^2 D_v^2 \left(\kappa_A - D_v \kappa_B \right) \\
&+ 2.07\sqrt{\mu\rho_f u_r^3} D_{pB}^{3/2} D_v^{3/2} \left(\kappa_A - D_v^{3/2} \kappa_B \right) \\
&+ 9.05\mu u_r D_{pB} D_v \left(\kappa_A - D_v^2 \kappa_B \right) \} / (D_v^3 + 1
\end{aligned}
\tag{5.6}
$$

The above two expressions for the dispersion forces are assumed for single agglomerated particles. The effect of neighboring particles on the dispersion force has also been reported.[3]

Figure 5.2 shows the calculated result of dispersion force expressed by Equation 5.4.[4] In the calculation, the diameter of particle A was kept constant at 1 μm, and the diameter of particle B was varied. The difference of the diameters is expressed by the ratio of diameters D_{pB}/D_{pA} for the case of $D_{pB} < D_{pA}$ and D_{pA}/D_{pB} for the case of $D_{pA} < D_{pB}$. When the diameter ratio is unity (i.e., the diameter of particle B is the same as that of particle A), the dispersion force induced by the acceleration or deceleration is zero. In addition, when the ratio of D_{pA}/D_{pB} and D_{pB}/D_{pA} is zero (i.e., diameter of particle B is quite a lot bigger or smaller than that of A), the dispersion force induced by the acceleration or deceleration is also zero. In other words, the dispersion induced by the acceleration or deceleration of fluid is effective for the agglomeration that consists of primary particles having an adequate size ratio.

Dispersion Force Induced by Shear Flow Field

When agglomerated particles are dipped into a shear flow field having a velocity gradient γ, the dispersion force is induced by the shear flow. Bagster and Tomi[5] studied the force induced by the flow theoretically and reported that the shear stress τ acting on the agglomerated particles has its maximum in the central plane of the agglomerated particle < and the maximum value is expressed by following equation:

$$
\tau = 8.5\mu\gamma
\tag{5.7}
$$

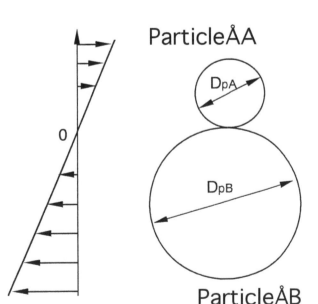

FIGURE 5.2 Estimation of dispersible region. (a) Comparison between dispersion force and adhesive force. (b) Dispersible region.

On the other hand, when an agglomerated particle consisting of two primary particles, shown in Figure 5.3, is dipped into the shear flow field having velocity gradient γ, it can be considered that the bending moment $\sigma\gamma$ is also induced. In the case where the diameter of particle B is quite a bit larger than that of particle A, the bending moment $\sigma\gamma$ is expressed by the following equation[1]:

$$\sigma_{\gamma} \approx \frac{93\mu\gamma}{\pi c^3} \qquad (5.8)$$

where c is a ratio of the diameter of the circular contact area to the diameter of the particle. Although the coefficient c depends on the contact configuration, the value is, in general, quite a bit smaller than unity. Thus, the value of the bending moment $\sigma\gamma$ calculated by Equation 5.8 is larger than the

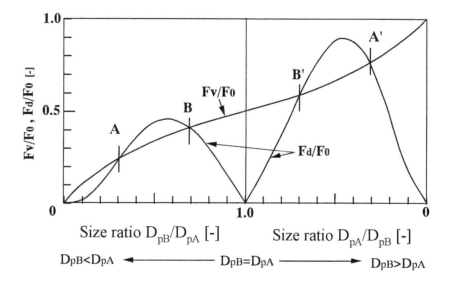

a) Comparison between dispersion force and adhesive force

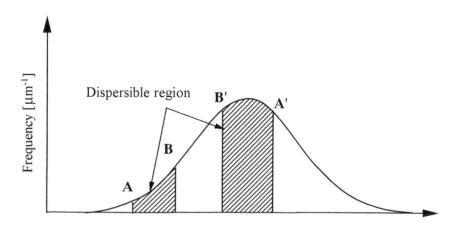

b) Dispersible region

FIGURE 5.3   Agglomerated particle in shear flow field.

value of the shear stress τ calculated by Equation 5.7. This fact means that the bending moment dominates the dispersion force in the shear flow field.

Impaction Force

When an agglomerated particle having mass m collides with an obstacle with velocity v_i, the impaction force F_i expressed by the following equation arises[1]:

$$F_i = mv_i / \Delta t \qquad (5.9)$$

where Δt is the duration time of the impaction.

On the other hand, compression stress σ_i is generated in the central plane of the agglomeration. If the agglomeration is modeled by a single spherical particle having a diameter of D_{pag}, the compression stress σ_i can be expressed by[1]

$$\sigma_i = \frac{2}{3} \rho_p D_{pag} \left(\frac{v_i}{\Delta t} \right) \qquad (5.10)$$

When we consider the dispersion by the impaction against an obstacle, the impaction probability η should be taken into consideration. The probability η is a function of the shape of the obstacle and of the inertia parameter Ψ or Stokes number S_k. The inertia parameter Ψ is defined by

$$\Psi = \frac{\rho_p u D_{pag}}{18 \mu D} = \frac{S_k}{2} \qquad (5.11)$$

where u and D are representative velocity and representative length, respectively. The dependence of the impaction probability η on the obstacle shape can be found in the literature describing the prediction of collection efficiency of particles on the impactor or filters. In general, the impaction probability η increases with the inertia parameter Ψ. It can be found by Equation 5.9 through Equation 5.11 that the impaction force F_i, the compression stress, and the inertia parameter increase with the diameter of the agglomerated particles. It means that dispersion by impaction is effective for the larger agglomeration, independently of the structure of the agglomeration.

The impaction force described above is the force induced by the collision of the agglomerated particle. The force induced by the collision of other obstacle with the agglomerated particle is also utilized for the dispersion. Two of the dispersers that utilize the collision force are a ball mill type and a fluidized bed type disperser. In these dispersers, not only the impaction force but also the friction or attrition force are induced and act as the dispersion force.

Other Dispersion Forces

The dispersion forces described above assume a simple flow field and simple particle behavior. However, the flow in a real disperser is complex, and in many case, the flow is turbulent. The turbulent flow induces dispersion force. Although the mechanism of the dispersion in the turbulent flow is not well known, it was reported that the mass median diameter D_{p50} of an aerosol passing through the disperser such as the orifice type[6] and the tube type[7] is proportional to the energy dissipation rate $\varepsilon^{-0.2}$, which is calculated by the pressure drop ΔP in the disperser. If the diameter of the agglomerated particles is smaller than the microscale of the vortex in the turbulence, the shear stress τ and the bending stress σ_g can be expressed by the flowing equations[1]:

$$\tau \cong 3.1 \sqrt{\rho_f \mu \varepsilon}$$
$$\sigma_\gamma \cong \frac{10.8 \sqrt{\rho_f \mu \varepsilon}}{c^3} \qquad (5.12)$$

Disperser in Gaseous State

Orifice

An orifice is a plate having a smaller hole than the diameter of the pipe, as shown in Figure 5.4. It makes a converging and expanding flow, inducing a rapid acceleration and a rapid deceleration. The mass median diameter D_{p50} of aerosol particles dispersed by the orifice correlates with the energy dissipation rate ε calculated by the pressure drop ΔP at the orifice[6]:

$$D_{p50}/D_{p50s} = 31.3\varepsilon^{-0.2}$$
$$\varepsilon = 0.4\Delta P\bar{u}/d_0$$

(5.13)

D_{p50s}: mass median diameter of primary particles
$\bar{u}$: average air velocity
d_0: diameter of orifice

Capillary Tube

When the flow in a capillary tube is turbulent, the tube can be used as a disperser. Yamamoto et al.[7] reported that the mass median diameter D_{p50} of aerosol particles dispersed by the capillary tube correlates with the energy dissipation rate ε calculated by the pressure drop ΔP.

$$D_{p50} = 15\varepsilon^{-0.2}$$
$$\varepsilon = \Delta P\bar{u}/2.5d$$

(5.14)

d: diameter of capillary tube

Ejector[1,2]

A schematic cross-sectional view of an ejector is shown in Figure 5.5. Pressurized primary air is introduced into the ejector. The air jet generated by the nozzle induces a pressure lower than the atmospheric pressure. The low pressure causes a secondary airflow. When a powder containing agglomerated particles is fed with the secondary airflow, the agglomerated particles are dispersed at the merging point of the primary and the secondary flow by the forces induced by the acceleration and shear flow field.

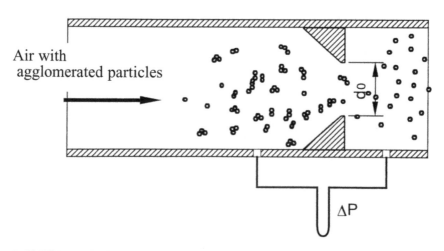

Air with agglomerated particles

ΔP

FIGURE 5.4 Orifice.

Venturi[1,2]

A venturi has a throat that causes converging and expanding flows, as shown in Figure 5.6. In the venturi, low pressure is induced at the throat, resulting in a secondary airflow through the inlet tube. If agglomerated particles are fed with the secondary airflow, the agglomerated particles suffer from the forces induced by the acceleration and shear flow field and are dispersed.

Fluidized Bed[8]

In a fluidized bed, a particle suffers mechanical forces such as impaction and attrition by the neighboring fluidizing particles. When the airflow velocity above the fluidized bed is higher than the terminal settling velocity of the primary particle fluidizing, the particles dispersed by the mechanical forces are entrained to the airflow. Thus, the fluidized bed can be utilized as a disperser.

When particles do not fluidize because of their adhesive force, mixing in large particles such as glass or metal beads as the fluidizing medium is effective for dispersion (Figure 5.7). The large particles are fluidized easily and generate impaction and attrition forces that act as dispersion forces on the adhesive particles.

Mixer-Type Disperser[4,8,9]

The mixer-type disperser shown in Figure 5.8 is one of the mechanical dispersers with a rotating obstacle. The disperser consists of the impeller rotating at high speed (from 4000 to 20,000, in general), an inlet tube attached to the top of the vessel, and outlet tube attached tangentially to the

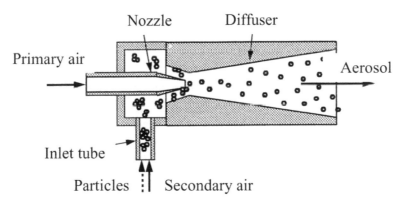

FIGURE 5.5   Ejector.

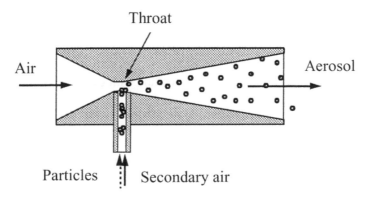

FIGURE 5.6   Venturi.

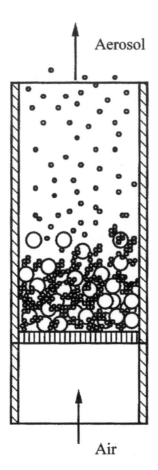

FIGURE 5.7   Fluidized bed.

vessel. By the rotation of the impeller, low pressure is generated above the impeller, and it causes suction flow through the inlet tube. The powder containing agglomerated particles is fed with the suction flow. The agglomerated particles are dispersed by several forces; these are the forces induced by acceleration and shear flow and/or impaction force to the impeller and the vessel wall.

4.5.2   PARTICLE DISPERSION IN LIQUID STATE

Particles in suspensions are thermodynamically unstable and tend to coagulate each other. Because well-dispersed suspensions are required in various particulate processes, it is important to establish the dispersion technology of particles in solutions. The term "dispersion" has been used in two different ways: the dispersion that will be discussed in this section represents the breakup of coagulated particles (flocs) by external forces, but it is often used to imply that a suspension is stable, such that particles do not coagulate even if they collide with each other. The dispersion of this meaning is discussed in Section 2.8.2.

Flow fields, collisions, high-pressure differences, and ultrasonic fields have been employed as the fundamental principle to deflocculate flocs in various commercial dispersers. In the defloccu-lation by flow fields, particles will be deflocculated by the hydrodynamic drag force exerted on coagulated particles. In the collision defloccuation flocs are broken by their collision to the solid surface or the other flocs, but the contribution of this deflocculation will be small because the inertia of microscopic particles in small in liquids. Particles are deflocculated when the suspension experi-ences either the high-pressure drop or the ultrasonic exposure. The cavitation, vibration, or turbu-

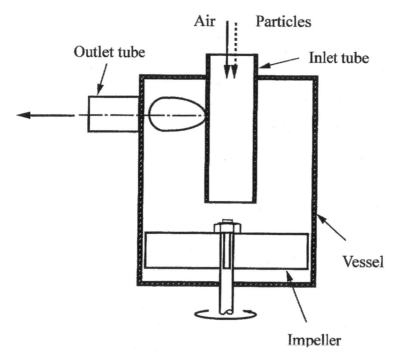

FIGURE 5.8   Mixer-type disperser.

lence has been considered as the possible mechanism for these deflocculations, but enough data to clarify the mechanism have not been reported.

Flow and ultrasonic fields have been popularly employed in many commercial devices to deflocculate flocs in suspensions. Here, the fundamentals of the deflocculation by the flow and ultrasonic fields are examined and their characteristics are compared.

Theoretical Background of Deflocculation by Flow Fields

The flow fields employed in various dispersers are usually very complicated, but they will be decomposed into a few fundamental flows: the shear, elongational, and turbulent flows. Because the turbulent flow is too complicated to know the details of the flow field, the mechanism of turbulent deflocculation is hard to clarify, but the features of the turbulent deflocculation can be estimated using the characteristics of the deflocculation by shear flow, because flocs will be broken by the shearing flow in turbulent eddies.

The hydrodynamic theory for the relative trajectory between a pair of particles of equal size in the shear and elongational flows was developed.[10] Applying this trajectory analysis to a pair of coagulated particles located in a shearing plane, the conditions under which particles remain coagulated or dispersed are calculated, as illustrated in Figure 5.9, where N_R is the dimensionless measure of the electrostatic repulsive force against the attractive force acting between particles $6\mu\pi a^3\gamma/A$, and N_F is the measure of the strength of flow field against the attractive force between particles, $\varepsilon a\psi_o^2/A$, where μ and ε are the viscosity and permitivity of the medium, respectively, a and ψ_o are the radius and surface potential of particles, respectively, γ is the shear rate, and A is the Hamaker constant.[11, 12] The boundary condition between coagulated and dispersed states can be expressed analytically also by the following equation when a pair of particles is in a shearing plane[12]:

$$N_P = \frac{(a/\delta)^2}{12} - \pi\kappa a N_R \qquad (5.15)$$

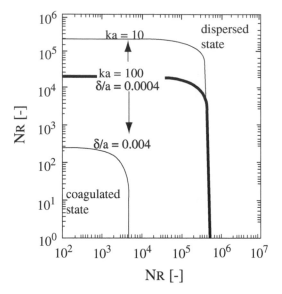

FIGURE 5.9   Boundary between coagulated and dispersed states for a pair of particles in a shearing plane.

where δ is the minimum gap between the surfaces of coagulated particles, $k = (2n_0z^2e^2/\varepsilon kT)^{0.5}$, n_0 and z are the number concentration and valency of ions, respectively, e is the elementary charge, k is the Boltzmann constant, and T is the temperature. This equation enables us to estimate on what conditions the shear-induced deflocculation depends. Figure 5.9 shows that δ/a and κa, especially δ/a, are the important parameters to control the deflocculation. The value of δ is often assumed to be 4Å for bare surfaces, which is attributable to the Born repulsion between the surfaces. When molecules are absorbed on the surface, the value of δ is closely correlated with the thickness of the adsorbed layer of molecules on the surface. The adsorbed layers of surfactants and polymers are often used to stabilize particles in solutions.

It is also possible to evaluate numerically the deflocculation process of a pair of particles tilted from a shearing plane and a pair of unequal deflocculated than the flocs described above.

Applying the trajectory analysis for a pair of flocs in an elongational flow, the coagulation-dispersion map similar to Figure 5.9 will be obtained. When the map for the elongational deflocculation is compared with Figure 5.9, it is found that flocs are deflocculated at the smaller dissipation energy of flow in the elongational deflocculation. This prediction confirms are the experimental finding that the elongational flow is more effective to deflocculate particles than the shear flow.[13]

Deflocculation by Flow Fields

It is difficult to experimentally realize the process where flocs are deflocculated solely by the simple shear or the elongational flow. Here, the characteristics of the deflocculation by the orifice contractile flow and rotational flow generated by a rotary disk are examined experimentally. These are the fundamental flows employed in the commercial devices such as dispersers with rotational blades; thus, it is important to understand the deflocculation by these flow fields.

Deflocculation by Orifice Contractile Flow[14]

The orifice contractile flow can be realized by placing a plate with a small hole (an orifice) in the Poisuille flow. The fluid is contracted and abruptly accelerated just before the orifice. The flow along the centerline may be regarded as an elongational flow and the flow near the wall inside the orifice is regarded as a high shear flow.

Figure 5.10 shows the dependence of the number-averaged diameter, D_{av}, of deflocculated particles on the energy dissipation of the orifice contractile flow, Φ, for three orifice diameters d, where flocs of monodispersed latex particles of diameter $D_o = 0.71, 0.91$, and $1.27\ \mu m$ coagulated in $1\ M$ KCl solution are employed. It is important to note that D_{av} depends on Φ but not on d, as for as the value of D_o is the same, and that the relation D_{av} versus Φ is expressed by $D_{av} \sim \Phi^{-0.035}$, irrespective of D_o. The maximum number of constituent particles in a floc, i_{max}, also shows the similar dependence on Φ, D_o, and d, and the relation i_{max} versus Φ is expressed by $i_{max} \sim \Phi^{-0.11}$. These equations are well supported by the data given elsewhere.[15,16] The results indicate that the energy dissipation of the flow and the size of constituent particles in the floc are the most important parameters to control the degree of deflocculation.

Deflocculation by Flow Generated by Rotary Disk[17]

Because dispersers with rotary blades are widely used, it is fundamentally important to know the characteristics of the deflocculation by the rotational flow generated by a rotary disk, as schematically shown in Figure 5.11. Because the fluid flows along the tube with the rotational motion generated by the rotor, the flow is a combined flow of the rotational, contractile, and shear flows. Figure 5.12a shows the dependence of the average diameter of broken flocs, D_{av}, on the rotational speed of the disk, n, for various gaps between the rotor and the inside wall of the cylinder, δ. The value of D_{av} decreases with n, as expected . It is interesting to note that D_{av} increases with δ when $\delta \geq 2.0$ mm, but it is independent of δ at $\delta \geq 2.0$ mm, whereas when the same data are plotted against the shear rate γ as in Figure 5.12b, the data of $\delta \geq 2.0$ mm fall on a single line. These results indicate the flocs are broken by the rotational and/or contractile flow at the upstream before the rotary disk but not by the shear flow generated within the gap when $\delta \geq 2.0$ mm. On the other hand, flocs are broken mainly by the shearing flow in the gap when the gap is widely open.

A high-shear flow has been considered to be effective to break flocs and a very small gap between the rotor and cylinder wall has been designed to gain a high-shear flow in many commercial devices. However, according to the above results that deflocculation is not attributable to the shear flow

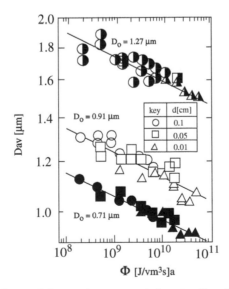

FIGURE 5.10 Dependence of the number-averaged diameter D_{av} of particles deflocculated by the orifice contractile flow on the energy dissipation Φ of the flow, d is the orifice diameter, and D_0 is the diameter of the constituent particle in the floc.

within the gap when the gap is sufficiently small, a sufficient deflocculation will be achieved by employing a thin disk, thus reducing the power required to rotate the disk considerably.

Deflocculation by Ultrasonic Field[18]

Ultrasonic dispersers are widely used, but there have been very few systematic investigations on the deflocculation by ultrasonication. Here, the characteristics of the ultrasonic deflocculation that the author has clarified are explained.

Figures 5.13a and 5.13b show the dependence of D_{av} and i_{max}, respectively, on the frequency and power W_s of sonic generator, the radiation time t_r, and the volume of suspensions V_f, where flocs of latex particles coagulated in 1 M KCl solution are employed. It is important to know that all the data are expressed by single curves irrespective of various experimental conditions. These results indicate that the degree of ultrasonic deflocculation is solely determined by the total sonic energy radiated to the floc solution of unit volume, $E_t (= W_s t_r / V_f)$, so long as the size of constituent particles is the same. This information is very important because the degree of deflocculation is easily controlled by adjusting the combination of the values of W_s, t_r, and V_f. It is also found that the ultrasonication is strong enough to defloculate flocs into particles with the initial size distribution when a sufficient amount of energy ($E \cong 4 \times 10^7$ J/m^3 in this case) is radiated.

It is known that the sonic vibration contributes not only to the deflocculation but also to the coagulation of suspending particles. Hence, it is necessary to make deflocculated particles free from coagulating again. This can be done by either diluting the suspension to minimize the collision frequency or dosing surfactants to form the adsorbed layer that stabilize the particles, just after flocs are deflocculated.

Comparison among Deflocculations by Flow and Ultrasonic Fields[18]

Here the features of the deflocculations examined above are compared. Figure 5.14 shows the size distributions of the flocs with nearly the same average size that are deflocculated by the orifice contractile flow, rotational flow, and ultrasonication. It is found that the size distributions of the flocs deflocculated by fluid flows are similar to each other, but they differ from that by ultrasonication. Fractions of single constituent particles and large coagulated particles are much higher in the sonic deflocculation than in the deflocculation by the fluid flows. This result implies that flocs are split into smaller agglomerates in the case of flow-induced deflocculation, whereas single particles are ripped off one by one from the floc surface in the case of sonic deflocculation as illustrated schematically in Figure 5.15. The former mechanism is called "splitting" and the latter, "erosion." This difference in deflocculation mechanism suggests that the flow-induced deflocculation with the subsequent sonic deflocculation will be the effective process to deflocculate particles.

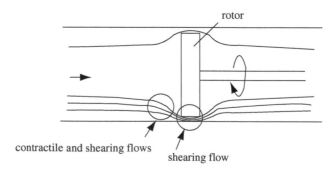

FIGURE 5.11 Schematic drawing of the flow field generated by the rotary disk. The contractile and rotational flow is predominant just before the rotor, but the shear flow is predominant within the gap.

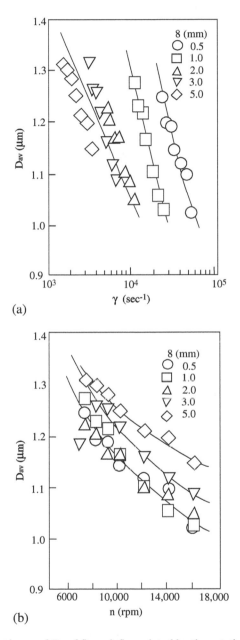

(a)

(b)

FIGURE 5.12 Dependence of D_{av} of flocs deflocculated by the rotational disk flow on various experimental conditions. (a) Dependence of D_{av} on γ; (b) dependence of D_{av} on n.

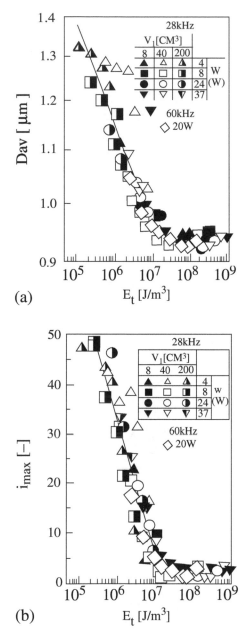

FIGURE 5.13 Dependence of D_{av} and i_{max} on $E_t \, (= W_s t_r / V_f)$ under various experimental conditions. (a) D_{av} versus E_t under various conditions; (b) i_{max} versus E_t under various conditions.

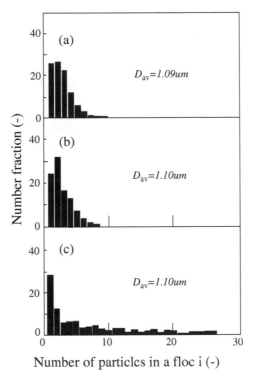

FIGURE 5.14 Comparison between size distributions of flocs dispersed by (a) orifice contractile flow, (b) rotational disk flow, and (c) ultrasonication.

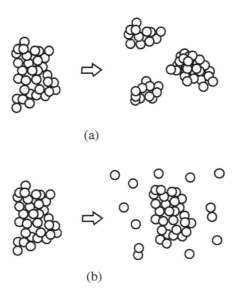

FIGURE 5.15 Schematics of two mechanisms of deflocculation: (a) splitting, (b) erosion.

REFERENCES

1. Kousaka, Y., Okuyama, K., Shimizu, A., and Yoshida, T., *J. Chem. Eng. Jpn.*, 12, 152–159, 1979.
2. Kousaka, Y., Endo, N., Horiuchi, T., and Niida, R., *Kagaku Kogaku Ronbunshu*, 18, 233–239, 1992.
3. Yuu, S. and Oda, T., *AIChE J.*, 29, 191–198, 1983.
4. Gotoh, K., Takahashi, M., and Masuda, H., *J. Soc. Powder Technol. Jpn.*, 29, 11–17, 1992.
5. Bagster, D. F. and Tomi, D., *Chem. Eng. Sci.*, 29, 1773–1783, 1974.
6. Yamamoto, H. and Suganuma, A., *Kagaku Kogaku Ronbunshu*, 9, 183–188, 1983.
7. Yamamoto, H., Suganuma, A., and Kunii, T., *Kagaku Kogaku Ronbunshu*, 3, 12–18, 1977.
8. Masuda, H., Fushiro, S., and Iinoya, K., *J. Res. Assoc. Powder Technol. Jpn.*, 14, 3–10, 1977.
9. Gotoh, K., Asaoka, H., and Masuda, H., *Adv. Powder Technol.*, 5, 323–337, 1994.
10. Batchelor, G. k. and Green, J.T., *J. Fluid Mech.*, 56, 375, 1972.
11. Higashitani, K., Tsutsumi, T., and Iimura, K., Proc. of China Japan Symposium on Particology, Beijing, 1996, p. 1.
12. Higashitani, K., Iimura, K., and Vakarelski, I.U. *KONA*, 18, 6, 2000.
13. van de Ven, T. G. M., *Colloidal Hydrodynamics*, Academic Press, London, 1989, p. 531
14. Higashitani, K., Tanise, N.,Yoshiba, A., Kondo, A., and Murata, H., *J. Chem. Eng. Japan*, 25, 502, 1992.
15. Sonntag, R. and Russel, W., *J. Colloid Interf. Sci.*, 115, 390, 1987.
16. Higashitani, K., Inada, N., and Ochi, T., *Colloid Surfaces*, 56, 13, 1991.
17. Higashitani, K. and Biryushi, K., Asakura Shoten, Tokyo, 1994, p. 43.
18. Higashitani, K., Yoshida, K., Tanise, N., and Murata, H., *Colloids Surfaces* A, 81, 167, 1993.

4.6 Electrical Charge Control

Motoaki Adachi
Osaka Prefecture University, Sakai, Osaka, Japan

Kikuo Okuyama
Hiroshima University, Higashi-Hiroshima, Hiroshima, Japan

Hiroaki Masuda, Shuji Matsusaka, and Ko Higashitani
Kyoto University, Katsura, Kyoto, Japan

4.6.1 IN GASEOUS STATE

The charging process of particles by gaseous ions depends on two physical mechanisms: field and diffusion charging. When ions drift along electric field lines and impinge on the particle, the charging process is referred to as field charging. Diffusion charging is due to thermal collisions between particles and ions. Photoelectron charging, in which electrons are emitted from particles by UV irradiation, is also used to control the particle charge.

Gaseous Ions

Production Methods

Gaseous ions are produced by three methods of irradiations: UV or other radiations, discharges in gas, and fissions of droplets, as shown in Table 6.1. In α-ray ($\cong$5 MeV) and β-ray ($\cong$0.7 MeV) irradiations, high-energy particles collide with air molecules and produce primary positive ions and electrons. The primary ions grow metastable positive ions by ion–molecule reactions and ion clustering. The free electrons attach molecules with high electron affinity, such as O_2 or H_2O, and form negative ions. In soft X-ray ($\cong$10 keV) and vacuum UV-ray ($\cong$3~10 eV) irradiations, gas atoms and molecules are excited by multiphoton adsorption and emit electrons. These primary positive ions and electrons change to metastable positive and negative ions at atmospheric and low pressures but do not change in vacuum. When UV-rays ($\cong$5 eV) irradiate a metal film, photoelectrons are emitted from the film. Photoelectrons become metastable negative ions by electron attachment and ion clustering at atmospheric and low pressures.

Corona discharge is produced in an inequality electric field between a needle electrode and a plate electrode or between a line electrode and a cylinder electrode. When energy higher than the ionization potential of gas is given to casual electrons, the accelerated electrons collide with gas molecules and produce pairs of a new electron and a primary positive ion near the needle or the line electrode (discharge electrode). Either the electrons or primary positive ions are drawn toward the plate or cylinder electrode (collection electrode) according to the polarity of the discharge voltage. During the travel from the discharge electrode to the collection electrode, the electrons and primary positive ions change to metastable negative and positive ions, respectively.

When water droplets are broken by a collision with an obstacle, like a waterfall, the space is filled by negative ions. This is called the Lenard effect. In a droplet suspended in air, anions generally form a layer near a surface of the droplet. By the collision of a droplet, many droplets with very small sizes are produced from a surface layer, and a few droplets with relatively large size are produced from the

465

TABLE 6.1   Methods for Generation of Ions

Method	Principle	Products	Maximum ion Concentration	Other Applications
Irradiation				
α-ray	Ionization of gas	Bipolar ions	10^7 cm^{-3}	Static charge elimination
β-ray	Ionization of gas	Bipolar ions	10^7 cm^{-3}	Static charge elimination
Soft X-ray	Photoionization of gas	Bipolar ions	10^8 cm^{-3}	Static charge elimination
UV-ray	Photoionization in vacuum	Positive ions and electrons	10^5 cm^{-3}	Static charge elimination
UV-ray	Photoelectron emission from metals	Photoelectrons or negative ions	10^5 cm^{-3}	Contamination control
Discharge				
DC corona	Ionization of gas	Unipolar ions and ozone	10^7 cm^{-3}	Electrostatic precipitator Indoor air cleaner
AC corona	Ionization of gas	Bipolar ions and ozone	10^7 cm^{-3}	Static charge elimination
Fission of droplets				
Collision	Lenard effect	Negative ions and charged droplets	10^4 cm^{-3}	Physical therapy Sterilization
Electrospray	Reyleigh fission	Unipolar ions and charged droplets	10^6 cm^{-3}	Spray pyrolysis Mass spectrometer

body of an original droplet. As a result, the very small droplets have a negative charge and the large droplets have a positive one. The very small droplets with negative charge evaporate and become negative cluster ions consisting of 10~30 water molecules. In electrospray, DC high voltage is applied to the nozzle and droplets have a high unipolar charge. When the repulsive force between charges accumulated on the droplet is stronger than the surface tension, that is Reyligh limit, and the droplets are broken into very small droplets. Two processes for ion formation are considered. One is that liquid molecules evaporate from the very small droplets and the sizes of the droplets reduce to ions. Another is that ions evaporate directly from the surfaces of droplets.

Formation Process and Reactivity

The primary ions and electrons produced by the high-energy irradiation and the discharge change with time by the following ion–molecule reactions.

$$e^- + A \rightarrow A^- \qquad \text{Electron attachment}$$
$$A^+ + B \rightarrow A + B^+ \qquad \text{Charge transfer}$$
$$A^- + B \rightarrow A + B^- \qquad \text{Charge transfer}$$
$$A^+B_n + B + M \rightarrow A^+B_{n+1} + M \qquad \text{Clustering}$$
$$A^-B_n + B + M \rightarrow A^-B_{n+1} + M \qquad \text{Clustering}$$

Ions that have high chemical reactivity, such as O^-, O_2^-, O_3^-, H^+, and N_2O^+, are generated generally in a first stage of growth processes. Furthermore, the ions and electrons often produce radicals and excited molecules by the following recombination reactions:

$$AB^+ + e^- \rightarrow A\cdot + B\cdot \qquad \text{Dissociation recombination}$$
$$AB^+ + e^- \rightarrow AB^* + h\nu \qquad \text{Radiative recombination}$$

$$AB^+ + e^- + M \rightarrow A\cdot + B\cdot + M \qquad \text{Three body recombination}$$
$$AB^+ + C^- \rightarrow AC^+ + B\cdot \qquad \text{Ion–ion recombination}$$

In air, $\cdot O$, $\cdot OH$, and O_3 are produced by the above process. These radicals and excited molecules induce further chain reactions with other chemical reactants. Effects of the radicals and excited molecules on the particle materials should be considered when the ions are used to control the electrical charge.

Physical Properties

The physical properties of ions depend on the ion–molecule reaction and ion-clustering processes. Table 6.2 shows ion properties in air. Electrical mobility was measured by Bricard,[9] and other properties were calculated from them. The positive ions are $H^+(H_2O)_n$ ($n = 5-8$) and negative ions are $O^-(H_2O)_n$, $O_2^-(H_2O)_n$, $OH^-(H_2O)_n$, $NO^-(H_2O)_n$, $CO_3^-(H_2O)_n$, and so forth.[10] The number concentrations N of unipolar and bipolar ions are obtained by

$$N^s = \frac{I}{B_{ei}^s eAE} \quad \text{for unipolar ion} \tag{6.1}$$

$$N^+ = N^- = \left(\frac{I_s}{\alpha_i eV} \right)^{1/2} \quad \text{for bipolar ion} \tag{6.2}$$

where I is the ion current in the electric field of intensity E, I_s is the saturation ion current, A is the area of the electrode, V is the volume of the ion-generation chamber, B_{ei} is the electrical mobility of ions, e is the elementary charge ($= 1.602 \times 10^{-19}$ C) and α_i is the recombination coefficient of a bipolar ion. The superscripts are the polarity of the ion.

Field Charging

Field charging tends to dominate the charging for particles lager than 2 μm in diameter under a strong electric field. The particle charge q can be evaluated from the equation introduced by White[11]:

$$q = \frac{q_\infty (eB_{ei}^s N^s t / 4\varepsilon_0)}{1 + (eB_{ei}^s N^s t / 4\varepsilon_0)} \tag{6.3}$$

$$q_\infty = \left(1 + \frac{2(\varepsilon_1 - 1)}{\varepsilon_1 + 2} \right) \frac{\pi \varepsilon E D_p^2}{e} \tag{6.4}$$

where q is the number of elementary charges captured by a particle, q_∞ is the saturation charge, D_p is the particle diameter, t is the charging time, ε_0 is the dielectric constant of gas ($= 8.855 \times 10^{-12}$ F/m), and ε_1 is the specific dielectric constant of particles. The average charges q_{av} calculated

TABLE 6.2 Physical Properties of Air Ions

	Property				
	Electrical mobility	Diffusion coefficient	Mean thermal velocity	Mean free path	Recombination coefficient
Polarity	B_{ei} (m^2 s^{-1} V^{-1})	D_i (m^2 s^{-1})	c_i (m s^{-1})	l_i (m)	α_i (m^2 s^{-1})
Positive ion	1.4×10^{-4}	3.53×10^{-6}	2.18×10^{2}	1.46×10^{-8}	1.6×10^{-12}
Negative ion	1.9×10^{-4}	4.80×10^{-6}	2.48×10^{2}	1.94×10^{-8}	

by Equation 6.3 and Equation 6.4 at $N^+ t = 10^{13}$ s/m³ are shown in Figure 6.1. The particle charge q decreases steeply as the particle size decreases. For the charging of high-resistivity particles, Equation 6.3 and Equation 6.4 cannot be applied, and numerical computation is necessary.[12]

Diffusion Charging

Diffusion charging is caused by the kinetic energy of an ion and the electrostatic energy between a particle and an ion in the absence of an external electric field. Charging is classified into bipolar charging, where particles are mixed with both negative and positive ions, and unipolar charging, where they are exposed to either ion. This mechanism dominates the charging in the size range $D_p < 0.2$ μm even in strong electric fields because field charging does not work.

Combination Probability of an Ion with a Particle

In the theoretical estimation for the diffusion charging, a combination probability β of an ion and a particle is necessary because diffusion charging is the coagulation process of ions and particles. The following equation, derived by Fuchs,[13] can be applied in the whole range of particle size:

$$\beta_q^s = \pi c_i^s \, \xi \lambda^2 \exp\left(-\frac{\phi(\lambda)}{kT}\right) \Bigg/ \left[1 + \exp\left(-\frac{\phi(\lambda)}{kT}\right) \frac{c_i^s \, \xi \lambda^2}{2D_i^s} \int_1^\infty \frac{\exp[-\phi(\lambda)/kT]}{r^2} dr\right] \qquad (6.5)$$

where λ is the radius of the image sphere ($\cong l_i^s + D_p/2$), k is the Boltzmann constant (= 1.380×10^{-23} J/K), T is the absolute temperature, D_i is the diffusion coefficient of ion, c_i is the mean thermal velocity, l_i is the mean free path of ions, and ξ is the correction coefficient of free molecule collision by the electrostatic force between a particle and an ion. $\phi(\lambda)$ is the electrostatic potential between a particle and an ion given by

$$\phi(\lambda) = \int_r^\infty F_e(r) dr = \frac{qe^2}{4\pi\varepsilon_0 r} - \frac{\varepsilon_1 - 1}{\varepsilon_1 + 1} \frac{e^2}{8\pi\varepsilon_0 r} \frac{D_p^3}{2r^2(4r^2 - D_p^2)} \qquad (6.6)$$

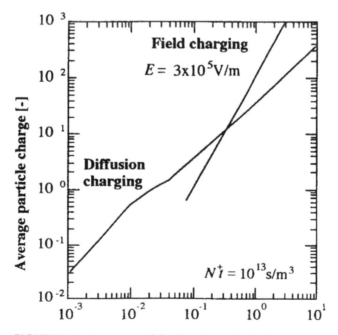

FIGURE 6.1   Average particle charge by unipolar ions.

Figure 6.2 shows the calculated results of Equation 6.5 using the ion properties listed in Table 6.2. The combination probability for a negative ion is larger than that of a positive ion because a negative ion has larger ion properties than a positive ion.

Diffusion Charging by Unipolar Ions

Diffusion charging can be evaluated theoretically by two approaches: deterministic and stochastic, using the combination probability β. In the deterministic approach, the time change in particle charges q is expressed by

$$\frac{dq}{dt} = \beta_q^s N^s \tag{6.7}$$

In the stochastic approach, the time-dependent change in charge distribution is expressed by the following birth–death equations:

$$\frac{dn_0}{dt} = -\beta_0^s n_0 N^s \tag{6.8}$$

$$\frac{dn_q}{dt} = \left(\beta_{q-1}^s n_{q-1} - \beta_q^s n_q \right) N^s, \quad q > 0 \text{ and } s = + \tag{6.9}$$

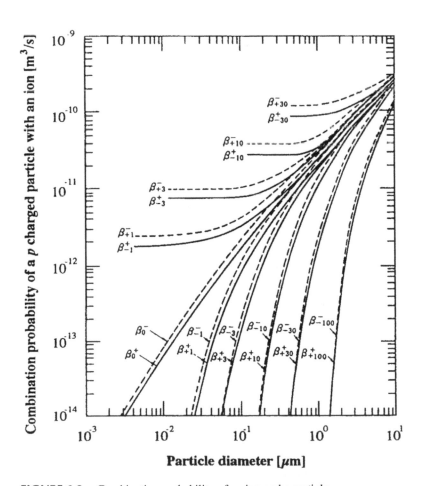

FIGURE 6.2 Combination probability of an ion and a particle.

or

$$\frac{dn_q}{dt} = \left(\beta_{q+1}^s \, n_{q+1} - \beta_q^{\,s} n_q \right) N^s, q < 0 \text{ and } s = - \tag{6.10}$$

where n is the particle number concentration and the subscripts are the numbers of elementary charges and their polarity. The above equations were solved analytically by Natanson[14] under the initial condition that particles were uncharged:

$$\frac{n_0}{n_T} = \exp\left(-\beta_0^s N^s t \right) \tag{6.11}$$

$$\frac{n_q}{n_T} = \left(\prod_{k=0}^{q-1} \beta_k^s N^s \right) \sum_{i=0}^{q} \frac{\exp(-\beta_j^s N^s t)}{\displaystyle\prod_{\substack{j=0 \\ j \ne i}}^{q} \left(\beta_i^s - \beta_j^s \right) N^s}, \quad q > 0 \text{ and } s = + \tag{6.12}$$

$$\frac{n_q}{n_T} = \left(\prod_{k=0}^{q+1} \beta_k^s N^s \right) \sum_{i=0}^{q} \frac{\exp\left(-\beta_j^s N^s t \right)}{\displaystyle\prod_{\substack{j=0 \\ j \ne i}}^{q} \left(\beta_i^s - \beta_j^s \right) N^s}, \quad q < 0 \text{ and } s = - \tag{6.13}$$

or

The average charges q_{av} calculated by Equation 6.11 and Equation 6.12 at $N^+ t = 10^{13}$ s/m^3 are shown in Figure 6.1.

A necessary condition to bring uncharged particles to the desired charge distribution is obtained by[15]

$$N^s t = \frac{\ln\left(1 - n_{pT}/n_T \right)}{\beta_0^s} \tag{6.14}$$

$$\beta_0^+ = \exp\left(-33.69 + 1.479 \ln D_p + 0.04214 \ln^2 D_p - 0.01001 \ln^3 D_p \right) \tag{6.15}$$

$$\beta_0^- = \exp\left(-33.49 + 1.389 \ln D_p + 0.08163 \ln^2 D_p - 0.01448 \ln^3 D_p \right) \tag{6.16}$$

where n_{pT}/n_T is the desired number ratio of total charged particles to total particles. Equation 6.15 and Equation 6.16 are approximate equations of the theoretical curves shown in Figure 6.2. D_p is measured in nanometers in these equations.

A numerical computation is necessary to solve Equation 6.5, Equation 6.11, and Equation 6.12. Although the following equation was derived under incorrect assumptions by White,[11] it is useful for making a rough estimate of the diffusion charging:

$$q = \frac{2\pi\varepsilon_0 D_p kT}{e^2} \ln\left(1 + \frac{D_p c_i^s e^2 N^s t}{8\varepsilon_0 kT} \right) \tag{6.17}$$

However, it should be noted that results given by Equation 6.17 are about two times higher than strict solutions.

In the range 0.2 μm$< D_p <$2 μm, particles are charged by both field and diffusion charging. Liu and Kapadia[16] solved this problem numerically. If errors of 50–100% are allowed, the sum of Equation 6.3 and Equation 6.17 is available as a simple estimation.

Diffusion Charging by Bipolar Ions (Neutralization)

Bipolar diffusion charging is expressed generally by birth–death equations because the deterministic approach cannot reflect the difference of ion properties:

$$\frac{dn_o}{dt} = \beta_{+1}^- n_{+1} N^- - \beta_0^- n_0 N^- + \beta_{-1}^+ n_{-1} N^+ - \beta_0^+ n_0 N^+ \tag{6.18}$$

$$\frac{dn_q}{dt} = \beta_{q+1}^- n_{q+1} N^- - \beta_q^- n_q N^- + \beta_{q-1}^+ n_{q-1} N^+ - \beta_q^+ n_q N^+ \quad |q| > 0 \tag{6.19}$$

When the number concentration of bipolar ions is sufficiently high under sufficient charging time, particle charge reaches the equilibrium state. The equilibrium distribution at steady state is obtained from the above equations.[17]

$$\frac{n_0}{n_T} = \frac{1}{\Sigma} \tag{6.20}$$

$$\frac{n_q}{n_T} = \prod_{p=+1}^{q} \left(\beta_{p-1}^+ / \beta_p^- \right) / \Sigma, \quad q \geq +1 \tag{6.21}$$

$$\frac{n_q}{n_T} = \prod_{p=-1}^{q} \left(\beta_{p+1}^- / \beta_p^+ \right) / \Sigma, \quad q \leq -1 \tag{6.22}$$

where

$$\Sigma = \sum_{P=+1}^{+\infty} \left[\prod_{P=+1}^{+\infty} \left(\frac{\beta_{P-1}^+}{\beta_P^-} \right) \right] + \sum_{P=-1}^{-\infty} \left[\prod_{P=-1}^{-\infty} \left(\frac{\beta_{P+1}^-}{\beta_P^+} \right) \right] + 1 \tag{6.23}$$

Figure 6.3 shows the equilibrium charge distribution in the size range of $0.001\ \mu m \leq D_p \leq 10\ \mu m$. Fractions of negatively charged particles are higher than those of positively charged particles due to differences in ion properties.

In the size range of $D_p \geq 0.5\ \mu m$, strict solutions agree with the Boltzmann charge distribution[18] within 30% error:

$$\frac{n_q}{n_T} = \frac{\exp\left(-q^2 e^2 / 4\pi\varepsilon_0 D_p kT\right)}{\sum_{P=-\infty}^{+\infty} \exp\left(-p^2 e^2 / 4\pi\varepsilon_0 D_p kT\right)} \tag{6.24}$$

The necessary condition for which q_0 charged particles attain the Boltzmann charge distribution is obtained by

$$N^b = -\varepsilon_0 kT \frac{\ln\left(0.1 / |q_0|\right)}{D_i e^2} \tag{6.25}$$

where N^b is the number concentration of ion pair $(= N^+ = N^-)$.

Photoelectron Charging

When a solid is irradiated by UV light, electrons are emitted from the surface. This phenomenon is called a photoelectron emission and is used as a principle of surface analysis by photoelectron

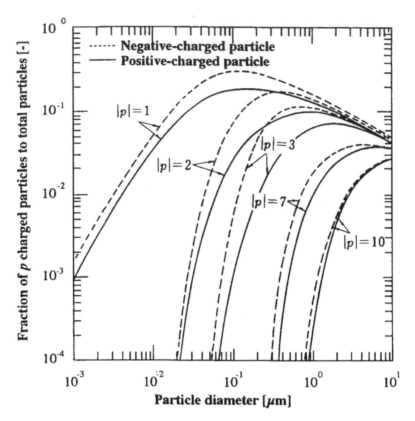

FIGURE 6.3 Equilibrium charge distribution by bipolar ions.

spectroscopy. The photoelectron emission is generally explained by the three-step model shown in Figure 6.4. Electrons in the solid are first optically excited into states of high energy; then they move to the surface of the solid and escape into vacuum or the gas phase. Therefore, electrons can be emitted from the surface when the photon energy $h\nu$ is higher than the work function ϕ_w of the solid. Photoelectron emission is also caused by UV irradiation to particles suspended in a gas stream (aerosol). In this case, aerosol particles are charged positively because of the escape of electrons. The potential barrier ϕ_q which must be overcome by an electron to escape from the q charged particle is expressed by[19]

$$\phi_q = \phi_w + \frac{e^2(q+1)}{2\pi\varepsilon_0 D_p} - \frac{5}{8}\frac{e^2}{2\pi\varepsilon_0 D_p} \tag{6.26}$$

When ϕ_q is higher than $h\nu$, the particle charges reach to a maximum value q_{max} obtained from Equation 6.26.

$$q_{max} = \frac{h\nu - \phi_w}{e^2}2\pi\varepsilon_0 D_p + \frac{5}{8} \tag{6.27}$$

The probability ζ_q which an electron is emitted from the q charged particle is obtained by

$$\zeta_q = C\left(h\nu - \phi_q\right)^m \frac{I\pi D_p^2}{4h\nu} \tag{6.28}$$

where C and m are material-dependent empirical constants, and I is the irradiation intensity.

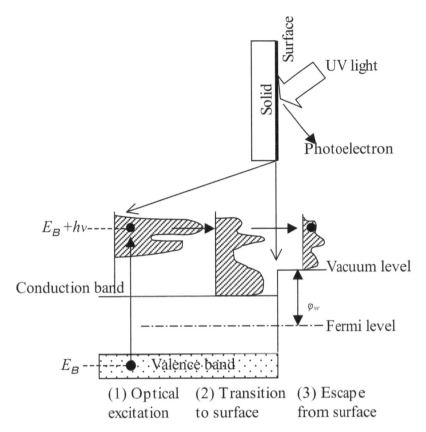

FIGURE 6.4   Illustration of three-step model for photoelectron emission from a solid.

4.6.2   CHARGE CONTROL BY CONTACT ELECTRIFICATION IN GASEOUS STATE

When two different materials come into contact, charge is transferred according to the contact potential difference, which depends on the work functions of the two materials; thus particle charging can be controlled by changing the wall material. If the work function of the wall is close to that of the particles, the particle charging will be repressed. Furthermore, it is possible to give the opposite polarity to the particles using a different material.

Since the charge transfer occurs in the contact area, the surface property is more significant than the bulk property. If the wall is covered with the oxidation layer or the layer is removed by particle collision, the surface property will be changed, and thus the particle charging will fluctuate. To avoid such instabilities, the wall surface is treated chemically or mechanically.

Surface treatment such as surface coating,[20] attaching a charge-control agent,[21] and chemical modification is effective to change the work function of particles. Figure 6.5 shows the contact potential difference of alumina particles coated with stearic acid.[20] The value of the contact potential difference varies according to the thickness of the coating layer but is almost constant where the thickness is larger. Therefore, particle charging can be controlled by the thickness of the coating layer.

The amount of charge transferred by elastic collision is proportional to the maximum contact area. When a spherical particle collides with a wall, the maximum contact area S can be approximated by[22]

$$S = 1.36 k_e^{2/5} \rho_p^{2/5} D_p^2 \, v_i^{4/5} \tag{6.29}$$

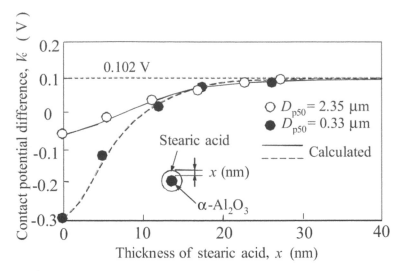

FIGURE 6.5   Contact potential difference of coated particles against the reference (Au). [From Yoshida, H., Fukuzono, T., Ami, H., Iguchi, Y., and Masuda, H., *J. Soc. Powder Technol. Jpn.,* 29, 504–510, 1992. With permission.]

where k_e is the elasticity parameter, ρ_p is the density of the sphere, D_p is the particle diameter, and v_i is the impact (incident) velocity. If the particle is not spherical, the particle shape should be taken into account.[23] The elasticity parameter k_e is given by

$$k_e = \frac{1-v_1^2}{E_1} + \frac{1-v_2^2}{E_2} \qquad (6.30)$$

where v is Poisson's ratio, E is Young's modulus, and subscripts 1 and 2 represent the particle and the wall, respectively. To reduce the charge transferred by impact, it is desirable to choose materials having a larger Young's modulus, that is, to use harder materials. The net contact area depends on the surface state: the net contact area decreases with increasing roughness.[24]

If there is no further electrification, the charge on the surface decreases in the course of time, which is called electric leakage. For conductive particles, the charge tends to leak out for a short time. It is more difficult for particles having high electric resistance to release their charge, but it is easier if they are mixed with conductive fine particles or fillers, because of the higher conductivity. Hydrophilic surface modification with a surfactant is also effective to release the charge. Charge reduction is prominent at higher humidity[25]; then the maximum charge on the particles is lower even though they are charged repeatedly.

If the wall is made of nonconductive material, the charge on the wall is also taken into consideration. The charge forms an electric field so as to reduce the effective contact potential difference, and also causes electric discharge or dust explosion; therefore, the charge on the wall should be released. One of the effective methods for this is to induce dielectric breakdown using a thin dielectric wall reinforced with a conductive material such as metal.[26]

4.6.3   IN LIQUID STATE

As described in Section 2.5.2, particles in solutions bear a surface charge as a result of the adsorption of ions or the dissociation of functional groups on the particle surface. Then, because of the electroneutrality principle, counterions are attracted toward the particle surface, as illustrated in Figure 6.6. Some of the ions are adsorbed directly on the surface to form the so-called Stern layer. When the counterions excessively adsorb on the surface with specific affinity, sign reversal of the surface charge may occur. The rest of the ions in the bulk are in thermal motion, balancing the electrical attractive force to form a diffuse double layer. The potential at the particle surface is called the surface potential, ψ_0, and the potential at the Stern layer is called the Stern potential, ψ_d. Because the structure within the Stern layer is not well known, the Stern potential is often regarded as the surface potential of particles for many engineering purposes. Hence the Stern potential is expressed by ψ_0 in the succeeding subsection unless the specific effect of the Stern layer is discussed.

Potential and Charge in the Diffuse Layer[27]

The Poisson equation holds between the volume density of charge ρ and the potential ψ in the diffuse double layer:

$$\nabla^2 \psi = -\frac{\rho}{\varepsilon} \tag{6.31}$$

where ε is the permitivity of the medium. The charge density is given by the sum of all the ionic contributions in the unit volume:

$$r = \sum n_i z_i e \tag{6.32}$$

where n_i and z_i are the number concentration and valency of ions of type i, respectively, and e is the elementary charge. The Boltzmann equation holds at the equilibrium where the electrostatic and chemical potentials of ions balance:

$$n_i = n_{0i} \exp\left(-\frac{z_i e \psi}{kT}\right) \tag{6.33}$$

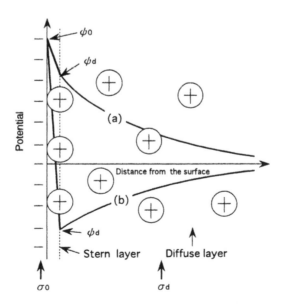

FIGURE 6.6 Model of Stern layer and diffuse layer: (a) a small amount of ions is adsorbed on the surface; (b) an excessive amount of ions is adsorbed because of their high affinity to the surface.

where n_{0i} is the bulk concentration of ions of type i, k is the Boltzmann constant, and T is the temperature. Substitution of Equation 6.32 and Equation 6.33 into Equation 6.31 results in the Poisson–Boltzmann equation:

$$\nabla^2\psi = -\Sigma\left(\frac{n_{0i}z_i e}{\varepsilon}\right)\exp\left(-\frac{z_i e\psi}{kT}\right) \tag{6.34}$$

This equation cannot be solved explicitly in general. However, when the Debye–Huckel approximation that ψ is small everywhere [i.e., $ze\psi < kT$ ($\psi < 25.7/z$ mV at 25°C)] holds, and the electrolyte is symmetric with valency $z_+ = -z_- = z$ and concentration $n_{0+} = n_{0-} = n_0$, the equation is simplified and the analytical solution is derived.

When the surface is a flat plate, the potential distribution is given by

$$\psi + = \psi_0 \exp(-\kappa x) \tag{6.35}$$

$$\kappa = \left(\frac{2n_0 z^2 e^2}{\varepsilon kT}\right)^{0.5} \tag{6.36}$$

where x is the distance from the plate surface, κ is a measure of the ionic concentration of the bulk solution, and $1/\kappa$ is called the thickness of the diffuse layer. The charge density of the diffuse layer, σ_d, is given by

$$\sigma_d = -\int_0^\infty \rho dx \tag{6.37}$$

Then, Equation 6.37 is rewritten as follows, using Equation 6.32 and Equation 6.34:

$$\sigma_d = \varepsilon\left(\frac{d\psi}{dx}\right)_{x=0} = -\left(\frac{4n_0 ze}{\kappa}\right)\sinh\left(\frac{ze\psi_0}{2kT}\right) \tag{6.38}$$

When the effect of the Stern layer is regarded as negligible, the surface charge, σ_d, is given by

$$\sigma_0 = -\sigma_d = \left(\frac{4n_0 ze}{\kappa}\right)\sinh\left(\frac{ze\psi_0}{2kT}\right) \tag{6.39}$$

This shows the correlation between the surface potential and the charge density of the diffuse layer.

A similar argument can be developed for the potential around a spherical particle of radius a, and the potential in the diffuse layer is given as

$$\psi = \left(\frac{\psi_0 a}{r}\right)\exp[-\kappa(r-a)] \tag{6.40}$$

The total charge of the particle Q is then written as

$$Q = -\int_a^\infty 4\pi r^2 \rho dr = 4p\varepsilon a(1+\kappa a)\psi_0 \tag{6.41}$$

Charge Control by Electrolytes

In order to control the charge of the particle surface, it is necessary to know not only the charging mechanism but also how the surface charge varies with the surrounding conditions. This is illustrated using a simple model for surface charging.[28]

When the particle surface is charged by the complete dissociation of "strong" functional groups, such as $-OSO_3H \rightarrow -OSO_3^- + H^+$, for polystyrene latex particles in an aqueous solution, the surface charge σ_0 depends only on the density of dissociation sites, $N (= [-OSO_3^-])$, as shown by Equation 6.42, where [ ] denotes the site density or solution concentration:

$$\sigma_0 = -FN \qquad (6.42)$$

where F is the Faraday constant.

When the particle surface is charged by the dissociation of "weak" functional groups, such as $-COOH$, the degree of dissociation is influenced greatly by the property of the aqueous solution. Suppose there is a weak-acid functional group described by $-SH$. Then it dissociates as

$$- SH \underset{\xrightarrow{\hspace{1cm}}}{\xleftarrow{\hspace{0.3cm} K \hspace{0.3cm}}} - S^- + Hs^+ \qquad (6.43)$$

where $K = [-S^-][Hs^+]/[-SH]$ and $[Hs^+]$ is the interfacial concentration of protons, which is related to the concentration in the bulk $[H^+]$ by $[H_s^+] = [H^+]\exp(-e\psi_0/kT)$. Because the density of the surface site, N, is given by $N = [-SH] + [-S^-]$, the surface charge is written as

$$\sigma_0 = -F[-S^-]$$

$$= -\frac{FN}{\left\{1 + [H^+]\exp(-e\psi_0/kT)/K\right\}} \qquad (6.44)$$

When $K = \infty$, Equation 6.44 coincides with Equation 6.42. Then the values of σ_0 and ψ_0 will be obtained by solving Equation 6.39 and Equation 6.44 simultaneously. Hence, it is known that σ_0 and ψ_0 are the functions of N, K, $[H^+]$, and κ, respectively. Because the values of N and K are usually fixed in a given solution, the surface charge may be controlled by $[H^+]$ and κ (i.e., the pH and ionic concentration of the solution).

The description of the surface charge of amphoteric materials, such as metal oxides and proteins, is more complicated than that of the monofunctional surface explained above. The dissociation of a metal oxide surface can be written as

$$- SH_{2+}^+ + OH^- \underset{\xrightarrow{\hspace{1cm}}}{\xleftarrow{\hspace{0.3cm} K_1 \hspace{0.3cm}}} - SH + H_2O \underset{\xrightarrow{\hspace{1cm}}}{\xleftarrow{\hspace{0.3cm} K_2 \hspace{0.3cm}}} - S^- + H^+ + H_2O \qquad (6.45)$$

It is clear that the surface charge varies from positive to negative as the value of pH increases. An argument similar to that for the monofunctional surface using a generalized $1-pK$ model enables us to evaluate the surface charge of amphoteric materials.[12] The predicted surface charge and potential are drawn in Figure 6.7. The estimated values might not be correct quantitatively, but the qualitative features are well predicted: (1) the positive charge decreases with pH at low pH, the charge increases negatively with pH at high pH, and there is the point of zero charge (p.z.c.) in between; and (2) the absolute value of the surface potential decreases, but that of the surface charge increases, as the ionic concentration in the bulk increases. It is clear that the pH and ionic concentration are the important parameters to control the charge and potential of particle surfaces.

Surface Charge and Specific Adsorption

There are ions that have a certain affinity for the surface that cannot be explained by the simple principles described above. They are called specifically adsorbing ions. When these ions exist, the surface charge or potential in any particular system is not easy to control, although it seems there exists

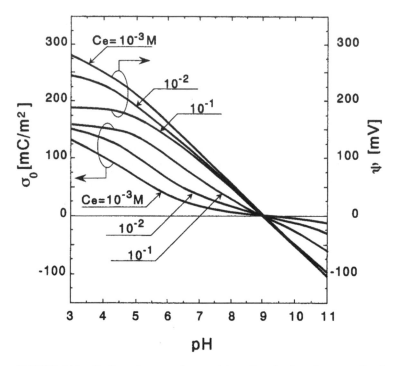

FIGURE 6.7   Surface charge and potential predicted using the generalized 11-pK model. Ce is the electrolyte concentration.[29] [From Koopal, L.L., in *Coagulation and Flocculation*, Bobias, B., Ed., Marcel Dekker, New York, 1993, p.101.]

some rule to govern the specific adsorption. It is known experimentally that the adsorption depends on various conditions, such as the pH, the electrolyte concentration, and the type of electrolyte in the solution. Some typical data are presented in Figure 6.8, showing the electrophoretic mobility of rutile particles as a function of pH in $Ca(NO_3)_2$ solutions and for various nitrates at the constant concentration.[13] When no specifically adsorbing ion exists, the mobility must decrease monotonically with pH, as predicted in Figure 5.10. The increase of the mobility at high pH in Figure 6.8 indicates the existence of specific adsorption. The data in Figure 6.8 a shows that cations adsorb specifically on the negatively charged surface, and the degree of specific adsorption increases with the magnitude of surface charge and the ionic concentration. Figure 6.8 shows that the adsorption depends also on the type of electrolyte, and the adsorbing ability of this system can be written as Ba > Sr > Ca > Mg. It is known that the ability of specific adsorption follows the Hofmeister or lyotropic series given below, if the valency of ions is the same. These series coincide with the order of the radii of ions that are given in parentheses (angstroms). Because the degree of hydration of ions decreases with the increasing radius of ions, the explanation of the higher adsorbing ability of larger ions is that less energy is needed for the ions to be dehydrated in their adsorption on the surface:

$$Cs^+(1.81) > Rb^+(1.66) > K^+(1.52) > Na^+(1.13) > Li^+(0.73)$$

$$Ba^{++}(1.49) > Sr^{++}(1.32) > Ca^{++}(1.14) > Mg^{++}(0.71)$$

$$I^-(2.06) > Br^-(1.82) > Cl^-(1.67) > F^-(1.19) \qquad (6.46)$$

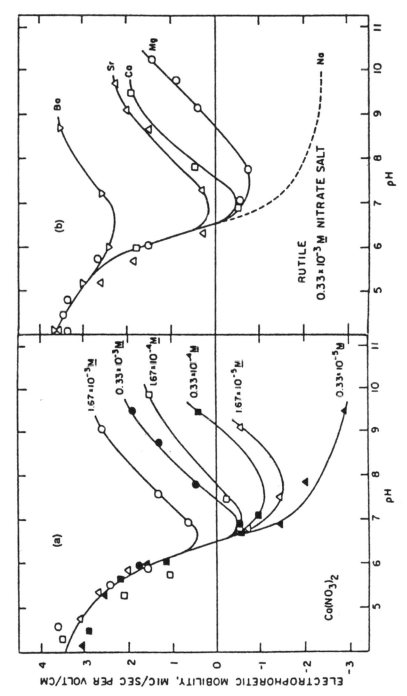

FIGURE 6.8 Dependence of electrophoretic mobility on pH for rutile (a) in solutions of various concentrations of $Ca(NO_3)_2$ and (b) in 0.33×10^{-3} M $Ca(NO_3)_2$ solutions with various types of cations. [From Furstenau, D.W., Manmotian, D., and Raghavan, in *Adsorption from Aqueous Solutions*, Tewari, P.H., Ed., Plenum Press, New York, 1981, p. 81.]

Charge Control by Surfactants

The surface charge of particles in aqueous solutions is controllable using surfactants. When the alkyl group of a surfactant has a high affinity to the surface, the aliphatic tail is adsorbed on the surface and the polar head is directed toward the solution, as shown in Figure 6.9. Then particles are dispersed and stable because of the interparticle repulsive force caused by the charge of the adsorbed surfactants.

When the bare surface is charged as shown in Figure 6.9 and surfactants of the opposite charge are dosed in the solution, the head groups are adsorbed on the surface because of the electrostatic attractive force between them. When the first adsorbed layer is formed completely, the surface becomes hydrophobic, and particles will be unstable because of the hydrophobic attractive force. When the surfactant is dosed further, the secondary adsorbed layer will be formed such that the polar heads are directed toward the solution. Then the surface charge is reversed and the suspension becomes stable again. An example is given in Figure 6.10, showing the change of the zeta potential of AgI sol with the concentration of anionic surfactants.[29] It is important to note that the longer the alkyl group is, the less molar concentration of surfactant is needed to vary the charge of the surface. Because the longer the aliphatic tail is, the more segregated from water surfactants are, surfactants with the long tail are adsorbed on the surface effectively.

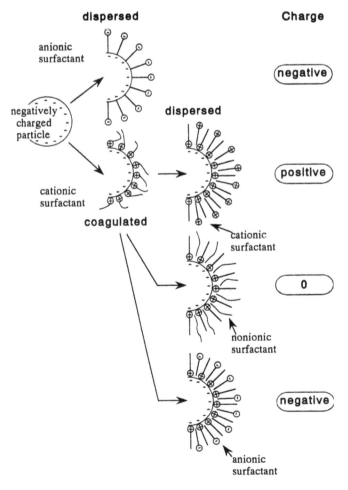

FIGURE 6.9   Schematic of the surfactant adsorption, the charge, and the stability of particles.

It is possible to control the surface charge in such a way that the secondary surfactants of different kinds are dosed after the first layer is formed. In this case, the final surface charge will be given by the charge of the secondary surfactants.

Charge Control by Polymers

Water-soluble polymers are very effective agents to control the surface charge of particles in aqueous solutions. Various polymers, which differ in charge density, hydrophobicity, molecular weight, and so forth, have been synthesized, and even polymers which have several different properties simultaneously are able to be synthesized these days. Hence, it is possible to prepare a polymer which has a high affinity to the particular surface and the property desired as the adsorbed layer at the same time.

High molecular weight polymers which are opposite to the particle surface in charge are popularly employed to control the charge of particles in solutions. When highly charged polymers are dosed, they adsorb on the particle surface with strong electrostatic attractive force, and the surface charge is altered even by a small dosage. Typical data are presented in Figure 6.11,

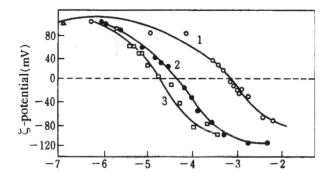

FIGURE 6.10 Dependance of the zeta potential of AgI sol on the surfactant concentration: (1) $CH_3(CH_2)_9SO_4Na$; (2) $CH_3(CH_2)_{11}SO_4Na$; (3) $CH_3(CH_2)_{13}SO_4Na$.

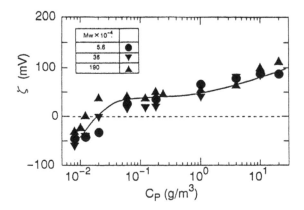

FIGURE 6.11 Zeta potential of latex particles versus the concentration of dosed cationic polymers. Cp and Mw are the concentration and molecular weight of dosed polymers, respectively.

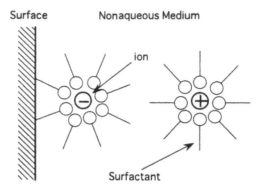

FIGURE 6.12 Schematic of the surface charge by ions in micelles in nonaqueous medium.

showing the change of the zeta potential of negatively charged polystyrene lattices with the concentration of dosed cationic polymers.[30] The strong coulombic force acts between the polymers and surface, and almost all polymers are adsorbed on the surface unless the amount of dosage is too large. It is interesting to note that the surface charge depends only on the total weight of dosed polymers but does not depend on the molecular weight of the polymer. This is because the charge density of the adsorbed layer is determined by total charge (i.e., the total weight) of adsorbed polymers, as far as the structure of polymer is the same and all the polymers dosed are adsorbed on the surface.

Charge in Nonaqueous Medium

The charge of particles in nonaqueous solutions varies sensitively with the property of the medium, water, and surfactants dissolved. There is no general rule to explain the charging mechanism in nonaqueous solutions at present. Nonaqueous solutions are classified into polar and nonpolar solutions. Because the characteristics of particles in polar solutions are more or less similar to those in aqueous solutions, the characteristics of the charge of particles in nonpolar solutions are considered here.

It is almost impossible to eliminate a small amount of water stuck on the particle surface even if the suspension is carefully prepared. The water dissociates as $H_2O \rightleftharpoons H^+ + OH^-$. This proton combines with either the particle surface ($-S$) or the medium molecules (M), depending on the relative acidic and basic balance between the surface and medium. When the basic strength of the surface is greater than that of the medium, the proton will combine with the particle surface and the surface will be charged positively, as shown below. But if it is the reverse, the surface will be negatively charged:

$$-SH^+ + MOH^- \rightleftharpoons -S + M + H^+ + OH^- \rightleftharpoons -SOH^- + MH^+ \qquad (6.47)$$

This mechanism was successfully confirmed by the experimental data in Table 5.2.[31]

If surfactants exist in nonaqueous solvents, ions are able to be solubilized into reversed micelles of surfactants, as shown in Figure 6.12. If these charged micelles are adsorbed on the surface, the surface will be charged. The charge in this case depends on the combination of the properties or the surfactant, the medium, and the surface, as illustrated in Table 6.3.[32] It is known that the existence of water in these systems affects sensitively the charge of the particle surface.[33]

TABLE 6.3 Particle Charge in Organic Solvents Containing a Small Amount of Water[a]

Particle	Solvent			
	2,2,4-Trimethylpentane (40)	Benzene (270)	Methyl ethyl ketone (650)	Ethylacetate (700)
Carbon black	+	−	−	−
Toluidine red	+	−	−	−
Titanium dioxide	+	+	+	−

[a] Values in parentheses are water content (ppm).

Source: Tamaribuchi, T. and Smith, M. L., *J. Colloid Interface Sci.,* 22, 404, 1966.

REFERENCES

1. Kousaka, Y., Adachi, M., Okuyama, K., Kitada, N., and Motouchi, T., *Aerosol Sci. Technol.,* 2, 421–427, 1983.
2. Lui, B. Y. H. and Pui, D. Y. H., *Aerosol Sci. Technol.,* 5, 465–472, 1974.
3. Han, B., Shimada, M., Choi, M., and Okuyama, K., *Aerosol Sci. Technol.,* 37, 330–341, 2003.
4. Inaba, H., Ohmi, T., Yoshida, T., and Okada, T., *J. Electrostatics,* 33, 15–42, 1994.
5. Shimada, M., Cho, S. C., Tamura, T., Adachi, M., and Fuhii, T., *J. Aerosol Sci.,* 4, 649–661, 1997.
6. Akasaki, M., in *Handbook of Electrostatics,* The Institute of Electrostatics, Japan, Ed., Ohm-Sha, Tokyo, 1981, pp. 217–218 (in Japanese).
7. Moore, A. D., in *Electrostatics and Its Applications,* John Wiley, New York, 1973, Chap. 4.
8. Kebarle, P. and Tang, L., *Anal. Chem.,* 65, A972–A986, 1993.
9. Bricard, J., in *Problems of Atmospheric and Space Electricity,* Coroniti, S. C., Ed., Elsevier, Amsterdam, 1965, pp. 82–117.
10. Mohnen, V. A., in *Electrical Processes in Atmospheres,* Dolezalk, H. and Reiter, R., Steinkopff Verlag, Darmstadt, 1977, pp. 1–16.
11. White, H. J., *AIEE Trans.,* 70, 1186–1191, 1951.
12. Masuda, S. and Washizu, M., *J Electrostat.,* 6, 57–67, 1979.
13. Fuchs, N. A., *Geofis. Pura Appl.,* 56, 185–193, 1963.
14. Natanson, G. L., *Sov. Phys.,* 5, 538–551, 1960.
15. Adachi, M., Okuyama, K., and Kousaka, Y., in *Proceedings of the Second International Conference on Electrostatic Precipitation,* Kyoto, 1984, pp. 698–701.
16. Liu, B. Y. H. and Kapadia, A., *J. Aerosol Sci.,* 9, 277–242, 1978.
17. Adachi, M., Kousaka, Y., and Okuyama, K., *J. Aerosol Sci.,* 16, 109–123, 1985.
18. Keefe, D., Nolan, P. J., and Rich, T. A., *Proc. R. Ir. Acad.,* 60-A, 27–45, 1959.
19. Maisels, A., Jordan, F., and Fissan, H., *J. Appl. Phys.,* 91, 3377–3383, 2002.
20. Yoshida, H., Fukuzono, T., Ami, H., Iguchi, Y., and Masuda, H., *J. Soc. Powder Technol., Jpn.,* 29, 504–510, 1992.
21. Itakura, T., Masuda, H., Ohtsuka, C., and Matsusaka, S., *J. Electrostat.,* 38, 213–226, 1996.
22. Timoshenko, S. P. and Goodier, J. N., *Theory of Elasticity,* 3rd Ed., McGraw-Hill, New York, 1970, pp. 409–422.
23. Masuda, H. and Iinoya, K., *AIChE J.,* 24, 950–956, 1978.
24. Ema, A., Tanoue, K., Maruyama, H., and Masuda, H., *J. Powder Technol. Jpn.,* 38, 695–701, 2001.
25. Nomura, T., Taniguchi, N., and Masuda, H., *J. Soc. Powder Technol. Jpn.,* 36, 168–173, 1999.
26. Matsusaka, S., Nishida, T., Gotoh, Y., and Masuda, H., *Adv. Powder Technol.,* 14, 127–138, 2003.
27. Hunter, R. J., *Foundations of Colloid Science,* Vol. 1., Clarendon Press, Oxford, 1987.
28. James, R. O., in *Polymer Colloids,* Buscall, R., Corner, T., and Stageman, J. R., Eds., Elsevier, London, 1985, p. 69.

29. Watanabe, A., *Bull. Inst. Chem. Res. Kyoto Univ.*, 38,179, 1960.
30. Higashitani, K., Kage, A., and Arao, E., *Dispersion and Aggregation,* Moudgil, B. M. and Somasun-daran, P., Eds., Engineering Foundation, New York, 1994, p. 191.
31. Tamaribuchi, T. and Smith, M. L., *J. Colloid Interface Sci.,* 22, 404, 1966.
32. Kitahara, A., Amano, M., Kawasaki, S., and Kon-no, K., *Colloid Polymer Sci.,* 255, 1118, 1977.
33. Cooper, W. D., *J. Chem. Soc. Faraday Trans.,* 1, 70, 864, 1974.

4.7 Surface Modification

Mamoru Senna
Keio University, Yokohama, Kanagawa, Japan

4.7.1 PURPOSE OF SURFACE MODIFICATION

The surface of powder particles is often modified to adapt them to their final uses. The purposes of surface modification are very diverse. They range from simple organophilication of inorganic substances by a coupling agent to very complicated complex formation with the guest species under well-controlled distribution and interactions with a substrate. Introduction of electronically or biologically functional groups belongs to an important part of a fast expanding palette of techniques.

Surface modification and formation of composite or complex particles are almost indistinguishable, since the change of the outermost surface almost always influences the physicochemical properties of the near-surface region. It is therefore more appropriate to discuss the method and consequences of surface modification together with those of complex or composite particles.

Modern surface modification techniques aim at finer particles down to nanosized ones under precisely controlled homogeneity. A number of elegant methods including *in situ* surface reactions are being developed. After a brief description of modern techniques, some case studies and examples are given. The significance of characterization and related techniques is also referred to.

4.7.2 METHODS OF SURFACE MODIFICATION

The technology of surface modification of fine particles has been traditionally developed in the pigment and paint industries. They have been dealing exclusively with fine particles. Similar techniques have been applied for rather traditional fillers, for example, calcium carbonate, kaolin, or carbon black. Soaking of powders in an appropriate solution containing inorganic salts, dispersants or surfactants, as well as coupling agents, is quite easy and popular. However, this is always followed by a drying process, which brings about a nuisance of agglomeration due to capillary pressure at the final stage of solvent evaporation.

In huge contrast to those traditional fields, many of the modern surface coating and modification methods are being applied to fine particles. Specific techniques are becoming more precision oriented under conditions as mild as, and as energy-saving as possible. One of the most important aspects for any kind of surface modification is the chemical affinity between the substrate and newly settled surface species. The weakest interaction is physical adsorption, which is by van der Waals forces, followed by various chemical adsorptions. Surface nucleation and graft polymerization root the active centers on the host surface, so that they result in stronger attachment to the substrate, as compared to any kind of adsorption. Frequently, those active centers must be artificially introduced by etching, adsorption of various activators, irradiation of electromagnetic beams, or mechanical stressing.

Methods of surface modification are conventionally divided into physical and chemical ones. Physical methods are subdivided into mechanical treatment, irradiation, sputtering, or similar techniques, combined with vacuum technology. Chemical methods begin with the above-mentioned soaking and resulting adsorption, and surface deposition with or without *in situ* surface reaction. In contrast to a posttreatment, that is, to put some new materials on the surface of already matured solid surfaces, there are also many chemical processes, where host particles and surface layers are synthesized simultaneously. The latter is generally called an *in situ* reaction. Organic polymers are very often subject to such an *in situ* synthesis to give desired properties by choosing an appropriate initiator[1] or adding different monomers at an interval.[2] A chemical process of surface modification is often regarded as immobilization of chemical species on the substrate surface. This subject is discussed at the end of this chapter (4.7.8).

To make uniform layers or films on the surface of the host particles is very common in the field of surface modification. They are generally categorized as microencapsulation.[3] Surface polymerization, coacervation, or even mechanical deposition with and without a subsequent heat or chemical treatment belong to this category. Electroless deposition is also a well-developed chemical method, leading to a similar result. The guest species are mostly metallic ones,[4] but not exclusively.

While mechanical routes are fairly free from choice of combination between the host and guest species,[5] most of the surface chemical methods are conditional with respect to interfacial affinity. On the other hand, the former has serious drawbacks of structural degradation and contamination after prolonged handling in a machine similar to mixers or grinding mills. The thickness of activated layers is distributed from a few atomic layers to several micrometers.[6] In the latter case, the activated near-surface region has its specific "bulk" properties as well.

Chemical affinity at the host–guest interface within the scope of conventional colloid and interface science is one of the fundamental prerequisites for a chemical route. Active centers are often provided on the host surface for coating with a film. The density and uniformity of such active centers are decisive for uniform nucleation and growth of the surface species, which covers the surface of the host particles continuously and uniformly. For the purpose of the controlled release of drugs or fertilizers, control of the properties of the surface film is very important.[7,8]

4.7.3   CONVENTIONAL TREATMENTS WITH SURFACTANTS, COUPLING AGENTS, AND SIMPLE HEATING

Every surface-active or coupling agent can be made suitable for surface modification, provided the agent firmly adsorbs on the surface of the substrate particles. This has long been done by soaking immersion or impregnation.[9] Most of the supported catalysts are prepared by soaking the carrier in an aqueous solution of the active species, followed by an appropriate heat treatment.[10] Since supported catalyses are quite sensitive whether and to what extent the chemical interaction takes place between the host surface and the guest species, it is important to learn from the entire preparation process of supported catalyses, not only the method of preparation but also characterization for better surface modification. For the control of chemical interaction at the host–guest interface, subsequent heat treatment is also of vital importance.

Some classical examples are given below for the surface modification of pigments.[11] Titania, as one of the most frequently used pigments, is normally surface treated by inorganic salts such as aluminum sulfate or sodium silicate.[12] By choosing subsequent heat treatment conditions, the microstructure, notably pores and fissures, is controllable. Organophilication is very usual and accomplished by using various organic reagents with polar groups, for example, n-buthanol, decyl amine, and tetramethyrol cyclohexanol.[13] Calcium carbonate and kaolin are frequently made organophilic by using, for example, poly(acrylic acid) or acetates of alkyl amine.[13] Various coupling agents are also used, for example, titanates, chromates, or silane coupling agents.

The surface of carbon black can be made hydrophilic by simple heating to give oxidation products. Carbon black can also be surface treated by chemicals such as dodecyl benzene sulfate.[13] Thermal treatments to introduce a controlled surface oxide layer are not restricted to carbon black but used on ultrafine metal particles to avoid self-ignition.[14]

4.7.4   MICROENCAPSULATION AND NANOCOATING

Microencapsulation is a kind of surface coating, developed mainly in the pharmaceutical industry for the specific purpose of a better drug delivery system and toxicity protection.[15] In this context, one

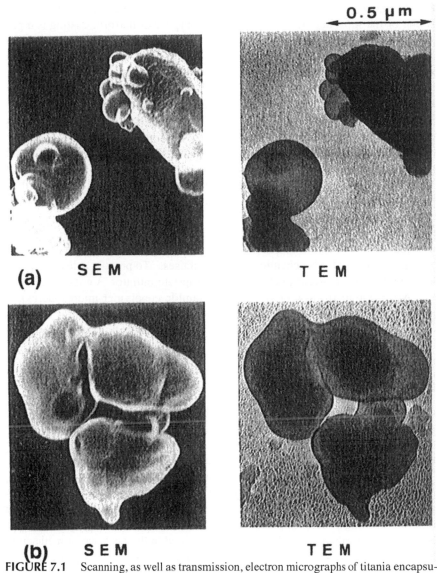

0.5 µm

(a) S E M   T E M

(b) S E M   T E M

FIGURE 7.1   Scanning, as well as transmission, electron micrographs of titania encapsulated by poly(methylmethacrylate). Without (a) and with (b) pretreatment by SDS. Kindly supplied by Professor Masahiro Hasagawa.

of the most popular ways of drug coating is coating the drug with gelatin sponges through coacervation. For successful coacervation, the core particles, being mostly drug particles, should not be too soluble in the coacervation solution. As long as an aqueous solution of gelatinlike substance is used, an initiator of phase separation should also be hydrophilic. Pretreatment of the core particles by polymeric ions often facilitates coacervation by attracting a larger amount of gelatin to form surface layers of high quality.

Encapsulation of inorganic particles is also becoming popular. Ultrafine silicon particles, for example, can be coated by carbon with a relatively simple method: exposing the core silicon particles in a vapor of benzene diluted by argon and fire at temperatures as high as 1000°C.[16] The obtained carbon layer is considerably graphitized. Graphite basal planes are found parallel to the surface of the silicon particles.

Surface treatment via an aqueous polymerization is particularly useful. Like most chemical processes, one of the most important factors for the purpose of uniform coating is a pretreatment. In the case of coating titania with poly(methacrylic acid), preadsorption of sodium dodecyl sulfate is indispensable. For the purpose of thickness control of the surface layer, successive addition of a monomer is recommended rather than applying the desired total amount at the beginning.[17]

When poly(methylmetacrylate) (PMMA) is to be coated on the surface of titania, it is necessary to use surfactants such as sodium dodecyl sulfate (SDS) prior to *in situ* polymerization. When surface pretreatment is insufficient, localized, clusterlike polymerization takes place, in contrast to the uniform polymeric surface layer achieved by the use of SDS pretreatments. A similar technique is applicable to coat ultrafine magnetic particles with PMMA. Successful and unsuccessful surface layers are easily distinguished on the transmission electron micrographs shown in Figure 7.1.

Those coated magnetic fine particles are not only used for a normal magnetic fluid, but also made bioactive after subsequent biochemical treatments. Ultrafine metallic iron with very thin protective oxide layers was successfully modified by adding acrolein for the purpose of further, higher-order surface modification, for example, by enzymes.[18] Polymeric particles or microspheres can be modified with various dissimilar polymeric species as well. They have been developed to the level of surface design for biochemical or medical uses.[19] To prepare a desired surface for subsequent treatment, it is important to choose an appropriate initiator. A comonomer may be most appropriate. After polymerization of styrene (acrylamide copolymer microspheres), the surface can be carboxylated under controlled hydrolysis. A Hofmann reaction brings about, in contrast, a series of amphiphilic surfaces with a regulated isoelectric point by controlling the temperature and time of the surface reaction.[20]

4.7.5   POLYMERIZATION AND PRECIPITATION *IN SITU*

An *in situ* synthesis of coated materials can be subdivided into several categories. By using a subtle difference in the polarity, a mixture of monomers often brings about various core-shell polymeric particles in one step. Core-shell polymerization is an example of *in situ* surface modification and results in a number of useful properties on the surface of microspherical particles.[21] Seeded polymerization is a similar technique but with wider variation and at the cost of an extra step.[22]

When inorganic precipitation is carried out in the presence of polymerizable species, fine inorganic particles are often coated *in situ* by polymer films. Magnetite fine particles are obtained, for example, in PVA solution, resulting in the precipitates of a well-dispersed nanometer regime with considerably large saturation magnetization.[23] Mn-doped ZnS is prepared in the presence of methacrylic acid in a methanolic medium to give nanoparticles with brilliant electroluminescent properties.[24] They show quantum dot effects with a blue shift of the absorption edge.

4.7.6   MECHANICAL ROUTES AND APPARATUS

Via mechanical routes, solid particles are covered relatively easily by various solids in the form of gels, fine particles, or films. Apart from traditional coating machines, there are a number of machines available on the market today for the purpose of mechanical surface modification or coating. There are numerous reports of using such commercial apparatuses to achieve, for example, acquired superplasticity after surface modification.[25,26] When the surface of the host materials is modified with another solid species, chemical interaction or solid-state reactions occur at the interface.[27,28] Such a mixture can be used for many practical purposes including graded functional materials.[29]

4.7.7   CHARACTERIZATION OF COATED PARTICLES

Morphological observation is by no means sufficient to analyze the results of surface modification. If the purpose is simple enough, for example, to prevent agglomeration due to a hygroscopic nature,

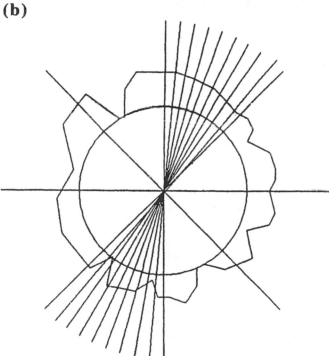

FIGURE 7.2   (a) Transmission electron micrograph of sliced polystyrene with electroplated Ni. (b) Scheme explaining the method of evaluating fluctuation of the thickness of the surface layer in the radial direction, to be evaluated from the variance of the local thickness. Subdivision up to 36 sections per particle suffices, in general, for a reliable result.

conventional measurements of various mechanical tests of the compacts could suffice. Rheological measurements are quite often used for the evaluation of surface modification.[30] Conversely, surface-modified particles can be used as dispersing particles for electrorheological fluids, which is gathering increasing interest.

Chemical interaction between the core and surface species, as well as in the surface layers, can conveniently be examined by various surface analyses including X-ray photoelectron spectroscopy.[31] If particles are embedded and cut into a very thin slice of a nanometer regime, transmission electron microscopy could reveal the microstructure at the interface, as well as the extent of mechanochemical short-range diffusion, provided the microscope is equipped with a tool for local elemental analysis. At the same time, microhomogeneity within a modified single particle can be quantitatively determined by using statistical variance obtained from repeated local analyses of the composition.[32]

Topochemical distribution of the deliberately introduced surface species, for example, by electroless deposition, can be made, combined with the above-mentioned ultrathin slice technique, by sectioning the particle into several radial elements and evaluating the statistical fluctuation of the thickness.[33] As shown in Figure 7.2, radial fluctuation and the average thickness of the surface layer can be determined from the repeated measurements of the thickness on the cross-sectional view of surface-coated particles.

4.7.8 REMARKS AND RECENT DEVELOPMENTS

To create a functional surface on powder particles is one of the basic issues for surface modification. Silica is one of the best examples in this context, for inorganic materials, as well as for inorganic–organic composites.[34] To immobilize the appropriate chemical species, their selectivity, irreversibility, and stability are of particular importance.[35]

Biomimetic methods are also used to modify nonbiological materials. Coating of polymeric microspheres by uniformly grown inorganic particles is only one example of such a tendency.[36] For tissue engineering including controlling cellular responses, various techniques of surface modification are applied.[37] There have been many attempts to modify natural organic products, such as leather powder, by copolymerization to gain stronger affinity with a matrix phase to obtain artificial leather.[38]

An entirely different aspect of modern surface modification methods is to carry out *in situ* methods, with or without passing through a dry state, when a process is wet and the end state of use is also wet. Drying after surface modification is not only energetically unfavorable but also brings about hazardous, undesired agglomeration.[39] An all-dry process can avoid undesirable agglomeration. One such technique is plasma-aided surface polymerization. It enables surface modification of fine particles with cross-linked, tough polymer layers from otherwise unpolymerizable species.

Thus, the technology of surface modification of powder particles is in the process of rapid expansion. A vast number of new technologies are introduced every year at symposia, meetings, and expositions, dealing with functional fine particles. The number of patents in this direction is also rapidly increasing. More systematic studies are required to impart those attractive technologies a sound scientific and theoretical basis for further development.

REFERENCES

1. Sugiyama, K., Ohga, K., and Kukukawa, K., *Macromol. Chem. Phys.,* 195, 1341, 1994.
2. Inomata, Y., Wada, T., and Sugi, Y., *J. Biomater. Sci. Polym. Ed.,* 5, 293, 1994.
3. Shiba, M., Tomioka, S., and Kondo, T., *Bull. Cem. Soc. Jpn.,* 46, 2584, 1973.
4. Jackson, R. L., *J. Electrochem. Soc.,* 137, 95, 1990.
5. Mizota, K., Fujiwara, S., and Senna, M., *Mater. Sci. Eng. B,* 10, 139, 1991.
6. Bernhardt, C. and Heegn, H., *Freiberger Forsch.-H.,* A602, 49, 1978.
7. Fukumori, Y., Ichikawa, H., Yamaoka, Y., Akino, E., Takeuchi, Y., Fukuda, T., Kanamori, R., and Osako, Y., *Chem. Pharm. Bull.,* 39, 164, 1991.
8. Nakahara, K., *Shikizai,* 59, 543, 1986.

9. Perfitt, G. D., *Dispersion of Powders in Liquids,* Applied Science Publ., London, 1981.
10. Delmon, B., in *Reactivity of Solids,* Barret, P. and Dufour, L., Eds., Elsevier, Amsterdam, 1984, p. 81.
11. Seino, M., *Shikizai,* 40, 163, 1967.
12. Perfitt, G. D. and Sing, K. S. W., *Characterization of Powder Surfaces,* Academic, London, 1976.
13. Moriyama, Y., in *Surface Modification,* Chem. Rev., No. 44, Nishiyama, Y., Ed., Gakkai Shuppan Center, Tokyo, 1984, p. 118.
14. Kashu, S., in *Ultrafine Particles, Science and Application,* Chem. Rev., Vol. 48, Chemical Society of Japan, Ed., Gakkai Shuppan Center, Tokyo, 1985, p. 135.
15. Nixon, J. R., Saleh, A. H., Khalil, J. R., and Carless, J. F., *J. Pharm. Pharmacol.,* 20, 528, 1968.
16. Iijima , S., *J. Surface Sci. Soc. Jpn.,* 8, 325, 1987.
17. Hasegawa, M., Arai, K., and Saito, S., *J. Polym. Sci. A Polym. Chem.,* 25, 3231, 1987.
18. Miyamoto, H., *J. Surface Sci. Soc. Jpn.,* 8, 345, 1987.
19. Kato, T., Fujimoto, K., and Kawaguchi, H., *Polym. Gels Networks,* 4, 237, 1993.
20. Kawaguchi, H., Hoshino, H., Amagasa, H., and Ohtsuka, Y., *J. Colloid Interface Sci.,* 97, 465, 1984.
21. Okubo, K. and Hattori, H., *Colloid Polym. Sci.,* 271, 1157, 1993.
22. Kobayashi, K. and Senna, M., *J. Appl. Polym. Sci.,* 46, 27, 1992.
23. Lee, J. W. and Senna, M., *Colloid Polym. Sci.,* 273, 76, 1995.
24. Yu, I., Isobe, T., and Senna, M., *Mater. Eng. Sci. B,* in press, 1995.
25. Yokoyama, T., Urayama, K., Naito, M., and Yokoyama, T., *Kona,* 5, 59, 1987.
26. Alonso, M., Satoh, M., and Miyanami, K., *Powder Technol.,* 59, 45, 1987.
27. Mizota, K., Fujiwara, S., and Senna, M., *Mater. Sci. Eng. B,* 10, 139, 1991.
28. Saito, I. and Senna, M., *Kona,* in press, 1995.
29. Tanno, K., Yokoyama, T., and Urayama, K., *J. Soc. Powd. Technol. Jpn.,* 27, 153, 1990.
30. Otsubo, Y., *Colloid Surf.,* 58, 73, 1991.
31. Saito, I. and Senna, M., *Kona,* in press, 1995
32. Komatsubara, S., Isobe, T., and Senna, M., *J. Am. Ceram. Soc.,* 77, 278, 1994.
33. Nakajima, S., Koga, T., Isobe, T., and Senna, M., *Mater. Sci. Eng. B,* in press, 1995.
34. Jal, P. K., Patel, S., and Mishra, B. K., *Talanta,* 62, 1005–1028, 2004.
35. Deorkar, N. V. and Tavlarides, L. L., *Ind. Eng. Chem. Res.,* 36, 39, 1997.
36. Tarasevich, B. J. and Rieke, P. C., *Mater. Res. Soc. Symp. Proc.,* 174, 51, 1991.
37. Shin, H., Jo, S., and Mikos, A. G., *Biomaterials,* 24, 4353–4364, 2003.
38. Nakajima, Y., Isobe, T., and Senna, M., *Mater. Sci. Eng. B,* 10, 139, 1996.
39. Nakajima, S., Koga, T., Isobe, T., and Senna, M., *Mater. Sci. Eng. B,* in press, 1995.

4.8 Standard Powders and Particles

Hideto Yoshida
Hiroshima University, Higashi-Hiroshima, Japan

Hiroaki Masuda
Kyoto University, Katsura, Kyoto, Japan

There are several standard particles currently in scientific use. For example, uniformly sized spherical polystyrene latex particles are used as a calibration standard in electron microscope studies. Natural pollens and spores are also monosized and nearly spherical, but they are sometimes swelled by moisture and have irregular surfaces. For example, the diameter of lycopodium particles measured by means of a microscope is about 30 μm, whereas it is about 24 μm when a liquid sedimentation method is used, because of surface irregularities. A specially made precipitated calcium carbonate is also monosized and is cubic in shape.

Standard powders for industrial use generally have fairly narrow size distributions. In the United States, AC fine and coarse dusts are defined by the Society of Automotive Engineers and are utilized for performance tests of automobile air cleaners. They have the same size distributions as Japanese Standard Powders (JIS Z 8901) No. 2 and No. 3 or No. 7 and No. 8, respectively.

Japan has also defined many types of standard powders and particles for industrial tests according to JIS Z 8901. They are sold by the Association of Powder Process Industry and Engineering (APPIE), Japan. Figure 8.1 shows some information on their size distributions, and Table 8.1 lists

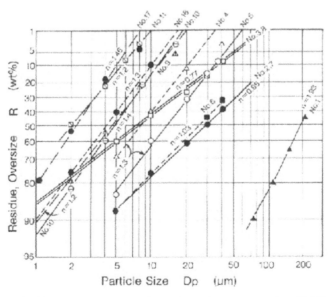

FIGURE 8.1 Size distributions of standard test powders (JIS Z 8901).

the materials they are prepared from and their applications. No. 8 Kanto loam powder is used mainly for the performance tests of automobile engine air cleaners in Japan, because it is easily dispersed in air by a mechanical or an air-jet-type powder disperser, shown in Figure 8.2 and Figure 8.3, respectively. Nearly monosized glass beads and white fused alumina are also defined according to JIS Z 8901, and their sizes are shown in Table 8.2 and Table 8.3.

TABLE 8.1   Materials and Application Examples of Standard Particles (JIS Z8901)

Type	Material	Test Application Examples
No. 1	Silica sand (coarse)	Abrasion or life test of machine; performance test of chemical plant
No. 2	Silica sand (medium)	Abrasion or life test of machine
No. 3	Silica sand (fine)	Abrasion or life test of machine
No. 4	Talc (fine)	Dust collector performance test
No. 5	Flyash (fine)	Dust collector performance test
No. 6	Portland cement	Airtightness test of car lamps
No. 7	Kanto loam (medium)	Dust collector performance test; abrasion or life test of machine
No. 8	Kanto loam (fine)	Dust collector performance test; abrasion or life test of machine
No. 9	Talc (ultrafine)	Performance test of high-efficiency dust collector air filter
No. 10	Flyash (ultrafine)	Performance test of high-efficiency dust collector air filter
No. 11	Kanto loam (ultrafine)	Performance test of high-efficiency dust collector air filter
No. 12	Carbon black (ultrafine)	Performance test of high-efficiency dust collector
No. 15	Mixed dust	Prefilter performance test
No. 16	Calcium (fine)	Carbonate air Classifier
No. 17	Calcium (ultrafine)	Carbonate air Classifier

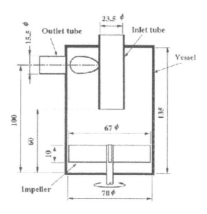

FIGURE 8.2   Mixer-type disperser.

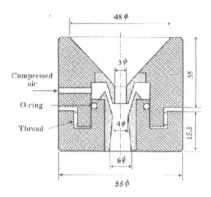

FIGURE 8.3   Nozzle-jet-type disperser.

TABLE 8.2   Size Distribution of Glass Beads Standard Powders in Japan (JIS Z 8901)

No.	Code[a]	Oversize in mass basis		
		90% Size (μm)	50% Median size (μm)	10% Size (μm)
1	GBL30	>26	30 ± 1.0	<34
2	GBL40	>37	41 ± 1.0	<45
3	GBL60	>55	59 ± 1.0	<63
4	GBL100	>95	0 + 1.0	<105
5	GBM20	>18	22 ± 1.0	<26
6	GBM30	>26	30 ± 1.0	<34
7	GBM40	>37	41 ± 1.0	<45

[a] GBL, sodalime silicate glass (ρ_p = 2.1–2.5 g/cm³); GBM, barium titanate glass (ρ_p = 4.0–4.2 g/cm³).

Table 8.4 presents several reference powders for certified size distributions, defined in Europe, and Table 8.5 presents several reference powders for the same purpose in the United States.

Recently, Japan also has defined two types of standard powders, MBP1-10 and MBP10-100. The particle size ranges are from 1 to 10 μm and 10 to 100 μm, respectively. They consist of barium titanate glass beads, and their sizes are controlled to nearly log-normal distributions within the specified size range. Table 8.6 shows some of their physical properties and Figure 8.4 shows photographs of the test particles.

Figure 8.5 and Figure 8.6 show particle size distributions that were carefully measured by a scanning electron microscope with a particle count of more than about 10,000 particles. The APPIE committee, Japan, recognizes these particles as a standard reference particle in Japan.[1,2]

TABLE 8.3 Size Distributions of White Fused Alumina Standard Powder in Japan (JIS Z 8901)

No.	92% Size (μm)	Oversize in Mass Basis 50% Median size (μm)	3% Size (μm)
1	>0.8	2 ± 0.4	<5
2	>2.0	4 ± 0.5	<11
3	>4.5	8 ± 0.6	<20
4	>9.0	14 ± 1	<31
5	>20	30 ± 2	<58
6	>40	57 ± 3	<103

Note: ρp = 3.9 4.0 g/cm^3

TABLE 8.4 IRMM—SMT in Europe 0BCR Before 1995

CRM No.	Description		Size Range (μm)	Unit Size	Price (ECU)
066	Quartz powder	Stokes diam.	0.35–3.50	10 g	125
067	Quartz powder	Stokes diam.	2.4–32	10 g	125
068	Quartz sand	Vol. diam.	160–630	100 g	125
069	Quartz powder	Stokes diam.	14–90	10 g	125
070	Quartz powder	Stokes diam.	1.2–20	10g	125
130	Quartz powder	Vol. diam.	50–220	50 g	125
131	Quartz powder	Vol. diam.	480–1800	200 g	125
132	Quartz gravel	Vol. diam.	1400–5000	700 g	125
165	Latex	Sphere.	2.223 ± 0.013	1 vial	100
166	Latex	Sphere.	4.821 ± 0.019	1 vial	100
167	Latex	Sphere.	9.475 ± 0.018	1 vial	100
169	Alpha	alumina	0.104 ± 0.012	60	100
170	Alpha	alumina	1.05 ± 0.05	50	100
171		Alumina	2.95 ± 0.13	50	100
172		Quartz	2.56 ± 0.10	10	100
173	Rutile	titania	8.23 ± 0.21	46	100

Note: Joint Research Centre Institute for Reference Materials and Measurements (IRMM), Retieseweb, B-2440 Geel, Belgium. Attention: Management of Reference Materials (MRM) Unit. Tel: 32 14 57 17 19, Fax: 32 14 59 04 06. Standards, Measurements, and Testing (SMT) Program.

TABLE 8.5   NIST-SRM (301) in the United States (1996)

SRM No.	Description	Particle Size (μm)	Unit issued	Price (US $)	Certif. Date
114p	Portland cement	45 μm oversize 8.24%	Set (20)	114.00	May 94
659	Silicon nitride (particle size)	0.2-10	Set (5)	207.00	Mar. 92
1003b	Glass (particle size)	10-60	25 g	159.00	Sept. 93
1004a	Glass (particle size)	40-170	70 g	154.00	Dec. 93
1017b	Glass (particle size)	100-310	70 g	252.00	Aug. 95
1018b	Glass (particle size)	225-780	74 g	In prep.	-
1019a	Glass (particle size)	760-2160	200 g	179.00	Oct. 84
1690	Polystylene (particle size)	0.895	5 ml	391.00	Dec. 82
1691	Polystylene (particle size)	0.269	5 ml	39.00	May 84
1692	Polystylene (particle size)	2.982	5 ml	384.00	May 91
1960	Polystylene (particle size)	9.89	5 ml	805.00	Apr. 85
1961	Polystylene (particle size)	29.64	5 ml	806.00	Jan. 87
1963	Polystylene (particle size)	0.1007	5 ml	504.00	Nov. 93
1965	Polystylene (on slide) (particle size)	9.94	1 slide	147.00	Jan. 87
1978	Zirconium oxide (particle size)	0.33-2.19	5 g	203.00	Oct. 93
8570	LGCGM calcined kaolin (surf. area)	10.3-10.9 m^2/g	10 g (25 g)	90.00	Sept. 94
8571	LGCGM alumina (surf.area)	153.2-158.7 m^2/g	10 g (25 g)	90.00	Sept. 94
8572	LGCGM silica (surf.area)	277.6-291.2 m^2/g	25 g	90.00	Sept. 94

Note: Purchase orders (in English) for all NIST SRMs/RMs should be directed to: National Institute of Standards and Technology (NIST), Standard Reference Materials Program (SRM), Room 204, Building 202, Gaithersburg, MD 20899-0001, USA; Tel.: (301) 975-6766, Fax: (301) 948-3730, e-mail: SRMINFO@enh.nist.gov.

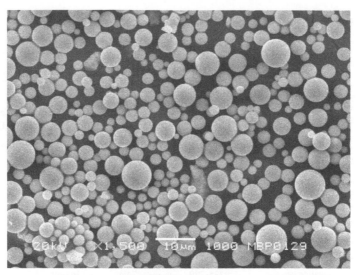

MBP 1-10

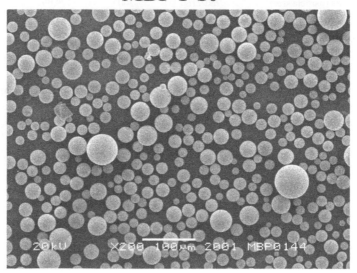

MBP 10-100

FIGURE 8.4   Photographs of test particles.

TABLE 8.6   Physical Properties of Test Particles

	MBP1-10 n=14806	MBP10-100 n=10515
Dp_{50} (μm) (mass base)	4.76	36.50
$\sigma = \ln\sigma_g$ (-)	0.312	0.395
σ_g (-)	1.37	1.48
ρ_p (g/cm3)	4.19	4.10

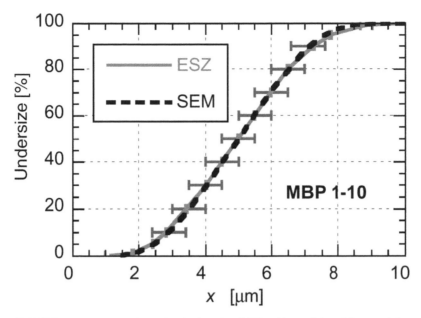

FIGURE 8.5 Particle size distribution for MBP1-10 particles. Measured by Electrical Sensing Zone (ESZ) and Scanning Electron Microscope (SEM) methods with the certified ranges as horizontal bars.

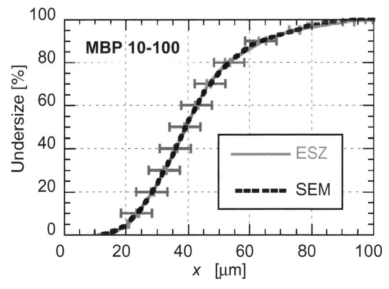

FIGURE 8.6 Particle size distribution for MBP10-100 particles. Measured by Electrical Sensing Zone (ESZ) and Scanning Electron Microscope (SEM) methods with the certified ranges as horizontal bars.

REFERENCES

1. Yoshida, H., Masuda, H., et al., *Adv. Powder Technol.,* 12, 79–94, 2001.
2. Yoshida, H., Masuda, H., et al., *Adv. Powder Technol.* 14, 17–31, 2003.

Index

Milton Keynes UK
Ingram Content Group UK Ltd.
UKHW052025071024
449327UK00027B/2435